Mathematics Higher Level

for the IB Diploma

Solutions Manual

**Paul Fannon, Vesna Kadelburg,
Ben Woolley and Stephen Ward**

CAMBRIDGE
UNIVERSITY PRESS

University Printing House, Cambridge CB2 8BS, United Kingdom

Cambridge University Press is part of the University of Cambridge.

It furthers the University's mission by disseminating knowledge in the pursuit of education, learning and research at the highest international levels of excellence.

www.cambridge.org
Information on this title: education.cambridge.org

© Cambridge University Press 2016

This publication is in copyright. Subject to statutory exception and to the provisions of relevant collective licensing agreements, no reproduction of any part may take place without the written permission of Cambridge University Press.

First published 2016

Printed in India by Multivista Global Pvt Ltd

A catalogue record for this publication is available from the British Library

ISBN 978-1-107-57937-8 Paperback

Cover image: Craig Jewell/Shutterstock

Cambridge University Press has no responsibility for the persistence or accuracy of URLs for external or third-party internet websites referred to in this publication, and does not guarantee that any content on such websites is, or will remain, accurate or appropriate. Information regarding prices, travel timetables, and other factual information given in this work is correct at the time of first printing but Cambridge University Press does not guarantee the accuracy of such information thereafter.

NOTICE TO TEACHERS IN THE UK
It is illegal to reproduce any part of this work in material form (including photocopying and electronic storage) except under the following circumstances:

(i)   where you are abiding by a licence granted to your school or institution by the Copyright Licensing Agency;
(ii)   where no such licence exists, or where you wish to exceed the terms of a licence, and you have gained the written permission of Cambridge University Press;
(iii)   where you are allowed to reproduce without permission under the provisions of Chapter 3 of the Copyright, Designs and Patents Act 1988, which covers, for example, the reproduction of short passages within certain types of educational anthology and reproduction for the purposes of setting examination questions.

Contents

	Introduction	iv
Chapter 1	Counting principles	1
Chapter 2	Exponents and logarithms	10
Chapter 3	Polynomials	21
Chapter 4	Algebraic structures	31
Chapter 5	The theory of functions	41
Chapter 6	Transformations of graphs	53
Chapter 7	Sequences and series	69
Chapter 8	Binomial expansion	84
Chapter 9	Circular measure and trigonometric functions	93
Chapter 10	Trigonometric equations and identities	100
Chapter 11	Geometry of triangles and circles	110
Chapter 12	Further trigonometry	124
Chapter 13	Vectors	135
Chapter 14	Lines and planes in space	148
Chapter 15	Complex numbers	179
Chapter 16	Basic differentiation and its applications	206
Chapter 17	Basic integration and its applications	223
Chapter 18	Further differentiation methods	233
Chapter 19	Further integration methods	248
Chapter 20	Further applications of calculus	266
Chapter 21	Summarising data	287
Chapter 22	Probability	293
Chapter 23	Discrete probability distributions	313
Chapter 24	Continuous distributions	328
Chapter 25	Mathematical induction	341
Chapter 26	Questions crossing chapters	372

Introduction

This book contains worked solutions for all the exam-style questions that, in the Higher Level coursebook, are colour-coded green, blue, red or gold. As a reminder:

Green questions should be accessible to students on the way to achieving a grade 3 or 4.

Blue questions are aimed at students hoping to attain a grade 5 or 6.

Red questions are the most difficult, and success with a good number of these would suggest a student is on course for a grade 7.

Gold questions are of a type not typically seen in the exam but are designed to provoke thinking and discussion in order to develop a better understanding of the topic.

Of course, these are just guidelines. If you are aiming for a grade 6, do not be surprised if occasionally you find a green question you cannot do; people are rarely equally good at all areas of the syllabus. Similarly, even if you are able to do all the red questions, that does not guarantee you will get a grade 7 – after all, in the examination you will have to deal with time pressure and exam stress! It is also worth remembering that these questions are graded according to our experience of the final examination. When you first start the course, you may well find the questions harder than you would do by the end of the course, so try not to get discouraged!

The solutions are generally written in the format of a 'model answer' that students should aim to produce in the exam, but sometimes extra lines of working are included (which wouldn't be absolutely necessary in the exam) in order to make the solution easier to follow and to aid understanding.

In many cases the approach shown is not the only option, and neither do we claim it to always be categorically the best or the approach that we would advise every student to take; this is clearly subjective, and different students will have different strengths and therefore preferences. Alternative methods are sometimes given, either in the form of a full worked solution or by way of a comment after the given worked solution.

Where the question has a calculator symbol next to it (indicating that it is for the calculator paper), the approach taken is intentionally designed to utilise the calculator and thereby to minimise work. Students should make sure they are familiar with all the calculator techniques used and, if not, the calculator skills sheets on the CD-ROM accompanying the coursebook should be consulted.

When there is no symbol (indicating that the question could appear on either the calculator or the non-calculator paper), the solution given usually assumes that a calculator is not available. Students should make sure they can cope in this situation but also that they can adapt and use features of the calculator to speed up the process. Perhaps the most common example of this is to use the calculator's equation solver rather than solving simultaneous or quadratic equations manually.

We strongly advise that these solutions be consulted only after having spent a good deal of time and effort working on a question. The process of thinking about the problems encountered, even if a full solution cannot ultimately be found, is really important in developing the skills and knowledge needed to succeed in the future.

We hope you find these solutions useful and that, when used wisely in conjunction with the coursebook, they lead to success in the IB exam.

Stephen Ward, Paul Fannon, Vesna Kadelburg, Ben Woolley

1 Counting principles

Exercise 1A

4 $12 \times 4 = 48$

5 a $5 \times 6 \times 3 = 90$

b $6 \times (5+3) = 48$

c $5 \times 6 + 5 \times 3 + 6 \times 3 = 63$

6 $5 \times 3 = 15$

7 a $58 \times 68 \times 61 \times 65 = 15\,637\,960$

b $15\,637\,960 \times \dfrac{42}{65} = 10\,104\,528$

8 $26^3 \times 9^4 = 115\,316\,136$

9 There are 4 options for the journey to the middle, then two possible directions (left or right), and then a single choice of path upwards. So the total number of possible routes is $4 \times 2 \times 1 = 8$

10 a $15 \times 4 \times 12 = 720$

b There are two cases:
- shirt is not pink $\Rightarrow$ no constraints on tie or waistcoat: $15 \times 4 \times 9 = 540$
- shirt is pink $\Rightarrow$ tie not red, waistcoat not red: $12 \times 3 \times 3 = 108$

Total number of possible outfits is $540 + 108 = 648$

11 a $5 \times 4 \times 3 = 60$

b $5^3 = 125$

12 a $3^4 = 81$

b $5^3 = 125$

Exercise 1B

4 a $7! = 5040$

b Number of arrangements with largest book at one end: $2 \times 6! = 1440$

So number of arrangements with largest not at either end:
$5040 - 1440 = 3600$

5 a $5! = 120$

b Require the final digit to be 5, so there are $4! = 24$ such numbers.

6 a $17! = 35\,568\,742\,809\,600$

b $16! = 20\,922\,789\,888\,000$

7 $5! \times 4! = 2880$

8 a $6! = 720$

b Require the first digit to be 1 or 2, so there are $2! \times 5! = 240$ such numbers.

9 $30! = 2.65 \times 10^{32}$ (3SF)

10 Because there are an odd number of red toys, the one in the middle must be red; then, on the left arrange two red and two blue, to be matched on the right. The following colour patterns are possible on the left:
RRBB
RBRB
RBBR
BRRB
BRBR
BBRR
There are $\dfrac{4!}{2! \times 2!} = 6$ colour patterns available.

Within each pattern, there are $5! \times 4!$ arrangements of the individual toys.

So total number of possible arrangements is $5! \times 4! \times 6 = 17\,280$

Exercise 1C

4 $\dfrac{n!}{(n-2)!} = n(n-1) = 20$

$n^2 - n - 20 = 0$

$(n-5)(n+4) = 0$

$\Rightarrow n = 5$ (as n is a positive integer)

5 $\dfrac{(n+1)!}{(n-2)!} = (n+1)(n)(n-1) = 990$

$\Rightarrow n^3 - n - 990 = 0$

From GDC, $n = 10$

6 $n! - (n-1)! = 16(n-2)!$

Dividing through by $(n-2)!$:

$n(n-1) - (n-1) = 16$

$n^2 - 2n - 15 = 0$

$(n-5)(n+3) = 0$

$\Rightarrow n = 5$ (for $n \in \mathbb{N}$)

Exercise 1D

4 Choose 9 from 15:

$\binom{15}{9} = 5005$

5 a Choose 3 from 7:

$\binom{7}{3} = 35$

b Choose 2 from 6:

$\binom{6}{2} = 15$

6 $\binom{39}{7} = 15\,380\,937$

7 Choose 3 boys from 16 and 2 girls from 12:

$\binom{16}{3} \times \binom{12}{2} = 560 \times 66 = 36\,960$

8 Choose 1 goalkeeper from 3, 4 defenders from 8, 4 midfielders from 6, and 2 forwards from 5:

$\binom{3}{1} \times \binom{8}{4} \times \binom{6}{4} \times \binom{5}{2}$

$= 3 \times 70 \times 15 \times 10 = 31\,500$

9 Choose 12 from 140, then 10 from 128, then 10 from the remaining 118:

$\binom{140}{12} \times \binom{128}{10} \times \binom{118}{10}$

$= 1.61 \times 10^{45}$ (3SF)

10 a Choose 3 boys from 14 and 2 girls from 16:

$\binom{14}{3} \times \binom{16}{2} = 364 \times 120 = 43\,680$

b Choose 3 boys from 14, and for the girls choose 2 from 15 (the 2 places not taken by Roberta or Priya) and choose 1 from 2 (Roberta or Priya):

$\binom{14}{3} \times \binom{15}{2} \times 2 = 364 \times 105 \times 2 = 76\,440$

11 a Exactly 6¢ is spent if you choose 3 of the 2¢ sweets from 7:

$\binom{7}{3} = 35$

b Exactly 7¢ is spent if you choose 1 of the 2¢ sweets from 7 and 1 of the 5¢ sweets from 5:

$\binom{7}{1} \times \binom{5}{1} = 35$

c Exactly 10¢ is spent if you choose 5 of the 2¢ sweets from 7 or 2 of the 5¢ sweets from 5:

$$\binom{7}{5}+\binom{5}{2}=21+10=31$$

d At most 5¢ is spent if you choose 1 or 2 of the 2¢ sweets from 7 or 1 of the 5¢ sweets from 5:

$$\binom{7}{1}+\binom{7}{2}+\binom{5}{1}=7+21+5=33$$

12 a Choose 4 from 9: $\binom{9}{4}=126$

  b Exclude the possibilities where all questions are from the same section.
  All questions from A: choose 4 from 5:

  $$\binom{5}{4}=5$$

  All questions from B: choose 4 from 4:

  $$\binom{4}{4}=1$$

  ∴ number of ways of choosing at least one from each section is $126 - 5 - 1 = 120$

13 To deliberately double-count: each point connects to 14 other points
∴ $2n = 15 \times 14$
⇒ $n = 15 \times 7 = 105$ lines

14

> **COMMENT**
>
> It is assumed that vertices of the triangles/quadrilaterals can only be at the original ten points and not at any intersections created by lines joining these ten points.

  a Each triangle is defined by three points.
  Number of different sets of 3 points =

  $$\binom{10}{3}=120$$

  b Each quadrilateral is defined by four points.
  Number of different sets of 4 points =

  $$\binom{10}{4}=210$$

15 With n people, each shakes hands with $n-1$ others. This double-counts the total number of handshakes.

∴ $n(n-1) = 2 \times 276$
$n^2 - n - 552 = 0$
$(n-24)(n+23) = 0$
⇒ $n = 24$

16 Once the rows are determined, there is only one arrangement for each row.
Choose 15 from 45 for the first row and 15 from 30 for the second row:

$$\binom{45}{15} \times \binom{30}{15} = 3.45 \times 10^{11} \times 1.55 \times 10^{8}$$

$$= 5.35 \times 10^{19} \text{ (3SF)}$$

Exercise 1E

1 Numbers divisible by 5 range from 21×5 to 160×5: 140 multiples of 5
So there are $700 - 140 = 560$ numbers not divisible by 5.

> **COMMENT**
>
> Alternatively, 4 out of every 5 of the numbers between 101 and 800 are not divisible by 5, so there are $\frac{4}{5} \times 700 = 560$ of them.

2 There are 6 different letters.
Total number of permutations: $6! = 720$
Number of permutations beginning with 'J': $5! = 120$
∴ total number not beginning with 'J': $720 - 120 = 600$

3 **a** Choose 3 from 10:
$$\binom{10}{3} = 120$$

b Choose 3 from 22, then exclude the choices of all chocolates: 3 from 12
$$\binom{22}{3} - \binom{12}{3} = 1540 - 220 = 1320$$

4 There are 7 different letters.

a Total permutations: $7! = 5040$
Permutations beginning with 'KI': $5! = 120$
∴ total not beginning with 'KI': $5040 - 120 = 4920$

b Permutations beginning with 'KI' or 'IK': $2 \times 5! = 240$
∴ total not beginning 'KI' or 'IK': $5040 - 240 = 4800$

5 Total committee possibilities:
$$\binom{21}{5} = 20349$$

Total all-male committees possible:
$$\binom{12}{5} = 792$$

∴ total committees that are not all male: $20349 - 792 = 19557$

6 Total possible selections of 7 tiles:
$$\binom{26}{7} = 657800$$

Total possible selections with no vowels:
$$\binom{21}{7} = 116280$$

Total possible selections with one vowel:
$$\binom{21}{6} \times 5 = 271320$$

∴ total possible selections with at least two vowels:
$657800 - 116280 - 271320 = 270200$

7 Total possibilities: $\binom{22}{6} = 74613$

Total possibilities with no women:
$$\binom{10}{6} = 210$$

Total possibilities with no men:
$$\binom{12}{6} = 924$$

Total possibilities with exactly one man:
$$\binom{12}{5} \times \binom{10}{1} = 7920$$

∴ total possibilities with at least 2 men and 1 woman:
$74613 - 210 - 924 - 7920 = 65559$

8 Among the integers 1–19, there are 10 odd numbers and 9 even numbers.

a No even numbers ⇒ choose 7 from 10:
$$\binom{10}{7} = 120$$

One even number ⇒ choose 6 from 10 and 1 from 9: $\binom{10}{6} \times \binom{9}{1} = 1890$

Two even numbers ⇒ Choose 5 from 10 and 2 from 9:
$$\binom{10}{5} \times \binom{9}{2} = 9072$$

∴ choices with at most two even numbers: $120 + 1890 + 9072 = 11082$

b Total possibilities ⇒ choose 7 from 19:
$$\binom{19}{7} = 50388$$

∴ choices with at least two even numbers: $50388 - 120 - 1890 = 48378$

9 Total permutations: $6! = 720$
Total beginning with 'D' and ending with 'L': $4! = 24$
∴ total not beginning with 'D' or not ending with 'L': $720 - 24 = 696$

Exercise 1F

5 $^{39}P_7 = 77\,519\,922\,480$

6 $^{24}P_4 = 255\,024$

7 $^9P_3 = 504$

8 $^8P_3 = 336$

9 $^{26}P_2 \times {}^{10}P_4 = 3\,276\,000$

10 $^nP_{n-1} = \dfrac{n!}{(n-(n-1))!} = \dfrac{n!}{1!} = n!$

$^nP_n = \dfrac{n!}{(n-n)!} = \dfrac{n!}{0!} = \dfrac{n!}{1} = n!$

∴ $^nP_{n-1} = {}^nP_n$

11 There are 7 different letters, consisting of 3 vowels and 4 consonants.
Total number of permutations: $^7P_3 = 210$
Total number of permutations with no vowels: $^4P_3 = 24$
∴ total permutations with at least one vowel: $210 - 24 = 186$

12 Other than James, choose 2 runners from 7: $\binom{7}{2} = 21$
Arrange 3 runners in medal positions: $3! = 6$
In two-thirds of these arrangements, James is in first or second.
Total valid arrangements: $21 \times 6 \times \dfrac{2}{3} = 84$

13 Case 1: Rajid not selected.
Choose 3 students from 17 and permute them: $\binom{17}{3} \times 3! = 4080$

Case 2: Rajid is selected.
Choose 2 students from 17, select one post from two for Rajid, then permute the other two candidates in the two remaining posts: $\binom{17}{2} \times 2 \times 2! = 544$
Total valid arrangements:
$4080 + 544 = 4624$

14 $^{2n}P_3 = {}^6P_n$

$\Rightarrow 2n(2n-1)(2n-2) = \dfrac{6!}{(6-n)!}$

Since ^6P_n is a real value, $n \leq 6$

Since $^{2n}P_3$ is a real value, $n \geq 2$
Trying all possible values:
$n = 2: \ ^4P_3 = 4 \neq {}^6P_2 = 30$
$n = 3: \ ^6P_3 = 120 = {}^6P_3$
$n = 4: \ ^8P_3 = 336 \neq 360 = {}^6P_4$
$n = 5: \ ^{10}P_3 = 720 = {}^6P_5$
$n = 6: \ ^{12}P_3 = 1320 \neq 720 = {}^6P_6$

So $n = 3$ or 5

Exercise 1G

1 Treating Joshua and Jolene as one unit, permute 13 units, then internally permute Joshua and Jolene:
$13! \times 2 = 12\,454\,041\,600$

2 Arrange the three blocks in $3! = 6$ ways.
Internally permute the members of each class in $6! \times 4! \times 4!$ ways.
Total arrangements:
$6 \times 6! \times 4! \times 4! = 2\,488\,320$

3 Treating the physics books as one unit, permute 7 units; then internally permute the 3 physics books:
$7! \times 3! = 30\,240$

4 The 13 Grays and Greens can be arranged in $13! = 6\,227\,020\,800$ ways.

There are 14 spaces in the line-up (including the ends), and one Brown must stand in each of 4 of these spaces.

There are $\binom{14}{4} = 1001$ possible space selections.

Then permute the Browns: $4! = 24$
Total possible arrangements:
$13! \times 1001 \times 4! = 1.50 \times 10^{14}$ (3SF)

5 a $^{15}P_7 = 32\,432\,400$

b There are 9 possible spaces for the leftmost friend to sit.
Total arrangements: $9 \times 7! = 45\,360$

6 a Treating the men as one unit, permute 6 units, then internally permute the 4 men:
$6! \times 4! = 17\,280$

b Treating all the men as one unit and all the women as one unit, permute the 2 units, then internally permute the 4 men and internally permute the 5 women:
$2! \times 4! \times 5! = 5760$

c Permute the 5 women: $5! = 120$
Into 4 of the 6 spaces, insert one man, and then permute the men: $^6P_4 = 360$
Total possible arrangements:
$120 \times 360 = 43\,200$

d Require a WMWMWMWMW arrangement.
Permute the 5 women and the 4 men:
$5! \times 4! = 2880$

Mixed examination practice 1
Short questions

1 Choose 3 from 7 and permute them:
$\binom{7}{3} \times 3! = {}^7P_3 = 210$

2 Permute 5 units: $5! = 120$

3 Permute 3 and permute 7: $3! \times 7! = 30\,240$

4 $9^3 = 729$

5 Total possible choices without restriction – choose 4 from 8: $\binom{8}{4} = 70$

Choices which contain the two oldest – choose 2 from the remaining 6:
$\binom{6}{2} = 15$

∴ choices not containing both of the oldest: $70 - 15 = 55$

6 $(n+1)! = 30(n-1)!$
$(n+1)(n)(n-1)! = 30(n-1)!$
$n(n+1) = 30$
$n^2 + n - 30 = 0$
$(n-5)(n+6) = 0$
∴ $n = 5$ (reject the negative solution $n = -6$)

7 There are 8 different letters, consisting of 4 vowels and 4 consonants.
Choose 1 of the 4 consonants to be the first letter, 1 of the other 3 consonants to be the last letter, and permute the remaining 6 letters for the centre:
$\binom{4}{1} \times \binom{3}{1} \times 6! = 8640$

8 $\binom{n}{2} = \frac{n(n-1)}{2} = 105$

$n^2 - n - 210 = 0$

$(n-15)(n+14) = 0$

$\Rightarrow n = 15$

(reject the negative solution $n = -14$)

9 Total possible choices without restriction:

choose 5 from 15: $\binom{15}{5} = 3003$

Choices which are all girls: choose 5 from 8: $\binom{8}{5} = 56$

∴ choices which contain at least one boy: $3003 - 56 = 2947$

10 Total permutations without restriction: $6! = 720$

Permutations containing BE or EB: permute 5 units and then internally permute B and E: $5! \times 2! = 240$

∴ permutations without B and E next to each other: $720 - 240 = 480$

11 Total possible choices without restriction:

choose 5 from 12: $\binom{12}{5} = 792$

Choices which contain the two youngest: choose 3 from the remaining 10:

$\binom{10}{3} = 120$

∴ choices not containing both of the youngest: $792 - 120 = 672$

12 Choose and permute 3 letters from 26:

$3! \times \binom{26}{3} = {}^{26}P_3 = 15\,600$

Choose 5 digits with repeats allowed: $9^5 = 59\,049$

∴ total possible registration numbers: $15\,600 \times 59\,049 = 921\,164\,400$

13 Choose 1 from 5 to be the driver and permute the remaining 7:

$\binom{5}{1} \times 7! = 25\,200$

COMMENT

Notice that the exact arrangement of people in each row of the other seats is irrelevant, since each seat is uniquely identified. The answer would be the same for a van with seats in a 2-2-2-2 arrangement, for example.

14 The drivers can be arranged in 2 ways; then choose 3 from 8 to go in the car:

$2 \times \binom{8}{3} = 112$

COMMENT

The specific people to go in the van need not be considered, since after choosing those to go in the car, the rest will go in the van. Because $\binom{8}{3} = \binom{8}{5}$, it makes no difference which vehicle is considered to have the first pick of passengers when making the calculation.

Long questions

1 a Choose 1 of 2 places for Anya, then permute the other 4: $2 \times 4! = 48$

b Total possible permutations: $5! = 120$

Permutations with Anya not at an end: $120 - 48 = 72$

c With Anya at the left, permute the other 4: $4! = 24$

With Elena on the right, permute the other 4: $4! = 24$

If both conditions hold, permute the other 3: $3! = 6$

To find the total number of possible permutations, remove the double-counted cases: $24 + 24 - 6 = 42$

2 a Unrestricted ways of sharing: $2^5 = 32$
Cases with all sweets to one person: 2
Cases with at least one sweet to each person: $32 - 2 = 30$

b $2^5 = 32$

c Choose 1 sweet from 5 to be split: $\binom{5}{1} = 5$

Choose 2 sweets from 4 for the first person: $\binom{4}{2} = 6$

(Then the second person gets the rest of the sweets.)

Total possible choices: $5 \times 6 = 30$

3 a Choose 8 seats from 14 and permute the 8 people: $\binom{14}{8} \times 8! = 121\,080\,960$

b Consider the people of the same family as a single unit.

Choose 6 seats from 12 and permute the 6 units, then internally permute the 3 people of the same family:

$\binom{12}{6} \times 6! \times 3! = 3\,991\,680$

c The person with the cough can sit at an end seat with one empty seat next to it or in a non-end seat with an empty seat on either side.

Case 1: Choose an end for the cougher, choose 7 seats from the remaining 12 and permute the 7 people $\Rightarrow$

$2 \times \binom{12}{7} \times 7! = 7\,983\,360$

Case 2: Choose 8 seats from 12 and permute the 8 $\Rightarrow$

$\binom{12}{8} \times 8! = 19\,958\,400$

Alternatively, choose 1 seat from 12 non-end seats for the cougher, choose 7 seats from the remaining 11 and permute the 7 people $\Rightarrow$

$\binom{12}{1} \times \binom{11}{7} \times 7! = 19\,958\,400$

Total possibilities:
$7\,983\,360 + 19\,958\,400 = 27\,941\,760$

4 a If the four slots are labelled 1, 2, 3 and 4, choose two of these to be the positions for the 'R's (then the other two will automatically be the positions for the 'D's); this is effectively choosing 2 slots from 4, which gives $\binom{4}{2}$.

b By the same logic as in (a), n 'R's and n 'D's can be arranged in $\binom{2n}{n}$ ways.

c To get from top left to bottom right in a 4×4 grid, the miner must make 3 moves to the right (R) and 3 moves down (D), i.e. in some ordering of 3 'R's and 3 'D's.

By (b), there are $\binom{6}{3} = 20$ ways.

d By the same argument as in (a), $(n-1)$ 'R's and $(m-1)$ 'D's can be arranged in $\binom{n+m-2}{n-1}$ ways.

8 Mixed examination practice 1

5 a Choose 6 people from 12 for one team: $\binom{12}{6}$ ways.

However, this will double-count, because choosing 6 people for one team is equivalent to choosing the other 6 people for the other team, since the two teams of equal size are not specified as A and B.
The total number of possible teams is therefore $\frac{1}{2}\binom{12}{6}$.

b Choose 1 from 12, 2 from 12, 3 from 12, 4 from 12, 5 from 12 or the previous answer:

$$\binom{12}{1}+\binom{12}{2}+\binom{12}{3}+\binom{12}{4}+\binom{12}{5}+\frac{1}{2}\binom{12}{6}=2047$$

c Choose 4 from 12, then 4 from 8. But this will over-count by the number of rearrangements of the 3 groups, i.e. 3!, so

$$\binom{12}{4}\times\binom{8}{4}\div 3!=5775$$

6 a Choose 3 from 31: $\binom{31}{3}=4495$

b Choose 3 from n: $\binom{n}{3}=1540$

$$\frac{n(n-1)(n-2)}{3!}=1540$$

$$n^3-3n^2+2n-9240=0$$

From GDC: $n=22$

c Choose 3 from n: $\binom{n}{3}=100n$

$$\frac{n(n-1)(n-2)}{3!}=100n$$

$$n(n^2-3n+2)=600n$$

$$n^2-3n-598=0 \quad \text{(reject } n=0\text{)}$$

$$(n-26)(n+23)=0$$

$\therefore n=26$ (reject negative solution $n=-23$)

2 Exponents and logarithms

Exercise 2A

8 n inputs are sorted in $k \times n^{1.5}$ microseconds.
1 million $= 10^6$ inputs are sorted in
0.5 seconds $= 0.5 \times 10^6$ microseconds.

$$\therefore k \times (10^6)^{1.5} = 0.5 \times 10^6$$
$$k \times 10^9 = 0.5 \times 10^6$$
$$k = 0.5 \times 10^{-3}$$
$$= 5 \times 10^{-4}$$

9 $V = xy^2$, and when $x = 2y$, $V = 128$.
$$\therefore (2y) \times y^2 = 128$$
$$y^3 = 64$$
$$y = 4$$
Hence $x = 2 \times 4 = 8$ cm.

10 a Substituting $A = 81$ and $V = 243$:
$$V = kA^{1.5}$$
$$243 = k \times (81)^{\frac{3}{2}}$$
$$243 = k \times \left(81^{\frac{1}{2}}\right)^3$$
$$243 = k \times 9^3$$
$$243 = 729k$$
$$k = \frac{1}{3}$$

b If $V = \dfrac{64}{3}$, then
$$\frac{64}{3} = \frac{1}{3} A^{1.5}$$
$$A^{\frac{3}{2}} = 64$$
$$A = 64^{\frac{2}{3}}$$
$$= \left(64^{\frac{1}{3}}\right)^2$$
$$= 4^2 = 16 \text{ cm}^2$$

11 $2^{350} = (2^7)^{50} = 128^{50} > 125^{50} = (5^3)^{50} = 5^{150}$

> **COMMENT**
> This question lends itself to a comparison where either bases (not possible in this case) or indices can be made to match. Alternatively, some standard approximations can be used which, if known, allow a different approach. You may find it useful to know that $2^{10} = 1024 \approx 10^3$ and $5^{10} = 9\,765\,625 \approx 10^7$. Then the following working gives a proof:
>
> $2^{10} = 1024 > 10^3$
> $\Rightarrow 2^{350} = (2^{10})^{35} > (10^3)^{35} = 10^{105}$
>
> $5^{10} = 9.77 \times 10^6 < 10^7$
> $\Rightarrow 5^{150} = (5^{10})^{15} < (10^7)^{15} = 10^{105}$
>
> $\therefore 2^{350} > 5^{150}$

12 $4^{ax} = b \times 8^x$

$ax \log_2 4 = \log_2 b + x \log_2 8$

$2ax = \log_2 b + 3x$

$x = \dfrac{\log_2 b}{2a - 3}$

Multiple solutions for x means that this is an undefined value, so $2a - 3 = 0 = \log_2 b$.

$\therefore a = \dfrac{3}{2},\ b = 1$

Exercise 2B

2 At 09:00 on Tuesday, $t = 0$ and $y = 10$.
Substituting:

$y = k \times 1.1^t$

$10 = k \times 1.1^0$

$k = 10$

At 09:00 on Friday, $t = 3$.

$\therefore y = 10 \times 1.1^3 = 13.31 \text{ m}^2$

3 $T = A + B \times 2^{-\frac{x}{k}}$

a Background temperature is 25°C, so as $x \to \infty,\ T \to 25$.

Since $2^{-\frac{x}{k}} \to 0,\ T \to A$.

Hence $A = 25$.

Temperature on surface of light bulb is 125°C, so when $x = 0,\ T = 125$.
Substituting:

$125 = 25 + B \times 2^0$

$\Rightarrow B = 100$

Air temperature 3 mm from surface of light bulb is 75°C, so when $x = 3$, $T = 75$. Substituting:

$75 = 25 + 100 \times 2^{-\frac{3}{k}}$

$\dfrac{1}{2} = 2^{-\frac{3}{k}}$

$2^{\frac{3}{k}} = 2$

$\dfrac{3}{k} = 1$

$k = 3$

b At 2 cm from the surface of the bulb, $x = 20$ (mm), so

$T = 25 + 100 \times 2^{-\frac{20}{3}}$

$= 26.0°\text{C}$

c

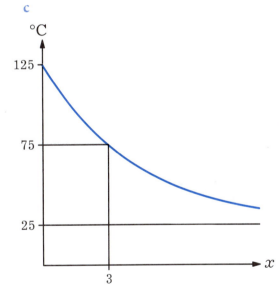

Figure 2B.3 Graph of $T = 25 + 100 \times 2^{-\frac{x}{3}}$

4 $h = 2 - 0.2 \times 1.6^{0.2m}$

a

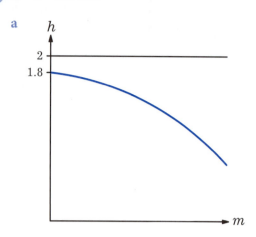

Figure 2B.4 Graph of $h = 2 - 0.2 \times 1.6^{0.2m}$

b When there is no fruit, $m = 0$, so
$h = 2 - 0.2 \times 1.6^0 = 1.8\,\text{m}$

c When $m = 7.5$,
$h = 2 - 0.2 \times 1.6^{0.2 \times 7.5} = 1.60\,\text{m}$ (3SF)

d The model was derived from data which, from (c), gave a height of 1.6 m above the ground at the harvest-time fruit load of 7.5 kg. Extrapolating so far beyond the model to reach $h = 0$ is unreliable and likely to be unrealistic; for example, the branch might simply break before being bent far enough to touch the ground.

5 a $y = 1 + 16^{1-x^2} = 1 + \dfrac{1}{16^{x^2-1}}$

As $x \to \infty$, $\dfrac{1}{16^{x^2-1}} \to 0$ and so $y \to 1$.

The maximum value for y must occur at the minimum value of $x^2 - 1$, which is when $x = 0$.

When $x = 0$, $y = 1 + \dfrac{1}{16^{0-1}} = 17$, so the maximum point is $(0, 17)$.

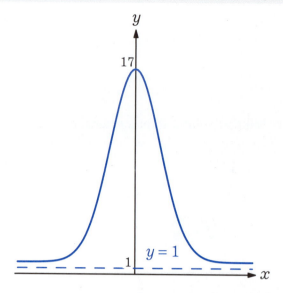

Figure 2B.5 Graph of $y = 1 + 16^{1-x^2}$

> **COMMENT**
> Some justification of how this graph could be constructed is given here, but you could just draw it on a GDC.

b When $y = 3$:

$3 = 1 + 16^{1-x^2}$

$2 = 16^{1-x^2}$

$16^{\frac{1}{4}} = 16^{1-x^2}$

$\dfrac{1}{4} = 1 - x^2$

$x^2 = \dfrac{3}{4}$

$x = \pm\dfrac{\sqrt{3}}{2} = \pm 0.866$ (3SF)

6 Temperature T of the soup decreases exponentially with time towards 20 (room temperature):
$T = 20 + A \times m^{kt}$

When the soup is served, $t = 0$ and $T = 55$,

$\therefore 55 = 20 + A \times m^0$

$A = 35$

Every 5 minutes the term $A \times m^{kt}$ must decrease by 30%, so $m=0.7$ and

$k = \dfrac{1}{5} = 0.2$.

$\therefore T = 20 + 35 \times 0.7^{0.2t}$

When $t=7$,

$T = 20 + 35 \times 0.7^{1.4} = 41.2°C$ (3SF)

7 $V = 40\left(1 - 3^{-0.1t}\right)$

  a When $t=0$,

  $V = 40(1-1) = 0 \text{ m s}^{-1}$

  b As $t \to \infty$, $3^{-0.1t} \to 0$ and so
  $V \to 40(1-0) = 40 \text{ m s}^{-1}$

Exercise 2C

2 This is a very close approximation to e (with less than a 7.5×10^{-7}% error).

> **COMMENT**
>
> It might at first seem that this is far too unlikely to be a mere coincidence and that there must be some underlying relationship, but this is in fact not the case. The so-called 'Strong Law of Small Numbers' gives some insight into the surprisingly regular occurrence of this type of coincidence.

Exercise 2D

6 $\log_{10}(9x+1) = 3$

$9x+1 = 10^3$

$9x+1 = 1000$

$x = 111$

7 $\log_8 \sqrt{1-x} = \dfrac{1}{3}$

$\sqrt{1-x} = 8^{\frac{1}{3}}$

$\sqrt{1-x} = 2$

$1-x = 4$

$x = -3$

8 $\ln(3x-1) = 2$

$3x - 1 = e^2$

$x = \dfrac{e^2 + 1}{3}$

9 $(\log_3 x)^2 = 4$

$\log_3 x = \pm 2$

$x = 3^{\pm 2} = 9 \text{ or } \dfrac{1}{9}$

10 $\begin{cases} \log_3 x + \log_5 y = 6 & \dots (1) \\ \log_3 x - \log_5 y = 2 & \dots (2) \end{cases}$

(1) + (2):

$2\log_3 x = 8$

$\log_3 x = 4$

$x = 3^4 = 81$

Substituting into (1):

$4 + \log_5 y = 6$

$\log_5 y = 2$

$y = 5^2 = 25$

i.e. $x=81$, $y=25$

11 $3(1 + \log x) = 6 + \log x$

$3 + 3\log x = 6 + \log x$

$2\log x = 3$

$\log x = \dfrac{3}{2}$

$x = 10^{\frac{3}{2}}$

$= 10\sqrt{10} = 31.6$ (3SF)

12 $\log_x 4 = 9$

$4 = x^9$

$x = 4^{\frac{1}{9}} = 1.17$ (3SF)

13 Let r be the Richter-scale value and s the strength of an earthquake.

Since an increase of one unit in r corresponds to an increase by a factor of 10 in s,

$s = C \times 10^r$ for some constant C.

Let t be the strength of an earthquake of Richter level 5.2:

$t = C \times 10^{5.2}$ ... (1)

For an earthquake twice as strong:

$2t = C \times 10^r$ ... (2)

$(2) \div (1)$: $\dfrac{2t}{t} = \dfrac{C \times 10^r}{C \times 10^{5.2}}$

$2 = 10^{r - 5.2}$

$r - 5.2 = \log 2$

$r = \log 2 + 5.2 = 5.50$

An earthquake twice as strong as a level-5.2 quake would measure 5.5 on the Richter scale.

> **COMMENT**
>
> The constant C is needed here as the precise relationship between r and s is not given, but it is not necessary to find the value of C to answer this particular question.

Exercise 2E

4 $2\ln x + \ln 9 = 3$

$\ln x = \dfrac{3 - \ln 9}{2}$

$= \dfrac{3 - 2\ln 3}{2}$

$= \dfrac{3}{2} - \ln 3$

$\therefore x = e^{\frac{3}{2} - \ln 3}$

$= e^{\frac{3}{2}} \times e^{-\ln 3}$

$= \dfrac{1}{3} e^{\frac{3}{2}}$

5 a $\ln 50 = \ln 2 + \ln 25$

$= \ln 2 + \ln 5^2$

$= \ln 2 + 2\ln 5$

$= a + 2b$

b $\ln 0.16 = \ln\left(\dfrac{4}{25}\right)$

$= \ln\left(\dfrac{2^2}{5^2}\right)$

$= \ln 2^2 - \ln 5^2$

$= 2\ln 2 - 2\ln 5$

$= 2a - 2b$

6 $\log_2 x = \log_x 2$

$\log_2 x = \dfrac{\log_2 2}{\log_2 x}$

$(\log_2 x)^2 = 1$

$\log_2 x = \pm 1$

$x = 2^{\pm 1} = 2$ or $\dfrac{1}{2}$

7 $a^x = (ab)^{xy}$

$x \ln a = xy \ln ab$

$x(\ln a - y \ln ab) = 0$

$\Rightarrow x = 0 \quad \text{or} \quad y = \dfrac{\ln a}{\ln ab} = \dfrac{\ln a}{\ln a + \ln b}$

$b^y = (ab)^{xy}$

$y \ln b = xy \ln ab$

$y(\ln b - x \ln ab) = 0$

$\Rightarrow y = 0 \quad \text{or} \quad x = \dfrac{\ln b}{\ln ab} = \dfrac{\ln b}{\ln a + \ln b}$

$x = 0 \Rightarrow a^x = 1$, so $b^y = 1$

and hence $y = 0$

(since $a, b > 1$)

$\therefore$ either $x = y = 0$ or $x + y = \dfrac{\ln b + \ln a}{\ln a + \ln b} = 1$

8 $\log \dfrac{1}{2} + \log \dfrac{2}{3} + \log \dfrac{3}{4} + \ldots + \log \dfrac{8}{9} + \log \dfrac{9}{10}$

$= \log \left(\dfrac{1}{\cancel{2}} \times \dfrac{\cancel{2}}{\cancel{3}} \times \dfrac{\cancel{3}}{\cancel{4}} \times \ldots \times \dfrac{\cancel{8}}{\cancel{9}} \times \dfrac{\cancel{9}}{10} \right)$

$= \log \left(\dfrac{1}{10} \right)$

$= -1$

9 $\log_a b = \log_b a$

Using change of base:

$\dfrac{\log b}{\log a} = \dfrac{\log a}{\log b}$

$(\log b)^2 = (\log a)^2$

$\log b = \pm \log a$

$\log b = \log(a^{\pm 1})$

$b = a^{\pm 1}$

Reject $b = a$

$\therefore b = a^{-1} = \dfrac{1}{a}$

Exercise 2F

2 The domains are different; $y = 2\log x$ has domain $x > 0$ whereas $y = \log(x^2)$ has domain $x \neq 0$ and is equivalent to $y = 2\log|x|$.

COMMENT

The rule of logarithms that $\log x^p = p \log x$ only applies to positive x.

Exercise 2G

2 $N = 100e^{1.03t}$

a When $t = 0$,

$N = 100e^0 = 100$

b When $t = 6$,

$N = 100e^{1.03 \times 6} = 48\,300$ (3SF)

c $N = 1000$ when

$1000 = 100e^{1.03t}$

$e^{1.03t} = 10$

$1.03t = \ln 10$

$t = \dfrac{1}{1.03} \ln 10$

$= 2.24$ hours (3SF)

The population will reach 1000 cells after approximately 2 hours and 14 minutes.

COMMENT

0.24 hours is $0.24 \times 60 = 14.4$ minutes.

3 Let N be the number of people who know the rumour t minutes after 9 a.m. Then N can be modelled by the exponential function $N = Ae^{kt}$

At 9 a.m. 18 people know the rumour, so when $t=0$, $N=18$:

$18 = Ae^0$
$\Rightarrow A = 18$

At 10 a.m. 42 people know the rumour, so when $t=60$, $N=42$:

$42 = 18e^{60k}$

$e^{60k} = \dfrac{42}{18}$

$60k = \ln\dfrac{7}{3}$

$k = \dfrac{1}{60}\ln\dfrac{7}{3}$

$= 0.0141216 = 0.0141\,(3\text{SF})$

Therefore $N = 18e^{0.0141t}$.

a At 10:30 a.m., $t = 90$,

$\therefore N = 18e^{90 \times 0.0141} = 64.2$

So 64 people know the rumour at 10:30 a.m.

b If 1200 people know the rumour, then

$1200 = 18e^{0.0141t}$

$e^{0.0141t} = \dfrac{1200}{18}$

$0.0141t = \ln\dfrac{200}{3}$

$t = \dfrac{1}{0.0141}\ln\dfrac{200}{3} = 297.4$

So after 297.4 minutes (4.96 hours), i.e. at 13:58, the whole school population will know the rumour.

4 $P = 32(e^{0.0012t} - 1)$

a 2 minutes = 120 seconds
$P(120) = 4.96$ units per second
(from GDC)

b $\dfrac{P}{32} + 1 = e^{0.0012t}$

$\therefore t = \dfrac{1}{0.0012}\ln\left(1 + \dfrac{P}{32}\right)$

$P = 7 \times 10^5 \Rightarrow t = \dfrac{1}{0.0012}\ln\left(1 + \dfrac{7 \times 10^5}{32}\right)$

$= 8327.6$

So the experiment can be run for 8328 seconds, equivalent to 2 hours, 18 minutes and 48 seconds.

5 a

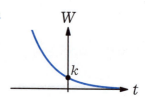

Figure 2G.5 Graph of $W = ke^{-0.01t}$

b 25% of the initial mass means $W = \dfrac{k}{4}$.

$\dfrac{k}{4} = ke^{-0.01t}$

$e^{-0.01t} = \dfrac{1}{4}$

$e^{0.01t} = 4$

$0.01t = \ln 4$

$t = 100\ln 4$

$= 138.6$ seconds

The block will be at 25% of its original weight after 2.31 minutes.

> **COMMENT**
>
> 138.6 seconds is $138.6 \div 60 = 2.31$ minutes.

6 $5 \times 4^{x-1} = \dfrac{1}{3^{2x}}$

$5 \times 4^{-1} \times 4^x = \dfrac{1}{(3^2)^x}$

$\dfrac{5}{4} = \dfrac{\frac{1}{9^x}}{4^x}$

$\dfrac{5}{4} = \dfrac{1}{36^x}$

$36^x = \dfrac{4}{5}$

$\ln 36^x = \ln \dfrac{4}{5}$

$x \ln 36 = \ln \dfrac{4}{5}$

$x = \dfrac{\ln \frac{4}{5}}{\ln 36}$

COMMENT

The answer in the back of the coursebook,

$x = \dfrac{\ln\left(\frac{5}{4}\right)}{\ln\left(\frac{1}{36}\right)}$, is an equivalent, though not fully simplified, form.

---

7 $\dfrac{1}{7^x} = 3 \times 49^{5-x}$

$7^{-x} = 3 \times (7^2)^{5-x}$

$7^{-x} = 3 \times 7^{10-2x}$

$\dfrac{7^{-x}}{7^{10-2x}} = 3$

$7^{x-10} = 3$

$x - 10 = \log_7 3$

$x = 10 + \log_7 3$

8 Let K (°C) be the temperature after t seconds; then

$K - 22 = A e^{-bt}$
$K(0) = 98 \Rightarrow A = 76$

$K(120) = 94 \Rightarrow 72 = 76 e^{-120b}$

$\therefore b = -\dfrac{1}{120} \ln\left(\dfrac{72}{76}\right) = 0.000451$ (3SF)

Hence $t = -\dfrac{1}{b} \ln\left(\dfrac{K-22}{A}\right)$

So $K = 78 \Rightarrow t = -\dfrac{1}{0.000451} \ln\left(\dfrac{56}{76}\right)$

$= 678$ (3SF)

It takes 678 seconds, equivalent to 11 minutes and 18 seconds, for the tea to cool to 78°C.

9 a $y = 3^x$ is always increasing and $y = 3 - x$ is always decreasing, so their two graphs can intersect at most once. Since $3^0 < 3 - 0$ and $3^1 > 3 - 1$, there must be an intersection in $[0, 1]$.

b From GDC, the intersection occurs at $x = 0.742$ (3SF)

Mixed examination practice 2

Short questions

1 $\log_5\left(\sqrt{x^2 + 49}\right) = 2$

$\sqrt{x^2 + 49} = 25$

$x^2 + 49 = 625$

$x^2 = 576$

$x = \pm 24$

2 a $\log \dfrac{x^2 \sqrt{y}}{z} = \log x^2 + \log \sqrt{y} - \log z$

$= \log x^2 + \log y^{\frac{1}{2}} - \log z$

$= 2\log x + \dfrac{1}{2}\log y - \log z$

$= 2a + \dfrac{b}{2} - c$

2 Exponents and logarithms 17

b $\log\sqrt{0.1x} = \log(0.1x)^{\frac{1}{2}}$

$= \frac{1}{2}\log 0.1x$

$= \frac{1}{2}(\log 0.1 + \log x)$

$= \frac{1}{2}(-1 + \log x)$

$= \frac{a-1}{2}$

c $\log_{100}\frac{y}{z} = \frac{\log\frac{y}{z}}{\log 100}$

$= \frac{\log y - \log z}{\log 100}$

$= \frac{b-c}{2}$

3 $\ln x + \ln y^2 = 8$

$\ln x^2 + \ln y = 6$

$\ln x + 2\ln y = 8 \quad \ldots(1)$

$2\ln x + \ln y = 6 \quad \ldots(2)$

$2 \times (2) - (1)$:

$3\ln x = 4$

$\ln x = \frac{4}{3}$

$x = e^{\frac{4}{3}} = 3.79 \ (3\text{SF})$

Substituting in (2):

$2 \times \frac{4}{3} + \ln y = 6$

$\ln y = \frac{10}{3}$

$y = e^{\frac{10}{3}} = 28.0 \ (3\text{SF})$

4 $y = \ln x - \ln(x+2) + \ln(4-x^2)$

$= \ln\left(\frac{x(4-x^2)}{x+2}\right)$

$= \ln\left(\frac{x(2-x)(2+x)}{x+2}\right)$

$= \ln(x(2-x))$

$= \ln(2x - x^2)$

$\therefore e^y = 2x - x^2$

$x^2 - 2x + e^y = 0$

$x = \frac{2 \pm \sqrt{(-2)^2 - 4 \times 1 \times e^y}}{2}$

$= \frac{2 \pm \sqrt{4 - 4e^y}}{2}$

$= \frac{2 \pm 2\sqrt{1 - e^y}}{2}$

$= 1 \pm \sqrt{1 - e^y}$

5

> **COMMENT**
> To approach this type of question, try to rewrite it so that all terms involving the unknown *x* have either a common base or a common exponent.

$2^{3x-2} \times 3^{2x-3} = 36^{x-1}$

$2^{-2} \times (2^3)^x \times 3^{-3} \times (3^2)^x = 36^{-1} \times 36^x$

$\frac{1}{4} \times 8^x \times \frac{1}{27} \times 9^x = \frac{1}{36} \times 36^x$

$\left(\frac{8 \times 9}{36}\right)^x = \frac{4 \times 27}{36}$

$2^x = 3$

$x = \frac{\ln 3}{\ln 2}$

6 Changing $\log_a$ and $\log_b$ into ln:

$\log_a b^2 = c$

$\Rightarrow 2\log_a b = c$

$\Rightarrow 2\dfrac{\ln b}{\ln a} = c \quad \ldots (1)$

$\log_b a = c - 1$

$\Rightarrow \dfrac{\ln a}{\ln b} = c - 1 \quad \ldots (2)$

$(1) - (2)$:

$2\dfrac{\ln b}{\ln a} - \dfrac{\ln a}{\ln b} = 1$

$2(\ln b)^2 - (\ln a)^2 = \ln a \ln b$

$(\ln a)^2 + \ln b(\ln a) - 2(\ln b)^2 = 0$

Treating this as a quadratic in $\ln a$ and factorising:

$(\ln a - \ln b)(\ln a + 2\ln b) = 0$

$\therefore \ln a = \ln b$ or $\ln a = -2\ln b$

i.e. $a = b$ or $a = e^{-2\ln b} = e^{\ln b^{-2}} = b^{-2}$

But we are given that $a < b$, so $a \neq b$ and hence $a = b^{-2}$.

7 $9\log_5 x = 25\log_x 5$

Using change-of-base formula:

$\dfrac{9\log x}{\log 5} = \dfrac{25\log 5}{\log x}$

$(\log x)^2 = \dfrac{25}{9}(\log 5)^2$

$\log x = \pm\dfrac{5}{3}\log 5 = \log\left(5^{\pm\frac{5}{3}}\right)$

$\therefore x = 5^{\pm\frac{5}{3}}$

8 $\ln x = 4\log_x e$

$\ln x = 4\dfrac{\ln e}{\ln x}$

$(\ln x)^2 = 4$

$\ln x = \pm 2$

$x = e^{\pm 2}$

Long questions

1 a As $t \to \infty$, $e^{-0.2t} \to 0$ and so $V \to 42$

When $t = 0$, $V = 42(1 - e^0) = 0$

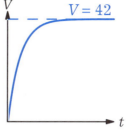

Figure 2ML.1 Graph of $V = 42(1 - e^{-0.2t})$

b When $t = 0$, $V = 42(1 - e^0) = 0\ \text{m s}^{-1}$

c As $t \to \infty$, $e^{-0.2t} \to 0$,

$\therefore V \to 42\ \text{m s}^{-1}$

d When $V = 22$,

$22 = 42(1 - e^{-0.2t})$

$1 - e^{-0.2t} = \dfrac{22}{42}$

$e^{-0.2t} = 1 - \dfrac{11}{21} = \dfrac{10}{21}$

$-0.2t = \ln\dfrac{10}{21}$

$0.2t = -\ln\dfrac{10}{21} = \ln\dfrac{21}{10}$

$t = 5\ln\dfrac{21}{10} = 3.71\ \text{s}\ (3\text{SF})$

> **COMMENT**
>
> The graph in (a) can be drawn immediately with a GDC; the answers to (b) and (c) can then simply be deduced from the graph without first calculating V when $t = 0$ and the limiting value of V as $t \to \infty$.

2 a $T = ka^n$

37 000 tigers in 1970, i.e. when $n = 0$,

$T = 37\,000$.

$\therefore 37\,000 = ka^0$

$k = 37\,000$

22 000 tigers in 1980, i.e. when $n = 10$,

$T = 22\,000$.

$\therefore 22\,000 = 37\,000 a^{10}$

$a^{10} = \dfrac{22}{37}$

$a = \sqrt[10]{\dfrac{22}{37}} = 0.949 \,(3\text{SF})$

Hence $T = 37\,000 \times 0.949^n$.

b In 2020, $n = 50$:

$T = 37\,000 \times 0.949^{50} = 2750$

c When $T = 1000$:

$1000 = 37\,000 \times 0.949^n$

$0.949^n = \dfrac{1}{37}$

$\ln 0.949^n = \ln \dfrac{1}{37}$

$n \ln 0.949 = \ln \dfrac{1}{37}$

$n = \dfrac{\ln \dfrac{1}{37}}{\ln 0.949} = 69.5$

so tigers will reach 'near extinction' in 2039.

d Under the initial model, in 2000 ($n = 30$) the number of tigers is

$T = 37\,000 \times 0.949^{30} = 7778$

The new model, $T = kb^m$, has $T = 7778$ when $m = 0$, so $k = 7778$.

Under this model ($T = 7778 b^m$), there are 10 000 tigers in 2010, i.e. when $m = 10$, $T = 10\,000$.

$\therefore 10\,000 = 7778 b^{10}$

$b^{10} = \dfrac{10\,000}{7778}$

$b = \sqrt[10]{\dfrac{10\,000}{7778}} = 1.025$

Therefore the new model is

$T = 7778 \times 1.025 m$.

e The growth factor each year is 1.025, equivalent to a 2.5% growth rate.

3 a $\ln y = 2 \ln x + \ln 3 = \ln(3x^2)$

$\Rightarrow y = 3x^2$

(Note that $x > 0$ for the original relationship to hold.)

b $\ln y = 4 \ln x + 6$

$= \ln(x^4) + \ln(e^6)$

$= \ln(x^4 e^6)$

$\Rightarrow y = e^6 x^4$

c $\ln y - 2 = 3(x - 1)$

$\ln y = 3x - 1$

$\Rightarrow y = e^{3x-1}$

d $e^y = 4x^2$

$\Rightarrow y = \ln(4x^2) = \ln 4 + 2 \ln x$

So the graph of y against $\ln x$ is a straight line with gradient 2.

3 Polynomials

Exercise 3A

6 Factorised form is more useful for finding roots; expanded form is more useful for evaluating the y-intercept and for comparing, adding or subtracting polynomials.

7 a Yes; the term in x^n is unaffected by adding a lower-order polynomial.

  b No; if the lead coefficients have a zero sum, then the sum of the polynomials will not have a term in x^n, so the resultant will be of lower order. For example, the sum of the third-order polynomials $f(x) = x^3 - 2x$ and $g(x) = 3 + x - x^3$ is $f(x) + g(x) = 3 - x$, a polynomial of order 1.

Exercise 3B

> **COMMENT**
> It can be helpful in questions to label a function as $f(x)$ or $g(x)$, so that evaluating at particular values of x can be clearly described.

4 $f(x) = 6x^3 + ax^2 + bx + 8$

By the factor theorem:
$f(-2) = 0 = -48 + 4a - 2b + 8$
$4a - 2b = 40$
$\Rightarrow b = 2a - 20 \quad \ldots (1)$

By the remainder theorem:
$f(1) = -3 = 6 + a + b + 8$
$\Rightarrow a + b = -17 \quad \ldots (2)$

Substituting (1) into (2):
$3a - 20 = -17$
$\therefore a = 1, b = -18$

5 $f(x) = x^3 + 8x^2 + ax + b$

By the factor theorem:
$f(2) = 0 = 8 + 32 + 2a + b$
$\Rightarrow b = -2a - 40 \quad \ldots (1)$

By the remainder theorem:
$f(3) = 15 = 27 + 72 + 3a + b$
$\Rightarrow 3a + b = -84 \quad \ldots (2)$

Substituting (1) into (2):
$a - 40 = -84$
$\therefore a = -44, \ b = 48$

6 $f(x) = x^2 + kx - 8k$

By the factor theorem:
$f(k) = 0 = k^2 + k^2 - 8k$
$2k^2 - 8k = 0$
$k(k - 4) = 0$
$k = 0 \ \text{or} \ k = 4$

7 $f(x) = x^2 - (k+1)x - 3$

By the factor theorem:
$f(k-1) = 0 = (k-1)^2 - (k+1)(k-1) - 3$
$k^2 - 2k + 1 - k^2 + 1 - 3 = 0$
$-2k - 1 = 0$

$\therefore k = -\dfrac{1}{2}$

8 $f(x) = x^3 - ax^2 - bx + 168$

a By the factor theorem:
$f(7) = 0 = 343 - 49a - 7b + 168$
$\Rightarrow b = -7a + 73$

Also,
$f(3) = 0 = 27 - 9a - 3b + 168$
$\Rightarrow 9a + 3b = 195$
$\therefore 9a + 3(-7a + 73) = 195$
$-12a = -24$
$\Rightarrow a = 2, b = 59$

b $f(x) = (x-3)(x-7)(x-k)$

Expanding:
$f(x) = x^3 - (10+k)x^2 + (21+10k)x - 21k$

Comparing coefficients:
$x^2 : 10 + k = a = 2 \Rightarrow k = -8$

$x^1 : 21 + 10k = -b = -59$, consistent with $k = -8$

$x^0 : -21k = 168$, consistent with $k = -8$
So the remaining factor is $(x+8)$.

COMMENT

Only one of the coefficient comparisons was needed to find the final factor; however, it is good practice to quickly verify – whether written down in an exam solution or not – that the other comparisons are valid, to check for errors in working.

9 $f(x) = x^3 + ax^2 + 9x + b$

a By the factor theorem:
$f(11) = 0 = 1331 + 121a + 99 + b$
$\Rightarrow b = -1430 - 121a$

By the remainder theorem:
$f(-2) = -52 = -8 + 4a - 18 + b$
$\Rightarrow 4a + b = -26$
$\therefore 4a - 1430 - 121a = -26$
$-117a = 1404$
$\Rightarrow a = -12, b = 22$

b By the remainder theorem, the remainder when divided by $(x-2)$ is $f(2)$:

$f(2) = 8 + 4a + 18 + b$
$= 8 - 48 + 18 + 22 = 0$

That is, $(x-2)$ is a factor of $f(x)$.

10 $f(x) = x^3 + ax^2 + 3x + b$

By the remainder theorem,
$f(-1) = 6 = -1 + a - 3 + b$
$\Rightarrow a + b = 10$

The remainder when divided by $(x-1)$ is $f(1)$:
$f(1) = 1 + a + 3 + b = 1 + 3 + 10 = 14$

11

COMMENT

There are two sensible approaches here. You may recognise that the given quadratic factorises readily into $(x-2)(x-3)$, so it would be possible to apply the factor theorem to the cubic and solve $f(2) = f(3) = 0$ to find a and b. Alternatively, propose a final factor $(2x - k)$, chosen to ensure that the lead coefficient will be correct, and then expand and compare coefficients. Both methods are shown below.

$f(x) = 2x^3 - 15x^2 + ax + b$

Method 1:

$x^2 - 5x + 6 = (x-3)(x-2)$, so both $(x-2)$ and $(x-3)$ are factors of $f(x)$.

By the factor theorem:
$f(2) = 0 = 16 - 60 + 2a + b$
$\Rightarrow b = 44 - 2a$

Also,
$f(3) = 0 = 54 - 135 + 3a + b$
$\Rightarrow 3a + b = 81$
$\therefore 3a + 44 - 2a = 81$
$\Rightarrow a = 37, \ b = -30$

Method 2:
$f(x) = (2x - k)(x^2 - 5x + 6)$
$\quad = 2x^3 + (-k - 10)x^2 + (5k + 12)x - 6k$

Comparing coefficients:

$x^3: \ 2 = 2$

$x^2: \ -15 = -10 - k \Rightarrow k = 5$

$x^1: \ a = 5k + 12 = 37$

$x^0: \ b = -6k = -30$

> **COMMENT**
> Although the two methods are of similar difficulty, the first requires that you spot the factors of the quadratic quickly, which is not needed for the second method. The second method also produces the final factor, which may be useful in a multi-part question.

Exercise 3C

7 Repeated root at $x = -2$, single root at $x = 3$; y-intercept at 24. Negative cubic shape.

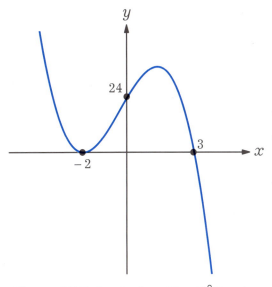

Figure 3C.7 Graph of $y = 2(x+2)^2(3-x)$

8 a Repeated root at $x = 3 \Rightarrow$ factor of $(x-3)^2$
Single root at $x = -2 \Rightarrow$ factor of $(x+2)$
$\therefore y = k(x-3)^2(x+2)$
$\quad = k(x^3 - 4x^2 - 3x + 18)$
$y(0) = 36 = 18k \Rightarrow k = 2$
$\therefore y = 2x^3 - 8x^2 - 6x + 36$
i.e. $p = 2, \ q = -8, \ r = -6, \ s = 36$

b Repeated root at $x = 0 \Rightarrow$ factor of x^2
Single root at $x = 3 \Rightarrow$ factor of $(x-3)$
$\therefore y = kx^2(x-3)$
$y(2) = 4 = -4k \Rightarrow k = -1$
$\therefore y = -x^3 + 3x^2$
i.e. $p = -1, \ q = 3, \ r = s = 0$

3 Polynomials

9 a $x^4 - q^4 = (x^2 + q^2)(x^2 - q^2)$
$= (x^2 + q^2)(x + q)(x - q)$

b Roots at $x = \pm q$ only;
y-intercept at $-q^4$.
Positive quartic shape. Even function (reflective symmetry about the y-axis).

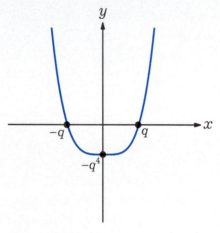

Figure 3C.9 Graph of $y = x^4 - q^4$

10 a Repeated root at $x = p$, single root at $x = q$; y-intercept at $-p^2 q$.
Positive cubic shape.

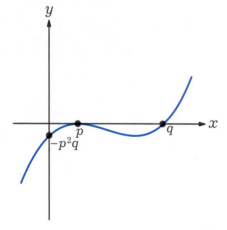

Figure 3C.10 Graph of $y = (x - p)^2(x - q)$

b From the graph, there will be only one solution for $y = k$ when $k > 0$.

Exercise 3D

5 Equal roots when discriminant is zero:
$\Delta = b^2 - 4ac = 0$
$(-4)^2 - 4 \times m \times 2m = 0$
$16 - 8m^2 = 0$
$m^2 = 2$
$m = \pm\sqrt{2}$

6 Tangent to the x-axis implies equal roots, so discriminant is zero:
$\Delta = b^2 - 4ac = 0$
$(2k+1)^2 - 4 \times (-3) \times (-4k) = 0$
$4k^2 + 4k + 1 - 48k = 0$
$4k^2 - 44k + 1 = 0$
$k = \dfrac{44 \pm \sqrt{44^2 - 4 \times 4 \times 1}}{2 \times 4}$
$= \dfrac{44 \pm \sqrt{1920}}{8}$
$= \dfrac{11}{2} \pm \sqrt{30}$

7 For a quadratic to be non-negative (≥ 0) for all x, it must have at most one root, so $\Delta \leq 0$ and $a > 0$.
$b^2 - 4ac \leq 0$
$(-3)^2 - 4 \times 2 \times (2c - 1) \leq 0$
$9 - 16c + 8 \leq 0$
$c \geq \dfrac{17}{16}$

> **COMMENT**
> Note that $\Delta \leq 0$ is not sufficient in general for a quadratic to be non-negative. The condition $a > 0$ is also necessary to ensure that the quadratic has a positive shape (opening upward) rather than a negative shape (opening downward), so that the curve remains above the x-axis and never goes below it, as would be the case if $a < 0$. In this question a was given as positive (2), so we did not need to use this condition at all.

> **COMMENT**
> The condition $a < 0$ ensures that the function is negative shaped and therefore remains below the x-axis. In this case $a = m$, and it followed from the condition on Δ that $a < 0$, as seen in the answer.

8 No real solutions when $\Delta < 0$:

$(-2k)^2 - 4 \times 1 \times 6k < 0$

$4k^2 - 24k < 0$

$k(k-6) < 0$

$0 < k < 6$

9 No real roots when $\Delta < 0$:

$(k+3)^2 - 4 \times k \times (-1) < 0$

$k^2 + 10k + 9 < 0$

$(k+9)(k+1) < 0$

$-9 < k < -1$

10 At least one root when $\Delta \geq 0$:

$m^2 - 4 \times m \times (-2) \geq 0$

$m(m+8) \geq 0$

$m \leq -8$ or $m \geq 0$

11 For a quadratic to be negative for all x, it must have no real roots, so $\Delta < 0$ and $a < 0$.

$b^2 - 4ac < 0$

$3^2 - 4 \times m \times (-4) < 0$

$9 + 16m < 0$

$m < -\dfrac{9}{16}$

12 The two zeros of $ax^2 + bx + c$ are

$\dfrac{-b + \sqrt{b^2 - 4ac}}{2a}$ and $\dfrac{-b - \sqrt{b^2 - 4ac}}{2a}$.

The positive difference between these zeros is

$\left| \dfrac{-b + \sqrt{b^2 - 4ac}}{2a} - \dfrac{-b - \sqrt{b^2 - 4ac}}{2a} \right|$

$= \left| \dfrac{2\sqrt{b^2 - 4ac}}{2a} \right|$

$= \left| \dfrac{\sqrt{b^2 - 4ac}}{a} \right|$

So, in this case,

$\dfrac{\sqrt{k^2 - 12}}{1} = \sqrt{69}$

$k^2 - 12 = 69$

$k^2 = 81$

$k = \pm 9$

> **COMMENT**
> Note that modulus signs were used in the general expression for the positive distance, as a could be negative. Here $a = 1$ and so the modulus was not required in the specific case in this question.

Mixed examination practice 3

Short questions

1 Roots at $x=k$ and $x=k+4 \Rightarrow$ line of symmetry is $x=k+2$ (midway between the roots).
So the x-coordinate of the turning point is $k+2$.

2
- Negative quadratic $\Rightarrow a$ is negative
- Negative y-intercept $\Rightarrow c$ is negative
- Single (repeated) root $\Rightarrow b^2 - 4ac = 0$
- Line of symmetry $x = -\dfrac{b}{2a}$ is positive
$\Rightarrow b$ is positive (as a is negative)

TABLE 3MS.2

Expression	Positive	Negative	Zero
a		✓	
c		✓	
$b^2 - 4ac$			✓
b	✓		

3 Repeated root at $x = -3 \Rightarrow$ factor of $(x+3)^2$

Single root at $x = 1 \Rightarrow$ factor of $(x-1)$

Single root at $x = 3 \Rightarrow$ factor of $(x-3)$

$\therefore y = k(x+3)^2(x-1)(x-3)$
$= k(x^4 + 2x^3 - 12x^2 - 18x + 27)$

$y(0) = 27 \Rightarrow k = 1$

$\therefore y = x^4 + 2x^3 - 12x^2 - 18x + 27$

So $a = 1$, $b = 2$, $c = -12$, $d = -18$, $e = 27$

4 $f(x) = (ax+b)^3$

By the remainder theorem:

$f(2) = 8 = (2a+b)^3$
$\Rightarrow 2a + b = 2$
$\Rightarrow b = 2 - 2a$

Also,

$f(-3) = -27 = (b-3a)^3$
$\Rightarrow b - 3a = -3$
$\therefore 2 - 2a - 3a = -3$
$-5a = -5$
$\Rightarrow a = 1, \ b = 0$

5 $f(x) = x^3 - 4x^2 + x + 6$

a $f(2) = 8 - 16 + 2 + 6 = 0$, so by the factor theorem, $(x-2)$ is a factor of $f(x)$.

b $f(x) = (x-2)(x^2 + ax + b)$
$= x^3 + (a-2)x^2 + (b-2a)x - 2b$

Comparing coefficients:
$x^2 : a - 2 = -4 \Rightarrow a = -2$

$x^1 : b - 2a = 1 \Rightarrow b = -3$

$x^0 : -2b = 6$ is consistent with the value found above.

> **COMMENT**
> The final coefficient comparison is useful for checking the validity of the solution. Always be thorough and compare all coefficients, even if you do not write down the check as part of your answer.

$\therefore f(x) = (x-2)(x^2 - 2x - 3)$
$= (x-2)(x-3)(x+1)$

c Roots at $x = -1, 2, 3$; y-intercept at 6. Positive cubic shape.

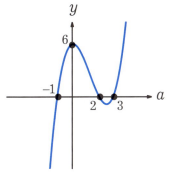

Figure 3MS.5 Graph of $f(x) = x^3 - 4x^2 + x + 6$

6 $f(x) = (ax+b)^4$

By the remainder theorem:
$f(2) = 16 = (2a+b)^4$
$2a + b = \pm 2$
$\Rightarrow b = \pm 2 - 2a$

Also,
$f(-1) = 81 = (b-a)^4$
$\Rightarrow b - a = \pm 3$

$\therefore \pm 2 - 3a = 3$ or $\pm 2 - 3a = -3$

Hence $(a,b) = \left(-\dfrac{1}{3}, \dfrac{8}{3}\right), \left(-\dfrac{5}{3}, \dfrac{4}{3}\right),$
$\left(\dfrac{5}{3}, -\dfrac{4}{3}\right), \left(\dfrac{1}{3}, -\dfrac{8}{3}\right)$

7 Repeated root at $x = a$, single roots at $x = b$ and $x = c$; y-intercept at $a^2bc < 0$.

Positive quartic shape.

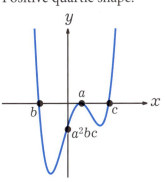

Figure 3MS.7 Graph of $y = (x-a)^2(x-b)(x-c)$ for $b < 0 < a < c$

8 Equal roots when $\Delta = 0$:
$b^2 - 4ac = 0$
$(k+1)^2 - 4 \times 2k \times 1 = 0$
$k^2 - 6k + 1 = 0$
$k = \dfrac{6 \pm \sqrt{32}}{2} = 3 \pm 2\sqrt{2}$

9 No real roots when $\Delta < 0$:
$k^2 - 4 \times 2 \times 6 < 0$
$k^2 < 48$
$\Rightarrow -4\sqrt{3} < k < 4\sqrt{3}$

10 At least one real root when $\Delta \geq 0$:
$(2k+1)^2 - 4 \times 1 \times 5 \geq 0$
$(2k+1)^2 \geq 20$

$\therefore 2k + 1 \geq 2\sqrt{5}$ or $2k + 1 \leq -2\sqrt{5}$

$\Rightarrow k \leq -\dfrac{1}{2} - \sqrt{5}$ or $k \geq -\dfrac{1}{2} + \sqrt{5}$

11 $f(x) = x^3 + ax^2 + 27x + b$
$= (x+k)(x^2 - 4x + 3)$
$= x^3 + (k-4)x^2 + (3-4k)x + 3k$

Comparing coefficients:
$x^3 : 1 = 1$
$x^2 : a = k - 4$
$x^1 : 27 = 3 - 4k \Rightarrow k = -6$
$x^0 : b = 3k = -18$

$\therefore a = -10, b = -18$

> **COMMENT**
>
> See Exercise 3B question 11 for an alternative method using the factor theorem.

12 a Roots of $x^2 - kx + (k-1) = 0$ are

$$\frac{k \pm \sqrt{k^2 - 4(k-1)}}{2} = \frac{k \pm \sqrt{k^2 - 4k + 4}}{2}$$

$$= \frac{k \pm \sqrt{(k-2)^2}}{2}$$

$$= \frac{k \pm (k-2)}{2}$$

$$= k-1 \text{ or } 1$$

$\therefore \alpha = k-1, \ \beta = 1$

b $\alpha^2 + \beta^2 = 17$

$(k-1)^2 + 1 = 17$

$k^2 - 2k + 2 = 17$

$k^2 - 2k - 15 = 0$

$(k-5)(k+3) = 0$

$k = 5$ or $k = -3$

13 Discriminant is

$\Delta = (k-2)^2 - 4 \times k \times (-2)$

$= k^2 - 4k + 4 + 8k$

$= k^2 + 4k + 4$

$= (k+2)^2$

$\Delta \geq 0$ for all values of k

$\therefore q(x)$ has at least one real root for any value of k.

14 Require $x^2 - kx + 2 \geq 0$ for all x, i.e. the quadratic $x^2 - kx + 2 = 0$ has at most one real root

$\therefore \Delta = k^2 - 8 \leq 0$

$k^2 \leq 8$

$-2\sqrt{2} \leq k \leq 2\sqrt{2}$

Long questions

1 a $y(0) = -a$, so y-intercept is $(0, -a)$

b Completing the square:

$$y = \left(x + \frac{b}{2}\right)^2 - a - \frac{b^2}{4}$$

$\therefore$ axis of symmetry is $x = -\frac{b}{2}$

c By the remainder theorem, the remainder is

$$y\left(\frac{a}{b}\right) = \left(\frac{a}{b}\right)^2 + b\left(\frac{a}{b}\right) - a = \frac{a^2}{b^2} > 0$$

d By the remainder theorem,

$y(a) = -9 = a^2 + ab - a$

$a^2 + (b-1)a + 9 = 0$

This quadratic must have at least one real solution for a, so discriminant $\Delta \geq 0$:

$(b-1)^2 - 36 \geq 0$

$(b-1)^2 \geq 36$

$b - 1 \leq -6$ or $b - 1 \geq 6$

$\therefore b \leq -5$ or $b \geq 7$

> **COMMENT**
>
> As an alternative, more direct method, rearrange the equation in a and b to find b in terms of a:
>
> $-9 = a^2 + ab - a$
>
> $\Rightarrow b = \dfrac{-9 + a - a^2}{a} = 1 - a - \dfrac{9}{a}$
>
> Plot the graph of this rational function on the GDC:
>
>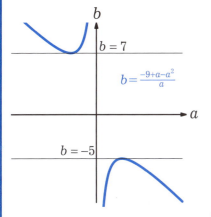
>
> **Figure 3 ML.1** Graph of $b = 1 - a - \dfrac{9}{a}$
>
> Then the range of b can be read from the graph.
> It would be appropriate to include a sketch in your solution to justify this method.

2 a $f(2) = 8 + 4(p-2) + 2(5-2p) - 10 = 0$

So by the factor theorem, $(x-2)$ is a factor of $f(x)$, irrespective of the value of p.

b $f(x) = (x-2)(x^2 + ax + b)$
$= x^3 + (a-2)x^2 + (b-2a)x - 2b$

Comparing coefficients:
$x^3 : 1 = 1$
$x^2 : p - 2 = a - 2 \Rightarrow a = p$
$x^1 : 5 - 2p = b - 2a \Rightarrow b = 5$
$x^0 : -10 = -2b$ is consistent with the value of b found above.

$\therefore f(x) = (x-2)(x^2 + px + 5)$

For exactly two roots, there is either a repeated root at 2 and a single root elsewhere or a single root at 2 and a repeated root elsewhere.

Let $g(x) = x^2 + px + 5$

If $g(2) = 0$ then $9 + 2p = 0$, so $p = -\dfrac{9}{2}$

If $g(x)$ has a repeated root then the discriminant is zero:

$p^2 - 20 = 0$

$\Rightarrow p = \pm 2\sqrt{5}$

$\therefore f(x)$ has exactly two roots when

$p = -\dfrac{9}{2}$ or $\pm 2\sqrt{5}$

c The middle value of p found in (b) is $-2\sqrt{5}$. In this case,

$f(x) = (x-2)\left(x^2 - 2\sqrt{5}x + 5\right)$

$= (x-2)\left(x - \sqrt{5}\right)^2$

Repeated root at $x = \sqrt{5}$, single root at $x = 2$; y-intercept at -10.
Positive cubic shape.

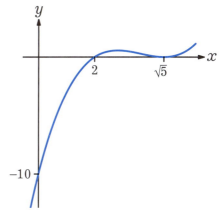

Figure 3ML.2 Graph of $f(x) = (x-2)\left(x - \sqrt{5}\right)^2$

3 a $x^2 + 4x + 5 = (x+2)^2 + 1 > 0$ for all values of x,
The numerator of the rational function is never equal to zero, hence $y \neq 0$.

b Vertical asymptote where denominator is zero: $x = -2$

c $(x+2)y = x^2 + 4x + 5$

$x^2 + (4-y)x + (5-2y) = 0$

$x = \dfrac{(y-4) \pm \sqrt{(4-y)^2 - 4(5-2y)}}{2}$

$= \dfrac{y \pm \sqrt{y^2 - 4}}{2} - 2$

d For real solutions of x, require $y^2 - 4 \geq 0$

i.e. $y^2 \geq 4$

$\therefore y \leq -2$ or $y \geq 2$

e Vertical asymptote at $x = -2$
Range excludes the interval $]-2, 2[$

$y = \dfrac{(x+2)^2 + 1}{x+2} = (x+2) + \dfrac{1}{x+2}$

As $x \to \pm\infty$, $y \to x+2$, so the line $y = x+2$ is an oblique asymptote.

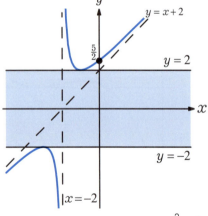

Figure 3ML.3 Graph of $y = \dfrac{x^2 + 4x + 5}{x+2}$

> **COMMENT**
>
> Finding oblique asymptotes is not expected within the syllabus; plotting rational functions is covered in depth in Section 5F.

4 $f(x) = x^4 + x^3 + x^2 + x + 1$

a $f(1) = 1 + 1 + 1 + 1 + 1 = 5$

b $(x-1)f(x) = (x-1)(x^4 + x^3 + x^2 + x + 1)$

$= x^5 + x^4 + x^3 + x^2 + x$
$\quad - (x^4 + x^3 + x^2 + x + 1)$
$= x^5 - 1$

c Root at $x = 1$, y-intercept at -1.
Quintic shape: graph of $y = x^5$ translated down one unit.

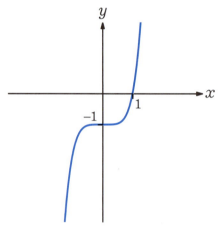

Figure 3ML.4 Graph of $y = x^5 - 1$

d From Figure 3ML.4, $y = x^5 - 1$ has a single real root at $x = 1$, and so $(x-1)f(x)$ has a single factor $(x-1)$.

Therefore $f(x)$ has no linear factors and hence, by the factor theorem, $f(x) \neq 0$ for any value x.

4 Algebraic structures

Exercise 4A

4 $(3x-1)^{x^2-4}=1$

$\therefore \begin{cases} 3x-1=1 \\ \text{or } 3x-1=-1 \text{ and } x^2-4 \text{ is even} \\ \text{or } x^2-4=0 \text{ and } 3x-1\neq 0 \end{cases}$

i.e. $x=\dfrac{2}{3}$ or $x=0$ or $x=\pm 2$

5 $x|x|=4x$

$x(|x|-4)=0$

$\Rightarrow x=0$ or $|x|=4$

$\therefore x=0$ or $x=\pm 4$

Exercise 4B

2 $9(1+9^{x-1})=10\times 3^x$

$9^x+9-10\times 3^x=0$

$(3^x)^2-10(3^x)+9=0$

Let $u=3^x$:

$u^2-10u+9=0$

$(u-1)(u-9)=0$

$u=1$ or $u=9$

$\therefore 3^x=1$ or $3^x=9$

$x=0$ or $x=2$

3 $a^x=-\dfrac{5}{a^x}+6$

$a^x-6+\dfrac{5}{a^x}=0$

$(a^x)^2-6a^x+5=0$

Let $u=a^x$:

$u^2-6u+5=0$

$(u-1)(u-5)=0$

$u=1$ or $u=5$

$\therefore a^x=1$ or $a^x=5$

$x=0$ or $x=\log_a 5$

4 $\log_2 x=6-5\log_x 2$

$\log_2 x=6-5\dfrac{\log_2 2}{\log_2 x}$

$(\log_2 x)^2-6\log_2 x+5=0$

Let $u=\log_2 x$:

$u^2-6u+5=0$

$(u-1)(u-5)=0$

$u=1$ or $u=5$

$\therefore \log_2 x=1$ or $\log_2 x=5$

$x=2^1=2$ or $x=2^5=32$

> **COMMENT**
> Instead of changing $\log_x$ into $\log_2$, the opposite could have been done, or both $\log_x$ and $\log_2$ could have been changed to log (base 10).

Exercise 4C

2 $y = x \ln x$

$\ln x$ ceases to be defined at $x = 0$; as $x \to 0$, $x \ln x \to 0$, so there is no vertical asymptote, but there is an empty circle at the origin and no graph for $x < 0$.

As x gets large, both x and $\ln x$ continue to increase, so there is no horizontal asymptote.

$x \ln x = 0$ when $x = 0$ or $\ln x = 0$. The first root has already been eliminated, so the only root is $x = 1$.

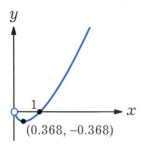

Figure 4C.2

3 $y = \dfrac{e^x}{\ln x}$

$\ln x = 0$ at $x = 1$, so there is a vertical asymptote at $x = 1$.

$\ln x$ ceases to be defined at $x = 0$, and as $x \to 0$ from above, $\ln x \to -\infty$, so the graph terminates with an empty circle at the origin.

For large x, e^x increases more rapidly than $\ln x$, so their ratio increases and there is no horizontal asymptote.

$y = 0$ when $e^x = 0$, which has no solutions, so there are no roots.

The exact value of the minimum is best found using a GDC, but from the above we can be confident that there must be a single minimum at some $x > 1$.

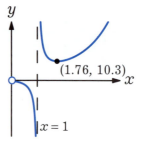

Figure 4C.3

4 $y = \dfrac{x^2(x^2 - 9)}{e^x} = \dfrac{x^2(x+3)(x-3)}{e^x}$

$e^x \neq 0$, so there is no vertical asymptote.

For large positive x, e^x increases more rapidly than any polynomial in x; so as $x \to \infty$, $y \to 0$.

For large negative x, $e^x \to 0$ and the numerator is positive, so $y \to \infty$.

$y = 0$ when $x = 0$ (double root) or $x = \pm 3$.

Exact values for local minima and maxima are best found using a GDC, but from the above we can be confident that there must be a minimum in $]-3, 0[$ and one in $]0, 3[$, and maxima at the origin and in $]3, \infty[$.

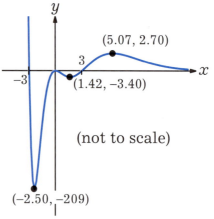

Figure 4C.4

Exercise 4D

2 $x \ln x = 3 - x^2$
The graphs of $y = x \ln x$ and $y = 3 - x^2$ intersect at one point.
From GDC: $x = 1.53$ (3SF)

3 For $\ln x = kx$ to have exactly one solution, the graph of $y = kx$ must be tangent to the graph of $y = \ln x$.

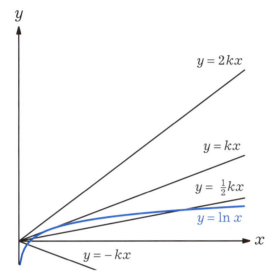

Figure 4D.3 Graphs of $y = kx$ and $y = \ln x$

a $\ln x^2 = 2 \ln x$, so $\ln x^2 = kx$ is equivalent to $\ln x = \dfrac{k}{2} x$, which will have two solutions since the line $y = \dfrac{k}{2} x$ has a smaller gradient than $y = kx$.

b $\ln \left(\dfrac{1}{x} \right) = \ln x^{-1} = -\ln x$, so
$\ln \left(\dfrac{1}{x} \right) = kx$ is equivalent to $\ln x = -kx$,
which will have one solution, since there will be an intersection in the lower right quadrant.

c $\ln \sqrt{x} = \ln x^{\frac{1}{2}} = \dfrac{1}{2} \ln x$, so $\ln \sqrt{x} = kx$
is equivalent to $\ln x = 2kx$, which will have no solutions since the line $y = 2kx$ has a greater gradient than $y = kx$.

Exercise 4E

3 Substitute $y = 8 - x$ into the circle equation to find the intersections:
$x^2 - 6x + (8 - x)^2 - 2(8 - x) + 2 = 0$
$2x^2 - 20x + 50 = 0$
$x^2 - 10x + 25 = 0$
$(x - 5)^2 = 0$
$\therefore x = 5$
There is a single solution (5, 3), so the line is a tangent to the circle.

> **COMMENT**
>
> Instead of solving the equation, it would be reasonable to show that the discriminant is zero and hence conclude that there is only one solution. In this case, it is easy enough to show that the quadratic is a perfect square, so solving the equation is just as efficient.

4 Substitute the equation of the line into the quadratic equation to find the equation governing intersections:
$3x^2 - x + 5 = mx + 3$
$\Rightarrow 3x^2 - (m+1)x + 2 = 0$
For the line to be tangent to the curve, this equation must have a single solution, so discriminant $\Delta = 0$:
$(m+1)^2 - 24 = 0$
$m + 1 = \pm 2\sqrt{6}$
$m = -1 \pm 2\sqrt{6}$

5 From the equation of the line,
$3y = k - 2x$, so $9y^2 = (k - 2x)^2$
Substitute this into the ellipse equation to find the equation governing intersections:
$4x^2 + (k - 2x)^2 = 36$
$8x^2 - 4kx + k^2 - 36 = 0$

4 Algebraic structures   33

For the line to be tangent to the ellipse, this equation must have a single solution, so discriminant $\Delta = 0$:

$(4k)^2 - 32(k^2 - 36) = 0$

$-16k^2 + 32 \times 36 = 0$

$k^2 = 2 \times 36$

$k = \pm 6\sqrt{2}$

> **COMMENT**
>
> When calculating values partway through solving an equation, it is often needless work to multiply out products if you are subsequently going to divide through by a common factor. In this case, rather than evaluating 32×36 it is more convenient to leave it in product form, ready to divide through by 16.

6 Intersections when

$kx + 5 = x^2 + 2$

$x^2 - kx - 3 = 0$

Discriminant $\Delta = k^2 + 12 > 0$ for all values of k,

so the quadratic has two roots,
i.e. there are always two intersection points, for any value of k.

7 Substituting $y = 4 - x$ from the second equation into the first equation:

$2^x + 2^{4-x} = 10$

$2^{2x} - 10 \times 2^x + 2^4 = 0$

This is a hidden quadratic. Let $u = 2^x$; then

$u^2 - 10u + 16 = 0$

$(u - 2)(u - 8) = 0$

$u = 2$ or $u = 8$

i.e. $2^x = 2$ or $2^x = 8$

$\Rightarrow x = 1$ or 3

$\therefore (x, y) = (1, 3)$ or $(3, 1)$

8 Substitute $y = x^5$ into $\log_3 x + \log_3 y = 3$:

$\log_3 x + \log_3(x^5) = 3$

$6 \log_3 x = 3$

$\log_3 x = \dfrac{1}{2}$

$x = 3^{\frac{1}{2}} = \sqrt{3}$

$\therefore (x, y) = (\sqrt{3}, 9\sqrt{3})$

Exercise 4F

4 $2x + y - 2z = 0$ …(1)

$x - 2y - z = 2$ …(2)

$3x + 4y - 3z = c$ …(3)

Eliminate y from (2) and (3):

(1) $2x + y - 2z = 0$ …(1)

(2) + 2×(1) $5x \quad -5z = 2$ …(2)

(3) − 4×(1) $-5x \quad +5z = c$ …(3)

For (4) and (5) to be consistent, require $c = -2$.

5 $x - 2y + 2z = 0$ …(1)

$2x + ky - z = 3$ …(2)

$x - y + 3z = -5$ …(3)

Eliminate x from (2) and (3):

(1) $x - 2y + 2z = 0$ …(1)

(2) − 2×(1) $(k+4)y - 5z = 3$ …(4)

(3) − (1) $y + z = -5$ …(5)

Eliminate z from (4):

(1) $x - 2y + 2z = 0$ …(1)

(4) + 5×(5) $(k+9)y = -22$ …(6)

(5) $y + z = -5$ …(5)

From (6), $y = -\dfrac{22}{k+9}$, so there is no valid unique solution when $k = -9$.

6 $2x + y - z = 2$ ...(1)
$x - 2y + 2z = 1$ ...(2)
$2x + y - 4z = a$ ...(3)

Eliminate y from (2) and (3):

(1) $2x + y - z = 2$ ...(1)
(2) + 2×(1) $5x = 5$ ...(4)
(3) − (1) $-3z = a - 2$ ...(5)

The solution is

$x = 1$, $z = \dfrac{2-a}{3}$ and $y = 2 - 2x + z = \dfrac{2-a}{3}$

i.e. $(x, y, z) = \left(1, \dfrac{2-a}{3}, \dfrac{2-a}{3}\right)$

7 a $x - 2y + z = 1$ ...(1)
$2x + y - z = a$ ...(2)
$4x + 7y - 5z = a^2$ ...(3)

Eliminate x from (2) and (3):

(1) $x - 2y + z = 1$ ...(1)
(2) − 2×(1) $5y - 3z = a - 2$ ...(4)
(3) − 4×(1) $15y - 9z = a^2 - 4$ ...(5)

Eliminate y from (5):

(1) $x - 2y + z = 1$ ...(1)
(4) $5y - 3z = a - 2$ ...(4)
(5) − 3×(4) $0 = a^2 - 4 - 3(a-2)$ ...(6)

For a consistent solution, require
$a^2 - 3a + 2 = 0$
$(a-1)(a-2) = 0$
$a = 1$ or 2

b With $a = 2$:
$x - 2y + z = 1$
$5y - 3z = 0$
Let $z = 5t$; then
$y = 3t$ and $x = 1 + 2y - z = 1 + t$
∴ general solution is
$(x, y, z) = (1 + t, 3t, 5t)$

> **COMMENT**
> When parameterising, it can be convenient to use judgement to avoid fractions in the end solution. If we parameterise as $z = k$ then the solution would be $\left(1 + \dfrac{k}{5}, \dfrac{3k}{5}, k\right)$, which is equally valid but less tidy.

8 a $x - 2y + z = 7$ ...(1)
$2x + y - 3z = b$ ...(2)
$x + y + kz = 4$ ...(3)

Eliminate x from (2) and (3):

(1) $x - 2y + z = 7$ ...(1)
(2) − 2×(1) $5y - 5z = b - 14$ ...(4)
(3) − (1) $3y + (k-1)z = -3$ ...(5)

Eliminate y from (5):

(1) $x - 2y + z = 7$ ...(1)
(4) $5y - 5z = b - 14$ ...(4)
5×(5) − 3×(4) $(5k + 10)z = 27 - 3b$ ...(6)

From (6), $z = \dfrac{3(9-b)}{5(k+2)}$, so for $k = -2$ there is no unique solution.

b For $k = -2$ the system will be consistent if $27 - 3b = 0$, i.e. $b = 9$.

c With $k = -2$ and $b = 9$:
$x - 2y + z = 7$
$5y - 5z = -5$
Let $z = t$; then $y = t - 1$ and
$x = 7 + 2y - z = t + 5$
∴ $(x, y, z) = (t + 5, t - 1, t)$

Exercise 4G

2 a

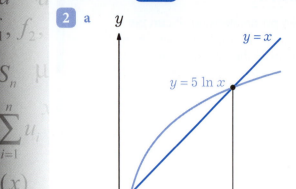

Figure 4G.2.1 Graphs of y = x and y = 5lnx

From the graph on GDC, $x > 5\ln x$ for $0 < x < 1.30$ or $x > 12.7$

b

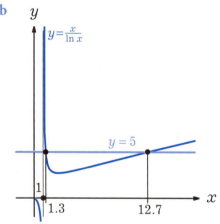

Figure 4G.2.2 Graphs of y = 5 and $y = \dfrac{x}{\ln x}$

From the graph on GDC, $\dfrac{x}{\ln x} > 5$ for $1 < x < 1.30$ or $x > 12.7$

The answers are different because for $x \in \,]0,1[$, $\ln x < 0$, so for $x < 1$ the conditions (a) and (b) are opposite, whereas for $x > 1$ the conditions are equivalent.

Exercise 4H

2
$$x^3 + px + q = (x-a)^2(x-b)$$
$$= (x^2 - 2ax + a^2)(x-b)$$
$$= x^3 + (-2a-b)x^2$$
$$\quad + (2ab + a^2)x - a^2 b$$

Comparing coefficients:

x^3: $1 = 1$

x^2: $0 = -2a - b \Rightarrow b = -2a$

x^1: $p = 2ab + a^2 = -3a^2$

x^0: $q = -a^2 b = -2a^3$

$\therefore q^2 = 4a^6 = -\dfrac{4}{27}p^3$

i.e. $4p^3 + 27q^2 = 0$

Mixed examination practice 4

Short questions

1 a $y = 2^x$: axis intercept at (0, 1), exponential shape.
$y = 1 - x^2$: axis intercepts at (0, 1) and (± 1, 0), negative quadratic.

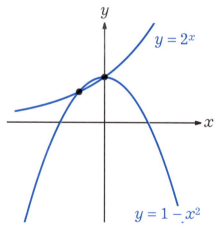

Figure 4MS.1

b Two intersection points $\Rightarrow$ two solutions of $2^x = 1 - x^2$.

2 Sketching the graph on GDC:

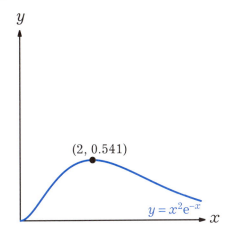

Figure 4MS.2
Maximum value of y is 0.541 (3SF)

3 Sketch the graphs $y_1 = x^3 - 4x$ and $y_2 = e^{-x}$ on GDC:

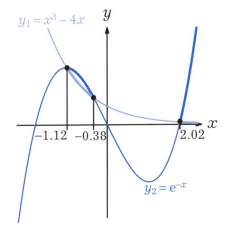

Figure 4MS.3
The intersections are at
$x = -1.12, -0.379, 2.02$ (3SF)

$y_1 > y_2$ for
$x \in \,]-1.12, -0.379[\, \cup \,]2.02, \infty[$

4 $3x - y + 2z = 2$ ...(1)
$x - 2y + z = 3$ ...(2)
$x - y + 3z = -5$ ...(3)

Eliminate x from (1) and (3):

(1)−3×(2) $\quad 5y - z = -7$ ...(4)
(2) $\quad\quad\quad x - 2y + z = 3$ ...(2)
(3)−(2) $\quad\quad y + 2z = -8$ ...(5)

Eliminate y from (4):

(4)−5×(5) $\quad -11z = 33$ ...(6)
(2) $\quad\quad\quad x - 2y + z = 3$ ...(2)
(5) $\quad\quad\quad y + 2z = -8$ ...(5)

So the solution is
$z = -3$
$y = -8 - 2z = -2$
$x = 3 + 2y - z = 2$

i.e. $(x, y, z) = (2, -2, -3)$

5 $e^x \ln x = 3e^x$

$e^x (\ln x - 3) = 0$

$e^x = 0$ (no solutions) or $\ln x = 3$

$\Rightarrow x = e^3$

6 a $x^4 + 36 = 13x^2$

$(x^2)^2 - 13x^2 + 36 = 0$

Substitute $u = x^2$:
$u^2 - 13u + 36 = 0$
$(u - 9)(u - 4) = 0$
$u = 9$ or $u = 4$
$\therefore x^2 = 4$ or 9
$x = \pm 2, \pm 3$

b $y = x^4 - 13x^2 + 36$ is a positive quartic with four roots, so $y \leq 0$ between the first and second roots and between the third and fourth roots.
That is, $x^4 + 36 \leq 13x^2$ for
$x \in [-3, -2] \cup [2, 3]$

4 Algebraic structures

7 a Vertical asymptote where
$e^x = 2 \Rightarrow x = \ln 2$

As $x \to \infty$, denominator gets very large and positive, so $y \to 0$

As $x \to -\infty$, denominator tends to -2 so $y \to -\dfrac{1}{2}$

Numerator is never zero, so $y \neq 0$

At $x = 0$, $y = -1$

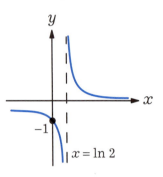

Figure 4MS.7

b $x = \ln 2$

8 a $2x + 2y - z = 1$ ...(1)
$x + y - z = -4$ ...(2)
$4x + 4y - 3z = p$ ...(3)

Eliminate x from (1) and (3):

$(1) - 2 \times (2)$ $z = 9$ ...(4)
(2) $x + y - z = -4$ ...(2)
$(3) - 4 \times (2)$ $z = p + 16$ ...(5)

For this to be a consistent system, need
$p + 16 = 9$
$p = -7$

b With $p = -7$, the system reduces to
$z = 9$
$x + y - z = -4$

Let $y = t$; then $x = z - 4 - y = 5 - t$

$\therefore (x, y, z) = (5 - t, t, 9)$

9 Substitute $y = 2x - k$ into the circle equation to find intersections:

$x^2 + (2x - k)^2 = 5$

$5x^2 - 4kx + k^2 - 5 = 0$

If the line is tangent to the circle, then there is a single solution to this quadratic equation, so the discriminant $\Delta = 0$:

$(-4k)^2 - 20(k^2 - 5) = 0$

$-4k^2 + 100 = 0$

$k^2 = 25$

$k = \pm 5$

10 On GDC, sketch graphs $y_1 = \dfrac{x}{\ln x}$ and $y_2 = 4$:

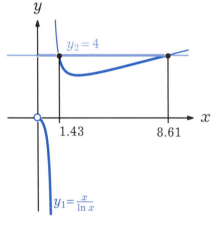

Figure 4MS.10

The intersections are at $x = 1.43$, 8.61

The graph of $y_1 = \dfrac{x}{\ln x}$ has a vertical asymptote at $x = 1$

$y_1 < y_2$ for $x \in\,]0, 1[\cup]1.43, 8.61[$

38   Mixed examination practice 4

11 $x \ln x + 4 \ln x = 0$

$(x+4) \ln x = 0$

$x = -4$ or $\ln x = 0$

$\therefore x = 1$

(as $\ln x$ has no real value for $x = -4$)

> **COMMENT**
> Always check the validity of algebraic solutions, especially in functions with restricted domains, such as rational functions and those containing logarithms.

Long questions

1 a i Expanding:

$(x-a)^3 - b = x^3 - 3ax^2 + 3a^2 x - a^3 - b$

Comparing coefficients:

$x^3 : 1 = 1$

$x^2 : -9 = -3a \Rightarrow a = 3$

$x^1 : k = 3a^2 = 27$

$x^0 : -28 = -a^3 - b \Rightarrow b = 1$

$\therefore k = 27$

ii With $k = 27$, the equation $x^3 - 9x^2 + kx - 28 = 0$ is equivalent to

$(x-3)^3 = 1$

$\therefore x - 3 = 1$

$\Rightarrow x = 4$

b i Vertical asymptote at $x = 4$.
For values close to $x = 3$, y is very close to -1.
As $x \to \infty$, denominator gets large and positive so $y \to 0$ from above.
As $x \to -\infty$, denominator gets large and negative so $y \to 0$ from below.
At $x = 0$, $y = -\dfrac{1}{28}$.

Numerator is never zero so $y \neq 0$.

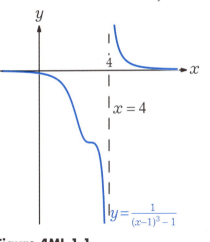

Figure 4ML.1.1

ii Vertical asymptote $x = 4$, horizontal asymptote $y = 0$.

c

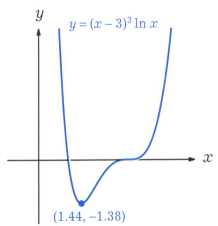

Figure 4ML.1.2

From GDC graph, the minimum point is $(x, y) = (1.44, -1.38)$.

2 a Let $u = x^2$; then equation becomes

$u^2 + u - 6 = 0$

$(u-2)(u+3) = 0$

$u = 2$ or $u = -3$

$\therefore x^2 = 2$ (reject $x^2 = -3$)

so $x = \pm\sqrt{2}$

b i $x^4 - 4x^3 + 7x^2 - 6x - 4$
$= f(x+k)$
$= (x+k)^4 + (x+k)^2 - 6$
$= x^4 + 4kx^3 + 6k^2x^2 + 4k^3x + k^4$
$\quad + x^2 + 2kx + k^2 - 6$
$= x^4 + 4kx^3 + (6k^2 + 1)x^2$
$\quad + (4k^3 + 2k)x + (k^4 + k^2 - 6)$

> **COMMENT**
> To expand $(x+k)^4$ quickly, use the binomial theorem; see Chapter 8.

Comparing coefficients:
x^4: $1 = 1$
x^3: $-4 = 4k \Rightarrow k = -1$
x^2: $7 = 6k^2 + 1$ is consistent with $k = -1$
x^1: $-6 = 4k^3 + 2k$ is consistent with $k = -1$
x^0: $-4 = k^4 + k^2 - 6$ is consistent with $k = -1$
So for $k = -1$ the two equations are equivalent.

ii From (b)(i), $x^4 + 7x^2 = 4x^3 + 6x + 4$ is equivalent to $f(x-1) = 0$.
From (a), $f(x-1) = 0 \Rightarrow x - 1 = \pm\sqrt{2}$
$\therefore x = 1 \pm \sqrt{2}$

c $x^4 + 7x^2 > 4x^3 + 6x + 4$ is equivalent to $f(x-1) > 0$.
$y = f(x-1)$ is a positive quartic with two roots at $x = 1 \pm \sqrt{2}$; it is positive on either side of those two roots, i.e. for $x < 1 - \sqrt{2}$ or $x > 1 + \sqrt{2}$.

3 a $x + y + z = 3$ ...(1)
$x + ky + 2z = 4$ ...(2)
$x - y + 3z = b$ ...(3)

Eliminate x from (2) and (3):

(1) $x + y + z = 3$ ...(1)
(2)−(1) $(k-1)y = 1$ ...(4)
(3)−(1) $-2y + 2z = b - 3$ ...(5)

Eliminate z from (5):

(1) $x + y + z = 3$ ...(1)
(4) $(k-1)y = 1$ ...(4)
(5)−2×(4) $-2ky = b - 5$ ...(6)

From (6), if $k = 0$ then there is no unique solution.

b If $b = 5$ then equation (6) is valid as $0 = 0$ and the system is consistent.

c With $k = 0$ and $b = 5$, the system reduces to
$x + y + z = 3$
$-y + z = 1$
Let $z = t$;
Then $y = -1$ $x = 3 - y - z = 4 - t$
So $(x, y, z) = (4 - t, t - 1, t)$

40 Mixed examination practice 4

5 The theory of functions

Exercise 5C

5 $f(x) = \sqrt{\ln(x-4)}$

Square root can have only non-negative values in its domain, so require $\ln(x-4) \geq 0$:

$x - 4 \geq e^0$

$\Rightarrow x \geq 5$

Domain of $f(x)$ is $x \geq 5$

6 $f(x) = \dfrac{4^{\sqrt{x-1}}}{x+2} - \dfrac{1}{(x-3)(x-2)} + x^2 + 1$

Cannot have division by zero, so $x \neq -2, 2, 3$

Square root can only have non-negative values in its domain, so require $x - 1 \geq 0$, i.e. $x \geq 1$

Domain of $f(x)$ is $x \geq 1, x \neq 2, x \neq 3$

> **COMMENT**
> Note that the restriction $x \neq -2$ is not needed in the final answer as it is already covered by the restriction $x \geq 1$.

7 Require that the boundary at $x = 2$ be consistent in the two parts of the function:

$3 \times 2^2 - 1 = a - 2^2$

$11 = a - 4$

$\therefore a = 15$

8 $g(x) = \ln(x^2 + 3x + 2)$

$\ln x$ can have only positive values in its domain, so require $x^2 + 3x + 2 > 0$:

$x^2 + 3x + 2 > 0$

$(x+1)(x+2) > 0$

$x < -2$ or $x > -1$

Domain of $g(x)$ is $x < -2$ or $x > -1$

> **COMMENT**
> It may be helpful to draw a graph of $y = x^2 + 3x + 2$ to solve the quadratic inequality $x^2 + 3x + 2 > 0$.

9 $f(x) = \sqrt{\dfrac{8x-4}{x-12}}$

Cannot have division by zero $\Rightarrow x \neq 12$

Square root can have only non-negative values in its domain, so require either $8x - 4 \geq 0$ and $x - 12 > 0$ or $8x - 4 \leq 0$ and $x - 12 < 0$.

$8x - 4 \geq 0$ and $x - 12 > 0$

$\Rightarrow x \geq \dfrac{1}{2}$ and $x > 12$

$\therefore x > 12$

$8x - 4 \leq 0$ and $x - 12 < 0$

$\Rightarrow x \leq \dfrac{1}{2}$ and $x < 12$

$\therefore x \leq \dfrac{1}{2}$

So domain of $f(x)$ is $x \leq \dfrac{1}{2}$ or $x > 12$

10 $f(x) = \sqrt{x-a} + \ln(b-x)$

a Square root can have only non-negative values in its domain, so require $x \geq a$

$\ln x$ can have only positive values in its domain, so require $x < b$

i $a < b \Rightarrow$ domain is $a \leq x < b$

ii $a > b \Rightarrow$ function has empty domain

b $f(a) = \begin{cases} \sqrt{a-a} + \ln(b-a) & \text{if } a < b \\ \text{undefined} & \text{if } a \geq b \end{cases}$

$= \begin{cases} \ln(b-a) & \text{if } a < b \\ \text{undefined} & \text{if } a \geq b \end{cases}$

Exercise 5D

3 $fg(x) = (3x+2)^2 + 1$
$= 9x^2 + 12x + 5$

$gf(x) = 3(x^2 + 1) + 2$
$= 3x^2 + 5$

$fg(x) = gf(x)$
$9x^2 + 12x + 5 = 3x^2 + 5$
$6x^2 + 12x = 0$
$6x(x+2) = 0$
$x = 0$ or $x = -2$

4 $gf(x) = 0$

$\dfrac{3x+1}{(3x+1)^2 + 25} = 0$

$3x + 1 = 0$

$\therefore x = -\dfrac{1}{3}$

5 a

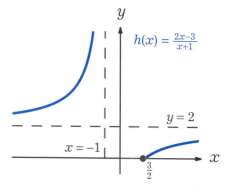

Figure 5D.5.1 Graph of $h(x) = \dfrac{2x-3}{x+1}$

Horizontal asymptote is $y = 2$, so range is $y \neq 2$

b $h(x) = 0$

$\dfrac{2x-3}{x+1} = 0$

$2x - 3 = 0$

$x = \dfrac{3}{2}$

c To define $g \circ h$, the range of h must be a subset of the domain of g. Domain of $g(x)$ is $x \geq 0$, so need to restrict the domain of h so that the range of $h(x)$ is $y \geq 0$. (Without restriction, the domain of h is $x \neq -1$.)

Figure 5D.5.2 Graph of $h(x) = \dfrac{2x-3}{x+1}$ with domain restricted so that the range is $y \geq 0$

42 Topic 5D Composite functions

Hence the domain, D, of $g \circ h(x)$ is
$x < -1$ or $x \geq \dfrac{3}{2}$

Range of h over domain D is $y \geq 0$, $y \neq 2$

Range of g over domain $x \geq 0$, $x \neq 2$ is $y \geq 0$, $y \neq \sqrt{2}$

∴ range of $g \circ h$ over domain D is $y \geq 0$, $y \neq \sqrt{2}$

6 a $fg(x) = 2x + 3$
$[g(x)]^3 = 2x + 3$
$\Rightarrow g(x) = \sqrt[3]{2x+3}$

b $gf(x) = 2x + 3$
$g(x^3) = 2x + 3$
$\Rightarrow g(x) = 2\sqrt[3]{x} + 3$

7 a $fg(x) = \sqrt{(3x+4)^2 - 2(3x+4)}$

$f \circ g$ is undefined for $x \in]a, b[$, so require $(3x+4)^2 - 2(3x+4) < 0$ (since square root is undefined for negative values).

$(3x+4^2) - 2(3x+4) < 0$
$(3x+4)[(3x+4) - 2] < 0$
$(3x+4)(3x+2) < 0$

$x \in \left]-\dfrac{4}{3}, -\dfrac{2}{3}\right[$

∴ $a = -\dfrac{4}{3}$, $b = -\dfrac{2}{3}$

b Over the domain $x \notin]a, b[$, $(3x+4)^2 - 2(3x+4)$ takes all non-negative values and so $fg(x) = \sqrt{(3x+4)^2 - 2(3x+4)}$ takes all non-negative values, i.e. the range of $f \circ g$ is $y \geq 0$.

8 a The range of f is $y > 2$; this lies within the domain of g, so $g \circ f$ is a valid composition.

The range of g is $y \geq 0$; values from $[0, 3]$ lie within the range of g but not within the domain of f, so $f \circ g$ is not a valid composition for the full domain of g.

b For $f \circ g$ to be defined, we require the range of g to be limited to $]3, \infty[$, so restrict the domain to $x \notin \left[-\sqrt{3}, \sqrt{3}\right]$.

9 By observation,

$g\left(\dfrac{x}{2} - 3\right) = 2\left(\dfrac{x}{2} - 3\right) + 5$
$= x - 6 + 5$
$= x - 1$

∴ $f(x-1) = fg\left(\dfrac{x}{2} - 3\right)$

$= \dfrac{\dfrac{x}{2} - 3 + 2}{3}$

$= \dfrac{x}{6} - \dfrac{1}{3}$

Alternatively, given that $fg(x) = \dfrac{x+2}{3}$ and $g(x) = 2x + 5$, we have

$f(2x+5) = \dfrac{x+2}{3}$.

Let $2x + 5 = u - 1$, so that $x = \dfrac{u-6}{2} = \dfrac{u}{2} - 3$. Then

$f(u-1) = \dfrac{\dfrac{u}{2} - 3 + 2}{3} = \dfrac{u}{6} - \dfrac{1}{3}$

∴ $f(x-1) = \dfrac{x}{6} - \dfrac{1}{3}$

Exercise 5E

4 For $x \geq 0$, $f(x) = x$ and is therefore an identity function.

However, this is not the case for $x < 0$, where $f(x) = -x$.

5 a $ff(2) = f(0) = -1$

b $f(1) = 3$ so $f^{-1}(3) = 1$

6 $y = \sqrt{3-2x}$

$3 - 2x = y^2$

$\Rightarrow x = \dfrac{3-y^2}{2}$

$\therefore f^{-1}(x) = \dfrac{3-x^2}{2}$

Hence $f^{-1}(7) = \dfrac{3-7^2}{2} = -23$

7 $y = 3e^{2x}$

$e^{2x} = \dfrac{y}{3}$

$2x = \ln \dfrac{y}{3}$

$\Rightarrow x = \dfrac{1}{2}\ln\left(\dfrac{y}{3}\right) = \ln\sqrt{\dfrac{y}{3}}$

$\therefore f^{-1}(x) = \ln\sqrt{\dfrac{x}{3}}$

The range of f is $y > 0$, so the domain of f^{-1} is $x > 0$.

8 $fg(x) = 2(x^3) + 3$

$y = 2x^3 + 3$

$x^3 = \dfrac{y-3}{2}$

$\Rightarrow x = \sqrt[3]{\dfrac{y-3}{2}}$

$\therefore (fg)^{-1}(x) = \sqrt[3]{\dfrac{x-3}{2}}$

9 a To find the inverse function of $f(x) = e^{2x}$:

$y = e^{2x}$

$2x = \ln y$

$\Rightarrow x = \dfrac{1}{2}\ln y$

$\therefore f^{-1}(x) = \dfrac{1}{2}\ln x = \ln\sqrt{x}$

To find the inverse function of $g(x) = x+1$:

$y = x+1$

$\Rightarrow x = y-1$

$\therefore g^{-1}(x) = x-1$

So

$f^{-1}(3) \times g^{-1}(3) = \left(\ln\sqrt{3}\right) \times (3-1)$

$= 2\ln\sqrt{3}$

$= \ln 3$

b $(fg)^{-1}(x) = g^{-1}f^{-1}(x) = \ln\sqrt{x} - 1$

$\therefore (fg)^{-1}(3) = \ln\sqrt{3} - 1$

10 $f^{-1}(x) = x^2$ $(x \geq 0)$

$\therefore f^{-1} \circ g(x) = (2^x)^2 = 4^x$

$4^x = 0.25 \Rightarrow x = -1$

11 $y = \dfrac{x^2 - 4}{x^2 + 9}$

$(x^2 + 9)y = x^2 - 4$

$x^2 y + 9y = x^2 - 4$

$x^2 y - x^2 = -4 - 9y$

$x^2(y-1) = -4 - 9y$

$x^2 = -\dfrac{4+9y}{y-1}$

$= \dfrac{4+9y}{1-y}$

$\therefore x = -\sqrt{\dfrac{4+9y}{1-y}}$ (as domain of f is $x \leq 0$)

Hence $f^{-1}(x) = -\sqrt{\dfrac{4+9x}{1-x}}$

The graph of f has a horizontal asymptote at $y = 1$ (as $x \to -\infty$) and is decreasing for all $x \leq 0$, so the range of f is $y \in \left[-\dfrac{4}{9}, 1\right[$.

44   Topic 5E Inverse functions

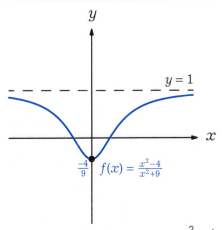

Figure 5E.11 Graph of $f(x) = \dfrac{x^2-4}{x^2+9}$ for real x

Hence the domain of f^{-1} is $x \in \left[-\dfrac{4}{9}, 1\right[$.

12 a $f(x) = x^2$, $x \leq k$

Taking $k = 0$,

$y = x^2 \Rightarrow x = -\sqrt{y}$ (choose the negative root since the domain of f is $x \leq 0$)

$\therefore f^{-1}(x) = -\sqrt{x}$

b $f(x) = (x+1)^2 + 2$, $x \geq k$

Taking $k = -1$,

$y = (x+1)^2 + 2$

$(x+1)^2 = y - 2$

$x + 1 = \sqrt{y-2}$

(take positive root since domain of f is $x \geq -1$)

$\Rightarrow x = \sqrt{y-2} - 1$

$\therefore f^{-1}(x) = \sqrt{x-2} - 1$

c $f(x) = |x|$, $x \leq k$

Taking $k = 0$,

$y = |x| = -x$ (since $x \leq 0$)

$y = -x \Rightarrow x = -y$

$\therefore f^{-1}(x) = -x$

13 a $y = \ln(x-1) + \ln 3$

$= \ln(3x-3)$

$3x - 3 = e^y$

$x = \dfrac{e^y + 3}{3}$

$= \dfrac{e^y}{3} + 1$

$\therefore f^{-1}(x) = \dfrac{e^x}{3} + 1$

The range of f is $\mathbb{R}$, so the domain of f^{-1} is also $\mathbb{R}$.

b $f(x) = \ln(x-1) + \ln(3)$

$= \ln[3(x-1)]$

$= \ln(3x-3)$

$\therefore gf(x) = e^{\ln(3x-3)}$

$= 3x - 3$

14 $f(x) = \begin{cases} 2 + (x-1)^2, & x < 1 \\ k - (x-1)^2, & x \geq 1 \end{cases}$

a Range for $x < 1$ is $]2, \infty[$.

For f to be one-to-one, require that for $x \geq 1$, $f(x) \leq 2$.

Maximum value of $f(x)$ for $x \geq 1$ is k

$\therefore k = 2$

b i When $k = 0$, range of f is

$]-\infty, 0] \cup]2, \infty[$

ii In the upper part of the range, $]2, \infty[$:

$y = 2 + (x-1)^2$

$(x-1)^2 = y - 2$

$x - 1 = -\sqrt{y-2}$

(choose the negative root since this part of the range comes from $x < 1$)

$\therefore x = 1 - \sqrt{y-2}$

5 The theory of functions 45

In the lower part of the range, $]-\infty, 0]$:

$y = -(x-1)^2$

$x - 1 = \sqrt{-y}$

(choose the positive root since this part of the range comes from $x \geq 1$)

$\therefore x = 1 + \sqrt{-y}$

Hence $f^{-1}(x) = \begin{cases} 1 + \sqrt{-x}, & x \leq 0 \\ 1 - \sqrt{x-2}, & x > 2 \end{cases}$

15 a Finding the inverse of $f(x) = \dfrac{1}{x}$:

$y = \dfrac{1}{x}$

$\Rightarrow x = \dfrac{1}{y}$

$\therefore f^{-1}(x) = \dfrac{1}{x}$

So $f^{-1}(x) = \dfrac{1}{x} = f(x)$, i.e. f is self-inverse.

b Finding the inverse of $g(x) = \dfrac{3x-5}{x+k}$:

$y = \dfrac{3x-5}{x+k}$

$y(x+k) = 3x-5$

$xy + ky = 3x - 5$

$3x - xy = 5 + ky$

$x(3-y) = 5 + ky$

$\Rightarrow x = \dfrac{5+ky}{3-y}$

$\therefore g^{-1}(x) = \dfrac{5+kx}{3-x}$

Require that $g(x) = g^{-1}(x)$ for all x:

$\dfrac{3x-5}{x+k} = \dfrac{5+kx}{3-x}$

$\Rightarrow \dfrac{3x-5}{x+k} = \dfrac{-kx-5}{x-3}$

Comparing these, it is evident that $k = -3$.

> **COMMENT**
>
> If it is difficult to see that multiplying the numerator and denominator as above enables a straightforward comparison to determine k, then the following (more lengthy!) process can be undertaken instead:
>
> $\dfrac{3x-5}{x+k} = \dfrac{5+kx}{3-x}$
>
> $(3x-5)(3-x) = (5+kx)(x+k)$
>
> $9x - 3x^2 - 15 + 5x = 5x + 5k + kx^2 + k^2 x$
>
> $-3x^2 + 14x - 15 = kx^2 + (k^2 + 5)x + 5k$
>
> Comparing coefficients of the two sides:
>
> $x^2: -3 = k$
>
> $x^1: 14 = k^2 + 5$
>
> $x^0: -15 = 5k$
>
> These three equations consistently give the unique solution $k = -3$.

Exercise 5F

5 $y = \dfrac{3x-1}{4-5x}$

Vertical asymptote where denominator equals zero: $x = \dfrac{4}{5}$

Horizontal asymptote as $x \to \pm\infty$: $y = -\dfrac{3}{5}$

6 $f(x) = \dfrac{1}{x+3}$

a Cannot have division by zero, so domain is $x \neq -3$

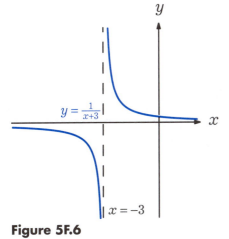

Figure 5F.6

Range is $y \neq 0$

b $y = \dfrac{1}{x+3}$

$y(x+3) = 1$

$xy = 1 - 3y$

$\Rightarrow x = \dfrac{1-3y}{y}$

$\therefore f^{-1}(x) = \dfrac{1-3x}{x}$

7 $y = \dfrac{3x-1}{x-5}$

Vertical asymptote where denominator equals zero: $x = 5$

Horizontal asymptote as x gets large: $y = 3$

Axis intercepts: $\left(0, \dfrac{1}{5}\right)$ and $\left(\dfrac{1}{3}, 0\right)$

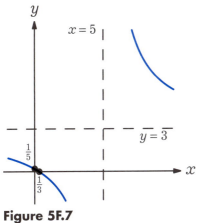

Figure 5F.7

8 $f(x) = \dfrac{ax+3}{2x-8}, \; x \neq 4$

a Horizontal asymptote as x gets large: $y = \dfrac{a}{2}$

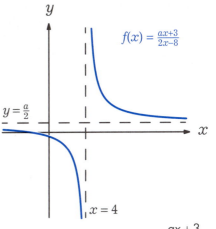

Figure 5F.8 Graph of $f(x) = \dfrac{ax+3}{2x-8}$ for positive a

Range of f is $y \in \mathbb{R}, \; y \neq \dfrac{a}{2}$

b $y = \dfrac{ax+3}{2x-8}$

$y(2x-8) = ax+3$

$2xy - 8y = ax + 3$

$2xy - ax = 8y + 3$

$x(2y-a) = 8y + 3$

$\Rightarrow x = \dfrac{8y+3}{2y-a}$

$\therefore f^{-1}(x) = \dfrac{8x+3}{2x-a}, \; x \neq \dfrac{a}{2}$

(The domain of f^{-1} is the range of f.)

c For f to be self-inverse, require that $f^{-1}(x) = f(x)$ for all x.

The vertical asymptote of f is $x = 4$; this must be the same as the vertical asymptote of f^{-1}, which is $x = \dfrac{a}{2}$:

$\dfrac{a}{2} = 4 \Rightarrow a = 8$

Mixed examination practice 5

Short questions

1 a $y = \log_3(x+3)$

$3^y = x+3$

$\Rightarrow x = 3^y - 3$

$\therefore f^{-1}(x) = 3^x - 3$

(Range of f is $\mathbb{R}$, so domain of f^{-1} is also $\mathbb{R}$.)

b $y = 3e^{x^3-1}$

$\dfrac{y}{3} = e^{x^3-1}$

$\ln\left(\dfrac{y}{3}\right) = x^3 - 1$

$\Rightarrow x = \left(1 + \ln\left(\dfrac{y}{3}\right)\right)^{\frac{1}{3}}$

$\therefore g^{-1}(x) = \left(1 + \ln\left(\dfrac{x}{3}\right)\right)^{\frac{1}{3}}$

(Range of g is $y > 0$, so domain of g^{-1} is $x > 0$.)

2 a Reflection of $f(x)$ in the line $y = x$ gives the graph of $f^{-1}(x)$, so C is $y = \log_2 x$.

b C cuts the x-axis where $y = 0$:

$\log_2 x = 0$

$\Rightarrow x = 2^0 = 1$

i.e. intersection at (1, 0).

3 a Vertical asymptote where denominator equals zero: $x = 5$

Horizontal asymptote for large x:

$y = \dfrac{4}{-1} = -4$

b $y = \dfrac{4x-3}{5-x}$

$(5-x)y = 4x - 3$

$5y - xy = 4x - 3$

$4x + xy = 5y + 3$

$x(4+y) = 5y + 3$

$\Rightarrow x = \dfrac{5y+3}{y+4}$

$\therefore f^{-1}(x) = \dfrac{5x+3}{x+4}$

4 a $f(x) = x^2 - 6x + 10$

$= (x-3)^2 - 9 + 10$

$= (x-3)^2 + 1$

b $y = (x-3)^2 + 1$

$x - 3 = \sqrt{y-1}$

(the positive square root is needed as $x \geq 3$)

$x = 3 + \sqrt{y-1}$

$\therefore f^{-1}(x) = 3 + \sqrt{x-1}$

c The minimum point of f is (3, 1), so the range of f is $y \geq 1$ and hence the domain of f^{-1} is $x \geq 1$.

5 a $h(x) = x^2 - 6x + 2$

$= (x-3)^2 - 9 + 2$

$= (x-3)^2 - 7$

b

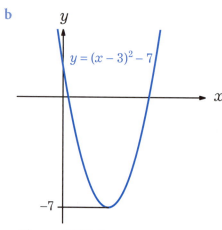

Figure 5MS.5

The domain of h is $x > 3$, so the range of h is $y > -7$

c $h(x) = (x-3)^2 - 7$, $x > k$

For the function to be one-to-one, take $k = 3$.

$y = (x-3)^2 - 7$

$(x-3)^2 = y + 7$

$x - 3 = \sqrt{y+7}$

(choose the positive root since the domain is $x > 3$)

$\Rightarrow x = 3 + \sqrt{y+7}$

$\therefore h^{-1}(x) = 3 + \sqrt{x+7}$

6 a Horizontal asymptote is $y = -1$, so the range is $y \neq -1$.

b Vertical asymptote: $x = -1$

Axis intercepts: $(3, 0)$ and $(0, 3)$

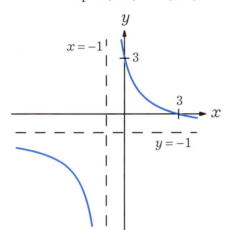

Figure 5MS.6 Graph of $f(x) = \dfrac{3-x}{x+1}$

c $y = \dfrac{3-x}{x+1}$

$(x+1)y = 3 - x$

$xy + y = 3 - x$

$xy + x = 3 - y$

$x(y+1) = 3 - y$

$\Rightarrow x = \dfrac{3-y}{y+1}$

$\therefore f^{-1}(x) = \dfrac{-x+3}{x+1}$

Domain and range of f are $x \neq -1$ and $y \neq -1$, so domain and range of f^{-1} are $x \neq -1$ and $y \neq -1$.

7 $f(x) = \begin{cases} 5 - x, & x < 0 \\ pe^{-x}, & x \geq 0 \end{cases}$

a i With $p = 3$:
lower part has range $]5, \infty[$
upper part has range $]0, 3]$
$\therefore f(x)$ has range $]0, 3] \cup]5, \infty[$

ii For $x < 0$:
$y = 5 - x \Rightarrow x = 5 - y$
$\therefore f^{-1}(x) = 5 - x$

For $x \geq 0$:
$y = 3e^{-x}$
$e^{-x} = \dfrac{y}{3}$
$e^x = \dfrac{3}{y}$
$\Rightarrow x = \ln\left(\dfrac{3}{y}\right)$
$\therefore f^{-1}(x) = \ln\left(\dfrac{3}{x}\right)$

So $f^{-1}(x) = \begin{cases} \ln\left(\dfrac{3}{x}\right), & 0 < x \leq 3 \\ 5 - x, & x > 5 \end{cases}$

5 The theory of functions 49

The domain of $f^{-1}(x)$ is the range of $f(x)$, i.e. $]0,3] \cup]5, \infty[$

b For continuous f, the value at the boundary $x = 0$ must be consistent:

$5 - 0 = pe^0$

$\Rightarrow p = 5$

8 a $f(x) = \sqrt{x-2}, g(x) = x^2 + x$

$\therefore fg(x) = \sqrt{x^2 + x - 2}$

$f \circ g$ is undefined when $x^2 + x - 2 < 0$ (since the square root is undefined for negative values):

$x^2 + x - 2 < 0$

$(x+2)(x-1) < 0$

$x \in]-2, 1[$

$\therefore a = -2, b = 1$

b

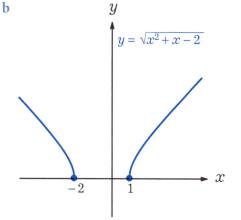

Figure 5MS.8 Graph of $fg(x) = \sqrt{x^2 + x - 2}$

Range of g on the domain $\mathbb{R} -]-2, 1[$ is $y \geq 2$

Range of f on the domain $x \geq 2$ is $y \geq 0$

$\therefore$ range of $f \circ g$ on the domain $\mathbb{R} -]-2, 1[$ is $y \geq 0$

Long questions

1 a $f(3) = 3^2 + 1 = 10$

b $gf(x) = g(x^2 + 1)$

$= 5 - (x^2 + 1)$

$= 4 - x^2$

c The graphs of a function and its inverse are reflections of each other in the line $y = x$.

d i $y = x^2 + 1$

$x^2 = y - 1$

$x = \sqrt{y-1}$

(the positive square root is needed as $x > 3$)

$\therefore f^{-1}(x) = \sqrt{x-1}$

ii Domain of f is $x > 3$, so range of f^{-1} is $y > 3$.

iii Range of f is $y > 10$, so domain of f^{-1} is $x > 10$.

e $f(x) = g(3x)$

$x^2 + 1 = 5 - 3x$

$x^2 + 3x - 4 = 0$

$(x+4)(x-1) = 0$

$x = -4$ or $x = 1$

However, the domain of f is $x > 3$ so there are no solutions to this equation.

2 a i $f(7) = 2 \times 7 + 1 = 15$

ii Range of f is $\mathbb{R}$

iii $fg(x) = f\left(\dfrac{x+3}{x-1}\right)$

$= 2\left(\dfrac{x+3}{x-1}\right) + 1$

$= \dfrac{2x+6}{x-1} + \dfrac{x-1}{x-1}$

$= \dfrac{3x+5}{x-1}$

50 Mixed examination practice 5

    iv $ff(x) = 2(2x+1)+1$
$$= 4x+3$$

b The value $f(0)=1$ is in the range of f but not in the domain of g, so $gf(0)$ is not defined.

c i $y = \dfrac{x+3}{x-1}$

$(x-1)y = x+3$

$xy - y = x+3$

$xy - x = y+3$

$x(y-1) = y+3$

$\Rightarrow x = \dfrac{y+3}{y-1}$

$\therefore g^{-1}(x) = \dfrac{x+3}{x-1}$

> **COMMENT**
>
> Note that $g(x)$ is self-inverse.

    ii $g(x)$ is self-inverse, so the domain and range of $g(x)$ and the domain and range of $g^{-1}(x)$ must all be the same.

$\therefore$ domain of $g^{-1}(x)$ is $x \neq 1$.

    iii Range of $g^{-1}(x)$ is $y \neq 1$.

3 a $f(x) = x^2 + 4x + 9$
$$= (x+2)^2 - 4 + 9$$
$$= (x+2)^2 + 5$$

b Symmetry line at $x = -2$, vertex at $(-2, 5)$.

Positive quadratic shape; y-intercept at $(0, 9)$.

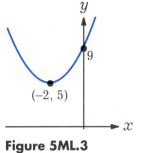

Figure 5ML.3

c Range of $f(x)$ is $]5, \infty[$

Range of $g(x) = e^x$ is $]0, \infty[$

d $h(x) = f \circ g(x)$

Range of $h(x)$ is the range of $f(x)$ with restricted domain $]0, \infty[$

$\therefore$ range of $h(x)$ is $]9, \infty[$

4 a $(2x+3)(4-y) = 12$

$8x + 12 - y(2x+3) = 12$

$y(2x+3) = 8x$

$\Rightarrow y = \dfrac{8x}{2x+3}$

b Vertical asymptote where denominator equals zero: $x = -\dfrac{3}{2}$

Horizontal asymptote: $y = \dfrac{8}{2} = 4$

Single axis intercept at $(0, 0)$

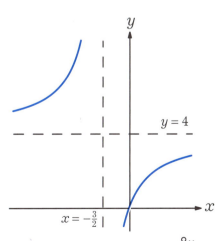

Figure 5ML.4 Graph of $y = \dfrac{8x}{2x+3}$

c i $hg(x) = h(2x+k)$

$$= \frac{8(2x+k)}{2(2x+k)+3}$$

$$= \frac{16x+8k}{4x+2k+3}$$

ii Vertical asymptote where denominator equals zero:

$$x = -\frac{2k+3}{4}$$

Horizontal asymptote: $y = \frac{16}{4} = 4$

iii For $k = -\frac{19}{2}$,

$$hg(x) = \frac{16x + 8\left(-\frac{19}{2}\right)}{4x + 2\left(-\frac{19}{2}\right) + 3}$$

$$= \frac{16x - 76}{4x - 16}$$

Finding the inverse:

$$y = \frac{16x - 76}{4x - 16}$$

$(4x-16)y = 16x - 76$

$4xy - 16y = 16x - 76$

$4xy - 16x = 16y - 76$

$(4y-16)x = 16y - 76$

$$\Rightarrow x = \frac{16y - 76}{4y - 16}$$

$$\therefore (hg)^{-1}(x) = \frac{16x - 76}{4x - 16} = hg(x)$$

i.e. $hg(x)$ is self-inverse.

5 a $gg(x) = g(g(x))$

$$= g(x^{-1})$$

$$= (x^{-1})^{-1}$$

$$= x$$

b $f(x) + 2f\left(\frac{1}{x}\right) = 2x + 1$ ...(1)

Replacing x with $\frac{1}{x}$:

$f\left(\frac{1}{x}\right) + 2f(x) = \frac{2}{x} + 1$ ...(2)

c $(1) - 2 \times (2)$:

$$-3f(x) = 2x - \frac{4}{x} - 1$$

$$\Rightarrow f(x) = \frac{1 + \frac{4}{x} - 2x}{3}$$

$$= \frac{4 + x - 2x^2}{3x}$$

52 Mixed examination practice 5

6 Transformations of graphs

Exercise 6D

6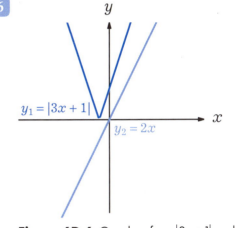

Figure 6D.6 Graphs of $y = |3x + 1|$ and $y = 2x$

From the graph, $y_1 > y_2$ for all x, so the solution is $x \in \mathbb{R}$.

7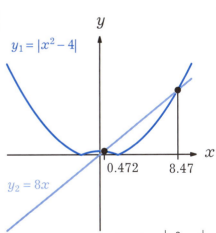

Figure 6D.7 Graphs of $y = |x^2 - 4|$ and $y = 8x$

Intersection of $y = 4 - x^2$ and $y = 8x$ in the interval $[0, 2]$:

$4 - x^2 = 8x$

$x^2 + 8x - 4 = 0$

$x = \dfrac{-8 \pm \sqrt{64 + 16}}{2}$

$= \dfrac{-8 \pm 4\sqrt{5}}{2}$

$= 2\sqrt{5} - 4$ (discard negative root)

Intersection of $y = x^2 - 4$ and $y = 8x$ in the interval $[2, \infty[$:

$x^2 - 4 = 8x$

$x^2 - 8x - 4 = 0$

$x = \dfrac{8 \pm \sqrt{64 + 16}}{2}$

$= \dfrac{8 \pm 4\sqrt{5}}{2}$

$= 2\sqrt{5} - 4$ (discard negative root)

∴ the solutions are $x = 2\sqrt{5} \pm 4 = 0.472$ and 8.47

8 $|x^2 - 7x + 10| = -(x^2 - 7x + 10)$

$\Rightarrow x^2 - 7x + 10 \leq 0$

$\Rightarrow (x - 5)(x - 2) \leq 0$

A positive quadratic is negative between its two roots

$\therefore x \in [2, 5]$

9 $x|x| = 4x$

$x(|x| - 4) = 0$

$x = 0$ or $|x| = 4$

$\therefore x = 0$ or ± 4

10 $|x+q^2| = |x-2q^2|$

> **COMMENT**
> This question can be solved either using a graph and algebra by intervals or by direct algebraic calculation. Both methods are given here, and either would be acceptable in an examination.

Graphically:

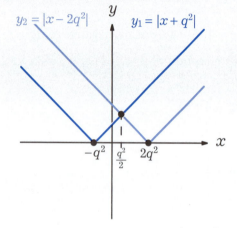

Figure 6D.10 Graphs of $y = |x+q^2|$ and $y = |x-2q^2|$

Intersection of $y = x+q^2$ and $y = 2q^2 - x$ in the interval $[-q^2, 2q^2]$:

$x + q^2 = 2q^2 - x$

$\Rightarrow x = \dfrac{q^2}{2}$

Algebraically:

$(x+q^2)^2 = (x-2q^2)^2$

$x^2 + 2xq^2 + q^4 = x^2 - 4xq^2 + 4q^4$

$6xq^2 = 3q^4$

$\Rightarrow x = \dfrac{q^2}{2}$

11 $y = f(x) + |f(x)|$

$\therefore y = 0$ wherever $f(x) \leq 0$ and $y = 2f(x)$ wherever $f(x) \geq 0$

Therefore, the graph of $y = f(x) + |f(x)|$ is:

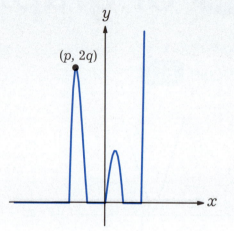

Figure 6D.11

Exercise 6E

10 a $f_1(x) = ax + b$

Translation by $\begin{pmatrix} 1 \\ 2 \end{pmatrix}$: replace x with $(x-1)$, add 2 $\Rightarrow f_2(x) = a(x-1) + b + 2$

Reflection in $y = 0$: multiply through by $-1 \Rightarrow f_3(x) = a(-x+1) - b - 2$

Horizontal stretch with scale factor $\dfrac{1}{3}$: replace x with $3x \Rightarrow$

$f_4(x) = a(-3x+1) - b - 2$
$= -3ax + a - b - 2$

$\therefore g(x) = 4 - 15x = a - b - 2 - 3ax$

Comparing coefficients of x^1:
$-3a = -15 \Rightarrow a = 5$

Comparing coefficients of x^0:
$a - b - 2 = 4 \Rightarrow b = -1$

b $f(x) = ax^2 + bx + c$

Reflection in $x = 0$: replace x with $-x \Rightarrow$
$f_2(x) = ax^2 - bx + c$

Translation by $\begin{pmatrix}-1\\3\end{pmatrix}$: replace
x with $(x+1)$, add $3 \Rightarrow$
$f_3(x) = a(x+1)^2 - b(x+1) + c + 3$

Horizontal stretch with scale factor 2:
replace x with $\dfrac{x}{2} \Rightarrow$

$f_4(x) = a\left(\dfrac{x}{2}+1\right)^2 - b\left(\dfrac{x}{2}+1\right) + c + 3$

$= \dfrac{a}{4}x^2 + ax + a - \dfrac{b}{2}x - b + c + 3$

$= \dfrac{a}{4}x^2 + \left(a - \dfrac{b}{2}\right)x + a - b + c + 3$

$\therefore g(x) = 4x^2 + ax - 6$

$= \dfrac{a}{4}x^2 + \left(a - \dfrac{b}{2}\right)x + a - b + c + 3$

Comparing coefficients of x^2:
$\dfrac{a}{4} = 4 \Rightarrow a = 16$

Comparing coefficients of x^1:
$a - \dfrac{b}{2} = a \Rightarrow b = 0$

Comparing coefficients of x^0:
$a - b + c + 3 = -6 \Rightarrow c = -25$

11 $f(x) = 2^x + x$

Vertical stretch with scale factor 8:
multiply through by $8 \Rightarrow$
$f_2(x) = 8(2^x + x)$

Translation by $\begin{pmatrix}1\\4\end{pmatrix}$: replace x with $(x-1)$,
add $4 \Rightarrow f_3(x) = 8(2^{x-1} + x - 1) + 4$

Horizontal stretch with scale factor $\dfrac{1}{2}$:
replace x with $2x \Rightarrow$

$f_4(x) = 8(2^{2x-1} + 2x - 1) + 4$

$= 2^3 \times 2^{2x-1} + 16x - 4$

$= 2^{2x+2} + 16x - 4$

$= (2^2)^{x+1} + 16x - 4$

$= 4^{x+1} + 16x - 4$

So $h(x) = 4^{x+1} + 16x - 4$

12 a Graph of $y = \ln x$:
vertical asymptote $x = 0$; intercept $(1, 0)$

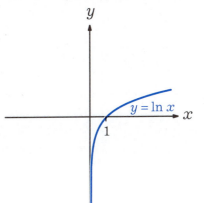

Figure 6E.12.1

b Graph of $y = 3\ln(x+2)$ is obtained
from the graph of $y = \ln x$ by:

translation $\begin{pmatrix}-2\\0\end{pmatrix}$ and vertical stretch
with scale factor $3 \Rightarrow$ vertical
asymptote $x = -2$; intercept $(-1, 0)$

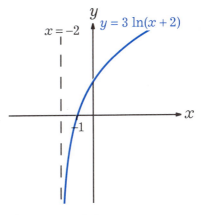

Figure 6E.12.2

c Graph of $y = \ln(2x-1)$ is obtained from the graph of $y = \ln x$ by:

translation $\begin{pmatrix} 1 \\ 0 \end{pmatrix}$ and horizontal stretch with scale factor $\frac{1}{2}$ ⇒ vertical asymptote $x = \frac{1}{2}$, intercept $(1, 0)$.

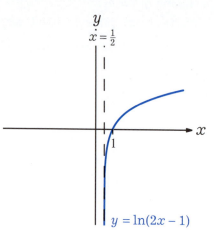

Figure 6E.12.3

Exercise 6F

3 a $f(x)$ to $f(x-3)$ is a translation by $\begin{pmatrix} 3 \\ 0 \end{pmatrix}$

∴ $p = 3$, $q = 0$

b i This is the graph of $y = x^2$ translated by $\begin{pmatrix} 3 \\ 0 \end{pmatrix}$:

x-intercept (and vertex) at $(3, 0)$; y-intercept at $(0, 9)$

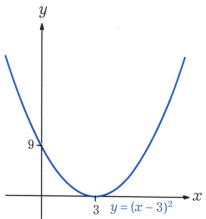

Figure 6F.3.1

ii This is the reciprocal transformation of (i):
- vertical asymptote at $x = 3$
- $y \to 0$ as $x \to \pm\infty$
- y-intercept at $\left(0, \frac{1}{9}\right)$

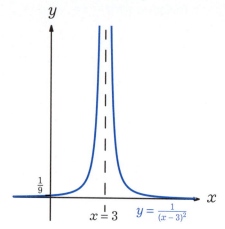

Figure 6F.3.2

4 $y = \dfrac{1}{f(x)}$ will have:
- vertical asymptotes where $f(x) = 0$: $x = -2$ and $x = 1$
- a maximum at the minimum of $f(x)$: approximately at $\left(1.8, -\dfrac{1}{3}\right)$
- a minimum at the maximum of $f(x)$, i.e. $\left(0, \dfrac{1}{3}\right)$

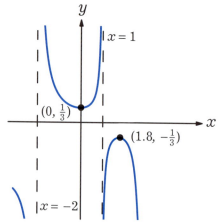

Figure 6F.4

5 Reciprocal graph followed by translation $\begin{pmatrix} 0 \\ -2 \end{pmatrix}$:
   - no vertical asymptotes as $g(x) \neq 0$
   - maximum at $(1, -1)$
   - roots of $\dfrac{1}{g(x)}$ at asymptotes of $g(x)$, i.e. at $(0, 0)$ and $(2, 0)$, so new curve passes through $(0, -2)$ and $(2, -2)$

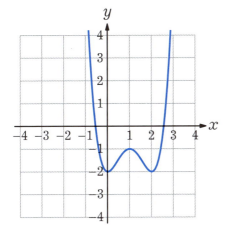

Figure 6F.5

6 Vertical asymptotes where denominator equals zero:
$$x^2 e^{x^2} - 4x^2 = 0$$
$$x^2 \left(e^{x^2} - 4 \right) = 0$$
$$x = 0 \text{ or } e^{x^2} = 4$$
$$\therefore x = 0 \text{ or } x = \pm\sqrt{\ln 4}$$

7 Roots of $\dfrac{f(x)}{g(x)}$ are at roots of $f(x)$: $(1.5, 0)$

Asymptotes of $\dfrac{f(x)}{g(x)}$ are at roots of $g(x)$: $x = 0$ and $x = 5$

As $x \to \infty$, $f(x)$ is negative and $g(x)$ gets large and negative, so $\dfrac{f(x)}{g(x)} \to 0$ from above.

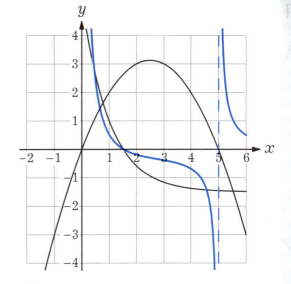

Figure 6F.7

Exercise 6G

2 **a** The graph of an odd function has two-fold rotational symmetry about the origin.

b i Even function $\Rightarrow$ reflective symmetry about $x = 0$

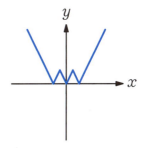

Figure 6G.2.1

ii Odd function $\Rightarrow$ two-fold rotational symmetry about the origin

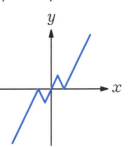

Figure 6G.2.2

3 $f(-x) = e^{(-x)^2}(1+(-x)^4)$

$\qquad = e^{x^2}(1+x^4)$

$\qquad = f(x)$

So by the definition of an even function, $f(x)$ is even.

4 For an even function, $f(-x) = f(x)$, so it cannot be one-to-one as each value in the range has at least two corresponding values in the domain.

5 Suppose that $f(x)$ and $g(x)$ are both even functions, and let $h(x) = f(x) + g(x)$ for $x \in \text{Dom}_h = \text{Dom}_f \cap \text{Dom}_g$. Then

$h(-x) = f(-x) + g(-x)$

$\qquad = f(x) + g(x) \quad \text{because } f \text{ and } g \text{ are even}$

$\qquad = h(x)$

$\therefore h(-x) = h(x)$ for all x in its domain, so by definition $h(x)$ is even.

6 Suppose that $f(x)$ and $g(x)$ are both odd functions, and let $h(x) = f(x) \times g(x)$ for $x \in \text{Dom}_h = \text{Dom}_f \cap \text{Dom}_g$. Then

$h(-x) = f(-x) \times g(-x)$

$\qquad = (-f(x)) \times (-g(x)) \quad \text{because } f \text{ and } g \text{ are odd}$

$\qquad = f(x) \times g(x)$

$\qquad = h(x)$

$\therefore h(-x) = h(x)$ for all x in its domain, so by definition $h(x)$ is even.

7 For $f(x)$ to be odd, require that $f(-x) = -f(x)$ for all $x \in \text{Dom}_f$.

If $f(x) = ax^n$ is odd, then $a(-x)^n = -ax^n$

One solution is $a = 0$, but this gives a trivial function $f(x) = 0$, which is not of interest.

$\therefore (-x)^n = -x^n$

For real-valued f, it must be the case that n is an integer.

Then the above condition becomes

$(-1)^n x^n = -x^n$ for all $x \in \text{Dom}_f$

$(-1)^n = -1$

which is true when n is an odd integer and false when n is an even integer.

$\therefore n$ is odd.

8 Suppose that $f(x)$ is an odd function and $g(x)$ is an even function, and let $h(x) = gf(x)$.
Then for all $x \in \mathrm{Dom}_h$,

$$h(-x) = g(f(-x))$$
$$= g(-f(x)) \text{ since } f \text{ is odd}$$
$$= g(f(x)) \text{ since } g \text{ is even}$$
$$= h(x)$$

$\therefore h(-x) = h(x)$ for all x in its domain, so by definition $h(x)$ is even.

9 a $f(x) = x^2 + 6x + 7$
$= (x+3)^2 - 2$

has line of symmetry $x = -3$.

b $f(x-a)$ is the function $f(x)$ after a translation by $\begin{pmatrix} a \\ 0 \end{pmatrix}$.

Require a symmetry line at $x = 0$ for $f(x-a)$ to be even.

$\therefore a = 3$

10 The graph has line of symmetry $x = 5$ since $f(5+a) = f(10-(5+a)) = f(5-a)$

11 If a function $f(x)$ is symmetrical in $y=x$, then $f(x) = f^{-1}(x)$ since the graph of $f^{-1}(x)$ is the graph of $f(x)$ reflected through $y=x$.

$\therefore ff(x) = x$ for all x

and hence $ff(4) = 4$

12 a Let $g(x) = \frac{1}{2}(f(x) - f(-x))$. Then

$$g(-x) = \frac{1}{2}(f(-x) - f(x))$$
$$= -g(x)$$

$\therefore g(-x) = -g(x)$ for all x in its domain, so by definition $g(x)$ is odd.

b By similar reasoning,
$h(x) = \frac{1}{2}(f(x) + f(-x))$ is an even function:

$$h(-x) = \frac{1}{2}(f(-x) + f(x)) = h(x)$$

c For any function f,

$$f(x) = \frac{1}{2}(f(x) - f(-x))$$
$$+ \frac{1}{2}(f(x) + f(-x))$$
$$= g(x) + h(x)$$

i.e. $f(x)$ can be written as the sum of an odd function $g(x)$ and an even function $h(x)$.

d i $g(x) = \frac{1}{2}(e^x - e^{-x})$

Axis intercept at $(0, 0)$ only.

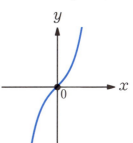

Figure 6G.12.1

ii $h(x) = \frac{1}{2}(e^x + e^{-x})$

Axis intercept at $(0, 1)$ only.

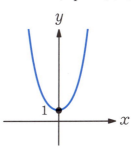

Figure 6G.12.2

> **COMMENT**
>
> These are the hyperbolic sine and cosine functions, $g(x) = \sinh x$ and $h(x) = \cosh x$, which you can find on your calculator.

13 If f is an odd function, then $f(-x) = -f(x)$ for all x in its domain.

If f is also a polynomial, then since any polynomial is defined at all real values of x, it must be defined at $x = 0$.

$f(-0) = -f(0)$ because f is odd,

but also $f(-0) = f(0)$.

So $f(0) = -f(0)$ and hence $f(0) = 0$, which means that the graph of $f(x)$ passes through the origin.

This must always be the case when 0 is within the domain of $f(x)$.

An example of an odd function not defined at 0 would be $f(x) = x^{-1}$, or indeed $f(x) = x^{-n}$ for any positive odd integer n.

$f(x) = \cot x$ and $f(x) = \csc x$ are other examples encountered in this course.

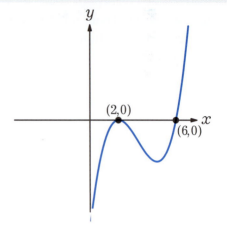

Figure 6MS.1.1

b Asymptotes of $\dfrac{1}{f(x)}$ at roots of $f(x)$:

$x = 0$ and $x = 4$.

$y \to 0$ from below as $x \to -\infty$ and
$y \to 0$ from above as $x \to \infty$.

Maximum value where $f(x)$ has minimum.

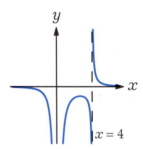

Figure 6MS.1.2

Mixed examination practice 6

Short questions

1 a Graph of $y = 3f(x-2)$ is obtained from the graph of $y = f(x)$ by:

translation $\begin{pmatrix} 2 \\ 0 \end{pmatrix}$ and vertical stretch with scale factor 3

$\Rightarrow$ asymptote becomes $x = 0$;
x-intercepts become $(2, 0)$ and $(6, 0)$

2 $f(x) = x^3 - 1$

Translation by $\begin{pmatrix} 2 \\ 0 \end{pmatrix}$: replace x with
$(x-2) \Rightarrow f_2(x) = (x-2)^3 - 1$

Vertical stretch with scale factor 2: multiply by $2 \Rightarrow f_3(x) = 2\left[(x-2)^3 - 1\right]$

So new graph is

$y = 2\left[(x-2)^3 - 1\right]$

$= 2\left[x^3 - 6x^2 + 12x - 8 - 1\right]$

$= 2x^3 - 12x^2 + 24x - 18$

3

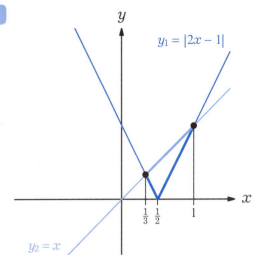

Figure 6MS.3 Graphs of $y = |2x - 1|$ and $y = x$

Intersection of $y = |2x - 1|$ and $y = x$ in interval $x < \dfrac{1}{2}$:

$1 - 2x = x$

$\Rightarrow x = \dfrac{1}{3}$

Intersection of $y = |2x - 1|$ and $y = x$ in interval $x > \dfrac{1}{2}$:

$2x - 1 = x$

$\Rightarrow x = 1$

$\therefore |2x - 1| < x$ for $\dfrac{1}{3} < x < 1$

4 a $y = |f(x)| = \begin{cases} -f(x), & x < a \\ f(x), & x \geq a \end{cases}$

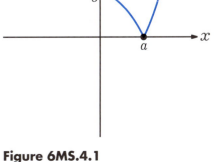

Figure 6MS.4.1

b $y = f(|x|) - 1 = \begin{cases} f(-x) - 1, & x < 0 \\ f(x) - 1, & x \geq 0 \end{cases}$

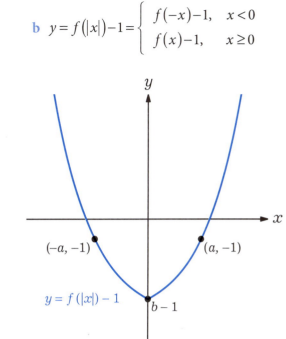

Figure 6MS.4.2

5 a The graph of $y = -\dfrac{3}{x}$ is the reciprocal graph of $y = \dfrac{1}{x}$, reflected through $y = 0$ and vertically stretched by scale factor 3.

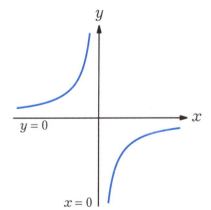

Figure 6MS.5

b Reflection through $y = 0$ followed by a vertical stretch with scale factor 3.

Alternatively, the reflection could be through $x = 0$ or the stretch could be horizontal with scale factor 3.

c $y = -\dfrac{3}{x}$

$\Rightarrow x = -\dfrac{3}{y}$

$\therefore f^{-1}(x) = -\dfrac{3}{x}$

(The function is self-inverse.)

6 a Asymptotes of $y = \dfrac{1}{f(x)}$ at roots of $f(x)$: $x = -5$, $x = 0$ and $x = 5$.

$y \to 0$ from above as $x \to \pm\infty$.

Maximum points where $f(x)$ has minimum points: $x = \pm 3$, at which $y = -\dfrac{1}{5}$.

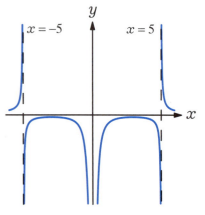

Figure 6MS.6

b Maximum points are $\left(-3, -\dfrac{1}{5}\right)$ and $\left(3, -\dfrac{1}{5}\right)$.

7 Let $f(x) = (x-1)^2$, $g(x) = 3(x+2)^2$.

Then $g(x) = 3f(x+3)$, which represents a translation by $\begin{pmatrix} -3 \\ 0 \end{pmatrix}$ and a vertical stretch with scale factor 3.

Note: there are alternative valid answers. For instance,

$g(x) = \left(\sqrt{3}x + 2\sqrt{3}\right)^2 = f\left(\sqrt{3}x + 1 + 2\sqrt{3}\right)$,

representing a translation $\begin{pmatrix} -1 - 2\sqrt{3} \\ 0 \end{pmatrix}$

followed by a horizontal stretch with scale factor $\dfrac{1}{\sqrt{3}}$.

Also, $g(x) = \left(\sqrt{3}x + 2\sqrt{3}\right)^2$

$= f\left(\sqrt{3}x + 2\sqrt{3} + 1\right)$

$= f\left(\sqrt{3}\left(x + 2 + \dfrac{1}{\sqrt{3}}\right)\right)$,

which represents a horizontal stretch with scale factor $\dfrac{1}{\sqrt{3}}$ followed by a translation

$\begin{pmatrix} -2 - \dfrac{1}{\sqrt{3}} \\ 0 \end{pmatrix}$.

COMMENT

In questions on transformations, it is often the case with simple curves that several possible transformations will lead to the same effective change, as here. In such cases, any single valid answer is acceptable, but one is usually simpler than the others.

8 a $y = 3f\left(\dfrac{x}{2}\right)$

In $y = f(x)$, x is replaced by $\dfrac{x}{2}$, corresponding to a horizontal stretch with scale factor 2.

Multiplication by 3 corresponds to a vertical stretch with scale factor 3.

b Graph of $y = \ln x$ has a vertical asymptote $x = 0$ and an axis intercept $(1, 0)$.

New graph still has asymptote at $x = 0$, but the intercept shifts to $(2, 0)$.

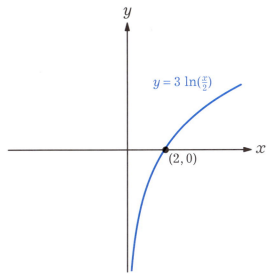

Figure 6MS.8.1

c The transformation from $y = 3\ln\left(\dfrac{x}{2}\right)$ to $y = 3\ln\left(\dfrac{x}{2} + 1\right) = 3\ln\left(\dfrac{x+2}{2}\right)$ is a translation by $\begin{pmatrix} -2 \\ 0 \end{pmatrix}$.

New graph shows the answer in (b) shifted 2 units to the left; the asymptote is at $x = -2$ and the axis intercept is $(0, 0)$.

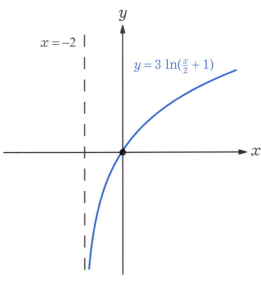

Figure 6MS.8.2

9 a Asymptote of $y = \dfrac{1}{f(x)}$ at root of $f(x)$: $x = -2$.

Horizontal asymptote at reciprocal of horizontal asymptote of $f(x)$: $y = \dfrac{1}{2}$, approached from below as $x \to \infty$.

$f(x)$ appears fairly linear until close to the maximum; that line has equation $y = x + 2$, so for an equivalent domain $\dfrac{1}{f(x)}$ will closely approximate $y = \dfrac{1}{x+2}$.

Minimum of $\dfrac{1}{f(x)}$ where $f(x)$ has a maximum.

y-intercept at reciprocal of y-intercept of $f(x)$: $\left(0, \dfrac{1}{2}\right)$.

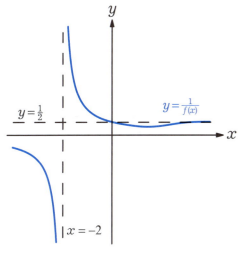

Figure 6MS.9.1

b Roots of $y = xf(x)$ at $x = 0$ and the root of $f(x)$: $x = -2$.

$f(x)$ appears fairly linear until close to the maximum; that line has equation $y = x + 2$, so for an equivalent domain $xf(x)$ will closely approximate $y = x^2 + 2x$.

6 Transformations of graphs   63

$f(x) \to 2$ as $x \to \infty$ so $xf(x) \to 2x$ as $x \to \infty$.

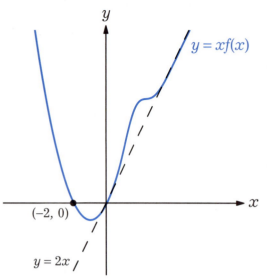

Figure 6MS.9.2

10

> **COMMENT**
>
> This problem could be solved graphically, by plotting for an arbitrary k the graphs of $y = |x+k|$ and $y = |x|+k$ and looking for intervals of $\mathbb{R}$ over which the two coincide. The question can, however, be approached more swiftly by direct algebra. To eliminate modulus signs, square a modulus; check for false solutions where appropriate.

$|x+k| = |x| + k$
$(x+k)^2 = (|x|+k)^2$
$x^2 + 2kx + k^2 = x^2 + 2k|x| + k^2$
$2kx = 2k|x|$
$x = |x|$
$\Rightarrow x \geq 0$

Long questions

1 a $y = 3x^2 - 12x + 12 = 3(x-2)^2$

Transformation from $y = f(x)$ to $y = 3f(x-2)$: translation by $\begin{pmatrix} 2 \\ 0 \end{pmatrix}$ and vertical stretch with scale factor 3.

b $y = x^2 + 6x - 1 = (x+3)^2 - 10$

Transformation from $y = f(x+3) - 10$ to $y = f(x)$: translation by $\begin{pmatrix} 3 \\ 10 \end{pmatrix}$, which is equivalent to translation by $\begin{pmatrix} 3 \\ 0 \end{pmatrix}$ and translation by $\begin{pmatrix} 0 \\ 10 \end{pmatrix}$.

> **COMMENT**
>
> It is usual to go from $y = f(x)$ to $y = f(x+3) - 10$, in which case the transformation would be a translation by $\begin{pmatrix} -3 \\ -10 \end{pmatrix}$. However, this question asks for the transformation in the opposite direction, hence the translation by $\begin{pmatrix} 3 \\ 10 \end{pmatrix}$.

c Transformation of $y = x^2 + 6x - 1$ to $y = 3x^2 - 12x + 12$ can be achieved by the transformation in (b) followed by the transformation in (a):

translation by $\begin{pmatrix} 3 \\ 10 \end{pmatrix}$ and then translation by $\begin{pmatrix} 2 \\ 0 \end{pmatrix}$ and vertical stretch with scale factor 3, which is equivalent to translation by $\begin{pmatrix} 5 \\ 10 \end{pmatrix}$ and vertical stretch with scale factor 3.

d

> **COMMENT**
>
> By factorising the denominator of the function and applying knowledge of roots and asymptotes, this question could be approached as though starting with no prior working. However, having determined a series of transformations mapping x^2 to $3x^2 - 12x + 12$, it is faster to use these same transformations to map $\dfrac{1}{x^2}$ to $\dfrac{1}{3x^2 - 12x + 12}$.

> **COMMENT**
>
> In both parts (a) and (d) the transformation could also be categorised as $x^2 \to \left(\sqrt{3}(x-2)\right)^2$, which would be a horizontal stretch with scale factor $\dfrac{1}{\sqrt{3}}$ followed by a translation $\begin{pmatrix} 2 \\ 0 \end{pmatrix}$. While this appears more complicated, it means that all the transformations are horizontal and are therefore exactly the same for both $f(x)$ and $\dfrac{1}{f(x)}$.

From (a), x^2 is mapped to $3x^2 - 12x + 12$ by a horizontal translation $\begin{pmatrix} 2 \\ 0 \end{pmatrix}$ and a vertical stretch with scale factor 3.

Therefore $\dfrac{1}{x^2}$ is transformed to $\dfrac{1}{x^2 - 12x + 12}$ by a horizontal translation $\begin{pmatrix} 2 \\ 0 \end{pmatrix}$ and a vertical stretch with scale factor $\dfrac{1}{3}$.

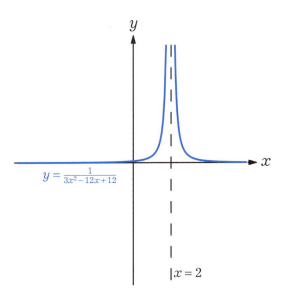

$y = \dfrac{1}{3x^2 - 12x + 12}$

$x = 2$

Figure 6ML.1

2 a As $x \to \infty$, $f(x) \to \dfrac{3x}{x} = 3$

∴ horizontal asymptote is $y = 3$

b $f(x) = \dfrac{3x - 5}{x - 2}$

$= \dfrac{3(x-2) + 1}{x - 2}$

$= \dfrac{3(x-2)}{x-2} + \dfrac{1}{x-2}$

$= 3 + \dfrac{1}{x - 2}$

∴ $p = 3$, $q = 1$

c If $g(x) = \dfrac{1}{x}$, then $f(x) = 3 + g(x - 2)$

Transformation from g to f is a translation by $\begin{pmatrix} 2 \\ 3 \end{pmatrix}$.

d $y = \dfrac{3x - 5}{x - 2}$

$(x - 2)y = 3x - 5$

$xy - 2y = 3x - 5$

$xy - 3x = 2y - 5$

$x(y - 3) = 2y - 5$

$\Rightarrow x = \dfrac{2y - 5}{y - 3}$

∴ $f^{-1}(x) = \dfrac{2x - 5}{x - 3}$

6 Transformations of graphs 65

Range of $f(x)$ is $y \neq 3$, so domain of $f^{-1}(x)$ is $x \neq 3$.

e The graph of $y = f^{-1}(x)$ is obtained from the graph of $y = f(x)$ by reflecting in the line $y = x$.

3 a Translation by $\begin{pmatrix} -2 \\ 0 \end{pmatrix}$

b i $y = \ln(x+2)$ is the graph of $y = \ln x$ after translation by $\begin{pmatrix} -2 \\ 0 \end{pmatrix}$

ii $y = \dfrac{1}{\ln(x+2)}$ is the reciprocal graph of $y = \ln(x+2)$
- vertical asymptote where $\ln(x+2) = 0$: $x = -1$
- as $x \to -2$, $y \to 0$ from below
- as $x \to \infty$, $y \to 0$ from above
- y-intercept at $\dfrac{1}{\ln 2}$, no x-intercept

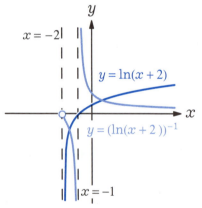

Figure 6ML.3.1 Graphs of $y = \ln(x+2)$ and $y = \dfrac{1}{\ln(x+2)}$

c i The turning point on the right has shifted from $(1, 4)$ to $(3, -4)$. Reflection in the x-axis has switched the sign of the y-coordinate, so the translation must be $\begin{pmatrix} 2 \\ 0 \end{pmatrix}$.

ii $g(x)$ has been translated by $\begin{pmatrix} 2 \\ 0 \end{pmatrix}$ and reflected in the x-axis, so $h(x) = -g(x-2)$.

$$h(x) = -g(x-2)$$
$$= -\left((x-2)^3 - 2(x-2) + 5\right)$$
$$= -\left(x^3 - 6x^2 + 12x - 8 - 2x + 4 + 5\right)$$
$$= -x^3 + 6x^2 - 10x - 1$$

$\therefore a = -1$, $b = 6$, $c = -10$, $d = -1$

d Roots of $y = (k(x))^2$ are at the same values as shown for $y = k(x)$.

$k(x)$ appears linear close to the roots, so the square of the curve should look quadratic close to the roots.

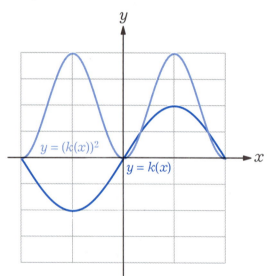

Figure 6ML.3.2

> **COMMENT**
>
> If you recognise the given curve as $k(x) = 2\sin\left(\dfrac{\pi x}{2}\right)$, you will have a shortcut to working out the shape of the result: $(k(x))^2 = 4\sin^2\left(\dfrac{\pi x}{2}\right) = 2 - \cos \pi x$. This material is covered in Chapter 12; although recognising the curve is helpful, it is not necessary.

4 a $f(x) = x^2 - 7x + 10 = (x-2)(x-5)$

Positive quadratic with roots at 2 and 5; y-intercept (0, 10).

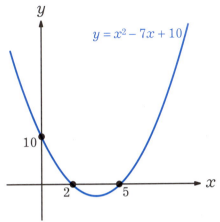

Figure 6ML.4.1

b $f(|x|) = (|x|)^2 - 7|x| + 10$
$= x^2 - 7|x| + 10$
$= g(x)$

c The modulus transformation replaces x by $|x|$, so the graph for negative x is just the mirror image (reflection in the y-axis) of the graph for positive x.

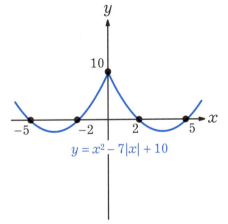

Figure 6ML.4.2

d $x^2 - 7|x| + 10 = x^2$

$-7|x| + 10 = 0$

$|x| = \dfrac{10}{7}$

$x = \pm \dfrac{10}{7}$

e $x^2 - 7|x| + 10 = -2$

$x^2 - 7|x| + 12 = 0$

$(|x| - 3)(|x| - 4) = 0$

$|x| = 3$ or 4

$x = \pm 3, \pm 4$

5 a Line of symmetry at $x = 3 \Rightarrow$
$f(x) = a(x-3)^2 + c$

Expanding:
$ax^2 - 6ax + 9a + c = 3x^2 + bx + 10$

Comparing coefficients:

x^2: $a = 3$

x^1: $-6a = b \Rightarrow b = -18$

x^0: $9a + c = 10 \Rightarrow c = -17$

$\therefore b = -18$

b $f(x)$ is symmetrical about $x = 3$, so
$f(3+k) = f(3-k)$

Replacing k with $x - 3$ gives the other form of the symmetry condition:

$f(3 + (x-3)) = f(3 - (x-3))$
$f(x) = f(6-x)$
$\therefore d = 6$

c $g(x) = f(x+p) + q$ is a quadratic; it is even and goes through the origin, so its vertex is at the origin.

$g(x)$ has symmetry line at $x = 0$ and y-value of vertex raised from $c = -17$ to 0, so it is obtained from $f(x)$ by a translation $\begin{pmatrix} -3 \\ 17 \end{pmatrix}$

$\Rightarrow g(x) = f(x+3) + 17$

$\therefore p = 3, \ q = 17$

d Since $g(x)$ is an even function, by definition $g(x) = g(|x|)$ for all $x \in \mathbb{R}$.

6 **a** The graph of $y = e^x - 2$ is the graph of $y = e^x$ translated by $\begin{pmatrix} 0 \\ -2 \end{pmatrix}$.

Horizontal asymptote at $y = -2$ for large negative x; intercepts at $(0, -1)$ and $(\ln 2, 0)$.

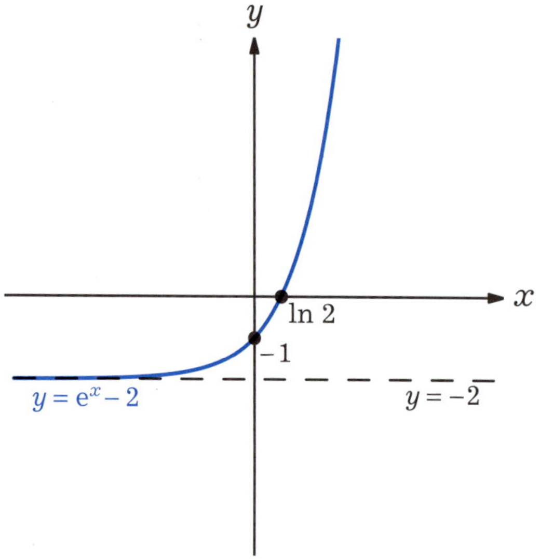

Figure 6ML.6.1

b i Vertical modulus transformation: graph in (a) has its negative-y part reflected in the x-axis.

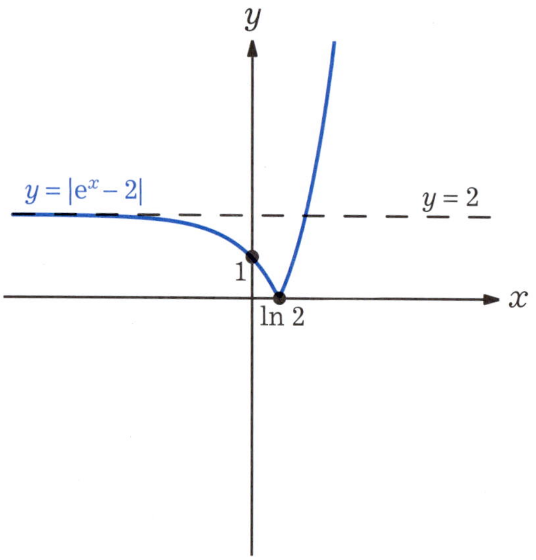

Figure 6ML.6.2

ii Horizontal modulus transformation: the graph for negative x is the mirror image (reflection in the y-axis) of the graph in (a) for positive x.

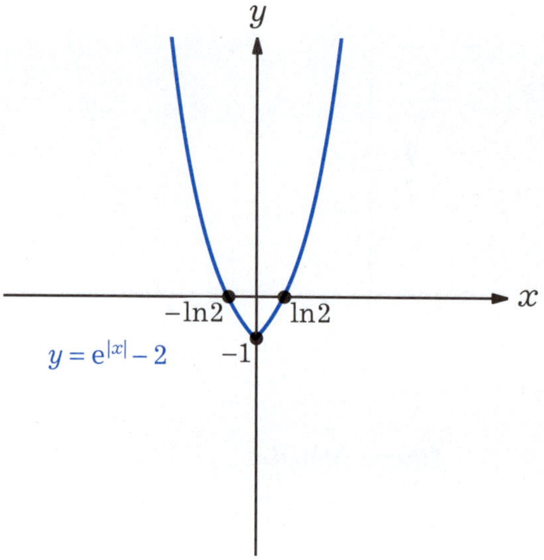

Figure 6ML.6.3

c Require those values of x for which the graphs in (b)(i) and (b)(ii) coincide:

The graphs are the same for $x \geq \ln 2$

They intersect for the negative value of x where $e^{-x} - 2 - e^x$

$e^{2x} - 4e^{-x} + 1 = 0$

$e^x = 2 \pm \sqrt{3}$

$x = \ln\left(2 \pm \sqrt{3}\right)$

Since this should be a negative value of x, the intersection is $x = \ln\left(2 - \sqrt{3}\right)$

7 Sequences and series

Exercise 7A

5 a $u_2 = 3(2) - 2(1) = 4$
$u_3 = 3(4) - 2(2) = 8$
$u_4 = 3(8) - 2(4) = 16$

b i It appears that $u_n = 2^n$

ii If $u_n = 2^n$ then $u_{n-1} = 2^{n-1}$ and $u_{n+1} = 2^{n+1}$, so

$3u_n - 2u_{n-1} = 3(2^n) - 2(2^{n-1})$
$= 3 \times 2^n - 2^n$
$= 2^n(3-1)$
$= 2^n \times 2$
$= 2^{n+1}$
$= u_{n+1}$

i.e. $u_n = 2^n$ satisfies the equation $u_{n+1} = 3u_n - 2u_{n-1}$.

> **COMMENT**
>
> This is an example of a method called proof by induction, which in this case can be used to establish that the result $u_n = 2^n$ is true for all $n \in \mathbb{Z}$. This method of proof is covered in Chapter 25.

Exercise 7C

3 a $a_1 = 5, a_2 = 13$
$d = a_2 - a_1$
$= 13 - 5 = 8$
$a_n = a_1 + (n-1)d$
$= 5 + 8(n-1)$
$= 8n - 3$

b $a_n < 400$
$8n - 3 < 400$
$n < \dfrac{403}{8} = 50.375$

So the first 50 terms are less than 400.

4 $u_{10} = 61$
$\therefore u_1 + (10-1)d = 61$
$\Rightarrow u_1 + 9d = 61 \quad \ldots(1)$

$u_{13} = 79$
$\therefore u_1 + (13-1)d = 79$
$\Rightarrow u_1 + 12d = 79 \quad \ldots(2)$

$(2) - (1)$:
$3d = 18 \Rightarrow d = 6$
$u_{20} = u_{10} + 10d$
$= 61 + 60 = 121$

> **COMMENT**
>
> Here $10d$ has been added to the tenth term to find the twentieth term, but this could also be calculated by first finding u_1 from equation (1) or (2) and then using the usual formula for u_n.

5 $u_8 = 74$

$\therefore u_1 + (8-1)d = 74$

$\Rightarrow u_1 + 7d = 74$ ...(1)

$u_{15} = 137$

$\therefore u_1 + (15-1)d = 137$

$\Rightarrow u_1 + 14d = 137$ ...(2)

(2) – (1):

$7d = 63 \Rightarrow d = 9$

Substituting in (1):

$u_1 + 7 \times 9 = 74 \Rightarrow u_1 = 11$

Let $u_n = 227$; then

$11 + (n-1)9 = 227$

$9n + 2 = 227$

$n = \dfrac{225}{9} = 25$

i.e. the 25th term is 227.

6 Third rung is 70 cm above ground:

$u_3 = 70$

$\therefore u_1 + (3-1)d = 70$

$\Rightarrow u_1 + 2d = 70$ ...(1)

Tenth rung is 210 cm above ground:

$u_{10} = 210$

$\therefore u_1 + (10-1)d = 210$

$\Rightarrow u_1 + 9d = 210$ ...(2)

(2) – (1):

$7d = 140 \Rightarrow d = 20$

Substituting in (1):

$u_1 + 2 \times 20 = 70 \Rightarrow u_1 = 30$

Top rung is 350 cm above ground, so let $u_n = 350$; then

$30 + (n-1)20 = 350$

$20n + 10 = 350$

$n = \dfrac{340}{20} = 17$

i.e. the ladder has 17 rungs.

7 $u_1 = 2$

$u_2 = a - b$

$\therefore 2 + d = a - b$ ...(1)

$u_3 = 2a + b + 7$

$\therefore 2 + 2d = 2a + b + 7$ ...(2)

$u_4 = a - 3b$

$\therefore 2 + 3d = a - 3b$ ...(3)

(3) – (1):

$2d = a - 3b - (a - b)$

$\Rightarrow d = -\dfrac{2b}{2} = -b$

Substituting $d = -b$ in (1):

$2 - b = a - b$

$\Rightarrow a = 2$

(3) – (2):

$d = a - 3b - (2a + b + 7)$

$\therefore -b = -a - 4b - 7$

$\Rightarrow b = \dfrac{-a - 7}{3}$

$= \dfrac{-2 - 7}{3} = -3$

8 a First 9 pages are numbered with single digits for a total of 9 digits.

10th and 11th pages are each numbered with two digits.

Total number of digits for the first 11 pages is $9 + 2 \times 2 = 13$.

b First 9 pages: 9 pages at 1 digit per page: total 9

Pages 10–99: 90 pages at 2 digits per page: total 180

So the total number of digits on the first 99 pages is 189.

Then, since pages 100–999 have 3 digits per page, define an arithmetic sequence with first term 192 (189 plus the 3 digits on page 100) and common difference 3.

Letting $u_n = 1260$:
$$192 + (n-1)3 = 1260$$
$$189 + 3n = 1260$$
$$n = 357$$

So there are $99 + 357 = 456$ pages in total.

Exercise 7D

3 $u_2 = 7$
$\Rightarrow u_1 + d = 7$ ...(1)
$S_4 = 12$
$\dfrac{4}{2}(2u_1 + (4-1)d) = 12$
$2(2u_1 + 3d) = 12$
$\Rightarrow 2u_1 + 3d = 6$ ...(2)

(2) − 2 × (1):
$d = 6 - 14 = -8$

Substituting in (1):
$u_1 = 7 - (-8) = 15$

4 $u_1 = 2$, $d = 3$

a $S_n = \dfrac{n}{2}(2u_1 + (n-1)d)$
$= \dfrac{n}{2}(4 + 3(n-1))$
$= \dfrac{n}{2}(3n + 1)$

b $S_n = 1365$
$\dfrac{n}{2}(3n+1) = 1365$
$3n^2 + n = 2730$
$3n^2 + n - 2730 = 0$
$(3n + 91)(n - 30) = 0$
$\therefore n = 30$ (as $n \in \mathbb{Z}^+$)

5 $u_1 = 85$, $d = -7$
$u_n < 0$
$85 + (n-1)(-7) < 0$
$-7n + 92 < 0$
$n > \dfrac{92}{7} = 13.14\ldots$

So the last positive term is u_{13}.
$S_{13} = \dfrac{13}{2}[2 \times 85 + (13-1)(-7)] = 559$

6 $u_2 = 6$
$\Rightarrow u_1 + d = 6$ ...(1)
$S_4 = 8$
$\dfrac{4}{2}(2u_1 + (4-1)d) = 8$
$2(2u_1 + 3d) = 8$
$\Rightarrow 2u_1 + 3d = 4$ ...(2)

(2) − 2 × (1):
$d = 4 - 12 = -8$

Substituting in (1):
$u_1 = 6 - (-8) = 14$

7 $u_1 = -6$, $d = 7$
$S_n > 10000$
$\dfrac{n}{2}[2(-6) + (n-1) \times 7] > 10000$
$\dfrac{n}{2}(7n - 19) > 10000$
$7n^2 - 19n - 20000 > 0$
$n < -52.1$ or $n > 54.8$ (roots from GDC)
So the smallest n such that $S_n > 10000$ is 55.

8 $S_n = 3n^2 - 2n$
$\dfrac{n}{2}(u_1 + u_n) = 3n^2 - 2n$
$\Rightarrow u_1 + u_n = 6n - 4$
$u_1 = S_1$
$= 3 - 2 = 1$
$\therefore 1 + u_n = 6n - 4$
$\Rightarrow u_n = 6n - 5$

7 Sequences and series

COMMENT

Remember that if you are given a formula for S_n, then by substituting $n = 1$ you can immediately calculate $S_1 = u_1$.

9 There are 12 angles in the sequence, so $n = 12$.
The angle of the largest sector is twice the angle of the smallest sector, so $u_{12} = 2u_1$.
Since the angles must add up to $360°$,

$$S_{12} = 360$$

$$\frac{12}{2}(u_1 + u_{12}) = 360$$

$$u_1 + 2u_1 = 60$$

$$3u_1 = 60$$

$$u_1 = 20$$

∴ smallest sector is $20°$.

10 $\dfrac{u_5}{u_{12}} = \dfrac{6}{13}$

$$\frac{u_1 + 4d}{u_1 + 11d} = \frac{6}{13}$$

$$13(u_1 + 4d) = 6(u_1 + 11d)$$

$$7u_1 = 14d$$

$$\Rightarrow u_1 = 2d \qquad \ldots(1)$$

$$u_1 u_3 = 32$$

$$\Rightarrow u_1(u_1 + 2d) = 32 \quad \ldots(2)$$

Substituting (1) into (2):

$$2d(2d + 2d) = 32$$

$$8d^2 = 32$$

$$d = \pm 2$$

From (1):

$$u_1 = 2(\pm 2) = \pm 4$$

As all terms are positive, must have $a = 4$ and $d = 2$. So

$$S_{100} = \frac{100}{2}(2a + 99d)$$

$$= 50(2 \times 4 + 99 \times 2)$$

$$= 10\,300$$

11 We need to find $112 + 140 + 154 + 182 + 196 + \ldots + 980 + 994$
This can be considered as the sum of two series:
$S_U = 112 + 154 + 196 + \ldots + 994$, series with $a_U = 112$, $d_U = 42$, $n_U = 22$
$S_V = 140 + 182 + 224 + \ldots + 980$, series with $a_V = 140$, $d_V = 42$, $n_V = 21$

$$S_U + S_V = \frac{22}{2}\left[(2 \times 112) + (22-1) \times 42\right]$$

$$+ \frac{21}{2}\left[(2 \times 140) + (21-1) \times 42\right]$$

$$= 12\,166 + 11\,760$$

$$= 23\,926$$

Alternatively, the value can be calculated as the difference of two series:
$S_W = 112 + 126 + 140 + \ldots + 994$, the 64 three-digit multiples of 14
$S_X = 126 + 168 + \ldots + 966$, the 21 three-digit multiples of 42 (i.e. multiples of both 21 and 14), which are excluded

$$S_W - S_X = \frac{64}{2}\left[(2 \times 112) + (64-1) \times 14\right]$$

$$- \frac{21}{2}\left[(2 \times 126) + (21-1) \times 42\right]$$

$$= 35\,392 - 11\,466$$

$$= 23\,926$$

Exercise 7E

4 $u_5 = u_2 r^3$

∴ $u_5 = 162 \Rightarrow u_2 r^3 = 162$

i.e. $6r^3 = 162$

$r^3 = 27$

$r = 3$

$u_{10} = u_5 r^5$

$= 162 \times 3^5$

$= 39\,366$

5 $u_6 = u_3 r^3$

$\therefore u_6 = 7168 \Rightarrow u_3 r^3 = 7168$

i.e. $112 r^3 = 7168$

$r^3 = 64$

$r = 4$

$u_6 r^m = 1\,835\,008$

$7168 \times 4^m = 1\,835\,008$

$4^m = 256$

$m = 4$

$\therefore 1\,835\,008 = u_{6+4} = u_{10}$

6 Here $u_1 = \dfrac{2}{5}, r = \dfrac{2}{5}$

$\therefore u_n = u_1 r^{n-1} = \dfrac{2}{5} \times \left(\dfrac{2}{5}\right)^{n-1}$

$= \left(\dfrac{2}{5}\right)^n = 0.4^n$

$u_N < 10^{-6}$

$0.4^N < 10^{-6}$

$N \log 0.4 < -6$

$N > -\dfrac{6}{\log(0.4)}$

$N > 15.08$

The least such N is 16, so the 16th term is the first to be less than 10^{-6}.

7 $|u_4 - u_3| = \dfrac{75}{8} u_1$

$u_1 (r^3 - r^2) = \pm \dfrac{75}{8} u_1$

$\Rightarrow 8r^3 - 8r^2 = \pm \dfrac{75}{8}$ (discounting the trivial case $u_1 = 0$)

Using GDC to find the roots of the two possible cubics, there is one solution for each:

$r = 2.5$ or -1.82

8 $u_5 = u_3 r^2$

$\therefore u_5 = 48 \Rightarrow u_3 r^2 = 48$

i.e. $12 r^2 = 48$

$r^2 = 4$

$r = \pm 2$

$u_8 = u_5 r^3$

$= 48 \times (\pm 2)^3$

$= \pm 384$

9 $u_1 = a$

$u_3 = 9a$

$\therefore ar^2 = 9a$

$\therefore r^2 = 9 \quad (a \neq 0)$

hence $r = \pm 3$

$u_2 = a + 14$

$\therefore ar = a + 14$

$\pm 3a = a + 14$

$2a = 14$ or $-4a = 14$

$a = 7$ or $a = -3.5$

10

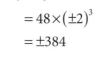

> **COMMENT**
> In sequences and series, the letters a, d and r usually have standard meanings, so take extra care in questions like this one which use these letters in other ways.

For the arithmetic progression:

$u_1 = a$

$u_2 = 1$

$\Rightarrow a + d = 1 \quad \ldots(1)$

$u_3 = b$

$\Rightarrow a + 2d = b \quad \ldots(2)$

$(2) - 2 \times (1)$:

$-a = b - 2$

$\Rightarrow b = 2 - a \quad \ldots(3)$

For the geometric progression:
$v_1 = 1$
$v_2 = a$
$\therefore r = a$
$v_3 = b$
$\therefore r^2 = b$
So $a^2 = b$
Hence, substituting in (3):
$a^2 = 2-a$
$a^2 + a - 2 = 0$
$(a+2)(a-1) = 0$
$a = -2$ or $a = 1$

$a = 1 \Rightarrow b = 1$, which contradicts the requirement that $a \neq b$.
$\therefore a = -2$ and $b = 4$

11 $S_n = 4n^2 - 2n$
$u_1 = S_1$
$= 4(1^2) - 2(1)$
$= 2$
$u_2 = S_2 - S_1$
$= (4(2^2) - 2(2)) - 2$
$= 12 - 2$
$= 10$
$\therefore d = u_2 - u_1 = 8$
$u_{32} = u_1 + (32-1)d$
$= 2 + 32 \times 8$
$= 250$
Since u_2, u_m, u_{32} (i.e. 10, u_m, 250) form a geometric sequence,
$250 = 10r^2$
$r^2 = 25$
$r = \pm 5$
$\therefore u_m = 10r = 50$
(Reject $10r = -50$ since this clearly does not lie in the arithmetic sequence.)

Returning to the arithmetic sequence,
$u_m = u_1 + (m-1)d$
$= 2 + (m-1) \times 8$
$= 8m - 6$
$\therefore 50 = 8m - 6$
$m = 7$

Exercise 7F

3 a $u_n = 3 \times 5^{n+2}$
$= 3 \times 5^3 \times 5^{n-1}$
$= 375 \times 5^{n-1}$

Comparing this with the standard formula $u_n = u_1 r^{n-1}$, we find $r = 5$.

b From (a), $u_1 = 375$
$S_n = \dfrac{u_1(r^n - 1)}{r - 1}$
$= \dfrac{375}{4}(5^n - 1)$

4 $S_3 = a(1 + r + r^2) = \dfrac{95}{4}$

$S_4 = a(1 + r + r^2 + r^3) = \dfrac{325}{8}$

Dividing gives
$\dfrac{1 + r + r^2 + r^3}{1 + r + r^2} = \dfrac{325}{8} \times \dfrac{4}{95} = \dfrac{325}{190} = \dfrac{65}{38}$

$38(1 + r + r^2 + r^3) = 65(1 + r + r^2)$
$38r^3 - 27r^2 - 27r - 27 = 0$
From GDC: $r = \dfrac{3}{2}$

$a(1 + r + r^2) = \dfrac{95}{4}$

$a\left(1 + \dfrac{3}{2} + \left(\dfrac{3}{2}\right)^2\right) = \dfrac{95}{4}$

$\therefore a = 5$

5 a $S_4 = 1 + x + x^2 + x^3$

b $S_6 = 1 + x + x^2 + x^3 + x^4 + x^5$

By the formula for geometric series,

$$S_6 = \frac{x^6 - 1}{x - 1}$$

$$\therefore x^6 - 1 = (x-1)S_6$$
$$= (x-1)(1 + x + x^2 + x^3 + x^4 + x^5)$$

b $S_\infty = \dfrac{27}{2}$

COMMENT

In part (b) the result $S_n = S_\infty(1 - r^n)$ was used, which enabled the answer to be read off immediately from part (a).

Exercise 7G

3 $u_1 = -18$, $u_2 = 12$

$$\Rightarrow r = \frac{12}{-18} = -\frac{2}{3}$$

$$S_\infty = \frac{u_1}{1 - r}$$

$$= \frac{-18}{\frac{5}{3}}$$

$$= -\frac{54}{5} = -10.8$$

4 a $u_1 = 18$

$u_1 = 18$, $u_4 = -\dfrac{2}{3}$

$$\therefore 18r^3 = -\frac{2}{3}$$

$$r^3 = \frac{-\frac{2}{3}}{18} = -\frac{1}{27}$$

$$r = -\frac{1}{3}$$

$$S_n = \frac{u_1(1 - r^n)}{1 - r}$$

$$= \frac{18\left(1 - \left(-\frac{1}{3}\right)^n\right)}{1 - \left(-\frac{1}{3}\right)}$$

$$= 18\left(1 - \left(-\frac{1}{3}\right)^n\right) \times \frac{3}{4}$$

$$= \frac{27}{2}\left(1 - \left(-\frac{1}{3}\right)^n\right)$$

5 a $S_2 = 15$

$$\Rightarrow u_1(1 + r) = 15 \quad \ldots(1)$$

$S_\infty = 27$

$$\Rightarrow \frac{u_1}{1 - r} = 27 \quad \ldots(2)$$

$(1) \div (2)$:

$$\frac{u_1(1 + r)}{\frac{u_1}{1 - r}} = \frac{15}{27}$$

$$(1 + r)(1 - r) = \frac{15}{27}$$

$$1 - r^2 = \frac{15}{27}$$

$$r^2 = \frac{4}{9}$$

$$r = \pm\frac{2}{3}$$

Each term of the series is positive,

so $r = \dfrac{2}{3}$

b From (2):

$$\frac{u_1}{1 - \frac{2}{3}} = 27$$

$$u_1 = 27 \times \frac{1}{3} = 9$$

6 $S_\infty = 32$

$$\Rightarrow \frac{u_1}{1 - r} = 32 \quad \ldots(1)$$

$S_4 = 30$

$$\Rightarrow \frac{u_1}{1 - r}(1 - r^4) = 30 \quad \ldots(2)$$

Substituting (1) into (2):
$32(1-r^4) = 30$

$1 - r^4 = \dfrac{15}{16}$

$r^4 = \dfrac{1}{16}$

$\therefore r = \dfrac{1}{2}$ ($r > 0$ as all terms are positive)

$S_\infty - S_8 = 32 - 32\left(1 - \left(\dfrac{1}{2}\right)^8\right)$

$= 32\left[1 - \left(1 - \left(\dfrac{1}{2}\right)^8\right)\right]$

$= 2^5 \times \dfrac{1}{2^8}$

$= \dfrac{1}{2^3} = \dfrac{1}{8}$

7 $S_\infty = 1 + \left(\dfrac{2x}{3}\right) + \left(\dfrac{2x}{3}\right)^2 + \ldots$ has $u_1 = 1$,

$r = \dfrac{2x}{3}$

a Convergence occurs when $-1 < r < 1$:

$-1 < \dfrac{2x}{3} < 1$

$-\dfrac{3}{2} < x < \dfrac{3}{2}$

b $x = 1.2 \Rightarrow r = \dfrac{2 \times 1.2}{3} = 0.8$

$S_\infty = \dfrac{u_1}{1 - r}$

$= \dfrac{1}{0.2} = 5$

8 $S_\infty = 13.5$

$\Rightarrow \dfrac{u_1}{1 - r} = 13.5 \quad \ldots(1)$

$S_3 = S_\infty(1 - r^3) = 13$

$\Rightarrow \dfrac{u_1}{1 - r}(1 - r^3) = 13 \quad \ldots(2)$

Substituting (1) into (2):
$13.5(1 - r^3) = 13$

$1 - r^3 = \dfrac{13}{13.5}$

$r^3 = \dfrac{1}{27}$

$r = \dfrac{1}{3}$

Substituting into (1):

$\dfrac{u_1}{1 - \dfrac{1}{3}} = 13.5$

$\Rightarrow u_1 = 13.5 \times \dfrac{2}{3} = 9$

9 This series has $u_1 = 2(4 - 3x)$, $r = 4 - 3x$

a Convergence occurs when $-1 < r < 1$:

$-1 < 4 - 3x < 1$

$1 < x < \dfrac{5}{3}$

b $x = 1.2$

$\Rightarrow r = 4 - 3.6 = 0.4$

and $u_1 = 2 \times (4 - 3.6) = 0.8$

$S_n > 1.328$

$\dfrac{u_1}{1 - r}(1 - r^n) > 1.328$

$\dfrac{0.8}{0.6}(1 - 0.4^n) > 1.328$

$1 - 0.4^n > 0.996$

$0.4^n < 0.004$

$n \log 0.4 < \log 0.004$

$n > 6.03$

Require at least 7 terms for a sum greater than 1.328.

10 $r = 2^x$

a Convergence occurs when $-1 < r < 1$:

$-1 < 2^x < 1$

$x < 0$

b $S_\infty = 40$

$\dfrac{35}{1-2^x} = 40$

$1-2^x = \dfrac{35}{40}$

$2^x = \dfrac{1}{8}$

$x = -3$

11 $f(x) = 1 + 2x + (2x)^2 + (2x)^3 + \ldots$ is a geometric series with $a = 1$ and $r = 2x$

a $x = \dfrac{1}{3} \Rightarrow r = \dfrac{2}{3}$, with $|r| < 1$

$S_\infty = \dfrac{u_1}{1-r}$

$= \dfrac{1}{\frac{1}{3}} = 3$

b $x = \dfrac{2}{3} \Rightarrow r = \dfrac{4}{3}$, with $|r| > 1$

$S_\infty = \infty$ since every term of the series is positive and it does not converge.

Exercise 7H

> **COMMENT**
> Be careful to define the terms you use; in finance questions it will often be critical whether you consider u_n to be the value at the start of year n or at the end of year n. If you are defining your own variables, always state the definitions clearly at the start of the question.

1 Let u_n represent the balance at the start of year n.
u_n follows a geometric sequence with $u_1 = 1000$, $r = 1.03$

a 6th year interest $= u_7 - u_6$

$= 1000 \times (1.03^6 - 1.03^5)$

$= 1000 \times 1.03^5 \times 0.03$

$= 34.78$

The interest for the sixth year is £34.78.

b The balance after six years is the balance at the start of the seventh year, u_7

$u_7 = 1000 \times 1.03^6 = 1194.05$

Balance after six years is £1194.05.

2 Let u_n be Lars's salary in the nth year.
u_n follows an arithmetic sequence with $u_1 = 32\,000$, $d = 1500$

a $u_{20} = u_1 + 19d$

$= 32\,000 + 19 \times 1500$

$= 60\,500$

In the twentieth year his salary will be $60\,500

b $S_n \geq 1\,000\,000$

$\dfrac{n}{2}(2u_1 + (n-1)d) \geq 1\,000\,000$

$n(62\,500 + 1500n) \geq 2\,000\,000$

$15n^2 + 625n - 20\,000 \geq 0$

$3n^2 + 125n - 4000 \geq 0$

Roots of this positive quadratic are 21.2 and -20.5 (from GDC)

$\therefore n > 21.2$

He will have earned more than $1 million after 22 years.

3 Let u_n be the balance at the start of year n.
u_n follows a geometric sequence with $u_1 = 5000$, $r = 1.063$

a After n full years the balance is the same as at the start of year $n+1$:

$u_{n+1} = 5000 \times 1.063^n$

b Balance at the end of 5 years:

$u_6 = 5000 \times 1.063^5 = 6786.35$

c i $5000 \times 1.063^n > 10000$

ii $5000 \times 1.063^n > 10000$

$1.063^n > 2$

$n \log 1.063 > \log 2$

$n > \dfrac{\log 2}{\log 1.063} = 11.3$

Balance will exceed $10 000 after 12 full years.

4 Let u_n be the number of seats in row n.
u_n follows an arithmetic sequence with $u_1 = 50, d = 200$

a $S_n \geq 8000$

$\dfrac{n}{2}(2u_1 + (n-1)d) \geq 8000$

$n(200n - 100) \geq 16000$

$2n^2 - n - 160 \geq 0$

Roots of this positive quadratic are 9.2 and −8.9 (from GDC)

$\therefore n > 9.2$

So 10 rows are required for there to be at least 8000 seats.

b $S_{10} = \dfrac{10}{2}(2 \times 50 + 9 \times 200) = 9500$

$S_5 = \dfrac{5}{2}(2 \times 50 + 4 \times 200) = 2250$

The percentage of seats in the front half (first 5 rows) is $\dfrac{2250}{9500} = 23.7\%$

5 a Balance at start of year n is $100 \times 1.05^{n-1}$

$\therefore V = 100 \times 1.05^{20} = \$265.33$

b Balance at the end of month m is

$100 \times \left(1 + \dfrac{0.05}{12}\right)^m$

$100 \times \left(1 + \dfrac{0.05}{12}\right)^m > 265.33$

$\left(1 + \dfrac{0.05}{12}\right)^m > 2.6533$

$m \log\left(1 + \dfrac{0.05}{12}\right) > \log 2.6533$

$m > \dfrac{\log 2.6533}{\log\left(1 + \dfrac{0.05}{12}\right)} = 234.6$

It takes 235 months, equivalent to 19 years and 7 months.

6 Let u_n be the number of miles run on day n.

u_n follows an arithmetic sequence with $u_1 = 1, d = \dfrac{1}{4}$

a $S_n \geq 26$

$\dfrac{n}{2}(2u_1 + (n-1)d) \geq 26$

$n\left(\dfrac{7}{4} + \dfrac{n}{4}\right) \geq 52$

$n^2 + 7n - 208 \geq 0$

Roots of this positive quadratic are 11.3 and −7.8 (from GDC)

$\therefore n > 11.3$

After 12 days the total distance exceeds 26 miles.

b $u_n > 26$

$u_1 + (n-1)d > 26$

$\dfrac{3}{4} + \dfrac{n}{4} > 26$

$n > 4 \times 26 - 3$

$n > 101$

On the 102nd day he runs more than 26 miles.

7 Let h_n be the height the ball rises on the nth bounce, i.e. after hitting the ground n times.

h_n follows a geometric sequence with $h_1 = 2 \times 0.8 = 1.6, r = 0.8$

a $h_4 = 1.6 \times 0.8^3 = 0.8192$ metres

78 Topic 7H Mixed questions

b Total distance travelled at the end of bounce n is

$$t_n = 2 + 2\sum_{k=1}^{n} h_k$$

$$= 2 + 2\frac{1.6(1-0.8^n)}{1-0.8}$$

$$= 2 + 16(1-0.8^n)$$

The ball hits the ground for the 9th time at the end of bounce 8.
$t_8 = 2 + 16(1-0.8^8) = 15.3$ metres

COMMENT

Note that the sum $\sum_{k=1}^{n} h_k$ is doubled because the ball goes up and down the same distance before it hits the ground again.

8 Let u_n be the account balance at the beginning of year n, where $n = 1$ is 2010.

a $u_1 = 1000$
At the beginning of 2011,
$u_2 = 1000 \times 1.04 + 1000$
At the beginning of 2012,
$u_3 = (1000 \times 1.04 + 1000) \times 1.04 + 1000$
$= 1000 + 1000 \times 1.04 + 1000 \times 1.04^2$

b The pattern in (a) shows that u_n is the sum of a geometric sequence with $u_1 = 1000$, $r = 1.04$
Hence

$$u_n = \frac{1000(1.04^n - 1)}{1.04 - 1}$$

$$= 25000(1.04^n - 1)$$

c $u_n \geq 50\,000$

$25\,000(1.04^n - 1) \geq 50\,000$

$1.04^n - 1 \geq 2$

$1.04^n \geq 3$

$n \log 1.04 \geq \log 3$

$n \geq \dfrac{\log 3}{\log 1.04} = 28.01$

In the 29th year of saving, Samantha will have accumulated at least $50 000.

Mixed examination practice 7

Short questions

1 $u_4 = 9.6$
$\Rightarrow u_1 + 3d = 9.6 \quad \ldots(1)$
$u_9 = 15.6$
$\Rightarrow u_1 + 8d = 15.6 \quad \ldots(2)$
(2) − (1):
$5d = 15.6 - 9.6$
$d = \dfrac{6}{5} = 1.2$

Substituting in (1):
$u_1 = 9.6 - 3 \times 1.2 = 6$

$S_9 = \dfrac{9}{2}(u_1 + u_9)$

$= \dfrac{9}{2}(6 + 15.6)$

$= 97.2$

2 $S_n = 2n^2 - n$

a $S_1 = 2 \times 1^2 - 1 = 1 \Rightarrow u_1 = 1$
$S_2 = 2 \times 2^2 - 2 = 6 = S_1 + u_2 \Rightarrow u_2 = 5$
$S_3 = 2 \times 3^2 - 3 = 15 = S_2 + u_3 \Rightarrow u_3 = 9$

b Arithmetic sequence, with $a = 1$, $d = 4$
$\therefore u_n = 1 + (n-1) \times 4 = 4n - 3$

7 Sequences and series

> **COMMENT**
>
> As an alternative approach, recognise that if S_n is a quadratic with a zero constant term, then the context is an arithmetic sequence. State this and rewrite S_n in the form
> $$S_n = \frac{n}{2}(2a + d(n-1)) = n\left(a - \frac{d}{2}\right) + \frac{dn^2}{2};$$
> compare coefficients to find $d = 4$ and $a = 1$, and then use these values to answer (a) and (b) directly.

3 Geometric sequence with $u_1 = \frac{1}{3}$, $r = \frac{1}{3}$

$$u_n = \frac{1}{3} \times \left(\frac{1}{3}\right)^{n-1} = \frac{1}{3^n}$$

$$u_n < 10^{-6}$$

$$\frac{1}{3^n} < 10^{-6}$$

$$3^n > 10^6$$

$$n \log 3 > \log 10^6$$

$$n > \frac{6}{\log 3} = 12.6$$

The least such n is 13.

4 $u_5 = u_1 + 4d$ and $u_2 = u_1 + d$

$u_5 = 3u_2$

$u_1 + 4d = 3(u_1 + d)$

$2u_1 = d$

$\Rightarrow \dfrac{d}{u_1} = 2$

5 Arithmetic sequence $\{u_n\}$ has $u_1 = 1$.
Geometric sequence $\{v_n\}$ has $v_1 = 1$.
$u_3 = v_2$
$\therefore 1 + 2d = r$ ...(1)
$u_4 = v_3$
$\therefore 1 + 3d = r^2$ ...(2)

Substituting (1) into (2):
$1 + 3d = (1 + 2d)^2$

$1 + 3d = 1 + 4d + 4d^2$

$4d^2 + d = 0$

$d(4d + 1) = 0$

So $d = 0$ (corresponding to $r = 1$ and both $\{u_n\}$ and $\{v_n\}$ being the constant sequence 1, 1, 1,...)

or $d = -\dfrac{1}{4}$ (corresponding to $r = \dfrac{1}{2}$).

6 This is the sum of two infinite geometric series:

$$u_i = \frac{2^i}{6^i} = \left(\frac{1}{3}\right)^i \text{ has sum to infinity}$$

$$U_\infty = \sum_{i=0}^{i=\infty} \left(\frac{1}{3}\right)^i = \frac{1}{1 - \frac{1}{3}} = \frac{3}{2}$$

$$v_i = \frac{4^i}{6^i} = \left(\frac{2}{3}\right)^i \text{ has sum to infinity}$$

$$V_\infty = \sum_{i=0}^{i=\infty} \left(\frac{2}{3}\right)^i = \frac{1}{1 - \frac{2}{3}} = 3$$

$\therefore$ the total value is $\dfrac{3}{2} + 3 = 4.5$

7 This is an arithmetic series with $u_1 = 301$ and $d = 7$.
To find the number of terms:
$u_n \leq 600$

$301 + (n-1) \times 7 \leq 600$

$7n - 7 \leq 299$

$n \leq \dfrac{306}{7} = 43.7$

$\therefore n = 43$

$S_{43} = \dfrac{43}{2}(2 \times 301 + (43-1) \times 7) = 19264$

80   Mixed examination practice 7

8 $u_1 = \ln\left(\dfrac{a^3}{b^{\frac{1}{2}}}\right) = 3\ln a - \dfrac{1}{2}\ln b$

$u_2 = \ln\left(\dfrac{a^3}{b}\right) = 3\ln a - \ln b$

$u_3 = \ln\left(\dfrac{a^3}{b^{\frac{3}{2}}}\right) = 3\ln a - \dfrac{3}{2}\ln b$

from which it can be seen that the sequence is arithmetic, with

$u_1 = 3\ln a - \dfrac{1}{2}\ln b$ and $d = -\dfrac{1}{2}\ln b$.

$\therefore S_{23} = \dfrac{23}{2}(2u_1 + 22d)$

$= \dfrac{23}{2}(6\ln a - \ln b - 11\ln b)$

$= 69\ln a - 138\ln b$

$= \ln\left(\dfrac{a^{69}}{b^{138}}\right)$

Long questions

1 a Let A_n be the amount in plan A after n years; then $\{A_n\}$ is an arithmetic sequence with $u_1 = 10\,800$, $d = 800$:
$A_n = 10\,000 + 800n$

b Let B_n be the amount in plan B after n years; then $\{B_n\}$ is a geometric sequence with $u_1 = 10\,500$, $r = 1.05$:
$B_n = 10\,000 \times 1.05^n$

c From GDC, intersection of the two graphs occurs at $n = 18.8$, so for the first 19 years $A_n > B_n$, i.e. plan A is better than plan B.

2 Let u_n be the number of bricks in row n, where row 1 is the top row.
Then $u_1 = 1$ and $u_{n+1} = u_n + 2$: this is an arithmetic sequence with $u_1 = 1$, $d = 2$.

a $u_n = 1 + (n-1) \times 2$
$= 2n - 1$

b $S_n = 36$

$\dfrac{n}{2}(2u_1 + (n-1)d) = 36$

$\dfrac{n}{2}(2 + 2n - 2) = 36$

$n^2 = 36$

$\therefore n = 6$

c $S_n = 4u_n + 4$

$\dfrac{n}{2}(2 + (n-1)\times 2) = 4(2n-1) + 4$

$n^2 = 8n$

$n^2 - 8n = 0$

$n(n-8) = 0$

$\therefore n = 8$ (reject $n = 0$)

Hence $S_n = n^2 = 64$

3 a There are n integers on the nth line

b

TABLE 7ML.3

Line	Final integer	Equals
1	1	
2	3	$= 1 + 2$
3	6	$= 1 + 2 + 3$
4	10	$= 1 + 2 + 3 + 4$

From the table it can be seen that the final integer on the nth line is the sum of the first n integers, i.e. S_n for an arithmetic sequence with $u_1 = 1$, $d = 1$:

$S_n = \dfrac{n}{2}(2 + (n-1))$

$= \dfrac{n(n+1)}{2}$

$= \dfrac{n^2 + n}{2}$

7 Sequences and series

c The first integer on the nth line must be $n-1$ less than the final integer:

$$\frac{n(n+1)}{2}-(n-1)=\frac{n^2+n-2n+2}{2}=\frac{n^2-n+2}{2}$$

d The integers on the nth line form an arithmetic sequence of n consecutive values from $\frac{n^2-n+2}{2}$ to $\frac{n^2+n}{2}$, so their sum is

$$S_n = \frac{n}{2}\left(\frac{n^2-n+2}{2}+\frac{n^2+n}{2}\right)=\frac{n}{2}\left(\frac{2n^2+2}{2}\right)=\frac{n}{2}(n^2+1)$$

e $\frac{n}{2}(n^2+1)=16\,400$

From GDC, $n = 32$

4 a Consider the mortgage as held in one account (A) and the payments in a separate account (B). The mortgage account just rises at its interest rate: $A_n = 15\,000 \times 1.06^n$.

$\{A_n\}$ is a geometric sequence with $A_1 = 15\,000 \times 1.06$ and $r = 1.06$.

At the end of three years the mortgage account stands at $A_3 = 150\,000 \times 1.06^3$

The payments account works as $B_{n+1} = 10\,000 + 1.06 B_n$, since each year interest is added to the previous payments and then a new £10 000 payment is made.

Therefore B_n is a geometric series of n terms with $a = 10\,000$ and $r = 1.06$.

At the end of three years, the payments account stands at $10\,000\left(1+1.06+1.06^2\right)$

So, after three years, the balance is $A_3 - B_3$:

$$150\,000 \times 1.06^6 - \left(10\,000 \times 1.06^2 + 10\,000 \times 1.06 + 10\,000\right)$$
$$= 150\,000 \times 1.06^6 - 10\,000 \times 1.06^2 - 10\,000 \times 1.06 - 10\,000$$

b Continuing the pattern:

Balance after n years $= A_n - B_n$

$$= 150\,000 \times 1.06^n - 10\,000 \frac{(1.06^n - 1)}{1.06 - 1}$$

$$= 150\,000 \times 1.06^n - 500\,000 \frac{(1.06^n - 1)}{3}$$

c For the balance at the end of n years to be ≤ 0, require

$$150\,000 \times 1.06^n - 500\,000 \frac{(1.06^n - 1)}{3} \leq 0$$

$$150\,000 \times 1.06^n \leq 500\,000 \frac{(1.06^n - 1)}{3}$$

$$0.9 \times 1.06^n \leq 1.06^n - 1$$

$$1.06^n \geq 10$$

$$n \geq \frac{1}{\log 1.06} = 39.5$$

So the mortgage will be paid off after 40 years.

7 Sequences and series 83

8 Binomial expansion

Exercise 8B

5 General term of $(x-2y)^5$ has the form $\binom{5}{r}x^{5-r}(-2y)^r$

Coefficient of this term is $\binom{5}{r}(-2)^r$

a $\binom{5}{r}(-2)^r = 80$

$\Rightarrow r = 4$

$\therefore$ term is $\binom{5}{2}x^1(-2y)^4 = 80xy^4$

b $\binom{5}{r}(-2)^r = -80$

$\Rightarrow r = 3$

Term is $\binom{5}{3}x^2(-2y)^3 = -80x^2y^3$

6 General term of $(3x+2y^2)^5$ has the form $\binom{5}{r}(3x)^{5-r}(2y^2)^r$

Require coefficient of x^2y^6, so $r = 3$.

Term is $\binom{5}{3}(3x)^2(2y^2)^3 = 10 \times 9x^2 \times 8y^6$

$= 720x^2y^6$

Coefficient is 720

7 General term of $(x^2-3x^{-1})^7$ has the form

$\binom{7}{r}(x^2)^{7-r}(-3x^{-1})^r = \binom{7}{r}(-3)^r x^{14-3r}$

Require $14-3r = 5$, so $r = 3$

Term is $\binom{7}{3}(-3)^3 x^5 = -945x^5$

8 General term of $(2x-5x^{-2})^{12}$ has the form

$\binom{12}{r}(2x)^{12-r}(-5x^{-2})^r$

$= \binom{12}{r}2^{12-r}(-5)^r x^{12-3r}$

Require $12-3r = 0$, so $r = 4$

Term independent of x is

$\binom{12}{4}2^8(-5)^4 x^0 = 79\,200\,000$

9 General form of a term in the expansion of $(1+3x)^n$ is $\binom{n}{r}(3x)^r$

$\therefore \binom{n}{1}(3x)^1 = 42x$

$3n = 42$

$n = 14$

10 General form of a term in the expansion

of $(1+2x)^n$ is $\binom{n}{r}(2x)^r$

$$\therefore \binom{n}{2}(2x)^2 = 264x^2$$

$$\frac{n(n-1)}{2} \times 4 = 264$$

$$n^2 - n - 132 = 0$$

$$(n-12)(n+11) = 0$$

$n = 12$ (reject negative solution $n = -11$)

11 General form of a term in the expansion

of $(1-5x)^n$ is $\binom{n}{r}(-5x)^r$

$$\therefore \binom{n}{3}(-5x)^3 = -10\,500x^3$$

$$\frac{n(n-1)(n-2)}{6} \times (-125) = -10\,500$$

$$n^3 - 3n^2 + 2n - 504 = 0$$

$n = 9$ (from GDC)

12 General form of a term in the expansion

of $(3+2x)^n$ is $\binom{n}{r}(3)^{n-r}(2x)^r$

$$\therefore \binom{n}{2}(3)^{n-2}(2x)^2 = 20\,412x^2$$

$$\frac{n(n-1)}{2} \times \frac{3^n}{9} \times 4 = 20\,412$$

$$n(n-1) \times 3^n = 91854$$

$n = 7$ (from GDC)

Exercise 8C

5 $(y + 3y^2)^6 = (y(1+3y))^6 = y^6(1+3y)^6$

General form of a term in the expansion

of $y^6(1+3y)^6$ is $y^6 \binom{6}{r}(3y)^r$

> **COMMENT**
>
> If a common factor can be taken outside the brackets, it is often simpler to do so before finding the general term.

First four terms are:

$$y^6 \left(\binom{6}{0}(3y)^0 + \binom{6}{1}(3y)^1 + \binom{6}{2}(3y)^2 + \binom{6}{3}(3y)^3 \right)$$

$$= y^6 \left(1 + 6(3y) + 15(9y^2) + 20(27y^3) \right)$$

$$= y^6 + 18y^7 + 135y^8 + 540y^9$$

6 $(1-x)^{10}(1+x)^{10} = ((1-x)(1+x))^{10} = (1-x^2)^{10}$

General form of a term in the expansion

of $(1-x^2)^{10}$ is $\binom{10}{r}(-x^2)^r$

> **COMMENT**
>
> If a product can be simplified before expanding, this will generally lead to a more rapid solution than expanding and then calculating the product. Always be alert for this kind of shortcut.

First three terms are:

$$\binom{10}{0}(-x^2)^0 + \binom{10}{1}(-x^2)^1 + \binom{10}{2}(-x^2)^2$$

$$= 1 - 10x^2 + 45x^4$$

7 $(1 - 2x + x^2)^{10} = ((1-x)^2)^{10} = (1-x)^{20}$

General form of a term in the expansion

of $(1-x)^{20}$ is $\binom{20}{r}(-x)^r$

First four terms are:

$$\binom{20}{0}(-x)^0 + \binom{20}{1}(-x)^1 + \binom{20}{2}(-x)^2 + \binom{20}{3}(-x)^3$$

$$= 1 - 20x + 190x^2 - 1140x^3$$

8 General form of a term in the expansion of $(1+x)^3$ is $\binom{3}{r}x^r$

∴ expansion of $(1+x)^3$ is
$(1+x)^3 = 1+3x+3x^2+x^3$

General form of a term in the expansion of $(1+mx)^4$ is $\binom{4}{r}(mx)^r$

∴ expansion of $(1+mx)^4$ is
$(1+mx)^4 = 1+4mx+6m^2x^2+\ldots$

Comparing coefficients of the product of these expansions with the given expression:

$x^0: 1=1$

$x^1: 3+4m=n$

$x^2: 3+12m+6m^2=93$

$m^2+2m-15=0$

$(m-3)(m+5)=0$

∴ $m=3,\ n=15$ or $m=-5, n=-17$

9 General form of a term in the expansion of $(1+kx)^4$ is $\binom{4}{r}(kx)^r$

∴ expansion of $(1+kx)^4$ is
$(1+kx)^4 = 1+4kx+6k^2x^2+\ldots$

General form of a term in the expansion of $(1+x)^n$ is $\binom{n}{r}x^r$

∴ expansion of $(1+x)^n$ is
$(1+x)^n = 1+nx+\dfrac{n(n-1)}{2}x^2+\ldots$

Comparing coefficients of the product of these expansions with the given expression:

$x^0: 1=1$

$x^1: 4k+n=13 \Rightarrow n=13-4k$

$x^2: 6k^2+4kn+\dfrac{n(n-1)}{2}=74$

$6k^2+4k(13-4k)+(13-4k)(6-2k)=74$

$-2k^2+2k+4=0$

$k^2-k-2=0$

$(k-2)(k+1)=0$

∴ $k=2, n=5$ or $k=-1, n=17$

Exercise 8D

2 a General form of a term in the expansion of $(3-5x)^4$ is $\binom{4}{r}3^{4-r}(-5x)^r$

The first 3 terms are

$\binom{4}{0}(3)^4(-5x)^0 + \binom{4}{1}(3)^3(-5x)^1 + \binom{4}{2}(3)^2(-5x)^2$

$= 1\times(3)^4 + 4\times(3)^3(-5x) + 6\times(3)^2(-5x)^2$

$= 81-540x+1350x^2$

b Require $3-5x=2.995$, so $x=0.001$

$x^0 = 1 \Rightarrow 81x^0 = 81$

$x^1 = 0.001 \Rightarrow -540x^1 = -0.54$

$x^2 = 0.000001 \Rightarrow 1350x^2 = 0.00135$

Hence $2.995^4 \approx 81 - 0.54 + 0.00135$
$= 80.46135$

> **COMMENT**
> In this type of question, find a value of x which makes the first part of the question relevant to finding the approximation. In more complicated questions this may require some ingenuity.

Rounding to 6SF: the truncated term will be negative, so this estimate should be rounded down:
$2.995^4 = 80.4613$ (6SF)

> **COMMENT**
>
> Although a value like 80.46135 would normally be rounded up, when using an expansion for approximation you should consider the next term in determining whether to round up or down if the value is exactly on the boundary.

3 a General form of a term in the expansion of $(2+5x)^7$ is $\binom{7}{r}(2)^{7-r}(5x)^r$

The first 3 terms are

$\binom{7}{0}(2)^7(5x)^0 + \binom{7}{1}(2)^6(5x)^1 + \binom{7}{2}(2)^5(5x)^2$

$= 128 + 7(64)(5x) + 21(32)(25x^2)$

$= 128 + 2240x + 16\,800x^2$

b Require $(2+5x)^7 = 2.005^7$, so $x = 0.001$

$x^0 = 1 \quad \Rightarrow \quad 128x^0 = 128$

$x^1 = 0.001 \quad \Rightarrow \quad 2240x^1 = 2.24$

$x^2 = 0.000001 \Rightarrow 16\,800x^2 = 0.0168$

$\therefore 2.005^7 \approx 128 + 2.24 + 0.0168 = 130.2568$

Rounding to 6SF: $2.005^7 \approx 130.257$

4 a $(2+3x)^7 = \binom{7}{0}(2)^7 + \binom{7}{1}(2)^6(3x)^1$

$+ \binom{7}{2}(2)^5(3x)^2 + \ldots$

$= 128 + 1344x + 6048x^2 + \ldots$

b i Require $2 + 3x = 2.3$, so $x = 0.1$.

$x^0 = 1 \quad \Rightarrow \quad 128x^0 = 128$

$x^1 = 0.1 \quad \Rightarrow \quad 1344x^1 = 134.4$

$x^2 = 0.01 \quad \Rightarrow \quad 6048x^2 = 60.48$

Hence, approximately,
$2.3^7 = 128 + 134.4 + 60.48 = 322.88$

ii Require $2 + 3x = 2.03$, so $x = 0.01$

$x^0 = 1 \quad \Rightarrow \quad 128x^0 = 128$

$x^1 = 0.01 \quad \Rightarrow \quad 1344x^1 = 13.44$

$x^2 = 0.0001 \Rightarrow 6048x^2 = 0.6048$

Hence, approximately,
$2.03^7 = 128 + 13.44 + 0.6048$

$= 142.0448$

b Approximation (ii) will be more accurate, on both an absolute and a relative basis, since the discarded terms (higher powers of x) reduce more rapidly in this case and are less significant to the total.

Mixed examination practice 8
Short questions

1 General term of $(2-x)^{12}$ has the form

$\binom{12}{r} 2^{12-r}(-x)^r$

Term in x^5 is

$\binom{12}{5}(2)^7(-x)^5 = 792 \times 128 \times (-x^5) = -101376x^5$

Coefficient is -101376

2

$(2-\sqrt{2})^5 = \binom{5}{0}(2)^5 + \binom{5}{1}(2)^4(-\sqrt{2})^1$

$+ \binom{5}{2}(2)^3(-\sqrt{2})^2 + \binom{5}{3}(2)^2(-\sqrt{2})^3$

$+ \binom{5}{4}(2)^1(-\sqrt{2})^4 + \binom{5}{5}(-\sqrt{2})^5$

$= 32 + 5(16)(-\sqrt{2}) + 10(8)(2)$

$+ 10(4)(-2\sqrt{2}) + 5(2)(4) - 4\sqrt{2}$

$= 232 - 164\sqrt{2}$

3 a General form of a term in the expansion of $(2+x)^5$ is $\binom{5}{r}(2)^{5-r}x^r$

∴ expansion of $(2+x)^5$ is

$\binom{5}{0}(2)^5 x^0 + \binom{5}{1}(2)^4 x^1 + \binom{5}{2}(2)^3 x^2 + \binom{5}{3}(2)^2 x^3 + \binom{5}{4}(2)^1 x^4 + \binom{5}{5}(2)^0 x^5$

$= 32 + 5(16)x + 10(8)x^2 + 10(4)x^3 + 5(2)x^4 + x^5$

$= 32 + 80x + 80x^2 + 40x^3 + 10x^4 + x^5$

b Require $(2+x)^5 = 2.01^5$, so $x = 0.01$

$x^0 = 1$	$\Rightarrow$	$32x^0 =$	32
$x^1 = 0.01$	$\Rightarrow$	$80x^1 =$	0.8
$x^2 = 0.0001$	$\Rightarrow$	$80x^2 =$	0.008
$x^3 = 0.000001$	$\Rightarrow$	$40x^3 =$	0.00004
$x^4 = 0.00000001$	$\Rightarrow$	$10x^4 =$	0.0000001
$x^5 = 0.0000000001$	$\Rightarrow$	$1x^5 =$	0.0000000001

∴ $2.01^5 = 32.8080401001$

4 General form of a term in the expansion of $(1-2x)^3$ is $\binom{3}{r}(-2x)^r$

∴ expansion of $(1-2x)^3$ is

$\binom{3}{0}(-2x)^0 + \binom{3}{1}(-2x)^1 + \binom{3}{2}(-2x)^2 + \ldots$

$= 1 + 3(-2x) + 3(4x^2) + \ldots$

$= 1 - 6x + 12x^2 + \ldots$

General form of a term in the expansion of $(3+4x)^5$ is $\binom{5}{r}(3)^{5-r}(4x)^r$

∴ expansion of $(3+4x)^5$ is

$\binom{5}{0}(3)^5(4x)^0 + \binom{5}{1}(3)^4(4x)^1 + \binom{5}{2}(3)^3(4x)^2 + \ldots$

$= 243 + 5(81)(4x) + 10(27)(16x^2) + \ldots$

$= 243 + 1620x + 4320x^2 + \ldots$

So the first 3 terms in the product are

$243 + (1620 - 6 \times 243)x + (4320 - 6 \times 1620 + 12 \times 243)x^2$

$= 243 + 162x - 2484x^2$

5 $(x^2-2x^{-1})^4 = \binom{4}{0}(x^2)^4 + \binom{4}{1}(x^2)^3(-2x^{-1})$

$\qquad + \binom{4}{2}(x^2)^2(-2x^{-1})^2$

$\qquad + \binom{4}{3}(x^2)^1(-2x^{-1})^3$

$\qquad + \binom{4}{4}(-2x^{-1})^4$

$\qquad = x^8 - 8x^5 + 24x^2 - 32x^{-1} + 16x^{-4}$

6 General form of a term in the expansion of $\left(x + \dfrac{1}{ax^2}\right)^7$ is

$\binom{7}{r}(x)^{7-r}(a^{-1}x^{-2})^r = \binom{7}{r}a^{-r}x^{7-3r}$

The term in x^1 corresponds to $r = 2$

$\therefore \binom{7}{2}a^{-2} = \dfrac{7}{3}$

$\dfrac{21}{a^2} = \dfrac{7}{3}$

$a^2 = 9$

$a = \pm 3$

7 General form of a term in the expansion of $(1+x)^6$ is $\binom{6}{r}x^r$

$\therefore$ expansion of $(1+x)^6$ is
$(1+x)^6 = 1 + 6x + 15x^2 + \ldots$

General form of a term in the expansion of $(1+mx)^5$ is $\binom{5}{r}(mx)^r$

$\therefore$ expansion of $(1+mx)^5$ is
$(1+mx)^5 = 1 + 5mx + 10m^2x^2 + \ldots$

Comparing coefficients of the product of these expansions with the given expression:

$x^0 : 1 = 1$

$x^1 : 6 + 5m = n$

$x^2 : 15 + 30m + 10m^2 = 415$

$m^2 + 3m - 40 = 0$

$(m-5)(m+8) = 0$

$\therefore m = 5,\ n = 31\ $ or $\ m = -8,\ n = -34$

Long questions

1 a The graph of $y = (x+2)^3$ is the graph of $y = x^3$ after a translation by $\begin{pmatrix}-2\\0\end{pmatrix}$. Axis intercepts are at $(-2, 0)$ and $(0, 8)$.

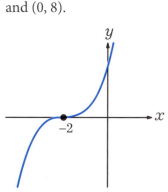

Figure 8ML.1 Graph of $y = (x+2)^3$

b $(x+2)^3 = \binom{3}{0}(x)^3 + \binom{3}{1}(x)^2(2)$

$\qquad + \binom{3}{2}(x)(2)^2 + \binom{3}{3}(2)^3$

$\qquad = x^3 + 6x^2 + 12x + 8$

c Require that $x + 2 = 2.001$, so take $x = 0.001$

$$x^0 = 1 \quad \Rightarrow \quad 8x^0 = 8$$
$$x^1 = 0.001 \quad \Rightarrow \quad 12x^1 = 0.012$$
$$x^2 = 0.000001 \quad \Rightarrow \quad 6x^2 = 0.000006$$
$$x^3 = 0.000000001$$

Hence, $2.001^3 = 8 + 0.012 + 0.000006 + 0.000000001 = 8.012006001$

d $x^3 + 6x^2 + 12x + 16 = 0$

$(x+2)^3 + 8 = 0$

$(x+2)^3 = -8$

$x + 2 = -2$

$x = -4$

2 Let $h(x) = \dfrac{f(x)}{g(x)} = \dfrac{(1+x)^5}{(2+x)^4}$

a Vertical asymptote where denominator is zero: $x = -2$

Axis intercepts at $(0, h(0))$ and where numerator is zero: $\left(0, \dfrac{1}{16}\right)$ and $(-1, 0)$

b General form of a term in the expansion of $(1+x)^5$ is $\binom{5}{r} x^r$

$\therefore$ expansion of $f(x)$ is $1 + 5x + 10x^2 + 10x^3 + 5x^4 + x^5$

General form of a term in the expansion of $(2+x)^4$ is $\binom{4}{r}(2)^{4-r} x^r$

$\therefore$ expansion of $g(x)$ is

$\binom{4}{0}(2)^4 x^0 + \binom{4}{1}(2)^3 x^1 + \binom{4}{2}(2)^2 x^2 + \binom{4}{3}(2)^1 x^3 + \binom{4}{4}(2)^0 x^4$

$= 16 + 32x + 24x^2 + 8x^3 + x^4$

c i Using the expansions in (b),

$$\dfrac{f(x)}{g(x)} = \dfrac{x^5 + 5x^4 + 10x^3 + 10x^2 + 5x + 1}{x^4 + 8x^3 + 24x^2 + 32x + 16}$$

$$= \dfrac{x^5 + 8x^4 + 24x^3 + 32x^2 + 16x - (3x^4 + 14x^3 + 22x^2 + 11x - 1)}{x^4 + 8x^3 + 24x^2 + 32x + 16}$$

$$= x + \dfrac{-3x^4 - 24x^3 - 72x^2 - 96x - 48 + (10x^3 + 50x^2 + 85x + 49)}{x^4 + 8x^3 + 24x^2 + 32x + 16}$$

$$= x - 3 + \dfrac{10x^3 + 50x^2 + 85x + 49}{x^4 + 8x^3 + 24x^2 + 32x + 16}$$

$\therefore k = 3$, $a = 10$

ii As $x \to \pm\infty$, the rational function
$$\frac{10x^3 + 50x^2 + 85x + 49}{x^4 + 8x^3 + 24x^2 + 32x + 16}$$ tends
to zero, since the denominator has higher order than the numerator.

Therefore $\dfrac{f(x)}{g(x)} \to x - 3$ as $x \to \pm\infty$

d

Figure 8ML.2 Graph of $y = \dfrac{(1+x)^5}{(2+x)^4}$

3 a $(1+\sqrt{2})^3 = \binom{3}{0} + \binom{3}{1}(\sqrt{2})^1 + \binom{3}{2}(\sqrt{2})^2$
$+ \binom{3}{3}(\sqrt{2})^3$

$= 1 + 3\sqrt{2} + 3 \times 2 + 2\sqrt{2}$

$= 7 + 5\sqrt{2}$

b General form of a term in the expansion of $(1+\sqrt{2})^n$ is $\binom{n}{r}(\sqrt{2})^r$

c $(1+x\sqrt{2})^n = \sum_{r=0}^{n} \binom{n}{r}(x\sqrt{2})^r$

$(1-x\sqrt{2})^n = \sum_{r=0}^{n} \binom{n}{r}(-x\sqrt{2})^r$

So $(1+x\sqrt{2})^n + (1-x\sqrt{2})^n$

$= \sum_{r=0}^{n} \binom{n}{r}\left[(x\sqrt{2})^r + (-x\sqrt{2})^r\right]$

$= \sum_{r=0}^{\frac{n}{2}} \binom{n}{2r}\left[2(x\sqrt{2})^{2r}\right]$

because the odd powers of x cancel while the even powers double up.

$\therefore (1+x\sqrt{2})^n + (1-x\sqrt{2})^n = \sum_{r=0}^{\frac{n}{2}} 2\binom{n}{2r}(2x^2)^r$

Taking $x = 1$:

$(1+\sqrt{2})^n + (1-\sqrt{2})^n = \sum_{r=0}^{\frac{n}{2}} 2\binom{n}{2r}(2)^r$

Since the sum of integer values must be an integer, it follows that $(1+\sqrt{2})^n + (1-\sqrt{2})^n$ is always an integer.

COMMENT

The above argument is more formal than strictly necessary, but you should be aware that talking about cancelling 'terms' in a sum lacking powers of x is problematic because there is no obvious ordering for the terms. By introducing x and then evaluating at $x = 1$ as shown above, this problem can be completely avoided.

d Since $|1-\sqrt{2}| < 0.5$, the distance between $(1+\sqrt{2})^n$ and the nearest whole number must in fact be $\left|(1-\sqrt{2})^n\right| = \left|1-\sqrt{2}\right|^n$.

Require $\left|1-\sqrt{2}\right|^n \leq 10^{-9}$

$\therefore n \log(\sqrt{2}-1) \leq -9$

$\Rightarrow n \geq \dfrac{-9}{\log(\sqrt{2}-1)} = 23.5$

So the least such n is 24.

4 **a** General form of a term in the expansion of $(a+x)^n$ is $\binom{n}{k}(a)^{n-k}x^k$

The ratio of coefficients of the rth and $(r+1)$th terms is $\dfrac{\binom{n}{r}a^{n-r}}{\binom{n}{r+1}a^{n-r-1}} = \dfrac{\alpha}{\beta}$

So $\dfrac{\alpha}{\beta} = \dfrac{\left(\dfrac{n!}{r!(n-r)!}\right)a}{\left(\dfrac{n!}{(r+1)!(n-r-1)!}\right)}$

$= a\dfrac{(r+1)!(n-r-1)!}{r!(n-r)!}$

$= a\dfrac{(r+1)}{(n-r)}$

b If $a=1$ and n is odd, say $n=2k+1$, then

$\dfrac{\alpha}{\beta} = \dfrac{r+1}{2k+1-r}$

When $r=k$,

$\dfrac{\alpha}{\beta} = \dfrac{k+1}{k+1} = 1$

i.e. the ratio of consecutive coefficients is 1.

This means that two consecutive terms have the same coefficient.

c Replacing r with $r+1$ in the answer to (a):

$\dfrac{\beta}{\gamma} = a\dfrac{(r+2)}{(n-r-1)}$

d For three consecutive terms to have the same coefficient, require that

$\dfrac{a(r+1)}{n-r} = 1 = \dfrac{a(r+2)}{n-r-1}$

i.e. $a(r+1) = n-r$ and $a(r+2) = n-r-1$

$\therefore a(r+1) - 1 = a(r+2)$

$\Rightarrow a = -1$

But then $-(r+1) = n-r$, so $n=-1$, which is invalid since $n \in \mathbb{N}$.

This proves that there cannot be three consecutive terms with the same coefficient in a binomial expansion.

COMMENT

There are many ways to establish this from the equations set up in (b) and (c). There is no 'best' method in a question of this sort, so you should seek the quickest way to find a contradiction.

9 Circular measure and trigonometric functions

Exercise 9B

13 $\cos(\pi+x)+\cos(\pi-x)$
$= \cos\pi\cos x - \sin\pi\sin x$
$\quad + \cos\pi\cos x + \sin\pi\sin x$
$= -\cos x - 0 - \cos x + 0$
$= -2\cos x$

14 $\sin x + \sin\left(x+\dfrac{\pi}{2}\right) + \sin(x+\pi) + \sin\left(x+\dfrac{3\pi}{2}\right)$
$\quad + \sin(x+2\pi)$
$= \sin x + \sin x \cos\dfrac{\pi}{2} + \cos x \sin\dfrac{\pi}{2}$
$\quad + \sin x \cos\pi + \cos x \sin\pi$
$\quad + \sin x \cos\dfrac{3\pi}{2} + \cos x \sin\dfrac{3\pi}{2} + \sin x$
$= \sin x + 0 + \cos x - \sin x + 0 + 0 - \cos x + \sin x$
$= \sin x$

> **COMMENT**
> Note that $\sin(x+2\pi) = \sin x$ by the periodicity of the sine function, so it isn't necessary to expand the last term in the expression.

Exercise 9E

4 $y = p\sin(qx)$ has amplitude p, i.e. y ranges from $-p$ to p.

From the graph, $p = 5$

$y = p\sin(qx)$ has period $\dfrac{2\pi}{q}$, so the second positive zero occurs at $x = \dfrac{2\pi}{q}$

From the graph, $\dfrac{2\pi}{q} = \pi \Rightarrow q = 2$

5 $y = a\cos(x-b)$ has amplitude a, i.e. y ranges from $-a$ to a.

From the graph, $a = 2$

$y = a\cos(x-b)$ has a zero at $x = 90° + b$

From the graph, the smallest positive solution is $110°$, so $b = 20°$

6 a $y = 1 + \sin 2x$: amplitude 1; centre $y = 1$; period π; axis intercepts
$(0, 1)$, $\left(\dfrac{3\pi}{4}, 0\right)$, $\left(\dfrac{7\pi}{4}, 0\right)$

$y = 2\cos x$: amplitude 2; centre $y = 0$; period 2π; axis intercepts
$(0, 2)$, $\left(\dfrac{\pi}{2}, 0\right)$, $\left(\dfrac{3\pi}{2}, 0\right)$

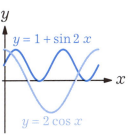

Figure 9E.6

b From Figure 9E.6, there are two points of intersection in $[0, 2\pi]$,
∴ two solutions.

c The pattern of the two curves in Figure 9E.6 repeats every 2π.

Since there are 2 solutions in an interval of length 2π, there must be 8 solutions in an interval of 8π.

7 a $y = 2\cos(x + 60°)$: amplitude 2; centre $y = 0$; period π; axis intercepts $(0, 1)$, $(30°, 0)$, $(210°, 0)$

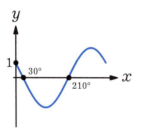

Figure 9E.7

b At maximum point:
$\cos(x + 60°) = 1$
$\Rightarrow x + 60° = 0°, 360°$
∴ $x = 300°$
At minimum point:
$\cos(x + 60°) = -1$
$\Rightarrow x + 60° = 180°$
∴ $x = 120°$

The minimum and maximum points are $(120°, -2)$ and $(300°, 2)$

c The graph of $y = 2\cos(x + 60°) - 1$ is the graph of $y = 2\cos(x + 60°)$ after a translation by $\begin{pmatrix} 0 \\ -1 \end{pmatrix}$, so its minimum and maximum points are $(120°, -3)$ and $(300°, 1)$.

Exercise 9F

1 $y = a\cos(bt) + m$: amplitude a, period $\frac{2\pi}{b}$, centre $y = m$

From the graph:

amplitude $= \frac{3}{2} \Rightarrow a = \frac{3}{2}$

period $= 12 \Rightarrow \frac{2\pi}{b} = 12 \Rightarrow b = \frac{\pi}{6}$

centre $= \frac{3+6}{2} \Rightarrow m = \frac{9}{2}$

2 a High tide will occur when d is a maximum, which occurs when

$\sin\left(\frac{\pi t}{12}\right) = 1$

$\frac{\pi t}{12} = \frac{\pi}{2}$

$t = 6$

Low tide will occur when d is a minimum, which occurs when

$\sin\left(\frac{\pi t}{12}\right) = -1$

$\frac{\pi t}{12} = \frac{3\pi}{2}$

$t = 18$

At high tide, $d = 16 + 7 = 23$ metres

At low tide, $d = 16 - 7 = 9$ metres

b Require $16 + 7\sin\left(\frac{\pi}{12}t\right) \geq 19$

i.e. $\sin\left(\frac{\pi}{12}t\right) \geq \frac{3}{7}$

From GDC: $t \in [1.69, 10.3]$

This is equivalent to the period of time between 01:42 and 10:18.

3 a $h = a\sin(kt)$: amplitude a, period $\frac{2\pi}{k}$

From the given information:

amplitude $= 5 \Rightarrow a = 5$

period $= 10 \Rightarrow \frac{2\pi}{k} = 10 \Rightarrow k = \frac{\pi}{5}$

b 3 cm below the x-axis corresponds to $h = -3$

$$5\sin\left(\frac{\pi t}{5}\right) = -3$$

$$\sin\left(\frac{\pi t}{5}\right) = -\frac{3}{5}$$

$$\frac{\pi t}{5} = 3.79, 5.64 \text{ (3SF, from GDC)}$$

$$\therefore t = 6.02, 8.98$$

The point is 3 cm below the x-axis 6.02 seconds and 8.98 seconds after starting.

4 $h = 120 - 10\cos 400t$: amplitude 10, centre $h = 120$, period $\frac{2\pi}{400} = \frac{\pi}{200}$

a Greatest height is $120 - (-10) = 130$ cm; least height is $120 - 10 = 110$ cm

b The time required to complete one full oscillation is the period, $\frac{\pi}{200} = 0.0157$ seconds

c Greatest height occurs when $\cos 400t = -1$

$$400t = \pi$$

$$t = \frac{\pi}{400} = 0.00785 \text{ (3SF)}$$

i.e. 0.00785 seconds after release.

Exercise 9G

8 a For example: $x = 1$ gives

$$\arctan x = \frac{\pi}{4}$$

$$\arcsin x = \frac{\pi}{2}$$

$$\arccos x = 0$$

Clearly in this case $\arctan x \neq \dfrac{\arcsin x}{\arccos x}$

COMMENT

The false idea being disproved by counter-example here is that division 'passes through' the inversion of a function – that for a function $h(x) = \dfrac{f(x)}{g(x)}$ it should follow that $h^{-1}(x) = \dfrac{f^{-1}(x)}{g^{-1}(x)}$; this is generally not the case!

b $\cos\theta = \sin\left(\dfrac{\pi}{2} - \theta\right)$

Let $\theta = \arccos x$, so $x = \cos\theta$; then

$$x = \sin\left(\frac{\pi}{2} - \theta\right)$$

$$\therefore \arcsin x = \frac{\pi}{2} - \arccos x$$

COMMENT

This is easily verified using a compound angle identity, as seen in Section 12B.

c From (b), $\arcsin x + \arccos x = \dfrac{\pi}{2}$, so the given equation becomes

$$2\arctan x = \frac{\pi}{2}$$

$$\arctan x = \frac{\pi}{4}$$

$$\therefore x = \tan\left(\frac{\pi}{4}\right) = 1$$

9 a $\sin x + \cos y = 0.6$ ... (1)
$\cos x - \sin y = 0.2$ ... (2)

From (1):
$\cos y = 0.6 - \sin x$
$\therefore y = \arccos(0.6 - \sin x)$

From (2):
$\sin y = \cos x - 0.2$
$\therefore y = \arcsin(\cos x - 0.2)$

b Need to solve

arccos(0.6 − sin x) = arcsin(cos x − 0.2)

From GDC, the solution is $x = 0$, and hence $y = 0.927$ (3SF)

Mixed examination practice 9
Short questions

1 $y = a \sin b(x+c) + d$ has amplitude a and period/wavelength $\dfrac{2\pi}{b}$

a $a = 1.4$, so amplitude is 1.4 metres

b Distance between consecutive peaks is the wavelength, which is $\dfrac{2\pi}{3} = 2.09$ metres (3SF)

2

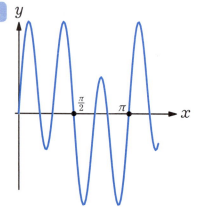

Figure 9MS.2 Graph of $\sin(2x) + 2\sin(6x)$

From the equation, it is clear that the period is at most π, which is 2π divided by the GCD of the coefficients of x in the trig functions.

From the graph, the period is not less than π, and so the period is exactly π.

3 $y = a \cos(bt)$ has amplitude a and period $\dfrac{2\pi}{b}$

a Period is $\dfrac{2\pi}{0.08} = 25\pi = 78.5$ seconds (3SF)

b The amplitude is equivalent to the radius of the track, i.e. 60 metres.

∴ Track length $= 2\pi \times 60$
$= 120\pi$
$= 377$ metres (3SF)

c Speed $= \dfrac{\text{Distance}}{\text{Time}}$

$= \dfrac{120\pi}{25\pi}$

$= 4.8 \text{ m s}^{-1}$

4 $f(x) = a \sin b(x+c)$ has amplitude a and period $\dfrac{2\pi}{b}$

a $b = 2$, so period is $\dfrac{2\pi}{2} = \pi$

b $x \in [0, 2\pi] \Rightarrow 2\left(x - \dfrac{\pi}{3}\right) \in \left[-\dfrac{2\pi}{3}, \dfrac{10\pi}{3}\right]$

$f(x) = 0$

$3 \sin 2\left(x - \dfrac{\pi}{3}\right) = 0$

$2\left(x - \dfrac{\pi}{3}\right) = 0, \pi, 2\pi, 3\pi$

$x = \dfrac{\pi}{3}, \dfrac{5\pi}{6}, \dfrac{4\pi}{3}, \dfrac{11\pi}{6}$

∴ zeros are $\left(\dfrac{\pi}{3}, 0\right), \left(\dfrac{5\pi}{6}, 0\right),$

$\left(\dfrac{4\pi}{3}, 0\right), \left(\dfrac{11\pi}{6}, 0\right)$

c Graph of $y = 3\sin 2\left(x - \dfrac{\pi}{3}\right)$

is maximum when

$2\left(x - \dfrac{\pi}{3}\right) = \dfrac{\pi}{2} + 2k\pi$, i.e. $x = \dfrac{7\pi}{12}, \dfrac{19\pi}{12}$;

minimum when

$2\left(x - \dfrac{\pi}{3}\right) = -\dfrac{\pi}{2} + 2k\pi$, i.e. $x = \dfrac{\pi}{12}, \dfrac{13\pi}{12}$;

amplitude 3; y-intercept $\left(0, -\dfrac{3\sqrt{3}}{2}\right)$

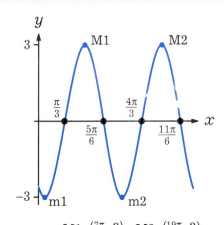

M1: $\left(\frac{7\pi}{12}, 3\right)$ M2: $\left(\frac{19\pi}{12}, 3\right)$
m1: $\left(\frac{\pi}{12}, -3\right)$ m2: $\left(\frac{13\pi}{12}, -3\right)$

Figure 9MS.4 Graph of $y = 3\sin 2\left(x - \frac{\pi}{3}\right)$

5 $y = a\sin(bx)$ has maximum point at
$\left(\frac{\pi}{2b}, a\right)$

From the graph, the maximum is at (2, 5)

$\therefore \frac{\pi}{2b} = 2 \Rightarrow b = \frac{\pi}{4}$ and $a = 5$.

Long questions

1 a i $y = \sin(x-k) + c$ has a maximum at
$\left(\frac{\pi}{2} + k, c+1\right)$

By symmetry, point A is midway horizontally between the first two zeros, i.e. at $x = \frac{2\pi}{3}$,

$\therefore \frac{\pi}{2} + k = \frac{2\pi}{3}$

$\Rightarrow k = \frac{\pi}{6}$

The graph goes through the origin,

$\therefore \sin\left(0 - \frac{\pi}{6}\right) + c = 0$

$-\frac{1}{2} + c = 0$

$\Rightarrow c = \frac{1}{2}$

So the coordinates of A are $\left(\frac{2\pi}{3}, \frac{3}{2}\right)$

ii From (i), $k = \frac{\pi}{6}, c = \frac{1}{2}$

b Period of $y = \sin(x-k) + c$ is 2π, so the zeros are those shown in the question and the same at intervals of 2π.

Within $[-4\pi, 0]$, these are
$-4\pi, -\frac{8\pi}{3}, -2\pi, -\frac{2\pi}{3}, 0$

c i For the equation $\sin\left(x - \frac{\pi}{6}\right) + \frac{1}{2} = k$,

as $k < 0$, the first pair of solutions will be in the interval $\left[\frac{4\pi}{3}, 2\pi\right]$,

and subsequent solutions will be at multiples of 2π further on, i.e. in the intervals $\left[\frac{10\pi}{3}, 4\pi\right], \left[\frac{16\pi}{3}, 6\pi\right]$

and $\left[\frac{22\pi}{3}, 8\pi\right]$. So there are only 8 solutions in $[0, 9\pi]$.

> **COMMENT**
> It is important to check that the next such interval is not needed too: in this case $\left[\frac{28\pi}{3}, 10\pi\right]$ is wholly outside $[0, 9\pi]$, so all the relevant intervals have been found.

ii Given that the smallest positive solution is α, the next solution, by symmetry about $x = \frac{5\pi}{3}$, must be

$2\pi - \left(\alpha - \frac{4\pi}{3}\right) = \frac{10\pi}{3} - \alpha$.

The following solution, by periodicity, must be $2\pi + \alpha$.

So the next two solutions after α are $\frac{10\pi}{3} - \alpha$ and $2\pi + \alpha$.

9 Circular measure and trigonometric functions

2 **i and ii**

Using GDC as necessary:

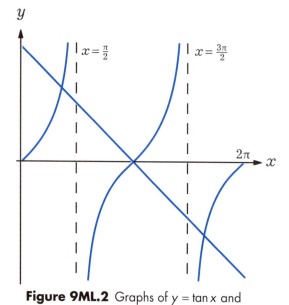

Figure 9ML.2 Graphs of $y = \tan x$ and $y = \pi - x$

b i $x + \tan x = \pi \Leftrightarrow \tan x = \pi - x$, so solutions of $x + \tan x = \pi$ are intersections of the graphs in Figure 9ML.2.

Given that x_0 is the first positive solution, by symmetry the other solutions in $[0, 2\pi]$ must be π and $2\pi - x_0$.

ii Since each period of $y = \tan x$ extends infinitely in the positive and negative y directions, the line $y = \pi - x$ must intersect each period of $y = \tan x$, so there are infinitely many solutions.

c i
$$\cos\left(\frac{\pi}{2} - A\right) = \cos\left(\frac{\pi}{2}\right)\cos A + \sin\left(\frac{\pi}{2}\right)\sin A$$
$$= 0 + \sin A$$
$$= s$$

$$\sin\left(\frac{\pi}{2} - A\right) = \sin\left(\frac{\pi}{2}\right)\cos A - \cos\left(\frac{\pi}{2}\right)\sin A$$
$$= \cos A - 0$$
$$= c$$

ii $\tan\left(\dfrac{\pi}{2} - A\right) = \dfrac{\sin\left(\dfrac{\pi}{2} - A\right)}{\cos\left(\dfrac{\pi}{2} - A\right)}$

$$= \frac{c}{s} \text{ from (i)}$$
$$= \frac{1}{s/c}$$
$$= \frac{1}{\tan A}$$

iii Let $\tan A = t$:

$$\tan A + \tan\left(\frac{\pi}{2} - A\right) = \frac{4}{\sqrt{3}}$$

$$t + \frac{1}{t} = \frac{4}{\sqrt{3}}$$

$$t^2 - \frac{4}{\sqrt{3}}t + 1 = 0$$

$$\left(t - \sqrt{3}\right)\left(t - \frac{1}{\sqrt{3}}\right) = 0$$

$\therefore \tan A = \sqrt{3}$ or $\dfrac{1}{\sqrt{3}}$

iv If $A \in \left]0, \dfrac{\pi}{2}\right[$,

$\tan A = \sqrt{3} \Rightarrow A = \dfrac{\pi}{3}$

$\tan A = \dfrac{1}{\sqrt{3}} \Rightarrow A = \dfrac{\pi}{6}$

$\therefore$ the values are $A = \dfrac{\pi}{3}$ or $\dfrac{\pi}{6}$

3 a Minimum value of $\cos x$ is -1, and the smallest positive value of x for which this occurs is $x = \pi$.

b i $f(x)$ to $2f\left(x + \dfrac{\pi}{6}\right)$: translation by $\begin{pmatrix} -\dfrac{\pi}{6} \\ 0 \end{pmatrix}$ and vertical stretch with scale factor 2.

ii Applying the two transformations in (i) to the minimum of $\cos x$:

the minimum point of

$$y = 2\cos\left(x + \frac{\pi}{6}\right) \text{ is}$$

$\left(\pi - \frac{\pi}{6}, -1 \times 2\right) = \left(\frac{5\pi}{6}, -2\right)$, i.e. the minimum value is -2, and it occurs at $x = \frac{5\pi}{6}$.

c i Vertical asymptotes occur where the denominator is zero.

The minimum value of the denominator $2\cos\left(x + \frac{\pi}{6}\right) + 3$ is $-2 + 3 = 1$, so there are no vertical asymptotes.

ii The maximum denominator value is $2 + 3 = 5$, so the range of the denominator is $[1, 5]$. Hence the range of $f(x)$ is $\left[\frac{5}{5}, \frac{5}{1}\right] = [1, 5]$.

> **COMMENT**
>
> Given the denominator is always strictly positive, the minimum of $f(x)$ occurs when the denominator is at a maximum, and the maximum of $f(x)$ occurs when the denominator is at a minimum.

10 Trigonometric equations and identities

Exercise 10A

8 $2\sin x + 1 = 0$

$\sin x = -\dfrac{1}{2}$

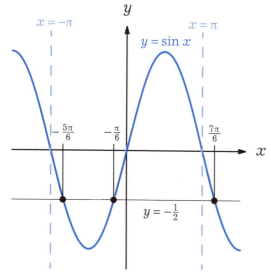

Figure 10A.8 Solutions to $\sin x = -\dfrac{1}{2}$ in $]-\pi, \pi[$

$\sin x = -\dfrac{1}{2}$ has the 2 solutions

$x_1 = \arcsin\left(-\dfrac{1}{2}\right) = -\dfrac{\pi}{6}$ and $x_2 = \pi - x_1 = \dfrac{7\pi}{6}$

But x_2 is outside the interval $(-\pi, \pi)$, so subtract 2π: $\dfrac{7\pi}{6} - 2\pi = -\dfrac{5\pi}{6}$

∴ the solutions are $x = -\dfrac{5\pi}{6}, -\dfrac{\pi}{6}$

Exercise 10B

6 $3\cos x = \tan x$

$3\cos x = \dfrac{\sin x}{\cos x}$

$3\cos^2 x - \sin x = 0$

$3 - 3\sin^2 x - \sin x = 0$

$3\sin^2 x + \sin x - 3 = 0$

$\sin x = \dfrac{-1 \pm \sqrt{1+36}}{6}$

$\sin x = 0.847$ or -1.18 (reject as < -1)

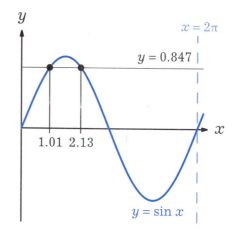

Figure 10B.6 Solutions to $\sin x = 0.847$ in $[0, 2\pi]$

There are 2 solutions to $\sin x = 0.847$ in $[0, 2\pi]$:

$x_1 = \arcsin 0.847 = 1.01$

$x_2 = \pi - 1.01 = 2.13$

∴ $x = 1.01, 2.13$

7 a $2\sin^2 x - 3\sin x = 2$

$2\sin^2 x - 3\sin x - 2 = 0$

$(2\sin x + 1)(\sin x - 2) = 0$

$\sin x = -\dfrac{1}{2}$ or 2 (reject as >1)

$\therefore \sin x = -\dfrac{1}{2}$

b

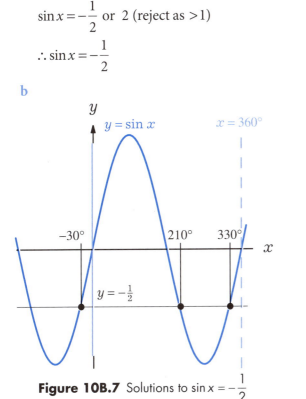

Figure 10B.7 Solutions to $\sin x = -\dfrac{1}{2}$ in $]0, 360°[$

$\sin x = -\dfrac{1}{2}$ has the 2 solutions

$x_1 = \arcsin\left(-\dfrac{1}{2}\right) = -30°$ and

$x_2 = 180° - x_1 = 210°$

But x_1 is outside the interval $]0, 360°[$, so add $360°$: $-30° + 360° = 330°$

$\therefore$ the solutions are $x = 210°, 330°$

8 $\sin x \tan x = \sin^2 x$

$\dfrac{\sin^2 x}{\cos x} = \sin^2 x$

$\sin^2 x = \sin^2 x \cos x$

$\sin^2 x(\cos x - 1) = 0$

$\sin x = 0$ or $\cos x = 1$

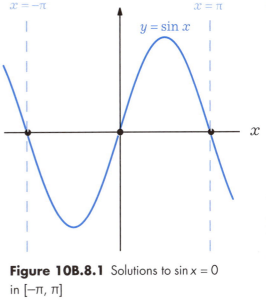

Figure 10B.8.1 Solutions to $\sin x = 0$ in $[-\pi, \pi]$

$\sin x = 0$

$\Rightarrow x = -\pi, 0, \pi$

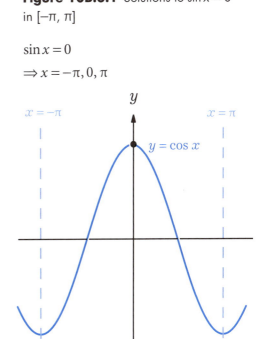

Figure 10B.8.2 Solutions to $\cos x = 1$ in $[-\pi, \pi]$

$\cos x = 1 \Rightarrow x = 0$

$\therefore$ the solutions are $x = -\pi, 0, \pi$

9 $x \in \left]-\sqrt{\pi}, \sqrt{\pi}\right[\Rightarrow x^2 \in [0, \pi[$

To solve $\sin(x^2) = \dfrac{1}{2}$, first consider $\sin\theta = \dfrac{1}{2}$:

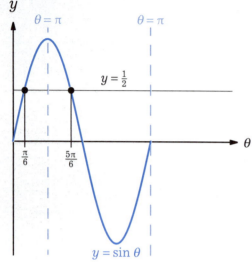

Figure 10B.9 Solutions to $\sin\theta = \dfrac{1}{2}$ in $[0, \pi[$

The first 2 solutions for $\theta = x^2$ are $\theta_1 = \dfrac{\pi}{6}$ and $\theta_2 = \pi - \theta_1 = \dfrac{5\pi}{6}$, and these are the only solutions in $[0, \pi[$.

$\therefore x^2 = \dfrac{\pi}{6}, \dfrac{5\pi}{6}$

$\Rightarrow x = \pm\sqrt{\dfrac{\pi}{6}}, \pm\sqrt{\dfrac{5\pi}{6}}$

Exercise 10C

9 $3 - 2\tan^2 x = 3 - \dfrac{2\sin^2 x}{\cos^2 x}$

$= \dfrac{3\cos^2 x - 2\sin^2 x}{\cos^2 x}$

$= \dfrac{3\cos^2 x - 2(1 - \cos^2 x)}{\cos^2 x}$

$= \dfrac{5\cos^2 x - 2}{\cos^2 x}$

$= 5 - \dfrac{2}{\cos^2 x}$

10 $\dfrac{1 + \tan^2 x}{\cos^2 x} = \dfrac{1 + \dfrac{\sin^2 x}{\cos^2 x}}{\cos^2 x}$

$= \dfrac{\cos^2 x + \sin^2 x}{(\cos^2 x)^2}$

$= \dfrac{1}{(1 - \sin^2 x)^2}$

11 a $\dfrac{\sin\theta}{\cos\theta} + \dfrac{\cos\theta}{\sin\theta} = \dfrac{\sin^2\theta + \cos^2\theta}{\cos\theta\sin\theta}$

$= \dfrac{1}{\sin\theta\cos\theta}$

(using $\sin^2\theta + \cos^2\theta = 1$)

b $\dfrac{1}{\cos\theta} + \tan\theta = \dfrac{1 + \sin\theta}{\cos\theta}$

$= \dfrac{(1 + \sin\theta)(1 - \sin\theta)}{\cos\theta(1 - \sin\theta)}$

$= \dfrac{1 - \sin^2\theta}{\cos\theta(1 - \sin\theta)}$

$= \dfrac{\cos^2\theta}{\cos\theta(1 - \sin\theta)}$

(using $\cos^2\theta = 1 - \sin^2\theta$)

$= \dfrac{\cos\theta}{(1 - \sin\theta)}$

Exercise 10D

> **COMMENT**
>
> Always check solutions to ensure that their values lie within the required interval, especially if the rearrangement has involved any division or multiplication. Sketching the graph is a good way of finding out how many solutions you need.

6 $5\sin^2\theta = 4\cos^2\theta$

$5\sin^2\theta = 4(1-\sin^2\theta)$

$9\sin^2\theta = 4$

$\sin^2\theta = \dfrac{4}{9}$

$\sin\theta = \pm\dfrac{2}{3}$

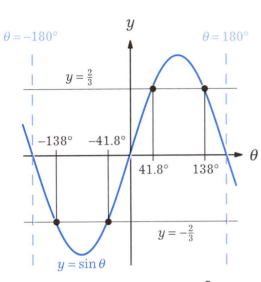

Figure 10D.6 Solutions to $\sin\theta = \pm\dfrac{2}{3}$ in $[-180°, 180°]$

There are 2 solutions to each of $\sin\theta = \dfrac{2}{3}$ and $\sin\theta = -\dfrac{2}{3}$ (positive and negative):

$\theta_1 = \arcsin\left(\pm\dfrac{2}{3}\right) = \pm 41.8°$

$\theta_2 = 180° - \theta_1 = 138.2°, 221.8°$

But 221.8° is outside the interval $-180° \le \theta \le 180°$, so subtract 360°:

$221.8° - 360° = -138.2°$

∴ the solutions are $\theta = \pm 41.8°, \pm 138°$ (3SF)

7 $2\cos^2 t - \sin t - 1 = 0$

$2(1-\sin^2 t) - \sin t - 1 = 0$

$2\sin^2 t + \sin t - 1 = 0$

$(2\sin t - 1)(\sin t + 1) = 0$

$\sin t = \dfrac{1}{2}$ or -1

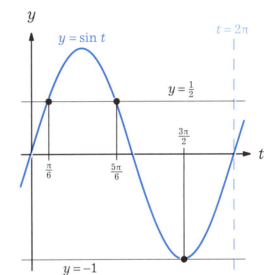

Figure 10D.7 Solutions to $\sin t = \dfrac{1}{2}$ and $\sin t = -1$ in $[0, 2\pi]$

There are 3 solutions in total.

For $\sin t = \dfrac{1}{2}$:

$t_1 = \arcsin\left(\dfrac{1}{2}\right) = \dfrac{\pi}{6}$

$t_2 = \pi - t_1 = \dfrac{5\pi}{6}$

For $\sin t = -1$:

$t_1 = \arcsin(-1) = -\dfrac{\pi}{2}$

But this is outside the interval $0 \le t \le 2\pi$, so add 2π: $-\dfrac{\pi}{2} + 2\pi = \dfrac{3\pi}{2}$

∴ the solutions are $t = \dfrac{\pi}{6}, \dfrac{5\pi}{6}, \dfrac{3\pi}{2}$

8 $4\cos^2 x - 5\sin x - 5 = 0$
$4(1-\sin^2 x) - 5\sin x - 5 = 0$
$4\sin^2 x + 5\sin x + 1 = 0$
$(4\sin x + 1)(\sin x + 1) = 0$
$\sin x = -\frac{1}{4}$ or -1

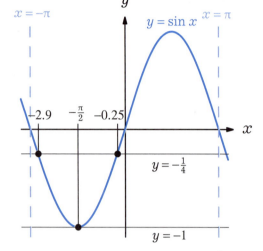

Figure 10D.8 Solutions to $\sin x = -\frac{1}{4}$ and $\sin x = -1$ in $[-\pi, \pi]$

There are 3 solutions in total.

For $\sin x = -\frac{1}{4}$:

$x_1 = \arcsin\left(-\frac{1}{4}\right) = -0.253$

$x_2 = -\pi - x_1 = -2.89$

For $\sin x = -1$:
$x_1 = \arcsin(-1) = -\frac{\pi}{2}$

∴ the solutions are $x = -2.89, -0.253, -\frac{\pi}{2}$

9 $\cos^2 t + 5\cos t = 2\sin^2 t$
$\cos^2 t + 5\cos t = 2(1-\cos^2 t)$
$3\cos^2 t + 5\cos t - 2 = 0$
$(3\cos t - 1)(\cos t + 2) = 0$
$\cos t = \frac{1}{3}$ or -2 (reject as < -1)
∴ $\cos t = \frac{1}{3}$

10 a $6\sin^2 x + \cos x = 4$
$6(1-\cos^2 x) + \cos x - 4 = 0$
$6\cos^2 x - \cos x - 2 = 0$
$(2\cos x + 1)(3\cos x - 2) = 0$
$\cos x = -\frac{1}{2}$ or $\frac{2}{3}$

b

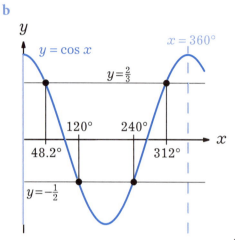

Figure 10D.10 Solutions to $\cos x = -\frac{1}{2}$ and $\cos x = \frac{2}{3}$ in $[0°, 360°]$

There are 2 solutions to each.

For $\cos x = -\frac{1}{2}$:

$x_1 = \arccos\left(-\frac{1}{2}\right) = 120°$

$x_2 = 360° - x_1 = 240°$

For $\cos x = \frac{2}{3}$:

$x_1 = \arccos\left(\frac{2}{3}\right) = 48.2°$

$x_2 = 360° - x_1 = 312°$

∴ the solutions are
$x = 48.2°, 120°, 240°, 312°$

11 a $2\sin^2 x - 3\sin x \cos x + \cos^2 x = 0$

$\frac{2\sin^2 x}{\cos^2 x} - \frac{3\sin x \cos x}{\cos^2 x} + \frac{\cos^2 x}{\cos^2 x} = 0$

$2\tan^2 x - 3\tan x + 1 = 0$

b $2\tan^2 x - 3\tan x + 1 = 0$

$(2\tan x - 1)(\tan x - 1) = 0$

$\tan x = \dfrac{1}{2}$ or 1

Figure 10D.11 Solutions to $\tan x = \dfrac{1}{2}$ and $\tan x = 1$ in $]-\pi, \pi[$

There are 2 solutions to each.

For $\tan x = \dfrac{1}{2}$:

$x_1 = \arctan\left(\dfrac{1}{2}\right) = 0.464$

$x_2 = x_1 - \pi = -2.68$

For $\tan x = 1$:

$x_1 = \arctan(1) = \dfrac{\pi}{4}$

$x_2 = x_1 - \pi = -\dfrac{3\pi}{4}$

∴ the solutions are

$x = -2.68, \ -\dfrac{3\pi}{4}, \ 0.464, \ \dfrac{\pi}{4}$

Mixed examination practice 10

Short questions

1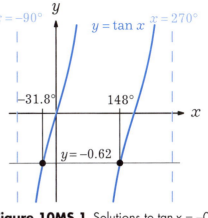

Figure 10MS.1 Solutions to $\tan x = -0.62$ in $]-90°, 270°[$

There are 2 solutions:

$x_1 = \arctan(-0.62) = -31.8°$

$x_2 = x_1 + \pi = 148°$

∴ $x = -31.8°, 148°$

2 $\dfrac{2}{\cos^2 x} - \tan^2 x = \dfrac{2}{\cos^2 x} - \dfrac{\sin^2 x}{\cos^2 x}$

$= \dfrac{2 - \sin^2 x}{\cos^2 x}$

$= \dfrac{2 - 2\sin^2 x + \sin^2 x}{\cos^2 x}$

$= \dfrac{2(1 - \sin^2 x) + \sin^2 x}{\cos^2 x}$

$= \dfrac{2\cos^2 x + \sin^2 x}{\cos^2 x}$

$= 2 + \dfrac{\sin^2 x}{\cos^2 x}$

$= 2 + \tan^2 x$

3 $5\sin^2\theta = 4\cos^2\theta$

$5\sin^2\theta = 4(1 - \sin^2\theta)$

$9\sin^2\theta = 4$

$\sin^2\theta = \dfrac{4}{9}$

$\sin\theta = \pm\dfrac{2}{3}$

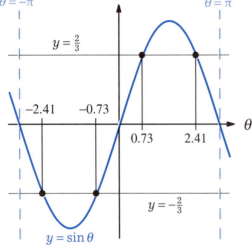

Figure 10MS.3 Solutions to $\sin\theta = \pm\dfrac{2}{3}$ in $[-\pi, \pi]$

There are 2 solutions to each (positive and negative):

$\theta_1 = \arcsin\left(\pm\dfrac{2}{3}\right) = \pm 0.730$

$\theta_2 = \pi - \theta_1 = 2.41, 3.87$

But 3.87 is outside the interval $-\pi \leq \theta \leq \pi$, so subtract 2π: $3.87 - 2\pi = -2.41$

$\therefore \theta = \pm 0.730, \pm 2.41$

4 $\dfrac{1}{1+\cos x} + \dfrac{1}{1-\cos x} = \dfrac{(1-\cos x)+(1+\cos x)}{(1+\cos x)(1-\cos x)}$

$= \dfrac{2}{1-\cos^2 x}$

$= \dfrac{2}{\sin^2 x}$

(using $\sin^2 x = 1 - \cos^2 x$)

5 Using $\sin^2\theta = 1 - \cos^2\theta$:

$\cos\theta - 2\sin^2\theta + 2 = 0$

$\cos\theta - 2(1-\cos^2\theta) + 2 = 0$

$\cos\theta - 2 + 2\cos^2\theta + 2 = 0$

$\cos\theta(1 + 2\cos\theta) = 0$

$\cos\theta = 0$ or $-\dfrac{1}{2}$

There are 2 solutions to each in the interval $[0°, 360°]$:

$\cos\theta = 0$

$\Rightarrow \theta = 90°, 270°$

$\cos\theta = -\dfrac{1}{2}$

$\Rightarrow \theta = 120°, 240°$

$\therefore$ the solutions are
$\theta = 90°, 120°, 240°, 270°$

6 $6\sin^2 x + \cos x = 4$

$6(1-\cos^2 x) + \cos x = 4$

$6\cos^2 x - \cos x - 2 = 0$

$(2\cos x + 1)(3\cos x - 2) = 0$

$\cos x = -\dfrac{1}{2}$ or $\dfrac{2}{3}$

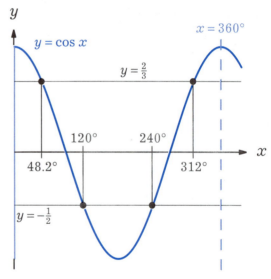

Figure 10MS.6 Solutions to $\cos x = -\dfrac{1}{2}$ and $\cos x = \dfrac{2}{3}$ in $[0, 360°]$

There are 2 solutions to each.

For $\cos x = -\dfrac{1}{2}$:

$x_1 = \arccos\left(-\dfrac{1}{2}\right) = 120°$

$x_2 = 360° - 120° = 240°$

106 Mixed examination practice 10

For $\cos x = \dfrac{2}{3}$:

$x_1 = \arccos\left(\dfrac{2}{3}\right) = 48.2°$

$x_2 = 360° - 48.2° = 312°$

$\therefore x = 48.2°, 120°, 240°, 312°$

7 Let $A = 2x + \dfrac{\pi}{3}$; then

$x \in [-\pi, \pi] \Rightarrow A \in \left[-\dfrac{5\pi}{3}, \dfrac{7\pi}{3}\right]$

$2\cos(A) = \sqrt{2}$

$\cos(A) = \dfrac{1}{\sqrt{2}}$

$A = \pm\dfrac{\pi}{4}, \dfrac{7\pi}{4}, \dfrac{9\pi}{4}$

$\therefore 2x + \dfrac{\pi}{3} = \pm\dfrac{\pi}{4}, \dfrac{7\pi}{4}, \dfrac{9\pi}{4}$

$2x = -\dfrac{7\pi}{12}, -\dfrac{\pi}{12}, \dfrac{17\pi}{12}, \dfrac{23\pi}{12}$

$x = -\dfrac{7\pi}{24}, -\dfrac{\pi}{24}, \dfrac{17\pi}{24}, \dfrac{23\pi}{24}$

8 a $\dfrac{1}{\sin^2 x} + \dfrac{1}{\cos^2 x} = \dfrac{16}{3}$

$\dfrac{\cos^2 x + \sin^2 x}{\sin^2 x \cos^2 x} = \dfrac{16}{3}$

$\dfrac{1}{\sin^2 x (1 - \sin^2 x)} = \dfrac{16}{3}$

Let $s = \sin x$; then the equation becomes

$16s^2(1 - s^2) = 3$

$16s^4 - 16s^2 + 3 = 0$

$(4s^2 - 1)(4s^2 - 3) = 0$

$s^2 = \dfrac{1}{4}$ or $\dfrac{3}{4}$

$\therefore \sin x = \pm\dfrac{1}{2}$ or $\pm\dfrac{\sqrt{3}}{2}$

b For $x \in \left]-\dfrac{\pi}{2}, \dfrac{\pi}{2}\right[$,

$\sin x = \pm\dfrac{1}{2} \Rightarrow x = \pm\dfrac{\pi}{6}$

$\sin x = \pm\dfrac{\sqrt{3}}{2} \Rightarrow x = \pm\dfrac{\pi}{3}$

$\therefore$ the solutions are $\pm\dfrac{\pi}{6}, \pm\dfrac{\pi}{3}$

Long questions

1

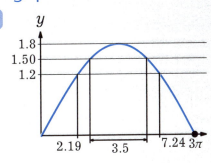

Figure 10ML.1 Graph of $y = 1.8\sin\left(\dfrac{x}{3}\right)$ between $x = 0$ and the first positive zero

a The width of the river is the x-coordinate of the first positive zero: $3\pi = 9.42$ metres (3SF)

b The maximum width of the barge is the distance between the first two positive solutions of $1.8\sin\left(\dfrac{x}{3}\right) = 1.2$.

From GDC, the solutions are $x = 2.19$ and $x = 7.24$, so the maximum width of the barge is $7.24 - 2.19 = 5.05$ metres.

c The centre of the bridge is at $x = \dfrac{3\pi}{2} = 4.71$ metres.

This barge of width 2.5 m, travelling along the centre of the river course, will be positioned in the interval $[4.71 - 1.75, 4.71 + 1.75] = [2.96, 6.46]$

In this interval, $y \geq 1.8\sin\left(\dfrac{2.96}{3}\right) = 1.50$, so the maximum height of the barge is 1.50 metres above water level.

2 a From GDC:

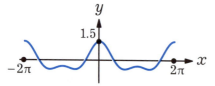

Figure 10ML.2 Graph of $C(x) = \cos x + \frac{1}{2}\cos 2x$ for $-2\pi \leq x \leq 2\pi$

b Since $\cos(a + 2k\pi) = \cos a$ for any integer k,

$$C(x + 2\pi) = \cos(x + 2\pi) + \frac{1}{2}\cos(2(x + 2\pi))$$
$$= \cos(x + 2\pi) + \frac{1}{2}\cos(2x + 4\pi)$$
$$= \cos x + \frac{1}{2}\cos 2x$$
$$= C(x)$$

So $C(x)$ is periodic with period a factor of 2π.

From Figure 10ML.2 it is clear that the period is no less than 2π, so the period equals 2π.

c From GDC, $C(x)$ has maximum points at $x = 0, \pm\pi, \pm 2\pi$

d $\cos x + \frac{1}{2}\cos 2x = 0$

$$\cos x + \frac{1}{2}(2\cos^2 x - 1) = 0$$

$$\cos^2 x + \cos x - \frac{1}{2} = 0$$

$$\cos x = \frac{-1 \pm \sqrt{1^2 - 4(1)\left(-\frac{1}{2}\right)}}{2}$$

$$= -\frac{1}{2} \pm \frac{\sqrt{3}}{2}$$

The smallest positive root is given by

$$x_0 = \arccos\left(-\frac{1}{2} + \frac{\sqrt{3}}{2}\right) = 1.2 \text{ (2SF)}.$$

e i $\cos x = \cos(-x)$ for all x

$$\therefore C(-x) = \cos(-x) + \frac{1}{2}\cos(-2x)$$
$$= \cos x + \frac{1}{2}\cos 2x$$
$$= C(x)$$

ii Using parts (e)(i) and (b):
$$C(x) = C(-x)$$
$$= C(2\pi - x)$$
$$\therefore x_1 = 2\pi - x_0$$

3 a Repeated root $\Rightarrow$ discriminant $= 0$:
$$k^2 - 16 = 0$$
$$\therefore k = \pm 4$$

b $4\sin^2\theta = 5 - k\cos\theta$

$$4(1 - \cos^2\theta) = 5 - k\cos\theta$$

$$4\cos^2\theta - k\cos\theta + 1 = 0$$

c i $f_4(\theta) = 4\cos^2\theta - 4\cos\theta + 1$

From (a), $4x^2 - kx + 1 = 0$ has a repeated root, so there is a single value of $\cos\theta$ which satisfies $f_4(\theta) = 0$.

ii $f_4(\theta) = 0$

$$4\cos^2\theta - 4\cos\theta + 1 = 0$$

$$(2\cos\theta - 1)^2 = 0$$

$$\cos\theta = \frac{1}{2}$$

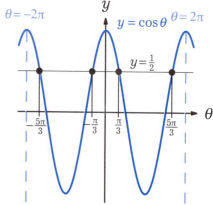

Figure 10ML.3.1 Solutions to $\cos\theta = \frac{1}{2}$ in $[-2\pi, 2\pi]$

108 Mixed examination practice 10

There are 4 solutions. The first 2 are:

$$\theta_1 = \arccos\left(\frac{1}{2}\right) = \frac{\pi}{3}$$

$$\theta_2 = -\theta_1 = -\frac{\pi}{3}$$

Then, adding/subtracting 2π gives

$$\frac{\pi}{3} - 2\pi = -\frac{5\pi}{3} \text{ and } -\frac{\pi}{3} + 2\pi = \frac{5\pi}{3}$$

$$\therefore \theta = \pm\frac{\pi}{3}, \pm\frac{5\pi}{3}$$

iii Substituting $x = 1$ into $4x^2 - kx + 1 = 0$:
$$4 - k + 1 = 0$$
$$k = 5$$

iv With $k = 5$,
$$f_5(\theta) = 0$$
$$4\cos^2\theta - 5\cos\theta + 1 = 0$$
$$(4\cos\theta - 1)(\cos\theta - 1) = 0$$
$$\cos\theta = \frac{1}{4} \text{ or } 1$$

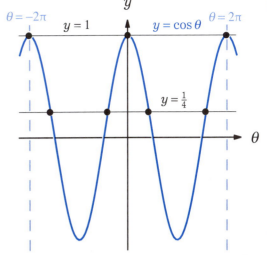

Figure 10ML.3.2 Solutions to $\cos\theta = \frac{1}{4}$ and $\cos\theta = 1$ in $[-2\pi, 2\pi]$

From the graph,

$\cos\theta = \frac{1}{4}$ has 4 solutions in $[-2\pi, 2\pi]$

and $\cos\theta = 1$ has 3 solutions.

In total, there are 7 solutions in $[-2\pi, 2\pi]$.

11 Geometry of triangles and circles

COMMENT

In questions on geometric shapes, if no diagram is given it is usually wise to draw a quick sketch. This reduces the opportunity for errors of interpretation and makes it easier to check for the sense of an answer.
When checking for sense, always remember that in a triangle, the widest angle lies opposite the longest side and the narrowest angle lies opposite the shortest side.

Exercise 11B

4

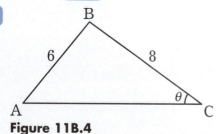

Figure 11B.4

By the sine rule:

$$\frac{\sin C\hat{A}B}{BC} = \frac{\sin A\hat{C}B}{AB}$$

$$\sin C\hat{A}B = \frac{BC \sin A\hat{C}B}{AB} = \frac{8\sin 35°}{6}$$

Two solutions for $C\hat{A}B$ are

$x_1 = \arcsin\left(\dfrac{8\sin 35°}{6}\right) = 49.9°$

$x_2 = 180° - 49.9° = 130.1°$

and hence $A\hat{B}C = 180° - 49.9° - 35° = 95.1°$
or $180° - 130.1° - 35° = 14.9°$ (both solutions are viable).

To find AC, use the sine rule:

$$\frac{\sin A\hat{C}B}{AB} = \frac{\sin A\hat{B}C}{AC}$$

$$AC = \frac{AB \sin A\hat{B}C}{\sin A\hat{C}B} = \frac{6 \sin A\hat{B}C}{\sin 35°}$$

∴ two possible triangles exist:

one with angles 35°, 49.9°, 95.1° and $AC = 10.4$ cm;
another with angles 35°, 130°, 14.9° and $AC = 2.69$ cm.

5 By sine rule in triangle ABD:

$$\frac{\sin A\hat{B}D}{AD} = \frac{\sin A\hat{D}B}{AB}$$

$$\Rightarrow A\hat{B}D = \arcsin\left(\frac{AD \sin A\hat{D}B}{AB}\right)$$

$$= \arcsin\left(\frac{5\sin 75°}{6}\right)$$

$$= 53.6°$$

By sine rule in triangle ABC:

$$\frac{\sin A\hat{C}B}{AB} = \frac{\sin A\hat{B}C}{AC}$$

$$\Rightarrow A\hat{C}B = \arcsin\left(\frac{AB \sin A\hat{B}C}{AC}\right)$$

$$= \arcsin\left(\frac{6 \sin 53.6°}{8}\right)$$

$$= 37.1°$$

Then $B\hat{A}C = 180° - 53.6° - 37.1° = 89.3°$

By sine rule in triangle ABC:

$$\frac{BC}{\sin B\hat{A}C} = \frac{AC}{\sin A\hat{B}C}$$

$$\Rightarrow BC = \frac{AC \sin B\hat{A}C}{\sin A\hat{B}C}$$

$$= \frac{8 \sin 89.3°}{\sin 53.6°}$$

$$= 9.94 \text{ cm}$$

6 By the sine rule:

$$\frac{\sin A\hat{C}B}{AB} = \frac{\sin A\hat{B}C}{AC}$$

$$\sin A\hat{C}B = \frac{AB \sin A\hat{B}C}{AC}$$

$$= \frac{12 \sin 47°}{8}$$

$$= 1.097$$

But since $\sin x < 1$ for any angle x in a triangle, this is not possible.

Exercise 11C

4

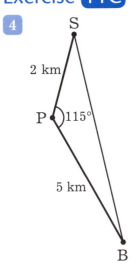

Figure 11C.4

$B\hat{P}S = 130 - 15 = 115°$

By the cosine rule:

$$BS = \sqrt{(BP)^2 + (PS)^2 - (BP)(PS)\cos B\hat{P}S}$$

$$= \sqrt{5^2 + 2^2 - 2(5)(2)\cos 115°}$$

$$= 6.12 \text{ km}$$

5 By cosine rule in triangle ACD:

$$A\hat{D}C = \arccos\left(\frac{(AD)^2 + (CD)^2 - (AC)^2}{2(AD)(CD)}\right)$$

$$= \arccos\left(\frac{6^2 + 7^2 - 10^2}{2(6)(7)}\right)$$

$$= 100.3°$$

$$\therefore B\hat{D}C = 180° - A\hat{D}C = 79.7°$$

By sine rule in triangle BCD:

$$\frac{BC}{\sin B\hat{D}C} = \frac{DC}{\sin D\hat{B}C}$$

$$\Rightarrow BC = \frac{DC \sin B\hat{D}C}{\sin D\hat{B}C}$$

$$= \frac{7 \sin 79.7°}{\sin 60°}$$

$$= 7.95$$

$$\therefore x = 7.95$$

6

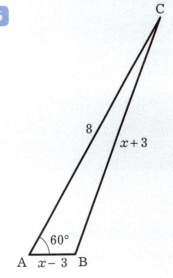

Figure 11C.6

By the cosine rule,
$BC^2 = AB^2 + AC^2 - 2(AB)(AC)\cos B\hat{A}C$

$$(x+3)^2 = (x-3)^2 + 8^2 - 2 \times 8 \times (x-3) \times \frac{1}{2}$$

$$x^2 + 6x + 9 = x^2 - 6x + 9 + 64 - 8x + 24$$

$$20x = 88$$

$$x = 4.4$$

7

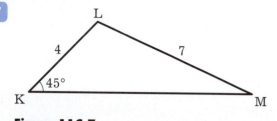

Figure 11C.7

By the cosine rule,
$(LM)^2 = (KL)^2 + (KM)^2 - 2(KL)(KM)\cos L\hat{K}M$

$\therefore 7^2 = 4^2 + x^2 - 2 \times 4x \times \dfrac{1}{\sqrt{2}}$

$49 = 16 + x^2 - \dfrac{8x}{\sqrt{2}}$

$x^2 - 4\sqrt{2}x - 33 = 0$

$x = \dfrac{4\sqrt{2} \pm \sqrt{32 - 4 \times 1 \times (-33)}}{2}$

$= 2\sqrt{2} \pm \sqrt{41}$

Since the length must be positive,
$KM = 2\sqrt{2} + \sqrt{41}$

Exercise 11D

3

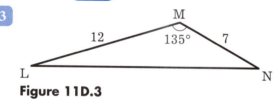

Figure 11D.3

By the cosine rule,
$LN = \sqrt{(LM)^2 + (MN)^2 - 2(LM)(MN)\cos L\hat{M}N}$
$= \sqrt{12^2 + 7^2 - 2(12)(7)\cos 135°}$
$= 17.7$ cm

$\text{Area} = \dfrac{1}{2}(LM)(MN)\sin L\hat{M}N$
$= \dfrac{1}{2} \times 12 \times 7 \sin 135°$
$= 29.7$ cm^2

4

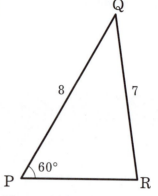

Figure 11D.4

Let $PR = x$. By the cosine rule,
$(RQ)^2 = (PQ)^2 + (PR)^2 - 2(PQ)(PR)\cos R\hat{P}Q$
$7^2 = 8^2 + x^2 - 2 \times 8x \cos 60°$
$49 = 64 + x^2 - 8x$
$x^2 - 8x + 15 = 0$
$(x-3)(x-5) = 0$
$x = 3$ or 5

Area of triangle $= \dfrac{1}{2}(PQ)x \sin R\hat{P}Q$,
where x can take the two possible values above.

Area difference $= \dfrac{1}{2}(PQ)\sin R\hat{P}Q \times (x_1 - x_2)$

$= \dfrac{1}{2} \times 8 \times \dfrac{\sqrt{3}}{2} \times (5-3)$

$= 4\sqrt{3}$ cm^2

Exercise 11E

2

> **COMMENT**
> There could be several possible answers to this question, depending on which cuboid edge (if any) is included in ABC. If the question had specified that no cuboid edges are included in ABC, then the solution narrows down to case (iv), which is the answer given in the back of the coursebook.

Case (i): If ABC includes a side of length 12.5 (right triangle), then the sides are 12.5,

$\sqrt{10^2 + 7.3^2} = 12.4$ and

$\sqrt{12.5^2 + 10^2 + 7.3^2} = 17.6$

From trigonometry, the angles are 90°,

$\arcsin\left(\dfrac{12.4}{17.6}\right) = 44.7°$ and 45.3°

Area $= \dfrac{1}{2} \times 12.4 \times 12.5 = 77.4 \text{ cm}^2$

Case (ii): If ABC includes a side of length 10 (right triangle), then the sides are 10,

$\sqrt{12.5^2 + 7.3^2} = 14.5$ and

$\sqrt{12.5^2 + 10^2 + 7.3^2} = 17.6$

From trigonometry, the angles are 90°,

$\arcsin\left(\dfrac{10}{17.6}\right) = 34.6°$ and 55.4°

Area $= \dfrac{1}{2} \times 10 \times 14.5 = 72.4 \text{ cm}^2$

Case (iii): If ABC includes a side of length 7.3 (right triangle), then the sides are 7.3,

$\sqrt{10^2 + 12.5^2} = 16.0$ and

$\sqrt{12.5^2 + 10^2 + 7.3^2} = 17.6$

From trigonometry, the angles are 90°,

$\arcsin\left(\dfrac{7.3}{17.6}\right) = 24.5°$ and 65.5°

Area $= \dfrac{1}{2} \times 7.3 \times 16.0 = 58.4 \text{ cm}^2$

Case (iv): If ABC includes no sides of the cuboid (oblique triangle), then the sides are $\sqrt{10^2 + 7.3^2} = 12.4$, $\sqrt{10^2 + 12.5^2} = 16.0$ and $\sqrt{12.5^2 + 7.3^2} = 14.5$.
By applying the cosine rule repeatedly, the angles are 47.6°, 59.7°, 72.7°

(For example, the angle between the sides of lengths 12.4 and 16.0 is

$\arccos\left(\dfrac{12.4^2 + 16.0^2 - 14.5^2}{2(12.4)(16.0)}\right) = 59.7°$.)

Area $= \dfrac{1}{2} \times 12.4 \times 14.5 \times \cos(72.7°)$

$= 85.6 \text{ cm}^2$

3 By trigonometry, the flagpole height BF is $12\tan 52° = 15.4$ m

$\therefore \tan\theta = \dfrac{\text{BF}}{8}$

$\theta = \arctan\left(\dfrac{\text{BF}}{8}\right) = 62.5°$

4

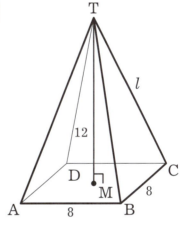

Figure 11E.4.1

Let M be the midpoint of the base. If one corner of the base is A and the apex of the pyramid is T, then by Pythagoras' Theorem,

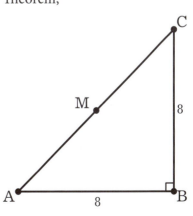

Figure 11E.4.2

$AM = \dfrac{\sqrt{8^2+8^2}}{2}$

$= \dfrac{8\sqrt{2}}{2}$

$= 4\sqrt{2}$

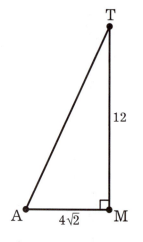

Figure 11E.4.3

$\therefore AT = \sqrt{12^2 + (4\sqrt{2})^2}$

$= \sqrt{144+32}$

$= \sqrt{176}$

$= 4\sqrt{11}$

$= 13.3$ cm

i.e. $l = 13.3$ cm

5 a By Pythagoras' Theorem,

$CM = \sqrt{17^2 - 12^2}$

$= \sqrt{145}$

$= 12.0$ cm

b CMB is an isosceles right triangle with $C\hat{M}B = 90°$ (the diagonals of a square meet at a right angle).

$\therefore CB = \sqrt{2(CM)^2}$

$= \sqrt{2}\, CM$

$= \sqrt{2}\sqrt{145}$

$= \sqrt{290}$

$= 17.0$ cm

6 a $QP = AQ \tan Q\hat{A}P$

$= 25 \tan 37°$

$= 18.8$

$\therefore h = 18.8$ m

b $QB = \dfrac{QP}{\tan Q\hat{B}P}$

$= \dfrac{25\tan 37°}{\tan 42°}$

$= 20.9$

By the cosine rule,

$AB = \sqrt{(QB)^2 + (QA)^2 - 2(QB)(QA)\cos A\hat{Q}B}$

$= \sqrt{20.9^2 + 25^2 - 2(20.9)(25)\cos 75°}$

$= 28.1$ m

7 a $\tan\alpha = \dfrac{h}{RA}$ and $\tan\beta = \dfrac{h}{RB}$

$\therefore RA = \dfrac{h}{\tan\alpha}$ and $RB = \dfrac{h}{\tan\beta}$

By Pythagoras' Theorem,

$(AB)^2 = (RA)^2 + (RB)^2$

i.e. $d^2 = \left(\dfrac{h}{\tan\alpha}\right)^2 + \left(\dfrac{h}{\tan\beta}\right)^2$

$= h^2 \left(\dfrac{1}{\tan^2\alpha} + \dfrac{1}{\tan^2\beta}\right)$

b $\alpha = 45° \Rightarrow \tan\alpha = 1 \Rightarrow \dfrac{1}{\tan^2\alpha} = 1$

$\beta = 30° \Rightarrow \tan\beta = \dfrac{1}{\sqrt{3}} \Rightarrow \dfrac{1}{\tan^2\beta} = 3$

$h^2 = \dfrac{d^2}{\left(\dfrac{1}{\tan^2\alpha} + \dfrac{1}{\tan^2\beta}\right)}$

$= \dfrac{26^2}{1+3}$

$= 169$

$\therefore h = 13$ m

Exercise 11F

3 $l = r\theta$
$= 10 \times 2.5$
$= 25$ cm

4 a $\theta = \dfrac{l}{r}$
$= \dfrac{7.5}{8}$
$= 0.9375$ radians

b 0.9375 radians $= 0.9375 \times \dfrac{180°}{\pi}$
$= 53.7°$

5 Let θ be the angle subtended by the major arc; then
$\theta = \dfrac{l}{r}$
$= \dfrac{15}{4}$
$= 3.75$
$\therefore \mathrm{M\hat{C}N} = 2\pi - 3.75$
$= 2.53$ radians

6 $r = \dfrac{l}{\theta}$
$= \dfrac{12}{1.6}$
$= 7.5$ cm

7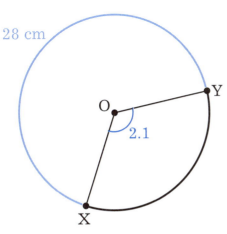

Figure 11F.7 Length of major arc XY is 28 cm

The angle subtended by the major arc is $2\pi - 2.1$, so
$r = \dfrac{l}{\theta}$
$= \dfrac{28}{2\pi - 2.1}$
$= 6.69$ cm

8 The perimeter p is composed of three arcs with radius 5 cm and angle $60° = \dfrac{\pi}{3}$
$\therefore p = 3\left(5 \times \dfrac{\pi}{3}\right)$
$= 5\pi = 15.7$ cm

9 Let l be the length of the arc; then
$l = r\theta$
$= 8 \times 0.7$
$= 5.6$
$\therefore p = 2(5+8) + 5.6$
$= 31.6$ cm

10 The angle between the two 5 cm sides is
$\theta = 180° - 2 \times 15°$
$= 150° = \dfrac{5\pi}{6}$
$\therefore p = 10 + 5 \times \dfrac{5\pi}{6}$
$= \left(10 + \dfrac{25\pi}{6}\right)$ cm

11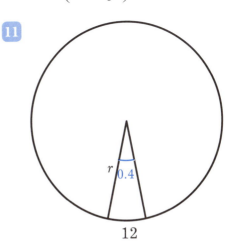

Figure 11F.11

The perimeter of the sector is made up of two radii and an arc:

$p = 2r + r\theta$

$12 = r(2 + 0.4)$

$r = \dfrac{12}{2.4} = 5$ cm

12 On the cone, the slant height is

$\sqrt{12^2 + \left(\dfrac{18}{2}\right)^2} = 15$, which is the radius r of the sector.

The perimeter of the base is 18π, which is the arc length l of the sector.

$\theta = \dfrac{l}{r}$

$= \dfrac{18\pi}{15}$

$= \dfrac{6}{5}\pi$

$= 3.77$ radians (or $216°$)

Exercise 11G

3 $A = \dfrac{1}{2}r^2\theta$

$\Rightarrow \theta = \dfrac{2A}{r^2}$

$= \dfrac{2 \times 40}{10^2}$

$= 0.8$ radians

4 The larger angle is

$\theta = \dfrac{2A}{r^2}$

$= \dfrac{2 \times 744}{21^2}$

$= 3.37$

∴ smaller angle $= 2\pi - 3.37$, which is equivalent to $(2\pi - 3.37) \times \dfrac{180°}{\pi} = 167°$

5 $A = \dfrac{1}{2}r^2\theta$

$r = \sqrt{\dfrac{2A}{\theta}}$

$= \sqrt{\dfrac{2 \times 54}{1.2}}$

$= \sqrt{90} = 9.49$ cm

6 $162° = 162 \times \dfrac{\pi}{180} = 2.827$ radians

$r = \sqrt{\dfrac{2A}{\theta}}$

$= \sqrt{\dfrac{2 \times 180}{2.827}}$

$= 11.3$ cm

7 $\theta = 45° = \dfrac{\pi}{4}$

Sector area $= \dfrac{r^2\theta}{2}$

$= \dfrac{6^2\pi}{8}$

$= 14.14$ cm^2

Triangle area $= \dfrac{1}{2} \times 6 \times 3 = 9$ cm^2

∴ shaded area $= 14.1 - 9 = 5.14$ cm^2

8 The perimeter of the sector is made up of two radii and an arc:

$p = 2r + r\theta$

$28 = r(2 + 1.6)$

$\Rightarrow r = \dfrac{28}{3.6} = 7.78$ cm

∴ $A = \dfrac{r^2\theta}{2} = \dfrac{1}{2}\left(\dfrac{28}{3.6}\right)^2 (1.6) = 48.4$ cm^2

9 $p = 2r + r\theta$

$7 = r(2+\theta)$ ...(1)

$A = \dfrac{r^2\theta}{2}$

$3 = \dfrac{r^2\theta}{2}$

$\Rightarrow \theta = \dfrac{6}{r^2}$ ...(2)

Substituting (2) into (1):

$r\left(2 + \dfrac{6}{r^2}\right) = 7$

$2r^2 - 7r + 6 = 0$

$(2r-3)(r-2) = 0$

$r = 1.5$ or 2

So the radius is 1.5 cm or 2 cm.

10 Let θ be the minor sector angle. Then:

major sector area $A_1 = \dfrac{r^2(2\pi - \theta)}{2}$

minor sector area $A_2 = \dfrac{r^2\theta}{2}$

$A_1 - A_2 = 15$

$\dfrac{r^2(2\pi - \theta)}{2} - \dfrac{r^2\theta}{2} = 15$

$\dfrac{5^2}{2}(2\pi - 2\theta) = 15$

$\pi - \theta = \dfrac{15}{25}$

$\therefore \theta = \pi - 0.6 = 2.54$ radians

Exercise 11H

4 a Minor segment area $= \dfrac{r^2}{2}(\theta - \sin\theta)$

$= \dfrac{5^2}{2}(\theta - \sin\theta)$

$= 12.5(\theta - \sin\theta)$ cm^2

b $12.5(\theta - \sin\theta) = 15$

$\theta - \sin\theta = 1.2$

$\Rightarrow \theta = 2.08$ radians (from GDC)

5 a By cosine rule in triangle PAQ:

$(PQ)^2 = (AP)^2 + (AQ)^2 - 2(AP)(AQ)\cos P\hat{A}Q$

$(PQ)^2 = 6^2 + 6^2 - 2\times 6\times 6\times \dfrac{1}{\sqrt{2}}$

$= 36(2 - \sqrt{2})$

$PQ = 6\sqrt{2 - \sqrt{2}}$

b By cosine rule in triangle PBQ:

$\cos P\hat{B}Q = \dfrac{(PB)^2 + (QB)^2 - (PQ)^2}{2(PB)(QB)}$

$= \dfrac{4^2 + 4^2 - 36(2-\sqrt{2})}{2(4)(4)}$

$= 0.341$

$\therefore P\hat{B}Q = \arccos 0.341 = 70.1°$

c Shaded area is the sum of two segments, one of radius 6 and angle $45° = \dfrac{\pi}{4}$ radians and the other of radius 4 and angle 1.22 radians.

Shaded area $= \left[\dfrac{1}{2}\times 6^2 \times \left(\dfrac{\pi}{4} - \sin\left(\dfrac{\pi}{4}\right)\right)\right]$

$+ \left[\dfrac{1}{2}\times 4^2 \times (1.22 - \sin(1.22))\right]$

$= 1.41 + 2.26$

$= 3.67$ cm^2

Mixed examination practice 11

Short questions

1 a $C\hat{O}Q + \dfrac{\pi}{2} + \dfrac{\pi}{6}$

$= \pi$ (angles on a straight line)

$\Rightarrow C\hat{O}Q = \dfrac{\pi}{3}$

b Total area = area COQ + area OABC + area OAP

$$= \frac{1}{2}\left(2^2 \times \frac{\pi}{3}\right) + (2\times 7) + \frac{1}{2}\left(7^2 \times \frac{\pi}{6}\right)$$

$$= \frac{57\pi}{12} + 14$$

$$= 28.9 \text{ cm}^2 \text{ (3SF)}$$

c

Perimeter = QC + CB + BA + AP + PO + OQ

$$= 2 \times \frac{\pi}{3} + 7 + 2 + 7 \times \frac{\pi}{6} + 7 + 2$$

$$= \frac{11\pi}{6} + 18 = 23.8 \text{ cm (3SF)}$$

2 $p = 2r + r\theta$

$36 = 2 \times 10 + 10\theta$

$\Rightarrow \theta = 1.6$

$$\therefore \text{Area} = \frac{r^2 \theta}{2} = \frac{10^2 \times 1.6}{2} = 80 \text{ cm}^2$$

3

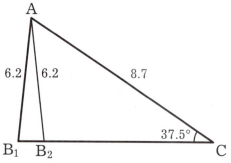

Figure 11MS.3

Using the sine rule,

$$\frac{\sin A\hat{B}C}{AC} = \frac{\sin A\hat{C}B}{AB}$$

$$\Rightarrow \sin A\hat{B}C = \frac{AC \times \sin A\hat{C}B}{AB}$$

$$= \frac{8.7 \sin 37.5°}{6.2}$$

$$= 0.854$$

$A\hat{B}C = \arcsin 0.854$

$= 58.7°$ or $180° - 58.7° = 121°$ (3SF)

4 a $\frac{12}{AB} = \tan 56°$

$$\therefore AB = \frac{12}{\tan 56°} = 8.09 \text{ m (3SF)}$$

b Triangle ABM is isosceles, with BM = AB.

Using the cosine rule:

$$AM = \sqrt{AB^2 + BM^2 - 2(AB)(BM)\cos A\hat{B}M}$$

$$= \sqrt{8.09^2 + 8.09^2 - 2 \times 8.09^2 \times \cos 48°}$$

$$= 6.58 \text{ m}$$

5 a Segment area $= \frac{1}{2}r^2(\theta - \sin\theta)$

$$= \frac{1}{2} \times 7^2 (1.4 - \sin 1.4)$$

$$= 10.2 \text{ cm}^2 \text{ (3SF)}$$

b By cosine rule in triangle OPQ,

$$PQ = \sqrt{OP^2 + OQ^2 - 2(OP)(OQ)\cos\theta}$$

$$= \sqrt{7^2 + 7^2 - 2 \times 7^2 \times \cos 1.4}$$

$$= 9.02$$

Perimeter = length of line PQ + length of arc PQ

$$= 9.02 + 1.4 \times 7$$

$$= 18.8 \text{ cm}$$

6

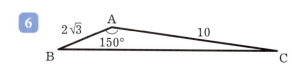

Figure 11MS.6

By the cosine rule,

$$BC^2 = AB^2 + AC^2 - 2(AB)(AC)\cos B\hat{A}C$$

$$BC = \sqrt{(2\sqrt{3})^2 + 10^2 - 2(10)(2\sqrt{3})\cos 150°}$$

$$= \sqrt{12 + 100 - 40\sqrt{3}\left(-\frac{\sqrt{3}}{2}\right)}$$

$$= \sqrt{112 + 60}$$

$$= \sqrt{172}$$

$$= 2\sqrt{43}$$

7 $p = 34$

$\therefore 2r + r\theta = 34$ ...(1)

Area = 52

$\therefore \frac{1}{2}r^2\theta = 52$ ...(2)

$(2) \Rightarrow \theta = \frac{104}{r^2}$

Substituting into (1):

$2r + \frac{104}{r} = 34$

$r^2 - 17r + 52 = 0$

$(r-13)(r-4) = 0$

$r = 13$ cm or 4 cm

8 $O\hat{T}A = 90°$ because AT is a tangent.
So, by Pythagoras' Theorem,

$AT = \sqrt{12^2 - 6^2}$

$= \sqrt{108} = 6\sqrt{3}$ cm

$\therefore$ Area of triangle OTA $= \frac{1}{2} \times 6\sqrt{3} \times 6$

$= 18\sqrt{3}$ cm^2

$\cos A\hat{O}T = \frac{6}{12} = \frac{1}{2}$

$\Rightarrow A\hat{O}T = \frac{\pi}{3}$

Shaded area = area of OTA − area of sector

$= 18\sqrt{3} - \frac{1}{2} \times 6^2 \times \frac{\pi}{3}$

$= 18\sqrt{3} - 6\pi$

$= 12.3$ cm^2

9 a Area of segment BDCP $= \frac{1}{2} \times 2^2 \left(\frac{\pi}{2} - \sin\frac{\pi}{2}\right)$

$= \pi - 2$

$= 1.14$ cm^2

b The semicircle with diameter BC has radius $\sqrt{2}$, so area of the region BECD is area of semicircle − area of segment BDCP

$= \frac{\pi(\sqrt{2})^2}{2} - (\pi - 2)$

$= \pi - (\pi - 2)$

$= 2$ cm^2

10 a Angle of sector 2 is $\frac{\pi}{2} - \theta$

$\therefore$ area of sector 2 $= \frac{2^2}{2}\left(\frac{\pi}{2} - \theta\right) = \pi - 2\theta$

b Total removed area is a semicircle with radius 2,

$\therefore$ remaining area $= \frac{9 \times 12}{2} - \frac{\pi \times 2^2}{2}$

$= 54 - 2\pi$

$= 47.7$ cm^2

11 By the sine rule, $\frac{\sin L\hat{K}M}{6.1} = \frac{\sin 42°}{4.2}$

$\Rightarrow L\hat{K}M = \arcsin\left(\frac{6.1 \sin 42°}{4.2}\right) = 76.37°$

or $180° - 76.37° = 103.63°$

$\therefore L\hat{M}K = 180° - 42° - L\hat{K}M$

$= 61.63°$ or $34.37°$

Since the triangle is obtuse,
$L\hat{K}M = 103.63°$ and $L\hat{M}K = 34.37°$

Area $= \frac{1}{2}(LM)(KM)\sin L\hat{M}K$

$= \frac{1}{2}(6.1)(4.2)\sin 34.37°$

$= 7.23$ cm^2 (3SF)

12

Figure 11MS.12

a By the cosine rule,

$$\cos A\hat{B}C = \frac{AB^2 + BC^2 - AC^2}{2(AB)(BC)}$$

$$= \frac{8^2 + 10^2 - 7^2}{2(8)(10)}$$

$$= \frac{115}{160}$$

$$= \frac{23}{32}$$

b $\sin A\hat{B}C = \sqrt{1 - \cos^2 A\hat{B}C}$

$$= \sqrt{1 - \left(\frac{23}{32}\right)^2}$$

$$= \frac{1}{32}\sqrt{1024 - 529}$$

$$= \frac{1}{32}\sqrt{495}$$

$$= \frac{3}{32}\sqrt{55}$$

c Area $= \frac{1}{2}(AB)(BC)\sin A\hat{B}C$

$$= \frac{1}{2} \times 10 \times 8 \times \frac{3}{32}\sqrt{55}$$

$$= \frac{15}{4}\sqrt{55} \text{ cm}^2$$

13 a Shaded area is the difference between two sectors:

$$\text{Area} = \frac{10^2 \theta}{2} - \frac{(10-x)^2 \theta}{2}$$

$$= \frac{\theta}{2}(100 - 100 + 20x - x^2)$$

$$= \frac{\theta x(20-x)}{2}$$

b $\theta = 1.2$:

Area $= 54.6$

$$\frac{1.2x(20-x)}{2} = 54.6$$

$0.6x(20-x) = 54.6$

$x^2 - 20x + 91 = 0$

$(x-13)(x-7) = 0$

$\therefore x = 7$ (since $x < 10$)

Long questions

1 a

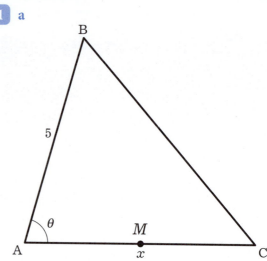

Figure 11ML.1

Using cosine rule in triangle AMB:

$$MB^2 = AM^2 + AB^2 - 2(AM)(AB)\cos M\hat{A}B$$

$$= \left(\frac{x}{2}\right)^2 + 5^2 - 2\left(\frac{x}{2}\right) \times 5\cos\theta$$

$$= \frac{x^2}{4} + 25 - 5x\cos\theta$$

b Using cosine rule in triangle ABC:

$$BC^2 = AC^2 + AB^2 - 2(AB)(BC)\cos B\hat{A}C$$

$$= x^2 + 25 - 10x\cos\theta$$

$BC = MB \Rightarrow BC^2 = MB^2$

i.e. $x^2 + 25 - 10x\cos\theta = \frac{x^2}{4} + 25 - 5x\cos\theta$

$$\frac{3x^2}{4} = 5x\cos\theta$$

$\therefore \cos\theta = \frac{3x}{20}$ (as $x \neq 0$)

c $x = 5 \Rightarrow \cos\theta = \frac{15}{20} = \frac{3}{4}$

$\therefore \theta = \arccos\left(\frac{3}{4}\right)$

$= 41.4°$ (3SF)

2 a

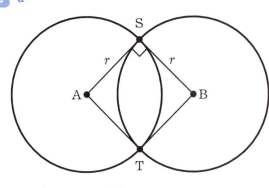

Figure 11ML.2

ASBT is a rhombus as each side has the same length r (the radius).

Since $A\hat{S}B = 90°$, ASBT is a square, and so $S\hat{A}T = 90°$.

b AB is the diagonal of a square with side r.
So, by Pythagoras' Theorem,
$AB^2 = 2r^2$
$AB = \sqrt{2}r$

c Sector AST is a quarter circle.
$\therefore$ Area AST $= \dfrac{\pi r^2}{4}$

d The overlap consists of two sectors minus the square ASBT.
$\text{Area} = 2 \times \dfrac{\pi r^2}{4} - r^2$
$= \left(\dfrac{\pi}{2} - 1\right)r^2$

3 a Area of minor segment
= area of minor sector AOB
− area of triangle AOB
$= \dfrac{r^2 \theta}{2} - \dfrac{1}{2}r^2 \sin\theta$
$= \dfrac{r^2}{2}(\theta - \sin\theta)$

b Area of major sector = area of circle
− area of minor sector
$= \pi r^2 - \dfrac{r^2 \theta}{2}$
$= \dfrac{r^2}{2}(2\pi - \theta)$

c $\dfrac{\dfrac{r^2}{2}(\theta - \sin\theta)}{\dfrac{r^2}{2}(2\pi - \theta)} = \dfrac{1}{2}$

$2(\theta - \sin\theta) = (2\pi - \theta)$
$2\theta - 2\sin\theta = 2\pi - \theta$
$\sin\theta = \dfrac{3\theta}{2} - \pi$

d From GDC, $\theta = 2.50$ (3SF)

4 a Area $= \dfrac{1}{2}ab\sin C$
$2.21 = \dfrac{1}{2} \times x \times 3x \times \sin\theta$
$\sin\theta = \dfrac{4.42}{3x^2}$

b $(x+3)^2 = x^2 + (3x)^2 - 2(x)(3x)\cos\theta$
$x^2 + 6x + 9 = x^2 + 9x^2 - 6x^2\cos\theta$
$\Rightarrow \cos\theta = \dfrac{9x^2 - 6x - 9}{6x^2}$
$= \dfrac{3x^2 - 2x - 3}{2x^2}$

c i $\cos^2\theta = 1 - \sin^2\theta$
$\Rightarrow \left(\dfrac{3x^2 - 2x - 3}{2x^2}\right)^2 = 1 - \left(\dfrac{4.42}{3x^2}\right)^2$

ii Using GDC: $x = 1.24,\ 2.94$ (3SF)

$\theta = \arccos\left(\dfrac{3x^2 - 2x - 3}{2x^2}\right)$
$= 1.86,\ 0.172$

5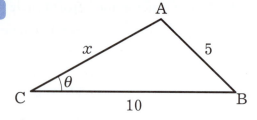

Figure 11ML.5

a By the cosine rule:
$$AB^2 = CA^2 + BC^2 - 2(CA)(BC)\cos A\hat{C}B$$
$$5^2 = x^2 + 10^2 - 2x \times 10\cos\theta = 0$$
$$25 - x^2 - 100 + 20x\cos\theta = 0$$
$$x^2 - 20x\cos\theta + 75 = 0$$

b Real solutions $\Rightarrow$ discriminant $\Delta \geq 0$:
$$(-20\cos\theta)^2 - 4 \times 1 \times 75 \geq 0$$
$$(20\cos\theta)^2 \geq 300$$
$$20\cos\theta \leq -10\sqrt{3} \quad \text{or} \quad 20\cos\theta \geq 10\sqrt{3}$$
$$\therefore -1 \leq \cos\theta \leq -\frac{\sqrt{3}}{2} \quad \text{or} \quad \frac{\sqrt{3}}{2} \leq \cos\theta \leq 1$$

COMMENT

Remember that $-1 \leq \cos\theta \leq 1$ always, so in the absence of other bounds these will still hold.

c $\arccos(-1) = 180°$

$\arccos\left(-\dfrac{\sqrt{3}}{2}\right) = 150°$

$\therefore -1 \leq \cos\theta \leq -\dfrac{\sqrt{3}}{2} \Rightarrow 150° \leq \theta < 180°$

$\arccos\left(\dfrac{\sqrt{3}}{2}\right) = 30°$

$\arccos(1) = 0°$

$\therefore \dfrac{\sqrt{3}}{2} \leq \cos\theta \leq 1 \Rightarrow 0° < \theta \leq 30°$

i.e. $0° < \theta \leq 30°$ or $150° \leq \theta < 180°$

6 a i $AP^2 = (8-x)^2 + (6-10)^2$
$$= 80 - 16x + x^2$$
$$\Rightarrow AP = \sqrt{x^2 - 16x + 80}$$

ii $OP = \sqrt{x^2 + 100}$

b By the cosine rule:
$$\cos O\hat{P}A = \frac{OP^2 + AP^2 - OA^2}{2(OP)(AP)}$$
$$= \frac{(x^2 + 100) + (x^2 - 16x + 80) - (8^2 + 6^2)}{2\sqrt{(x^2 + 100)(x^2 - 16x + 80)}}$$
$$= \frac{2x^2 - 16x + 80}{2\sqrt{(x^2 + 100)(x^2 - 16x + 80)}}$$
$$= \frac{x^2 - 8x + 40}{\sqrt{(x^2 + 100)(x^2 - 16x + 80)}}$$

c $x = 8 \Rightarrow O\hat{P}A = \arccos\left(\dfrac{40}{\sqrt{164 \times 16}}\right)$
$= 38.7°$ (3SF)

d $O\hat{P}A = 60° \Rightarrow \cos O\hat{P}A = \dfrac{1}{2}$

$\therefore \dfrac{x^2 - 8x + 40}{\sqrt{(x^2 + 100)(x^2 - 16x + 80)}} = \dfrac{1}{2}$

From GDC, $x = 5.63$ (3SF)

e i If $f(x) = \cos O\hat{P}A = 1$ then $O\hat{P}A = 0$, which happens when OAP is a straight line (so there is a solution).

ii When OAP forms a straight line, the gradients of OA and OP are equal:
$$\frac{6-0}{8-0} = \frac{10-0}{x-0}$$
$$\frac{6}{8} = \frac{10}{x}$$
$$x = \frac{80}{6} = \frac{40}{3}$$

7 a i $y = -16x^2 + 160x - 256$
$= -16(x^2 - 10x + 16)$
$= -16((x-5)^2 - 25 + 16)$
$= 144 - 16(x-5)^2$
Maximum value of y occurs at $x = 5$

ii Maximum value of y is 144

b i $x + z + 6 = 16$
$\Rightarrow z = 10 - x$

ii $z^2 = x^2 + 6^2 - 2 \times x \times 6 \cos Z$
$= x^2 + 36 - 12x \cos Z$

iii $\cos Z = \dfrac{x^2 + 36 - z^2}{12x}$ from (ii)
$= \dfrac{x^2 + 36 - (10-x)^2}{12x}$ from (i)
$= \dfrac{-64 + 20x}{12x}$
$= \dfrac{5x - 16}{3x}$

c $A = \dfrac{1}{2} \times 6 \times x \times \sin Z$
$= 3x \sin Z$
$\Rightarrow A^2 = 9x^2 \sin^2 Z$

d $A^2 = 9x^2 \sin^2 Z$
$= 9x^2 (1 - \cos^2 Z)$
$= 9x^2 \left(1 - \left(\dfrac{5x-16}{3x}\right)^2\right)$ by (b)(iii)
$= 9x^2 - 25x^2 + 160x - 256$
$= -16x^2 + 160x - 256$

e i From (a)(ii), the maximum value of A^2 is 144, so the maximum area is 12.

ii From (a)(i), the maximum occurs when $x = 5$, for which $z = 10 - 5 = 5$. Since $x = z$, the triangle is isosceles.

8 a $O_1\hat{A}B = \dfrac{\pi}{2}$, since AB is tangent to the circle.

b By the same reasoning, $O_2\hat{B}A = \dfrac{\pi}{2}$, and hence BAPO$_2$ is a rectangle, so PO$_2$ = AB (parallel sides in a rectangle have the same length).

c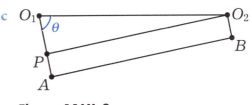

Figure 11ML.8

PO$_1$ = 8 − 3 = 5
$O_1O_2 = 25$
$\therefore PO_2 = \sqrt{25^2 - 5^2}$
$= \sqrt{600}$
$= 10\sqrt{6}$
AB = PO$_2$ = $10\sqrt{6}$ = 24.5 cm (3SF)

d $\sin \theta = \dfrac{PO_2}{25}$
$= \dfrac{10\sqrt{6}}{25}$
$\theta = \arcsin\left(\dfrac{10\sqrt{6}}{25}\right) = 1.369$ (4SF)

e Length of chain = arc AD + 2AB + arc BC
$= 8(2\pi - 2\theta) + 2 \times 10\sqrt{6}$
$+ 3(2\theta)$
$= 85.6$ cm (3SF)

12 Further trigonometry

Exercise 12A

7 Using $\cos 2\theta = 2\cos^2\theta - 1$, the equation becomes
$$\cos^2\theta + 2\cos^2\theta - 1 = 0$$
$$\cos^2\theta = \frac{1}{3}$$
$$\cos\theta = \pm\frac{1}{\sqrt{3}}$$
$\theta \in [-\pi, \pi]$, so
$$\cos\theta = \frac{1}{\sqrt{3}} \Rightarrow \theta = \pm 0.955$$
$$\cos\theta = -\frac{1}{\sqrt{3}} \Rightarrow \theta = \pm 2.19$$
$\therefore \theta = \pm 0.955, \pm 2.19$

8 $1 - \cos 2\theta = 2\sin^2\theta$ and $1 + \cos 2\theta = 2\cos^2\theta$, so
$$\frac{1-\cos 2\theta}{1+\cos 2\theta} = \frac{2\sin^2\theta}{2\cos^2\theta}$$
$$= \tan^2\theta$$

9 a Using $\tan 2\alpha = \dfrac{2\tan\alpha}{1-\tan^2\alpha}$, the equation becomes
$$\frac{2\tan^2\alpha}{1-\tan^2\alpha} = 6$$
$$2\tan^2\alpha = 6 - 6\tan^2\alpha$$
$$8\tan^2\alpha = 6$$
$$\tan^2\alpha = \frac{3}{4}$$
$$\tan\alpha = \pm\frac{\sqrt{3}}{2}$$

b As in (a), using $\tan 2\alpha = \dfrac{2\tan\alpha}{1-\tan^2\alpha}$, the equation becomes
$$\frac{2\tan^2\alpha}{1-\tan^2\alpha} = 1$$
$$2\tan^2\alpha = 1 - 1\tan^2\alpha$$
$$3\tan^2\alpha = 1$$
$$\tan^2\alpha = \frac{1}{3}$$
$$\tan\alpha = \pm\frac{1}{\sqrt{3}}$$
$$\alpha = \frac{\pm\pi}{6}$$
For $\alpha \in\,]0, \pi[$ the solutions are
$$\alpha = \frac{\pi}{6} \text{ and } \pi - \frac{\pi}{6} = \frac{5\pi}{6}$$

10 a $\cos 4\theta = 2\cos^2 2\theta - 1$
$$= 2(2\cos^2\theta - 1)^2 - 1$$
$$= 8\cos^4\theta - 8\cos^2\theta + 1$$

b $\cos 4\theta = 2\cos^2 2\theta - 1$
$$= 2(1 - 2\sin^2\theta)^2 - 1$$
$$= 8\sin^4\theta - 8\sin^2\theta + 1$$

> **COMMENT**
> In this case it is as fast to start again as to convert the answer from (a) using $\cos^2\theta = 1 - \sin^2\theta$.

11 a i $\cos^2\left(\dfrac{x}{2}\right) = \dfrac{1}{2}\left(2\cos^2\left(\dfrac{x}{2}\right)\right)$

$= \dfrac{1}{2}\left(1+\cos\dfrac{2x}{2}\right)$

$= \dfrac{1}{2}(1+\cos x)$

ii $\sin^2\left(\dfrac{x}{2}\right) = \dfrac{1}{2}\left(2\sin^2\left(\dfrac{x}{2}\right)\right)$

$= \dfrac{1}{2}\left(1-\cos\dfrac{2x}{2}\right)$

$= \dfrac{1}{2}(1-\cos x)$

b From (a): $\tan^2\left(\dfrac{x}{2}\right) = \dfrac{\tfrac{1}{2}(1-\cos x)}{\tfrac{1}{2}(1+\cos x)}$

$= \dfrac{1-\cos x}{1+\cos x}$

12 $a\sin 4x = 2a\sin 2x \cos 2x$, so
$a \sin 4x = b \sin 2x$
$2a \sin 2x \cos 2x = b \sin 2x$

$\cos 2x = \dfrac{b}{2a}$

$1 - 2\sin^2 x = \dfrac{b}{2a}$

$\sin^2 x = \dfrac{1}{2}\left(1-\dfrac{b}{2a}\right)$

$= \dfrac{2a-b}{4a}$

Exercise 12B

4 a $\sin\left(x+\dfrac{\pi}{3}\right) + \sin\left(x-\dfrac{\pi}{3}\right)$

$= \left(\sin x \cos\left(\dfrac{\pi}{3}\right) + \cos x \sin\left(\dfrac{\pi}{3}\right)\right)$

$+ \left(\sin x \cos\left(\dfrac{\pi}{3}\right) - \cos x \sin\left(\dfrac{\pi}{3}\right)\right)$

$= \dfrac{1}{2}\sin x + \dfrac{1}{2}\sin x$

$= \sin x$

b $\sin\left(x+\dfrac{\pi}{4}\right) + \cos\left(x+\dfrac{\pi}{4}\right)$

$= \left(\sin x \cos\left(\dfrac{\pi}{4}\right) + \cos x \sin\left(\dfrac{\pi}{4}\right)\right)$

$+ \left(\cos x \cos\left(\dfrac{\pi}{4}\right) - \sin x \sin\left(\dfrac{\pi}{4}\right)\right)$

$= \dfrac{1}{\sqrt{2}}\sin x + \dfrac{1}{\sqrt{2}}\cos x$

$+ \dfrac{1}{\sqrt{2}}\cos x - \dfrac{1}{\sqrt{2}}\sin x$

$= \sqrt{2}\cos x$

5 a $\tan\left(\theta - \dfrac{\pi}{4}\right) = \dfrac{\tan\theta - \tan\left(\dfrac{\pi}{4}\right)}{1 + \tan\theta \tan\left(\dfrac{\pi}{4}\right)}$

$= \dfrac{\tan\theta - 1}{1 + \tan\theta}$

b $\tan\left(\theta - \dfrac{\pi}{4}\right) = 6\tan\theta$

$\dfrac{\tan\theta - 1}{1 + \tan\theta} = 6\tan\theta$

$6\tan^2\theta + 6\tan\theta = \tan\theta - 1$

$6\tan^2\theta + 5\tan\theta + 1 = 0$

$(2\tan\theta + 1)(3\tan\theta + 1) = 0$

$\tan\theta = -\dfrac{1}{2}$ or $-\dfrac{1}{3}$

c For $\theta \in \,]0, \pi[$,

$\tan\theta = -\dfrac{1}{2} \Rightarrow \theta = -0.464 + \pi = 2.68$

$\tan\theta = -\dfrac{1}{3} \Rightarrow \theta = -0.322 + \pi = 2.82$

6 a $\sin x \cos\left(\dfrac{\pi}{4}\right) + \cos x \sin\left(\dfrac{\pi}{4}\right)$

$= \sin\left(x + \dfrac{\pi}{4}\right)$

Maximum value is 1, and the smallest positive x value at which this occurs is $x = \dfrac{\pi}{4}$.

b $2\cos x \cos 25° + 2\sin x \sin 25° = 2\cos(x - 25°)$

Maximum value is 2, and the smallest positive x value at which this occurs is $x = 25°$.

7 a $\sin(x + 0) = \sin x$

By the compound angle formula, $\sin(x + 0) = \sin x \cos 0 + \cos x \sin 0$

Assuming it is known that $\sin 0 = 0$,

$\sin x = \sin x \cos 0 + 0$

$\sin x (\cos 0 - 1) = 0$ for all values of x

$\therefore \cos 0 = 1$

b $\sin 2\theta = \sin(\theta + \theta)$

$\therefore$ using the compound angle formula: $\sin 2\theta = \sin \theta \cos \theta + \cos \theta \sin \theta = 2 \sin \theta \cos \theta$

c $\sin\left(\dfrac{\pi}{2} - \theta\right) = \sin\left(\dfrac{\pi}{2}\right)\cos\theta - \cos\left(\dfrac{\pi}{2}\right)\sin\theta$

Since $\sin\left(\dfrac{\pi}{2}\right) = 1$ and $\cos\left(\dfrac{\pi}{2}\right) = 0$, $\sin\left(\dfrac{\pi}{2} - \theta\right) = \cos\theta$

8 a $\cos 3A = \cos(A + 2A)$

$= \cos A \cos 2A - \sin A \sin 2A$

$= \cos A (2\cos^2 A - 1) - \sin A (2 \sin A \cos A)$

$= 2\cos^3 A - \cos A - 2\cos A \sin^2 A$

$= 2\cos^3 A - \cos A - 2\cos A (1 - \cos^2 A)$

$= 4\cos^3 A - 3\cos A$

b $\sin 3A = \sin(A + 2A)$

$= \sin A \cos 2A + \cos A \sin 2A$

$= \sin A (1 - 2\sin^2 A) + \cos A (2 \sin A \cos A)$

$= \sin A - 2\sin^3 A + 2\sin A \cos^2 A$

$= \sin A - 2\sin^3 A + 2\sin A (1 - \sin^2 A)$

$= 3\sin A - 4\sin^3 A$

$= 3\sin A - 3\sin A (1 - \cos^2 A) - \sin^3 A$

$= 3\sin A \cos^2 A - \sin^3 A$

Similarly, from (a):

$\cos 3A = \cos^3 A + 3\cos A (1 - \sin^2 A) - 3\cos A$

$= \cos^3 A - 3\cos A \sin^2 A$

Therefore $\tan 3A = \dfrac{\sin 3A}{\cos 3A}$

$= \dfrac{3\sin A \cos^2 A - \sin^3 A}{\cos^3 A - 3\cos A \sin^2 A}$

Dividing through by $\cos^3 A$ and writing $\dfrac{\sin^n A}{\cos^n A} = \tan^n A$:

$$\tan 3A = \dfrac{3\tan A - \tan^3 A}{1 - 3\tan^2 A}$$

COMMENT

You can derive this formula from a direct approach through $\tan(3A) = \tan(A + 2A)$, but this potentially generates a large and unwieldy nested fraction; finding identities for $\sin 3A$ and $\cos 3A$ is more standard, and you will encounter this technique again in connection with De Moivre's Theorem in Chapter 15.

9 a $\sin(A+B) + \sin(A-B)$
$= (\sin A \cos B + \cos A \sin B)$
$\quad + (\sin A \cos B - \cos A \sin B)$
$= 2 \sin A \cos B$

b By (a), $\sin\left(x + \dfrac{\pi}{6}\right) + \sin\left(x - \dfrac{\pi}{6}\right)$
$= 2 \sin x \cos \dfrac{\pi}{6}$

$\therefore 2 \sin x \cos\left(\dfrac{\pi}{6}\right) = 3 \cos x$

$\Rightarrow \dfrac{2\sqrt{3}}{2} \sin x = 3 \cos x$

$\tan x = \sqrt{3}$

$\therefore x = \dfrac{\pi}{3}$ for $x \in [0, \pi]$

10 a $\tan(A-B) = \dfrac{\sin(A-B)}{\cos(A-B)}$

$= \dfrac{\sin A \cos B - \cos A \sin B}{\cos A \cos B + \sin A \sin B}$

Dividing through by $\cos A \cos B$ gives

$\tan(A-B) = \dfrac{\tan A - \tan B}{1 + \tan A \tan B}$

b Let ϕ_1 be the angle of $y = 4x$ and ϕ_2 the angle of $y = 2x$ with the positive x-axis.

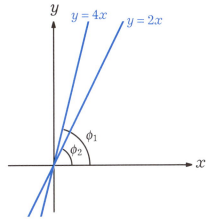

Figure 12B.10

Then $\theta = \phi_1 - \phi_2$
Since the gradient of a line is equal to the tangent of the angle under the line,
$\tan \phi_1 = 4, \quad \tan \phi_2 = 2$
$\therefore \tan \theta = \tan(\phi_1 - \phi_2)$
$= \dfrac{4 - 2}{1 + 4 \times 2} = \dfrac{2}{9}$

11 a $\cos(x+y) + \cos(x-y)$
$= (\cos x \cos y - \sin x \sin y)$
$\quad + (\cos x \cos y + \sin x \sin y)$
$= 2 \cos x \cos y$

b $\cos 3x + \cos x = \cos(2x + x) + \cos(2x - x)$
$\therefore \cos(2x + x) + \cos(2x - x) = 3 \cos 2x$
$\Rightarrow 2 \cos 2x \cos x = 3 \cos 2x$ by (a)
$\cos 2x (2 \cos x - 3) = 0$

$\Rightarrow \cos 2x = 0$ or $\cos x = \dfrac{3}{2}$ (no solutions)

For $x \in [0, 2\pi]$, i.e. $2x \in [0, 4\pi]$:
$\cos 2x = 0 \Rightarrow 2x = \dfrac{\pi}{2}, \dfrac{3\pi}{2}, \dfrac{5\pi}{2}, \dfrac{7\pi}{2}$

$\therefore x = \dfrac{\pi}{4}, \dfrac{3\pi}{4}, \dfrac{5\pi}{4}, \dfrac{7\pi}{4}$

12 a $\tan(\arctan 1.2 + \arctan 0.5) = \dfrac{1.2 + 0.5}{1 - 1.2 \times 0.5}$

$= \dfrac{1.7}{0.4}$

$= 4.25$

b $\tan\left(2\arctan\left(\dfrac{1}{3}\right)\right) = \dfrac{2 \times \dfrac{1}{3}}{1 - \left(\dfrac{1}{3}\right)^2}$

$= \dfrac{\dfrac{2}{3}}{\dfrac{8}{9}}$

$= \dfrac{3}{4} = 0.75$

13 a $\cos^2\left(\dfrac{A}{2}\right) = \dfrac{1}{2}\left(2\cos^2\left(\dfrac{A}{2}\right)\right)$

$= \dfrac{1}{2}\left(1 + \cos\left(2 \times \dfrac{A}{2}\right)\right)$

$= \dfrac{1}{2}(1 + \cos A)$

b Let $A = \arccos x$, so that $\cos A = x$; then

$\cos^2\left(\dfrac{\arccos x}{2}\right) = \dfrac{1}{2}(1 + x)$

$\cos\left(\dfrac{\arccos x}{2}\right) = \sqrt{\dfrac{1}{2}(1+x)}$

Note that arccos has range $[0, \pi]$, so $\dfrac{\arccos x}{2} \in \left[0, \dfrac{\pi}{2}\right]$; within this domain cosine is always non-negative, so there is no need to consider the negative square root in this case.

> **COMMENT**
>
> Always ensure that when you take a square root in a proof question, you either include both positive and negative roots or explicitly reject one option, giving your reason. Without this, your proof is incomplete.

Exercise 12C

5 a $5\sin x + 12\cos x = R\sin(x + \theta)$

$= R\cos\theta \sin x + R\sin\theta \cos x$

$R^2 = 5^2 + 12^2 = 13^2$

$\Rightarrow R = 13$

$\tan\theta = \dfrac{R\sin\theta}{R\cos\theta} = \dfrac{12}{5}$

$\Rightarrow \theta = \arctan\left(\dfrac{12}{5}\right) = 1.18$

$\therefore 5\sin x + 12\cos x = 13\sin(x + 1.18)$

b $f(x) \to 13f(x + 1.18)$

The transformations are a vertical stretch with scale factor 13 and a horizontal translation $\begin{pmatrix} -1.18 \\ 0 \end{pmatrix}$, in either order.

6 a $3\sin x - 7\cos x = R\sin(x - \theta)$

$= R\cos\theta\sin x - R\sin\theta\cos x$

$R^2 = 3^2 + 7^2 = 58$

$\Rightarrow R = \sqrt{58}$

$\tan\theta = \dfrac{R\sin\theta}{R\cos\theta} = \dfrac{7}{3}$

$\Rightarrow \theta = \arctan\left(\dfrac{7}{3}\right) = 1.17$

$\therefore 3\sin x - 7\cos x = \sqrt{58}\sin(x - 1.17)$

b Range of the function is $\left[-\sqrt{58}, \sqrt{58}\right]$

7 a $4\cos x - 5\sin x = R\cos(x + \alpha)$

$= R\cos\alpha\cos x - R\sin\alpha\sin x$

$R^2 = 4^2 + 5^2 = 41$

$\Rightarrow R = \sqrt{41}$

$\tan\alpha = \dfrac{R\sin\alpha}{R\cos\alpha} = \dfrac{5}{4}$

$\Rightarrow \alpha = \arctan\left(\dfrac{5}{4}\right) = 0.896$

$\therefore 4\cos x - 5\sin x = \sqrt{41}\cos(x + 0.896)$

b $4\cos x - 5\sin x = 0$ when
$\cos(x + 0.896) = 0$

The smallest positive solution occurs
where $x + 0.896 = \dfrac{\pi}{2}$

$\therefore x = 0.675$

8 a $\sqrt{3}\sin x + \cos x = R\cos(x - \theta)$
$= R\cos\theta\cos x + R\sin\theta\sin x$

$R^2 = \left(\sqrt{3}\right)^2 + 1^2 = 4 \Rightarrow R = 2$

$\tan\theta = \dfrac{R\sin\theta}{R\cos\theta} = \dfrac{\sqrt{3}}{1}$

$\Rightarrow \theta = \arctan\left(\sqrt{3}\right) = \dfrac{\pi}{3}$

$\therefore \sqrt{3}\sin x + \cos x = 2\cos\left(x - \dfrac{\pi}{3}\right)$

b For $x \in [0, 2\pi]$:

the maximum is at $\left(\dfrac{\pi}{3}, 2\right)$ and the

minimum is at $\left(\dfrac{4\pi}{3}, -2\right)$

9 $\sin 2x + \cos 2x = R\sin(2x + a)$
$= R\cos a \sin 2x + R\sin a \cos 2x$

$R^2 = 1^2 + 1^2 = 2 \Rightarrow R = \sqrt{2}$

$\tan a = \dfrac{R\sin a}{R\cos a} = \dfrac{1}{1}$

$\Rightarrow a = \arctan(1) = \dfrac{\pi}{4}$

$\therefore \sin 2x + \cos 2x = \sqrt{2}\sin\left(2x + \dfrac{\pi}{4}\right)$

$\sin 2x + \cos 2x = 1$

$\sqrt{2}\sin\left(2x + \dfrac{\pi}{4}\right) = 1$

$\sin\left(2x + \dfrac{\pi}{4}\right) = \dfrac{1}{\sqrt{2}}$

$x \in [-\pi, \pi] \Rightarrow A = 2x + \dfrac{\pi}{4} \in \left[-\dfrac{7\pi}{4}, \dfrac{9\pi}{4}\right]$

Primary solution for $\sin A = \dfrac{1}{\sqrt{2}}$ is $A_1 = \dfrac{\pi}{4}$

Secondary solution is $A_2 = \pi - A_1 = \dfrac{3\pi}{4}$

Periodic solutions $A_1 + 2n\pi$ and
$A_2 + 2n\pi$ within the specified interval:
$A = -\dfrac{7\pi}{4}, -\dfrac{5\pi}{4}, \dfrac{9\pi}{4}$

$\therefore 2x + \dfrac{\pi}{4} = -\dfrac{7\pi}{4}, -\dfrac{5\pi}{4}, \dfrac{\pi}{4}, \dfrac{3\pi}{4}, \dfrac{9\pi}{4}$

$\Rightarrow x = -\pi, -\dfrac{3\pi}{4}, 0, \dfrac{\pi}{4}, \pi$

Exercise 12D

7 $\sin^2\theta + \cot^2\theta \sin^2\theta = \sin^2\theta\left(1 + \cot^2\theta\right)$

$= \sin^2\theta\left(1 + \dfrac{\cos^2\theta}{\sin^2\theta}\right)$

$= \dfrac{\sin^2\theta\left(\sin^2\theta + \cos^2\theta\right)}{\sin^2\theta}$

$= \sin^2\theta + \cos^2\theta$

$= 1$

8 $\tan x + \sec x = 4, \quad x \in [0, 2\pi]$
From GDC, $x = 1.08$
Alternative algebraic solution, which
would be needed in a non-calculator
question specifying an exact solution:

$\tan x + \sec x = 4$

$\dfrac{\sin x + 1}{\cos x} = 4$

$\sin x + 1 = 4\cos x$

$4\cos x - \sin x = 1$

Let $4\cos x - \sin x = R\cos(x + c)$
$= R\cos c \cos x - R\sin c \sin x$. Then

$R^2 = 4^2 + 1^2 = 17 \Rightarrow R = \sqrt{17}$

$\tan c = \dfrac{R\sin c}{R\cos c} = \dfrac{1}{4} \Rightarrow c = \arctan\left(\dfrac{1}{4}\right)$

$\therefore \sqrt{17}\cos\left(x + \arctan\left(\dfrac{1}{4}\right)\right) = 1 \quad \ldots(*)$

$\Rightarrow x = \arccos\left(\dfrac{1}{\sqrt{17}}\right) - \arctan\left(\dfrac{1}{4}\right)$

$= 1.08 \text{ (3SF)}$

The secondary solution

$x = 2\pi - \arccos\left(\dfrac{1}{\sqrt{17}}\right) - \arctan\left(\dfrac{1}{4}\right)$ is not a valid solution to the original problem; from a right triangle with sides 1, 4 and $\sqrt{17}$ it is clear that $\arccos\left(\dfrac{1}{\sqrt{17}}\right) + \arctan\left(\dfrac{1}{4}\right) = \dfrac{\pi}{2}$, so this would be a false solution $x = \dfrac{3\pi}{2}$, for which both $\sec x$ and $\tan x$ are undefined.

> **COMMENT**
>
> In the absence of other instructions, it is clearly more efficient to use the GDC than to write an algebraic solution at great length.
>
> Also, note that there is only a single solution (clear if you plot the function), but the algebraic working suggests there should be two, with the other arising from the secondary solution to equation (*). However, as noted, this is a false solution, which arises from $\cos x = 0$ since the rearrangement involved multiplying through by $\cos x$, and this would need to be explicitly stated in an answer.

9 a $f(x) = \tan x + \csc x$, $x \in \left[-\dfrac{\pi}{2}, \dfrac{\pi}{2}\right]$

From GDC: local maximum is at $(-0.715, -2.39)$, local minimum is at $(0.715, 2.39)$

b Range of $f(x)$ is $]-\infty, -2.39] \cup [2.39, \infty[$

10 $\tan x + \cot x = \dfrac{\sin x}{\cos x} + \dfrac{\cos x}{\sin x}$

$= \dfrac{\sin^2 x + \cos^2 x}{\sin x \cos x}$

$= \dfrac{1}{\sin x \cos x}$

$= \dfrac{1}{\sin x} \times \dfrac{1}{\cos x}$

$= \csc x \sec x$

11 $\dfrac{\sin\theta}{1-\cos\theta} - \dfrac{\sin\theta}{1+\cos\theta} = \dfrac{\sin\theta\left((1+\cos\theta)-(1-\cos\theta)\right)}{(1-\cos\theta)(1+\cos\theta)}$

$= \dfrac{2\sin\theta\cos\theta}{1-\cos^2\theta}$

$= \dfrac{2\sin\theta\cos\theta}{\sin^2\theta}$

$= \dfrac{2\cos\theta}{\sin\theta}$

$= 2\cot\theta$

12 a $\sec^2 x - 3\tan x + 1 = 0$

But $\sec^2 x = 1 + \tan^2 x$

$\therefore 1 + \tan^2 x - 3\tan x + 1 = 0$

$\tan^2 x - 3\tan x + 2 = 0$

b $(\tan x - 1)(\tan x - 2) = 0$

$\Rightarrow \tan x = 1$ or 2

c For $x \in [0, 2\pi]$:

$\tan x = 1 \Rightarrow x = \dfrac{\pi}{4}, \dfrac{5\pi}{4}$

$\tan x = 2 \Rightarrow x = 1.11, 4.25$

$\therefore$ solutions are $x = \dfrac{\pi}{4}, 1.11, \dfrac{5\pi}{4}, 4.25$

13 $\csc 2x = \dfrac{1}{\sin 2x}$

$= \dfrac{1}{2\sin x \cos x}$

$= \dfrac{1}{2} \times \dfrac{1}{\sin x} \times \dfrac{1}{\cos x}$

$= \dfrac{1}{2}\csc x \sec x$

14 $\cot 2x = \dfrac{\cos 2x}{\sin 2x}$

$= \dfrac{\cos^2 x - \sin^2 x}{2\sin x \cos x}$

Dividing through by $\sin^2 x$ gives

$\cot 2x = \dfrac{\cot^2 x - 1}{2\cot x}$

15 Let $y = f(x) = \sec x$

Then $\dfrac{1}{y} = \cos x$

$\therefore x = \arccos\left(\dfrac{1}{y}\right)$

and so $f^{-1}(x) = \arccos\left(\dfrac{1}{x}\right)$

That is, the inverse function of $\sec x$ is the arccosine of the reciprocal.

Mixed examination practice 12
Short questions

1 $2\tan^2\theta - 5\sec\theta - 10 = 0$

$2(\sec^2\theta - 1) - 5\sec\theta - 10 = 0$

$2\sec^2\theta - 5\sec\theta - 12 = 0$

$(2\sec\theta + 3)(\sec\theta - 4) = 0$

$\sec\theta = -\dfrac{3}{2}$ or 4

For θ in the second quadrant, range of $\sec\theta$ is $]-\infty, -1]$, so only the first solution is valid.

$\therefore \sec\theta = -\dfrac{3}{2}$

2 a $\cos\left(x + \dfrac{\pi}{3}\right) = \cos\dfrac{\pi}{3}\cos x - \sin\dfrac{\pi}{3}\sin x$

$= \dfrac{1}{2}\cos x - \dfrac{\sqrt{3}}{2}\sin x$

b Similar to (a),

$\cos\left(x - \dfrac{\pi}{3}\right) = \dfrac{1}{2}\cos x + \dfrac{\sqrt{3}}{2}\sin x$

$\cos\left(x + \dfrac{\pi}{3}\right) = \cos\left(x - \dfrac{\pi}{3}\right)$

$\Rightarrow \dfrac{1}{2}\cos x - \dfrac{\sqrt{3}}{2}\sin x = \dfrac{1}{2}\cos x + \dfrac{\sqrt{3}}{2}\sin x$

$\sqrt{3}\sin x = 0$

$\therefore$ for $x \in [-2\pi, 2\pi]$,

$x = -2\pi, -\pi, 0, \pi, 2\pi$

3 a $\cos(\theta + \theta) = \cos\theta\cos\theta - \sin\theta\sin\theta$

$= \cos^2\theta - \sin^2\theta$

$= \cos^2\theta - (1 - \cos^2\theta)$

$= 2\cos^2\theta - 1$

b Taking $x = \dfrac{\theta}{2}$ in $\sin 2x = 2\sin x \cos x$:

$\sin\theta = 2\sin\dfrac{\theta}{2}\cos\dfrac{\theta}{2}$

Taking $x = \dfrac{\theta}{2}$ in $\cos 2x + 1 = 2\cos^2 x$:

$\cos\theta + 1 = 2\cos^2\left(\dfrac{\theta}{2}\right)$

$\therefore \dfrac{\sin\theta}{1 + \cos\theta} = \dfrac{2\sin\dfrac{\theta}{2}\cos\dfrac{\theta}{2}}{2\cos^2\dfrac{\theta}{2}} = \tan\dfrac{\theta}{2}$

Solving $\tan\dfrac{\theta}{2} = 3\cot\dfrac{\theta}{2}$:

$\tan^2\dfrac{\theta}{2} = 3$

$\Rightarrow \tan\dfrac{\theta}{2} = \pm\sqrt{3}$

For $\theta \in \,]0, 2\pi[$, $\dfrac{\theta}{2} \in \,]0, \pi[$

$\therefore \dfrac{\theta}{2} = \dfrac{\pi}{3}, \dfrac{2\pi}{3}$

and hence $\theta = \dfrac{2\pi}{3}, \dfrac{4\pi}{3}$

4 $\tan\theta + \cot\theta = 3$

Multiplying through by $\tan\theta$:

$\tan^2\theta - 3\tan\theta + 1 = 0$

$\Rightarrow \tan\theta = \dfrac{3 \pm \sqrt{3^2 - 4}}{2}$

$= \dfrac{3 \pm \sqrt{5}}{2}$

$\theta \in [0°, 90°] \Rightarrow \theta = 20.9°, 69.1°$

12 Further trigonometry 131

> **COMMENT**
> Notice that the two answers must add up to 90° because the original equation is symmetrical in $\tan\theta$ and $\cot\theta$.

5 a $\sqrt{15}\sin 2x + \sqrt{5}\cos 2x = R\sin(2x+\alpha)$
$= R\cos\alpha \sin 2x$
$+ R\sin\alpha \cos 2x$

$R^2 = 15+5 = 20 \Rightarrow R = \sqrt{20} = 2\sqrt{5}$

$\tan\alpha = \dfrac{R\sin\alpha}{R\cos\alpha} = \dfrac{\sqrt{5}}{\sqrt{15}} = \dfrac{1}{\sqrt{3}}$

$\Rightarrow \alpha = \arctan\left(\dfrac{1}{\sqrt{3}}\right) = \dfrac{\pi}{6}$

$\therefore \sqrt{15}\sin 2x + \sqrt{5}\cos 2x$
$= 2\sqrt{5}\sin\left(2x + \dfrac{\pi}{6}\right)$

b $f(x) = \dfrac{2}{5 + 2\sqrt{5}\sin\left(2x + \dfrac{\pi}{6}\right)}$

i Value of f is maximum when denominator is as small as possible:

$\dfrac{2}{5 - 2\sqrt{5}} = \dfrac{2(5+2\sqrt{5})}{5^2 - 20}$
$= \dfrac{2}{5}(5 + 2\sqrt{5})$
$= 2 + \dfrac{4}{5}\sqrt{5}$

$\therefore p = 2,\ q = \dfrac{4}{5}$

ii The maximum occurs when $\sin\left(2x + \dfrac{\pi}{6}\right) = -1$:

$2x + \dfrac{\pi}{6} = \dfrac{3\pi}{2}$

$2x = \dfrac{3\pi}{2} - \dfrac{\pi}{6} = \dfrac{4\pi}{3}$

$\therefore x = \dfrac{2\pi}{3}$

6 a $\sin(\arcsin x) = \begin{cases} x & |x| \le 1 \\ \text{undefined} & |x| > 1 \end{cases}$

b Let $x = \cos y$; then
$y = \arccos x$ and $\sin^2 y = 1 - x^2$
$\therefore \sin y = \sqrt{1 - x^2}$
i.e. $\sin(\arccos x) = \sqrt{1 - x^2}$
(Again, this is only valid over the natural domain of $\arccos x$: $[-1, 1]$.)

c $\arcsin x = \arccos x$
$\sin(\arcsin x) = \sin(\arccos x)$
$x = \sqrt{1 - x^2}$
$x^2 = 1 - x^2$
$x^2 = \dfrac{1}{2}$
$x = \dfrac{1}{\sqrt{2}}$

Long questions

1 a We know that $\hat{ABC} = 90°$ because $AC = 2r$ is the circle diameter and point B lies on the circumference.
$\therefore AB = 2r\sin\theta,\quad BC = 2r\cos\theta$

b Area of ABC $= \dfrac{1}{2}(AB)(BC) = 2r^2 \sin\theta\cos\theta$

Using the double angle formula:
Area of ABC $= r^2 \sin 2\theta$

c Triangle OBC is isosceles, so $\hat{BOC} = \pi - 2\theta$

Area of OBC $= \dfrac{1}{2}(OB)(OC)\sin\hat{BOC}$
$= \dfrac{1}{2}r^2 \sin(\pi - 2\theta)$

Since $\sin x = \sin(\pi - x)$,
Area of OBC $= \dfrac{1}{2}r^2 \sin(2\theta)$

d $\dfrac{\text{Area of OBC}}{\text{Area of ABC}} = \dfrac{\frac{1}{2}r^2 \sin(2\theta)}{r^2 \sin(2\theta)} = \dfrac{1}{2}$

$\therefore k = \dfrac{1}{2}$

2 a $\tan(A+B) = \dfrac{\tan A + \tan B}{1 - \tan A \tan B}$

$\Rightarrow \tan(A+A) = \dfrac{2\tan A}{1 - \tan^2 A}$

b $\tan 135° = \tan(-45°) = -1$

c Let $\tan 67.5° = t$; then

$\tan 135° = \dfrac{2t}{1-t^2}$ from (a)

$-1 = \dfrac{2t}{1-t^2}$ from (b)

$\therefore t^2 - 2t - 1 = 0$

$t = \dfrac{2 \pm \sqrt{2^2 + 4}}{2} = 1 \pm \sqrt{2}$

Since $67.5°$ lies in the first quadrant, $t > 0$

$\therefore t = 1 + \sqrt{2}$

i.e. $\tan 67.5° = 1 + \sqrt{2}$

3 a $y_1 = a\cos px$ has amplitude a and period $\dfrac{2\pi}{p}$

From the graph:

$y_1(0) = 1.2 \Rightarrow a = 1.2$

Half a period is 1.5, so

$\dfrac{\pi}{p} = 1.5 \Rightarrow p = \dfrac{2\pi}{3}$

b $y_2 = 0.9 \sin\left(\dfrac{2\pi}{3}x\right)$

Amplitude is 0.9; period is 3

c $y = 1.2\cos\left(\dfrac{2\pi}{3}x\right) + 0.9\sin\left(\dfrac{2\pi}{3}x\right)$

$= R\sin\left(\dfrac{2\pi}{3}x + \alpha\right)$

$= R\cos\alpha \sin\left(\dfrac{2\pi}{3}x\right) + R\sin\alpha \cos\left(\dfrac{2\pi}{3}x\right)$

$R^2 = 0.9^2 + 1.2^2 = 1.5^2 \Rightarrow R = 1.5$

$\tan\alpha = \dfrac{R\sin\alpha}{R\cos\alpha} = \dfrac{1.2}{0.9} = \dfrac{4}{3}$

$\Rightarrow \alpha = \arctan\left(\dfrac{4}{3}\right) = 0.927$ (3SF)

$\therefore y = 1.5\sin\left(\dfrac{2\pi}{3}x + 0.927\right)$

d Amplitude is 1.5; period is still 3.

e $y = 0 \Rightarrow \dfrac{2\pi}{3}x + 0.927 = \pi$

$\therefore x = \dfrac{3}{2\pi}(\pi - 0.927) = 1.06$ (3SF)

f $y = 1.3 \Rightarrow \sin\left(\dfrac{2\pi}{3}x + 0.927\right) = \dfrac{1.3}{1.5}$

Let $A = \dfrac{2\pi}{3}x + 0.927$

$x > 0 \Rightarrow A > 0.927$

Solving $\sin A = \dfrac{1.3}{1.5}$:

Primary solution

$A_1 = \arcsin\left(\dfrac{1.3}{1.5}\right) = 1.05$

Secondary solution $A_2 = \pi - A_1 = 2.09$

Hence the first two positive solutions are

$x = \dfrac{3}{2\pi}(A - 0.927) = 0.0580,\ 0.557$ (3SF)

4 a $\sqrt{3}\cos\theta - \sin\theta = r\cos(\theta + \alpha)$

$= r\cos\alpha\cos\theta - r\sin\alpha\sin\theta$

$r^2 = 3 + 1 = 4 \Rightarrow r = 2$

$\tan\alpha = \dfrac{r\sin\alpha}{r\cos\alpha} = \dfrac{1}{\sqrt{3}}$

$\Rightarrow \alpha = \arctan\left(\dfrac{1}{\sqrt{3}}\right) = \dfrac{\pi}{6}$

$\therefore \sqrt{3}\cos\theta - \sin\theta = 2\cos\left(\theta + \dfrac{\pi}{6}\right)$

b Over a complete period, the function $2\cos\left(\theta + \dfrac{\pi}{6}\right)$ has range $[-2, 2]$.

12 Further trigonometry 133

c $\sqrt{3}\cos\theta - \sin\theta = -1$

$2\cos\left(\theta + \dfrac{\pi}{6}\right) = -1$

$\cos\left(\theta + \dfrac{\pi}{6}\right) = -\dfrac{1}{2}$

Let $A = \theta + \dfrac{\pi}{6}$

$\theta \in [0, 2\pi] \Rightarrow A \in \left[\dfrac{\pi}{6}, \dfrac{13\pi}{6}\right]$

For $\cos A = -\dfrac{1}{2}$:

primary solution is $A_1 = \dfrac{2\pi}{3}$

secondary solution is $A_2 = 2\pi - A_1 = \dfrac{4\pi}{3}$

Hence $\theta = A - \dfrac{\pi}{6} = \dfrac{\pi}{2}, \dfrac{7\pi}{6}$

5 a $f(t) = t^3 - 3t^2 - 3t + 1$

By inspection, $f(-1) = 0$, so $(t+1)$ is a factor of $f(t)$ by the factor theorem.

> **COMMENT**
>
> Alternatively, plot the function on the GDC and try to find a recognisable rational root.

Hence $f(t) = (t+1)(t^2 + at + b)$
Expanding and comparing coefficients:

t^3: $1 = 1$

t^2: $a + 1 = -3 \Rightarrow a = -4$

t^1: $b + a = -3 \Rightarrow b = 1$

t^0: $b = 1$ is consistent with the value found above

$\therefore f(t) = (t+1)(t^2 - 4t + 1)$

b $\tan(3A) = \tan(A + 2A)$

$= \dfrac{\tan A + \tan 2A}{1 - \tan A \tan 2A}$

$= \dfrac{\tan A + \dfrac{2\tan A}{1 - \tan^2 A}}{1 - \tan A \left(\dfrac{2\tan A}{1 - \tan^2 A}\right)}$

Multiplying numerator and denominator by $1 - \tan^2 A$:

$\tan(3A) = \dfrac{\tan A(1 - \tan^2 A) + 2\tan A}{1 - \tan^2 A - 2\tan^2 A}$

$= \dfrac{3\tan A - \tan^3 A}{1 - 3\tan^2 A}$

c $\tan 45° = 1$

d Because $\tan(x + 180°) = \tan x$,

$\tan 405° = \tan 225° = \tan 45° = 1$

Let $t = \tan 15°$ or $\tan 75°$ or $\tan 135°$; in any of these cases, the following is true, using the formula in (b):

$1 = \dfrac{3t - t^3}{1 - 3t^2}$

$t^3 - 3t^2 - 3t + 1 = 0$

$(t+1)(t^2 - 4t + 1) = 0$ by (a)

$t = -1$ or $\dfrac{4 \pm \sqrt{4^2 - 4}}{2}$

$\therefore t = -1$ or $2 \pm \sqrt{3}$

Each of these three values corresponds to one of $\tan 15°$, $\tan 75°$ or $\tan 135°$.

$0 < \tan 15° < 1 \Rightarrow \tan 15° = 2 - \sqrt{3}$

$\tan 75° > 1 \Rightarrow \tan 75° = 2 + \sqrt{3}$

13 Vectors

Exercise 13A

4 a $\overrightarrow{AB} = \boldsymbol{b} - \boldsymbol{a}$

Figure 13A.4

$\overrightarrow{OC} = \overrightarrow{OA} + \dfrac{1}{2}\overrightarrow{AB}$

$= \boldsymbol{a} + \dfrac{1}{2}(\boldsymbol{b}-\boldsymbol{a})$

$= \dfrac{1}{2}(\boldsymbol{a}+\boldsymbol{b})$

c $\overrightarrow{AD} = -3\overrightarrow{AB}$

$\Rightarrow \overrightarrow{OD} = \overrightarrow{OA} + \overrightarrow{AD}$

$= \overrightarrow{OA} - 3\overrightarrow{AB}$

$= \boldsymbol{a} - 3(\boldsymbol{b}-\boldsymbol{a})$

$= 4\boldsymbol{a} - 3\boldsymbol{b}$

5 a $\overrightarrow{AB} = \begin{pmatrix}4\\2\end{pmatrix} - \begin{pmatrix}3\\0\end{pmatrix} = \begin{pmatrix}1\\2\end{pmatrix}$

$\overrightarrow{AC} = \dfrac{1}{2}\overrightarrow{AB} = \begin{pmatrix}0.5\\1\end{pmatrix}$

b $\overrightarrow{OD} = \overrightarrow{OA} + \overrightarrow{AD}$

$= \begin{pmatrix}3\\0\end{pmatrix} + \begin{pmatrix}7\\-2\end{pmatrix} = \begin{pmatrix}10\\-2\end{pmatrix}$

∴ the coordinates of D are (10, −2)

6 a $\overrightarrow{AB} = \overrightarrow{OB} - \overrightarrow{OA}$

$= \begin{pmatrix}4\\-2\\5\end{pmatrix} - \begin{pmatrix}3\\1\\-2\end{pmatrix}$

$= \begin{pmatrix}1\\-3\\7\end{pmatrix}$

b The position vector of the midpoint is the mean of the position vectors of the end points:

$\dfrac{1}{2}\left(\begin{pmatrix}3\\1\\-2\end{pmatrix} + \begin{pmatrix}4\\-2\\5\end{pmatrix}\right) = \begin{pmatrix}3.5\\-0.5\\1.5\end{pmatrix}$

7 $\overrightarrow{OD} = \overrightarrow{OA} + \overrightarrow{AD}$

$= (2\boldsymbol{i} - 3\boldsymbol{j}) + (\boldsymbol{i} - \boldsymbol{j})$

$= 3\boldsymbol{i} - 4\boldsymbol{j}$

8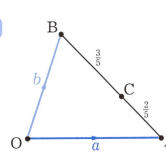

Figure 13A.8

$\overrightarrow{AC} = \dfrac{2}{5}\overrightarrow{AB}$

$\overrightarrow{AB} = \boldsymbol{b} - \boldsymbol{a} = \begin{pmatrix}1\\-1\\3\end{pmatrix} - \begin{pmatrix}2\\2\\1\end{pmatrix} = \begin{pmatrix}-1\\-3\\2\end{pmatrix}$

$\therefore \overrightarrow{OC} = \overrightarrow{OA} + \overrightarrow{AC}$

$= \begin{pmatrix} 2 \\ 2 \\ 1 \end{pmatrix} + \frac{2}{5}\begin{pmatrix} -1 \\ -3 \\ 2 \end{pmatrix}$

$= \begin{pmatrix} 1.6 \\ 0.8 \\ 1.8 \end{pmatrix}$

9 a M has position vector

$\frac{1}{2}(\mathbf{p}+\mathbf{q}) = \frac{1}{2}(2\mathbf{i}-\mathbf{j}-3\mathbf{k}+\mathbf{i}+4\mathbf{j}-\mathbf{k})$

$= \frac{3}{2}\mathbf{i} + \frac{3}{2}\mathbf{j} - 2\mathbf{k}$

b

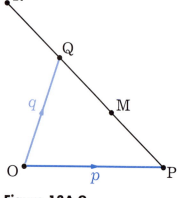

Figure 13A.9

$\overrightarrow{QR} = -\overrightarrow{QM}$

$\therefore \overrightarrow{OR} = \overrightarrow{OQ} + \overrightarrow{QR}$

$= \overrightarrow{OQ} - \overrightarrow{QM}$

$= \mathbf{q} - \frac{1}{2}(\mathbf{p}-\mathbf{q})$

$= \frac{1}{2}(3\mathbf{q}-\mathbf{p})$

Hence

$\overrightarrow{OR} = \frac{1}{2}(3(\mathbf{i}+4\mathbf{j}-\mathbf{k})-(2\mathbf{i}-\mathbf{j}-3\mathbf{k}))$

$= \frac{1}{2}\mathbf{i} + \frac{13}{2}\mathbf{j}$

$\therefore$ the coordinates of R are $\left(\frac{1}{2}, \frac{13}{2}, 0\right)$

10

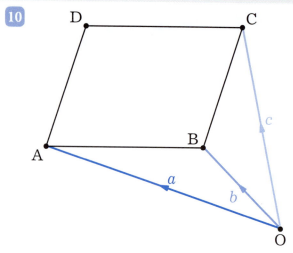

Figure 13A.10

$\overrightarrow{BA} = \mathbf{a} - \mathbf{b} = \begin{pmatrix} -3 \\ -2 \\ 2 \end{pmatrix}$

To form a parallelogram, $\overrightarrow{CD} = \overrightarrow{BA}$

$\therefore \overrightarrow{OD} = \overrightarrow{OC} + \overrightarrow{CD}$

$= \begin{pmatrix} 3 \\ 1 \\ 4 \end{pmatrix} + \begin{pmatrix} -3 \\ -2 \\ 2 \end{pmatrix}$

$= \begin{pmatrix} 0 \\ -1 \\ 6 \end{pmatrix}$

Exercise 13B

4 $3\mathbf{a} + 4\mathbf{x} = \mathbf{b}$

$\Rightarrow \mathbf{x} = \frac{1}{4}(\mathbf{b}-3\mathbf{a})$

$= \frac{1}{4}\left(\begin{pmatrix} 5 \\ 3 \\ 3 \end{pmatrix} - \begin{pmatrix} -3 \\ 3 \\ 6 \end{pmatrix}\right)$

$= \begin{pmatrix} 2 \\ 0 \\ -0.75 \end{pmatrix}$

5 $a + tb = c$
$tb = c - a$

$t\begin{pmatrix} 1 \\ -1 \\ 2 \end{pmatrix} = \begin{pmatrix} 1 \\ 0 \\ 1 \end{pmatrix} - \begin{pmatrix} 3 \\ -2 \\ 5 \end{pmatrix}$

$t\begin{pmatrix} 1 \\ -1 \\ 2 \end{pmatrix} = \begin{pmatrix} -2 \\ 2 \\ -4 \end{pmatrix}$

$\Rightarrow t = -2$

6 Require that $a + pb = k\begin{pmatrix} 3 \\ 2 \\ 3 \end{pmatrix}$ for some k

i.e. $\begin{pmatrix} 2+3p \\ p \\ 2+3p \end{pmatrix} = \begin{pmatrix} 3k \\ 2k \\ 3k \end{pmatrix}$

$\therefore p = 2k$ from the second component.

Substituting into $2 + 3p = 3k$:

$2 + 6k = 3k$

$\Rightarrow k = -\dfrac{2}{3}$

$\therefore p = -\dfrac{4}{3}$

7 Require that $\lambda x + y = \begin{pmatrix} 0 \\ k \\ 0 \end{pmatrix}$ for some k

i.e. $\begin{pmatrix} 2\lambda+4 \\ 3\lambda+1 \\ \lambda+2 \end{pmatrix} = \begin{pmatrix} 0 \\ k \\ 0 \end{pmatrix}$

$2\lambda + 4 = 0$ from the first component

$\therefore \lambda = -2$

8 Require that $pa + b = k\begin{pmatrix} 1 \\ 1 \\ 2 \end{pmatrix}$ for some k

i.e. $\begin{pmatrix} p+2q \\ -p+1 \\ 3p+q \end{pmatrix} = \begin{pmatrix} k \\ k \\ 2k \end{pmatrix}$

$\therefore p + 2q = k$ …(1)
$-p + 1 = k$ …(2)
$3p + q = 2k$ …(3)

(1) − (2):
$2p + 2q - 1 = 0$ …(4)

(3) − 2 × (2):
$5p + q - 2 = 0$ …(5)

Then 2 × (5) − (4):
$8p - 3 = 0$

$\Rightarrow p = \dfrac{3}{8}$

and hence, substituting into (4):

$q = \dfrac{1 - 2p}{2} = \dfrac{1}{8}$

Exercise 13C

7 $\left|\begin{pmatrix} 2c \\ c \\ -c \end{pmatrix}\right| = 12$

$\sqrt{4c^2 + c^2 + c^2} = 12$

$\sqrt{6c^2} = 12$

$6c^2 = 144$

$c^2 = 24$

$c = \pm 2\sqrt{6}$

8 $\overrightarrow{AB} = b - a = \begin{pmatrix} -2 \\ -2 \\ 1 \end{pmatrix}$

$AB = \sqrt{2^2 + 2^2 + 1^2} = 3$

$\therefore AC = \dfrac{3}{2}$

9

$$a + \lambda b = \begin{pmatrix} 2\lambda - 2 \\ -\lambda \\ 2\lambda - 1 \end{pmatrix}$$

$|a + \lambda b| = 5\sqrt{2}$

$\sqrt{(2\lambda - 2)^2 + \lambda^2 + (2\lambda - 1)^2} = 5\sqrt{2}$

$9\lambda^2 - 12\lambda + 5 = 50$

$3\lambda^2 - 4\lambda - 15 = 0$

$(3\lambda + 5)(\lambda - 3) = 0$

$\lambda = -\dfrac{5}{3}$ or 3

10 a Require $\left| k \begin{pmatrix} 4 \\ -1 \\ 1 \end{pmatrix} \right| = 6$ for some k

$\therefore \sqrt{(4^2 + 1^2 + 1^2)k^2} = 6$

$18k^2 = 36$

$k = \pm\sqrt{2}$

Possible vectors are $\pm\sqrt{2} \begin{pmatrix} 4 \\ -1 \\ 1 \end{pmatrix}$

b Require $\left| k \begin{pmatrix} 2 \\ -1 \\ 1 \end{pmatrix} \right| = 3$ for some $k > 0$

(same direction)

$\therefore \sqrt{(2^2 + 1^2 + 1^2)k^2} = 3$

$6k^2 = 9$

$k = \sqrt{\dfrac{3}{2}} = \dfrac{\sqrt{6}}{2}$ (choose positive root)

$\therefore$ the vector is $\dfrac{\sqrt{6}}{2} \begin{pmatrix} 2 \\ -1 \\ 1 \end{pmatrix}$

11 $\overrightarrow{AB} = \overrightarrow{OB} - \overrightarrow{OA}$

$= \begin{pmatrix} 2 + 2t \\ 4 + t \\ -9 - 5t \end{pmatrix}$

Require that $AB = 3$

$\therefore (2 + 2t)^2 + (4 + t)^2 + (9 + 5t)^2 = 9$

$30t^2 + 106t + 92 = 0$

$15t^2 + 53t + 46 = 0$

$(t + 2)(15t + 23) = 0$

$t = -2$ or $-\dfrac{23}{15} = -1.53 \,(3\text{SF})$

12 $\overrightarrow{PQ} = q - p$

$= \begin{pmatrix} 2 + t \\ 1 - t \\ 1 + t \end{pmatrix} - \begin{pmatrix} 1 \\ 1 \\ 3 \end{pmatrix} = \begin{pmatrix} t + 1 \\ -t \\ t - 2 \end{pmatrix}$

$(PQ)^2 = (t + 1)^2 + t^2 + (t - 2)^2$

$= 3t^2 - 2t + 5$

$= 3\left(t - \dfrac{1}{3}\right)^2 + \dfrac{14}{3}$

$\Rightarrow$ minimum $(PQ)^2$ is $\dfrac{14}{3}$, at $t = \dfrac{1}{3}$

$\therefore$ minimum PQ is $\sqrt{\dfrac{14}{3}}$, at $t = \dfrac{1}{3}$

Exercise 13D

5 $\overrightarrow{AB} = \begin{pmatrix} -3 \\ 5 \\ -1 \end{pmatrix}$, $\overrightarrow{OA} = \begin{pmatrix} 2 \\ 2 \\ 3 \end{pmatrix}$

$\theta = \arccos\left(\dfrac{\overrightarrow{AB} \cdot \overrightarrow{OA}}{|\overrightarrow{AB}||\overrightarrow{OA}|}\right)$

$= \arccos\dfrac{-6 + 10 - 3}{\sqrt{9 + 25 + 1}\sqrt{4 + 4 + 9}}$

$= \arccos\left(\dfrac{1}{\sqrt{35}\sqrt{17}}\right)$

$= 87.7° \,(3\text{SF})$

So the angle between $\overrightarrow{AB}$ and $\overrightarrow{OA}$ is 87.7° or 180° − 87.7° = 92.3°.

6 $\overrightarrow{AC} = \begin{pmatrix} 4 \\ 0 \\ -1 \end{pmatrix}$, $\overrightarrow{BD} = \begin{pmatrix} 6 \\ -4 \\ 1 \end{pmatrix}$

$\theta = \arccos\left(\dfrac{\overrightarrow{AC} \cdot \overrightarrow{BD}}{|\overrightarrow{AC}||\overrightarrow{BD}|}\right)$

$= \arccos\left(\dfrac{24+0-1}{\sqrt{16+0+1}\sqrt{36+16+1}}\right)$

$= \arccos\left(\dfrac{23}{\sqrt{17}\sqrt{53}}\right)$

$= 40.0°$ (3 SF)

7 a $\overrightarrow{AB} = \begin{pmatrix} k \\ 4 \\ 2k \end{pmatrix} - \begin{pmatrix} 2 \\ 4 \\ 1 \end{pmatrix} = \begin{pmatrix} k-2 \\ 0 \\ 2k-1 \end{pmatrix}$

$\overrightarrow{BC} = \begin{pmatrix} k+4 \\ 2k+4 \\ 2k+2 \end{pmatrix} - \begin{pmatrix} k \\ 4 \\ 2k \end{pmatrix} = \begin{pmatrix} 4 \\ 2k \\ 2 \end{pmatrix}$

$\overrightarrow{DC} = \begin{pmatrix} k+4 \\ 2k+4 \\ 2k+2 \end{pmatrix} - \begin{pmatrix} 6 \\ 2k+4 \\ 3 \end{pmatrix} = \begin{pmatrix} k-2 \\ 0 \\ 2k-1 \end{pmatrix}$

$\overrightarrow{AD} = \begin{pmatrix} 6 \\ 2k+4 \\ 3 \end{pmatrix} - \begin{pmatrix} 2 \\ 4 \\ 1 \end{pmatrix} = \begin{pmatrix} 4 \\ 2k \\ 2 \end{pmatrix}$

$\overrightarrow{AB} = \overrightarrow{DC}$ and $\overrightarrow{BC} = \overrightarrow{AD}$

∴ ABCD is a parallelogram.

b When $k = 1$,

$\overrightarrow{AB} = \begin{pmatrix} -1 \\ 0 \\ 1 \end{pmatrix}$, $\overrightarrow{AD} = \begin{pmatrix} 4 \\ 2 \\ 2 \end{pmatrix}$

$\theta = \arccos\left(\dfrac{\overrightarrow{AB} \cdot \overrightarrow{AD}}{|\overrightarrow{AB}||\overrightarrow{AD}|}\right)$

$= \arccos\left(\dfrac{-4+0+2}{\sqrt{1+0+1}\sqrt{16+4+4}}\right)$

$= \arccos\left(\dfrac{-2}{\sqrt{2}\sqrt{24}}\right)$

$= 107°$ (3 SF)

The angles of the parallelogram are 107° and $180° - \theta = 73.2°$

c For ABCD to be a rectangle, require $\overrightarrow{AB} \cdot \overrightarrow{AD} = 0$:

$\begin{pmatrix} k-2 \\ 0 \\ 2k-1 \end{pmatrix} \cdot \begin{pmatrix} 4 \\ 2k \\ 2 \end{pmatrix} = 0$

$4k - 8 + 0 + 4k - 2 = 0$

$8k = 10$

$k = \dfrac{5}{4}$

8 $\overrightarrow{AB} = \begin{pmatrix} 2 \\ 1 \\ 5 \end{pmatrix}$, $\overrightarrow{BC} = \begin{pmatrix} 2 \\ 1 \\ -7 \end{pmatrix}$, $\overrightarrow{CA} = \begin{pmatrix} -4 \\ -2 \\ 2 \end{pmatrix}$

a $\overrightarrow{AB} \cdot \overrightarrow{CA} = -8 - 2 + 10 = 0$

$\Rightarrow B\hat{A}C = 90°$

b $\overrightarrow{CB} = \begin{pmatrix} -2 \\ -1 \\ 7 \end{pmatrix}$

$B\hat{C}A = \arccos\left(\dfrac{\overrightarrow{CB} \cdot \overrightarrow{CA}}{|\overrightarrow{CB}||\overrightarrow{CA}|}\right)$

$= \arccos\left(\dfrac{8+2+14}{\sqrt{4+1+49}\sqrt{16+4+4}}\right)$

$= \arccos\left(\dfrac{24}{\sqrt{54}\sqrt{24}}\right)$

$= 48.2°$ (3 SF)

∴ $A\hat{B}C = 180° - 90° - 48.2° = 41.8°$

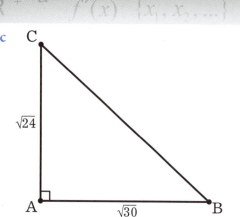

Figure 13D.8

$$\text{Area} = \frac{1}{2}|AB||CA|$$

$$= \frac{1}{2}\sqrt{30}\sqrt{24}$$

$$= 6\sqrt{5}$$

$$= 13.4 \, (3\text{SF})$$

Exercise 13E

7 a The vector equation can be written as the following system of equations:

$$4 + 2x = -2 + 3y \quad \ldots(1)$$
$$1 - 3x = 5 + y \quad \ldots(2)$$

From (2): $y = -3x - 4$

Substituting into (1):

$$4 + 2x = -2 + 3(-3x - 4)$$
$$11x = -18$$
$$x = -\frac{18}{11}, \text{ and so } y = \frac{10}{11}$$

b i $(4-3) + (2+9)x = (-2-15) + (3-3)y$

$$1 + 11x = -17 + 0y$$
$$x = -\frac{18}{11}$$

ii $(12+2) + (6-6)x = (-6+10) + (9+2)y$

$$14 + 0x = 4 + 11y$$
$$y = \frac{10}{11}$$

iii Vectors were picked which were perpendicular to each of the direction vectors of the two lines, $\begin{pmatrix} 2 \\ -3 \end{pmatrix}$ and $\begin{pmatrix} 3 \\ 1 \end{pmatrix}$.

c i

> **COMMENT**
>
> For a two-dimensional vector, find a perpendicular vector by exchanging the horizontal and vertical components and changing the sign of one of these.

$\begin{pmatrix} 1 \\ 2 \end{pmatrix}$ is perpendicular to $\begin{pmatrix} 2 \\ -1 \end{pmatrix}$

$\begin{pmatrix} -3 \\ 5 \end{pmatrix}$ is perpendicular to $\begin{pmatrix} 5 \\ 3 \end{pmatrix}$

ii $\begin{pmatrix} 1 \\ 1 \end{pmatrix} + x \begin{pmatrix} 2 \\ -1 \end{pmatrix} = \begin{pmatrix} 0 \\ 3 \end{pmatrix} + y \begin{pmatrix} 5 \\ 3 \end{pmatrix}$

Taking the scalar product with $\begin{pmatrix} 1 \\ 2 \end{pmatrix}$:

$$3 + 0x = 6 + 11y \Rightarrow y = -\frac{3}{11}$$

Taking the scalar product with $\begin{pmatrix} -3 \\ 5 \end{pmatrix}$:

$$2 - 11x = 15 + 0y \Rightarrow x = -\frac{13}{11}$$

8 a $\mathbf{a} \cdot (\mathbf{b} + \mathbf{c}) = \begin{pmatrix} 2 \\ -2 \\ 1 \end{pmatrix} \cdot \begin{pmatrix} 4 \\ -4 \\ 3 \end{pmatrix} = 19$

b $(\mathbf{b} - \mathbf{a}) \cdot (\mathbf{d} - \mathbf{c}) = \begin{pmatrix} -1 \\ 3 \\ 1 \end{pmatrix} \cdot \begin{pmatrix} 0 \\ 2 \\ 1 \end{pmatrix} = 7$

c $(\mathbf{b} + \mathbf{d}) \cdot (2\mathbf{a}) = \begin{pmatrix} 4 \\ -2 \\ 4 \end{pmatrix} \cdot \begin{pmatrix} 4 \\ -4 \\ 2 \end{pmatrix} = 32$

9 a $a \cdot b = 0$ and $a \cdot a = 1$

$\therefore a \cdot (2a - 3b) = 2a \cdot a - 3a \cdot b = 2$

b $\theta = 45° \Rightarrow \cos\theta = \dfrac{1}{\sqrt{2}}$

$p \cdot q = |p||q|\cos\theta$

$\therefore 3\sqrt{2} = 1 \times |q| \times \dfrac{1}{\sqrt{2}}$

$\Rightarrow |q| = 6$

10 a $\theta = 60° \Rightarrow \cos\theta = \dfrac{1}{2}$

$a \cdot b = |a||b|\cos\theta$

$= \dfrac{3|b|}{2}$

Also, $|a| = 3 \Rightarrow a \cdot a = |a|^2 = 9$

$a \cdot (a - b) = \dfrac{1}{3}$

$a \cdot a - a \cdot b = \dfrac{1}{3}$

$\therefore 9 - \dfrac{3|b|}{2} = \dfrac{1}{3}$

$\dfrac{3|b|}{2} = \dfrac{26}{3}$

$\Rightarrow |b| = \dfrac{52}{9}$

b $(3a + b) \cdot (a - 3b) = 0$

$3a \cdot a - 9a \cdot b + b \cdot a - 3b \cdot b = 0$

$3|a|^2 - 3|b|^2 - 8a \cdot b = 0$

Since $|a| = |b|$, the first two terms cancel and this becomes

$-8a \cdot b = 0$

$\therefore a \cdot b = 0$

which means that a and b are perpendicular.

11 a $\overrightarrow{BC} = \begin{pmatrix} -6 - 2\lambda \\ -\lambda - 17 \\ 5 \end{pmatrix}$, $\overrightarrow{AC} = \begin{pmatrix} -7 \\ 4 \\ 2 \end{pmatrix}$,

$\overrightarrow{AB} = \begin{pmatrix} 2\lambda - 1 \\ 21 + \lambda \\ -3 \end{pmatrix}$

$\overrightarrow{BC} \cdot \overrightarrow{AC} = 0$

$-7(-6 - 2\lambda) + 4(-\lambda - 17) + 10 = 0$

$10\lambda - 16 = 0$

$\lambda = 1.6$

b With $\lambda = 1.6$:

$\overrightarrow{BC} = \begin{pmatrix} -9.2 \\ -18.6 \\ 5 \end{pmatrix}$, $\overrightarrow{AC} = \begin{pmatrix} -7 \\ 4 \\ 2 \end{pmatrix}$,

$\overrightarrow{AB} = \begin{pmatrix} 2.2 \\ 22.6 \\ -3 \end{pmatrix}$

$B\hat{C}A = 90°$ from (a)

$B\hat{A}C = \arccos\left(\dfrac{\overrightarrow{AB} \cdot \overrightarrow{AC}}{|AB||AC|}\right)$

$= \arccos\left(\dfrac{69}{\sqrt{524.6}\sqrt{69}}\right)$

$= 68.7°$

$B\hat{A}C = 180° - 90° - 68.7° = 21.3°$

c Area $= \dfrac{1}{2}(BC)(AC)$

$= \dfrac{1}{2}\sqrt{455.6}\sqrt{69}$

$= 88.7$ (3SF)

12 **a**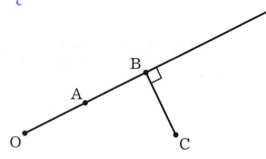

Figure 13E.12

$\overrightarrow{AC} = a + b$

$\overrightarrow{BD} = b - a$

b $(a+b) \cdot (b-a) = a \cdot b - a \cdot a + b \cdot b - a \cdot b$
$= |b|^2 - |a|^2$

c If ABCD is a rhombus, then $|a| = |b|$ and hence $(a+b) \cdot (b-a) = 0$.

This means that $\overrightarrow{AC} \cdot \overrightarrow{BD} = 0$, so the diagonals are perpendicular.

> **COMMENT**
>
> It is very important to know the defining qualities of various quadrilaterals: the square, rhombus, rectangle, parallelogram, trapezium and kite. In a question of this sort, you can assume all other properties without proof, and only need to demonstrate the property being requested.

13 **a** $\overrightarrow{OB} = \lambda \overrightarrow{OA}$, so $\overrightarrow{OB}$ and $\overrightarrow{OA}$ are parallel, hence B lies on (OA).

b $\overrightarrow{BC} = \begin{pmatrix} 12 - 2\lambda \\ 2 - \lambda \\ 4 - 4\lambda \end{pmatrix}$, $\overrightarrow{BA} = (1 - \lambda) \begin{pmatrix} 2 \\ 1 \\ 4 \end{pmatrix}$

$C\hat{B}A = 90°$

$\Rightarrow \overrightarrow{BC} \cdot \overrightarrow{BA} = 0$

$\Rightarrow (1-\lambda)(2(12-2\lambda)+(2-\lambda)+4(4-4\lambda)) = 0$

$\Rightarrow (1-\lambda)(42 - 21\lambda) = 0$

$\therefore \lambda = 2$ (since $\lambda = 1$ is the degenerate case where $\overrightarrow{OA} = \overrightarrow{OB}$)

c

Figure 13E.13

Since $C\hat{B}A = 90°$, B is the point on line (OA) that is closest to C.

Hence the distance from C to the line (OA) is equal to BC:

$\lambda = 2 \Rightarrow \overrightarrow{BC} = \begin{pmatrix} 8 \\ 0 \\ -4 \end{pmatrix}$

$\therefore BC = \left\| \begin{pmatrix} 8 \\ 0 \\ -4 \end{pmatrix} \right\| = \sqrt{64 + 0 + 16} = 4\sqrt{5}$

Exercise 13F

3 **a** $p = \overrightarrow{AB} \times \overrightarrow{AC}$

$= \begin{pmatrix} 4 \\ 12 \\ 1 \end{pmatrix} \times \begin{pmatrix} -4 \\ 6 \\ 2 \end{pmatrix}$

$= \begin{pmatrix} 24 - 6 \\ -4 - 8 \\ 24 + 48 \end{pmatrix}$

$= \begin{pmatrix} 18 \\ -12 \\ 72 \end{pmatrix}$

$q = \overrightarrow{BA} \times \overrightarrow{BC}$

$= \begin{pmatrix} -4 \\ -12 \\ -1 \end{pmatrix} \times \begin{pmatrix} -8 \\ -6 \\ 1 \end{pmatrix}$

$= \begin{pmatrix} -12-6 \\ 8+4 \\ 24-96 \end{pmatrix}$

$= \begin{pmatrix} -18 \\ 12 \\ -72 \end{pmatrix}$

b $p = -q$

4 a $\overrightarrow{BA} = \overrightarrow{CD}$

$\Rightarrow \overrightarrow{CD} = \begin{pmatrix} 4 \\ 0 \\ -3 \end{pmatrix}$

$\therefore \overrightarrow{OD} = \overrightarrow{OC} + \overrightarrow{CD}$

$= \begin{pmatrix} 7 \\ 2 \\ 3 \end{pmatrix} + \begin{pmatrix} 4 \\ 0 \\ -3 \end{pmatrix} = \begin{pmatrix} 11 \\ 2 \\ 0 \end{pmatrix}$

So the coordinates of D are (11, 2, 0).

b Area $= |\overrightarrow{BA} \times \overrightarrow{BC}|$

$= \left| \begin{pmatrix} 4 \\ 0 \\ -3 \end{pmatrix} \times \begin{pmatrix} 8 \\ 1 \\ -2 \end{pmatrix} \right|$

$= \left| \begin{pmatrix} 0+3 \\ -24+8 \\ 4-0 \end{pmatrix} \right|$

$= \left| \begin{pmatrix} 3 \\ -16 \\ 4 \end{pmatrix} \right|$

$= \sqrt{3^2 + 16^2 + 4^2}$

$= \sqrt{281} = 16.8$ (3SF)

5 a $D(0, 4, 0)$

$F(5, 0, 2)$

$G(5, 4, 2)$

$H(0, 4, 2)$

b Area $= \dfrac{1}{2} |\overrightarrow{BE} \times \overrightarrow{BG}|$

$= \dfrac{1}{2} \left| \begin{pmatrix} -5 \\ 0 \\ 2 \end{pmatrix} \times \begin{pmatrix} 0 \\ 4 \\ 2 \end{pmatrix} \right|$

$= \dfrac{1}{2} \left| \begin{pmatrix} 0-8 \\ 0+10 \\ -20-0 \end{pmatrix} \right|$

$= \left| \begin{pmatrix} -4 \\ 5 \\ -10 \end{pmatrix} \right|$

$= \sqrt{4^2 + 5^2 + 10^2}$

$= \sqrt{141} = 11.9$ (3 SF)

Exercise 13G

4 $|a \times b| = |a||b| \sin 30°$

$= 5 \times 7 \times \dfrac{1}{2}$

$= 17.5$

5 a Using $x \times x = 0$ and $x \times y = -y \times x$:

$(a-b) \times (a+b)$

$= a \times a + a \times b - b \times a - b \times b$

$= 0 + a \times b + a \times b - 0$

$= 2a \times b$

b $(2a - 3b) \times (3a + 2b)$

$= 6a \times a + 4a \times b - 9b \times a - 6b \times b$

$= 0 + 4a \times b + 9a \times b - 0$

$= 13a \times b$

6 a $a \times b$ is a vector perpendicular to a.
Therefore $(a \times b) \cdot a$ is the scalar product of two perpendicular vectors, so it is zero.

b $(a \times b) \cdot (a - b) = (a \times b) \cdot a - (a \times b) \cdot b$
$= 0 - 0$
$= 0$

7 If θ is the angle between vectors a and b, then by the properties of vector and scalar products:

$|a \times b|^2 + (a \cdot b)^2$
$= (|a||b|\sin\theta)^2 + (|a||b|\cos\theta)^2$
$= |a|^2|b|^2 \sin^2\theta + |a|^2|b|^2 \cos^2\theta$
$= |a|^2|b|^2 (\sin^2\theta + \cos^2\theta)$
$= |a|^2|b|^2$

Mixed examination practice 13

Short questions

1 $\left(\begin{pmatrix} 2 \\ 1 \\ -3 \end{pmatrix} \times \begin{pmatrix} -3 \\ 0 \\ 2 \end{pmatrix} \right) \cdot \begin{pmatrix} 1 \\ -3 \\ 4 \end{pmatrix}$

$= \begin{pmatrix} 2 \\ 5 \\ 3 \end{pmatrix} \cdot \begin{pmatrix} 1 \\ -3 \\ 4 \end{pmatrix}$

$= 2 - 15 + 12$
$= -1$

2 a $\overrightarrow{MD} = \overrightarrow{MC} + \overrightarrow{CD}$

$= \frac{1}{2}\overrightarrow{BC} + \overrightarrow{CD}$

$= \frac{1}{2}\overrightarrow{AD} - \overrightarrow{AB}$

b Since ABCD is a rectangle, $\overrightarrow{AB} \cdot \overrightarrow{AD} = 0$

$\overrightarrow{MD} \cdot \overrightarrow{MC} = \left(\frac{1}{2}\overrightarrow{AD} - \overrightarrow{AB} \right) \cdot \left(\frac{1}{2}\overrightarrow{AD} \right)$

$= \frac{1}{4}AD^2 - \frac{1}{2}\overrightarrow{AB} \cdot \overrightarrow{AD}$

$= \frac{1}{4} \times 4^2 - \frac{1}{2} \times 0$

$= 4$

3 a $\begin{pmatrix} 2 \\ -5 \\ 1 \end{pmatrix} \times \begin{pmatrix} 1 \\ 1 \\ -2 \end{pmatrix} = \begin{pmatrix} 10-1 \\ 1+4 \\ 2+5 \end{pmatrix} = \begin{pmatrix} 9 \\ 5 \\ 7 \end{pmatrix} = 9i + 5j + 7k$

b Area $= |\overrightarrow{ON} \times \overrightarrow{OL}| = |n \times l|$

$= \left| \begin{pmatrix} 9 \\ 5 \\ 7 \end{pmatrix} \right|$

$= \sqrt{9^2 + 5^2 + 7^2}$

$= \sqrt{155} = 12.4$ (3SF)

4 a $\begin{pmatrix} 3 \\ 0 \\ 1 \end{pmatrix} \times \begin{pmatrix} -1 \\ 5 \\ p \end{pmatrix} = \begin{pmatrix} -5 \\ -1-3p \\ 15 \end{pmatrix}$

b Require that $\begin{pmatrix} -5 \\ -1-3p \\ 15 \end{pmatrix} = k \begin{pmatrix} 1 \\ 4 \\ -3 \end{pmatrix}$ for some value k.

Clearly, by inspecting the first and third components, $k = -5$.

Then, from the second component:
$-1 - 3p = -20$
$-3p = -19$
$\Rightarrow p = \frac{19}{3}$

5 Choosing H to be the origin, $\overrightarrow{HG}$ as the positive x direction, $\overrightarrow{EH}$ as the positive y direction and $\overrightarrow{DH}$ as the positive z direction, we have

$$\overrightarrow{HA} = \begin{pmatrix} 0 \\ -4 \\ -3 \end{pmatrix} \text{ and } \overrightarrow{HC} = \begin{pmatrix} 6 \\ 0 \\ -3 \end{pmatrix}.$$

$$A\hat{H}C = \arccos\left(\frac{\overrightarrow{HA} \cdot \overrightarrow{HC}}{|\overrightarrow{HA}||\overrightarrow{HC}|}\right)$$

$$= \arccos\left(\frac{9}{\sqrt{25}\sqrt{45}}\right)$$

$$= 74.4° \text{ (3SF)}$$

6 $|a| = |b| = \sqrt{\sin^2\theta + \cos^2\theta} = 1$

$$\alpha = \arccos\left(\frac{a \cdot b}{|a||b|}\right)$$

$$= \arccos\left(\frac{\cos\theta\sin\theta + \sin\theta\cos\theta}{1 \times 1}\right)$$

$$= \arccos(2\cos\theta\sin\theta)$$

$$= \arccos(\sin 2\theta)$$

$$= \arccos\left(\cos\left(\frac{\pi}{2} - 2\theta\right)\right)$$

$$= \frac{\pi}{2} - 2\theta$$

7 $a = \frac{1}{2}((a+b)+(a-b))$,

$b = \frac{1}{2}((a+b)-(a-b))$

$a \cdot b = \frac{1}{4}((a+b)+(a-b)) \cdot ((a+b)-(a-b))$

$= \frac{1}{4}[(a+b) \cdot (a+b) - (a+b)(a-b)$

$\quad + (a-b)(a+b) - (a-b) \cdot (a-b)]$

$= \frac{1}{4}(|a+b|^2 - |a-b|^2)$

$= 0$

8 a $(b-a) \cdot (b-a) = b \cdot b - b \cdot a - a \cdot b + a \cdot a$

$= |b|^2 + |a|^2 - 2a \cdot b$

b

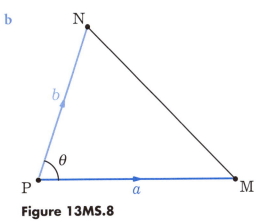

Figure 13MS.8

$b - a = \overrightarrow{MN}, \quad |a| = PM, \quad |b| = PN$

From (a):

$|b-a|^2 = |a|^2 + |b|^2 - 2a \cdot b$

$= |a|^2 + |b|^2 - 2|a||b|\cos\theta$

i.e. $MN^2 = PM^2 + PN^2 - 2(PM)(PN)\cos\theta$

Long questions

1 a $\overrightarrow{AD} = \begin{pmatrix} 3 \\ 1 \\ k \end{pmatrix} - \begin{pmatrix} 1 \\ 1 \\ 7 \end{pmatrix}$

$= \begin{pmatrix} 2 \\ 0 \\ k-7 \end{pmatrix}$

b (AD) is perpendicular to (AB)

$\therefore \overrightarrow{AD} \cdot \overrightarrow{AB} = 0$

$\begin{pmatrix} 2 \\ 0 \\ k-7 \end{pmatrix} \cdot \begin{pmatrix} -2 \\ 5 \\ -4 \end{pmatrix} = 0$

$-4 - 4(k-7) = 0$

$4k = 24$

$k = 6$

c With $k = 6$, $\overrightarrow{AD} = \begin{pmatrix} 2 \\ 0 \\ -1 \end{pmatrix}$

$\therefore \overrightarrow{BC} = \begin{pmatrix} 4 \\ 0 \\ -2 \end{pmatrix}$

$c - b = \begin{pmatrix} 4 \\ 0 \\ -2 \end{pmatrix}$

$c = \begin{pmatrix} 4 \\ 0 \\ -2 \end{pmatrix} + \begin{pmatrix} -1 \\ 6 \\ 3 \end{pmatrix} = \begin{pmatrix} 3 \\ 6 \\ 1 \end{pmatrix}$

i.e. the coordinates of C are (3, 6, 1)

d $\cos(A\hat{D}C) = \dfrac{\overrightarrow{DA} \cdot \overrightarrow{DC}}{(DA)(DC)}$

$= \dfrac{\begin{pmatrix} -2 \\ 0 \\ 1 \end{pmatrix} \cdot \begin{pmatrix} 0 \\ 5 \\ -5 \end{pmatrix}}{\sqrt{5}\sqrt{50}}$

$= -\dfrac{1}{\sqrt{10}}$

2 a $\overrightarrow{CD} = \overrightarrow{CB} + \overrightarrow{BD}$

$= \overrightarrow{CB} + \dfrac{k}{k+1}\overrightarrow{BA}$

$= \begin{pmatrix} 2 \\ 3 \\ -4 \end{pmatrix} + \dfrac{k}{k+1}\begin{pmatrix} -3 \\ -3 \\ 0 \end{pmatrix}$

b $\overrightarrow{CD} \cdot \overrightarrow{AB} = 0$

$\left(\begin{pmatrix} 2 \\ 3 \\ -4 \end{pmatrix} + \dfrac{k}{k+1}\begin{pmatrix} -3 \\ -3 \\ 0 \end{pmatrix}\right) \cdot \begin{pmatrix} 3 \\ 3 \\ 0 \end{pmatrix} = 0$

$15 - \dfrac{18k}{k+1} = 0$

$15k + 15 - 18k = 0$

$k = 5$

c With $k = 5$,

$\overrightarrow{CD} = \begin{pmatrix} 2 \\ 3 \\ -4 \end{pmatrix} + \dfrac{5}{6}\begin{pmatrix} -3 \\ -3 \\ 0 \end{pmatrix} = \begin{pmatrix} -\frac{1}{2} \\ \frac{1}{2} \\ -4 \end{pmatrix}$

i.e. $\mathbf{d} - \mathbf{c} = \begin{pmatrix} -\frac{1}{2} \\ \frac{1}{2} \\ -4 \end{pmatrix}$

$\Rightarrow \mathbf{d} = \begin{pmatrix} -\frac{1}{2} \\ \frac{1}{2} \\ -4 \end{pmatrix} + \begin{pmatrix} 2 \\ 1 \\ 6 \end{pmatrix} = \begin{pmatrix} \frac{3}{2} \\ \frac{3}{2} \\ 2 \end{pmatrix}$

$\therefore$ the coordinates of D are $\left(\dfrac{3}{2}, \dfrac{3}{2}, 2\right)$

d $CD = \sqrt{\dfrac{1}{4} + \dfrac{1}{4} + 16}$

$= \sqrt{\dfrac{33}{2}}$

$= 4.06\ (3\text{SF})$

3 a The coordinates of P are (a, a^2)

b $\overrightarrow{PO} = -\overrightarrow{OP} = \begin{pmatrix} -a \\ -a^2 \end{pmatrix}$

$\overrightarrow{PS} = \begin{pmatrix} 0 \\ 4 \end{pmatrix} - \begin{pmatrix} a \\ a^2 \end{pmatrix} = \begin{pmatrix} -a \\ 4 - a^2 \end{pmatrix}$

c $\overrightarrow{PO} \cdot \overrightarrow{PS} = 0$

$a^2 - 4a^2 + a^4 = 0$

$a^2(-3 + a^2) = 0$

$a = 0$ or $\pm\sqrt{3}$

Since $a > 0$, $a = \sqrt{3}$

146 Mixed examination practice 13

d With $a = \sqrt{3}$, we know that
$O\hat{P}S = 90°$, $OP = \sqrt{12}$ and $PS = 2$

$\therefore$ Area of OPS $= \dfrac{1}{2} \times \sqrt{12} \times 2 = 2\sqrt{3}$

4 a Area of base $= \dfrac{1}{2}|\boldsymbol{a} \times \boldsymbol{b}|$

b $h = |\boldsymbol{c}|\cos\theta$

c Volume $= \dfrac{1}{3}$ height $\times$ base area

$= \dfrac{1}{3}(|\boldsymbol{c}|\cos\theta)\left(\dfrac{1}{2}|\boldsymbol{a} \times \boldsymbol{b}|\right)$

$= \dfrac{1}{6}|\boldsymbol{a} \times \boldsymbol{b}||\boldsymbol{c}|\cos\theta$

Since [AE] is perpendicular to the base,
it is parallel to $\boldsymbol{a} \times \boldsymbol{b}$, and so the angle
between $\boldsymbol{a} \times \boldsymbol{b}$ and $\boldsymbol{c}$ is also θ.

$\therefore$ Volume $= \dfrac{1}{6}|(\boldsymbol{a} \times \boldsymbol{b}) \cdot \boldsymbol{c}|$

d $\boldsymbol{a} = \begin{pmatrix} -1 \\ 0 \\ -1 \end{pmatrix}$, $\boldsymbol{b} = \begin{pmatrix} 2 \\ -7 \\ 1 \end{pmatrix}$, $\boldsymbol{c} = \begin{pmatrix} -1 \\ -2 \\ -1 \end{pmatrix}$

Volume $= \dfrac{1}{6}\left|\left(\begin{pmatrix} -1 \\ 0 \\ -1 \end{pmatrix} \times \begin{pmatrix} 2 \\ -7 \\ 1 \end{pmatrix}\right) \cdot \begin{pmatrix} -1 \\ -2 \\ -1 \end{pmatrix}\right|$

$= \dfrac{1}{6}\left|\begin{pmatrix} -7 \\ -1 \\ 7 \end{pmatrix} \cdot \begin{pmatrix} -1 \\ -2 \\ -1 \end{pmatrix}\right|$

$= \dfrac{1}{6}|7 + 2 - 7|$

$= \dfrac{1}{3}$

14 Lines and planes in space

Exercise 14A

4 Equation of line l: $r = \begin{pmatrix} 2-t \\ 1+t \\ -4+2t \end{pmatrix}$

a At $t=-2$, $r = \begin{pmatrix} 4 \\ -1 \\ -8 \end{pmatrix} = \overrightarrow{OA}$

At $t=0$, $r = \begin{pmatrix} 2 \\ 1 \\ -4 \end{pmatrix} = \overrightarrow{OB}$

So A and B lie on l.

b At A, $t=-2$, and at B, $t=0$. Require AB=BC, so the point C must lie where $t=2$.
∴ the coordinates of C are $(0, 3, 0)$

5 a $\overrightarrow{PQ} = q - p = \begin{pmatrix} -4 \\ -2 \\ 3 \end{pmatrix}$

$\Rightarrow r = \begin{pmatrix} 7 \\ 1 \\ 2 \end{pmatrix} + \lambda \begin{pmatrix} -4 \\ -2 \\ 3 \end{pmatrix}$

b At P, $\lambda=0$, and at Q, $\lambda=1$. Require PR=3PQ, so the point R must lie where $\lambda=\pm 3$, since the distances are proportional to the differences in λ.
Hence

$\overrightarrow{OR} = \begin{pmatrix} 7 \\ 1 \\ 2 \end{pmatrix} + 3\begin{pmatrix} -4 \\ -2 \\ 3 \end{pmatrix}$ or $\begin{pmatrix} 7 \\ 1 \\ 2 \end{pmatrix} - 3\begin{pmatrix} -4 \\ -2 \\ 3 \end{pmatrix}$

i.e. the coordinates of R are $(-5, -5, 11)$ or $(19, 7, -7)$

6 a $2\mathbf{i} - 3\mathbf{j} + 6\mathbf{k}$ gives a direction vector of $\begin{pmatrix} 2 \\ -3 \\ 6 \end{pmatrix}$ for the line,

∴ $r = \begin{pmatrix} 2 \\ 1 \\ 4 \end{pmatrix} + \lambda \begin{pmatrix} 2 \\ -3 \\ 6 \end{pmatrix}$

b $\left\| \begin{pmatrix} 2 \\ -3 \\ 6 \end{pmatrix} \right\| = \sqrt{4+9+36} = 7$

c As P is a point on the line,

$AP = \left\| \lambda \begin{pmatrix} 2 \\ -3 \\ 6 \end{pmatrix} \right\| = |\lambda| \left\| \begin{pmatrix} 2 \\ -3 \\ 6 \end{pmatrix} \right\| = 7|\lambda|$

$AP = 35 \Rightarrow |\lambda| = 5 \Rightarrow \lambda = \pm 5$

∴ $\overrightarrow{OP} = \begin{pmatrix} 2 \\ 1 \\ 4 \end{pmatrix} + 5\begin{pmatrix} 2 \\ -3 \\ 6 \end{pmatrix}$ or $\begin{pmatrix} 2 \\ 1 \\ 4 \end{pmatrix} - 5\begin{pmatrix} 2 \\ -3 \\ 6 \end{pmatrix}$

i.e. the coordinates of P are $(12, -14, 34)$ or $(-8, 16, -26)$

Exercise 14B

COMMENT

In these problems, assign unknowns to the values that need to be determined, then use one or more standard equations which describe the geometry you are given. Solving the equations will give the values for the unknowns.

4 Since C lies on l, $\overrightarrow{OC} = \begin{pmatrix} 4+2\lambda \\ 2-\lambda \\ -1+2\lambda \end{pmatrix}$ for some λ

$\overrightarrow{CP} = \mathbf{p} - \mathbf{c} = \begin{pmatrix} 3-2\lambda \\ \lambda \\ 4-2\lambda \end{pmatrix}$

[PC] perpendicular to l

$\Rightarrow \overrightarrow{CP} \cdot \begin{pmatrix} 2 \\ -1 \\ 2 \end{pmatrix} = 0$

$6 - 4\lambda - \lambda + 8 - 4\lambda = 0$

$9\lambda = 14$

$\lambda = \dfrac{14}{9}$

∴ the coordinates of C are $\left(\dfrac{64}{9}, \dfrac{4}{9}, \dfrac{19}{9} \right)$

5 Let P be the point on the line that is closest to $A(-1, 1, 2)$

$\overrightarrow{OP} = \begin{pmatrix} 1-3t \\ t \\ 2+t \end{pmatrix}$ for some t

$\overrightarrow{PA} = \overrightarrow{OA} - \overrightarrow{OP} = \begin{pmatrix} -2+3t \\ 1-t \\ -t \end{pmatrix}$ and

we require that $\overrightarrow{PA} \cdot \begin{pmatrix} -3 \\ 1 \\ 1 \end{pmatrix} = 0$

$\begin{pmatrix} -2+3t \\ 1-t \\ -t \end{pmatrix} \cdot \begin{pmatrix} -3 \\ 1 \\ 1 \end{pmatrix} = 0$

$6 - 9t + 1 - t - t = 0$

$11t = 7$

$t = \dfrac{7}{11}$

∴ $\overrightarrow{PA} = \dfrac{1}{11} \begin{pmatrix} -1 \\ 4 \\ -7 \end{pmatrix}$

$= \dfrac{1}{11} \sqrt{1 + 16 + 49} = \dfrac{\sqrt{66}}{11}$

6 a At P, $\begin{pmatrix} -5-3\lambda \\ 1 \\ 10+4\lambda \end{pmatrix} = \begin{pmatrix} 3+\mu \\ \mu \\ -9+7\mu \end{pmatrix}$, so

$-5 - 3\lambda = 3 + \mu$ ...(1)
$1 = \mu$ ...(2)
$10 + 4\lambda = -9 + 7\mu$ ...(3)

From (2), $\mu = 1$
Substituting into (1) gives

$-5 - 3\lambda = 4 \Rightarrow \lambda = -3$

Then, substituting into (3):

$10 + 4\lambda = -2 = -9 + 7\mu$ is valid, so the lines do intersect.

Substituting, say, $\mu = 1$ into the equation for l_2 gives $(4, 1, -2)$ as the coordinates of P.

> **COMMENT**
> Always use all three of the equations to ensure that the solution obtained from two of them is valid. In this case, the question states that the lines intersect, so this check serves to reveal errors in working rather than determining whether or not the lines are skew, but it is nonetheless worthwhile.

b When $\mu = 2$, $\mathbf{r}_2 = \begin{pmatrix} 5 \\ 2 \\ 5 \end{pmatrix}$, so $Q(5, 2, 5)$ does lie on l_2.

c $\overrightarrow{QM} = \begin{pmatrix} -5-3\lambda \\ 1 \\ 10+4\lambda \end{pmatrix} - \begin{pmatrix} 5 \\ 2 \\ 5 \end{pmatrix} = \begin{pmatrix} -10-3\lambda \\ -1 \\ 5+4\lambda \end{pmatrix}$

for some value of λ

14 Lines and planes in space 149

Require that $\overrightarrow{QM} \cdot \begin{pmatrix} -3 \\ 0 \\ 4 \end{pmatrix} = 0$:

$\begin{pmatrix} -10-3\lambda \\ -1 \\ 5+4\lambda \end{pmatrix} \cdot \begin{pmatrix} -3 \\ 0 \\ 4 \end{pmatrix} = 0$

$30 + 9\lambda + 20 + 16\lambda = 0$

$25\lambda = -50$

$\lambda = -2$

∴ the coordinates of M are (1, 1, 2)

d

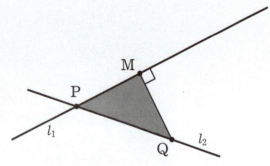

Figure 14B.6

$P\hat{M}Q = 90°$

∴ Area of PMQ $= \frac{1}{2}$(PM)(MQ)

$= \frac{1}{2} \left\| \begin{pmatrix} -3 \\ 0 \\ 4 \end{pmatrix} \right\| \left\| \begin{pmatrix} -4 \\ -1 \\ -3 \end{pmatrix} \right\|$

$= \frac{5\sqrt{26}}{2}$

$= 12.7 (3SF)$

7 a At t hours, the ship is at $r_1 = \begin{pmatrix} 3t \\ 4t \end{pmatrix}$

b At t hours, the second ship is at

$r_2 = \begin{pmatrix} 0 \\ 18 \end{pmatrix} + t \begin{pmatrix} 3 \\ -5 \end{pmatrix} = \begin{pmatrix} 3t \\ 18-5t \end{pmatrix}$

c $d = |r_2 - r_1|$

$= \left\| \begin{pmatrix} 0 \\ 18-9t \end{pmatrix} \right\|$

$= |18 - 9t|$

When $t = 0.5$, $d = 13.5$

d The ships meet if there is a positive value of t for which $d = 0$:

$18 - 9t = 0$

$\Rightarrow t = 2$

∴ the ships do meet, and this happens after 2 hours.

e $d = 18$ km when

$|18 - 9t| = 18$

$9t - 18 = 18$ (since $t > 2$)

$9t = 36$

$t = 4$

∴ the ships are 18 km apart a further 2 hours after meeting.

> **COMMENT**
>
> Note that the other solution to the modulus equation $|18-9t| = 18$ is $18 - 9t = 18 \Rightarrow t = 0$, which is already known since the ships start 18 km apart.

8 a At time t, the first aircraft is at

$r_1 = \begin{pmatrix} 0 \\ 5 \\ 0 \end{pmatrix} + t \begin{pmatrix} 3 \\ -4 \\ 1 \end{pmatrix} = \begin{pmatrix} 3t \\ 5-4t \\ t \end{pmatrix}$

b At time t, the second aircraft is at

$r_2 = \begin{pmatrix} 0 \\ 0 \\ 7 \end{pmatrix} + t \begin{pmatrix} 5 \\ 2 \\ -1 \end{pmatrix} = \begin{pmatrix} 5t \\ 2t \\ 7-t \end{pmatrix}$

$d = |r_1 - r_2|$

$= \sqrt{(2t)^2 + (6t-5)^2 + (7-2t)^2}$

$\Rightarrow d^2 = 4t^2 + 36t^2 - 60t + 25 + 49 - 28t + 4t^2$

$= 44t^2 - 88t + 74$

c Completing the square:

$d^2 = 44(t-1)^2 + 30 \geq 30$

i.e. $d \neq 0$ for all t and so the aircraft do not collide.

d From (c), the minimum d^2 is 30, so the minimum distance d is

$\sqrt{30} = 5.48$ km (3SF)

9 Let P be the closest point to the origin on the line.

$\overrightarrow{OP} = \begin{pmatrix} 1+2\lambda \\ -2+2\lambda \\ 2+\lambda \end{pmatrix}$ for some value of λ

Require that $\overrightarrow{OP} \cdot \begin{pmatrix} 2 \\ 2 \\ 1 \end{pmatrix} = 0$:

$\begin{pmatrix} 1+2\lambda \\ -2+2\lambda \\ 2+\lambda \end{pmatrix} \cdot \begin{pmatrix} 2 \\ 2 \\ 1 \end{pmatrix} = 0$

$2 + 4\lambda - 4 + 4\lambda + 2 + \lambda = 0$

$9\lambda = 0$

$\lambda = 0$

$\therefore$ the coordinates of P are $(1, -2, 2)$

$OP = \sqrt{1+4+4} = 3$

10 a At P, $\begin{pmatrix} \lambda \\ -1+5\lambda \\ 2+3\lambda \end{pmatrix} = \begin{pmatrix} 2-t \\ 2+t \\ 1+3t \end{pmatrix}$, so

$\lambda = 2 - t$ ...(1)

$-1 + 5\lambda = 2 + t$ ...(2)

$2 + 3\lambda = 1 + 3t$ ...(3)

(1)+(2):

$6\lambda - 1 = 4 \Rightarrow \lambda = \dfrac{5}{6}$

Substituting into (1):

$t = 2 - \lambda = \dfrac{7}{6}$

Substituting into (3):

$2 + 3\lambda = \dfrac{9}{2} = 1 + 3t$ is valid, so the lines do intersect.

Substituting, say, $\lambda = \dfrac{5}{6}$ into the equation for l_1 gives $\left(\dfrac{5}{6}, \dfrac{19}{6}, \dfrac{9}{2}\right)$ as the coordinates of P.

b $d_1 = \begin{pmatrix} 1 \\ 5 \\ 3 \end{pmatrix}$, $d_2 = \begin{pmatrix} -1 \\ 1 \\ 3 \end{pmatrix}$

Let θ be the angle between the lines. Then

$\theta = \arccos\left(\dfrac{d_1 \cdot d_2}{|d_1||d_2|}\right)$

$= \arccos\left(\dfrac{-1+5+9}{\sqrt{1+25+9}\sqrt{1+1+9}}\right)$

$= \arccos\left(\dfrac{13}{\sqrt{35}\sqrt{11}}\right)$

$= 48.5°$ (3SF)

c At $t = 3$, $r_2 = \begin{pmatrix} -1 \\ 5 \\ 10 \end{pmatrix} = \overrightarrow{OQ}$, so Q lies on l_2.

d $\overrightarrow{PQ} = \begin{pmatrix} -1 \\ 5 \\ 10 \end{pmatrix} - \dfrac{1}{6}\begin{pmatrix} 5 \\ 19 \\ 27 \end{pmatrix} = \dfrac{1}{6}\begin{pmatrix} -11 \\ 11 \\ 33 \end{pmatrix} = \dfrac{11}{6}\begin{pmatrix} -1 \\ 1 \\ 3 \end{pmatrix}$

$\therefore PQ = \dfrac{11}{6}\sqrt{1+1+9}$

$= \dfrac{11\sqrt{11}}{6} = 6.08$ (3SF)

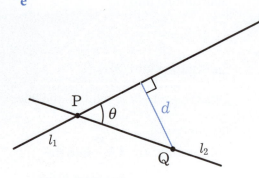

Figure 14B.10

Let R be the closest point to Q on line l_1. Then PQR is a triangle with

$$P\hat{R}Q = 90°, \quad PQ = \frac{11\sqrt{11}}{6} \text{ and } Q\hat{P}R = 48.5°$$

By trigonometry in PQR,

$$QR = PQ \sin Q\hat{P}R$$
$$= 4.55 \text{ (3SF)}$$

11 a $\overrightarrow{PM} = \begin{pmatrix} 5+2\lambda \\ 1-3\lambda \\ 2+3\lambda \end{pmatrix} - \begin{pmatrix} 21 \\ 5 \\ 10 \end{pmatrix} = \begin{pmatrix} -16+2\lambda \\ -4-3\lambda \\ -8+3\lambda \end{pmatrix}$

for some value of λ

Require $\overrightarrow{PM} \cdot \begin{pmatrix} 2 \\ -3 \\ 3 \end{pmatrix} = 0$

$\therefore \begin{pmatrix} -16+2\lambda \\ -4-3\lambda \\ -8+3\lambda \end{pmatrix} \cdot \begin{pmatrix} 2 \\ -3 \\ 3 \end{pmatrix} = 0$

$-32 + 4\lambda + 12 + 9\lambda - 24 + 9\lambda = 0$

$22\lambda = 44$

$\lambda = 2$

$\therefore$ the coordinates of M are $(9, -5, 8)$

b When $\lambda = 5$, $r = \begin{pmatrix} 15 \\ -14 \\ 17 \end{pmatrix} = \overrightarrow{OQ}$,

so Q lies on l.

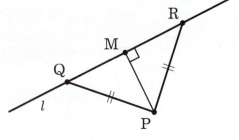

Figure 14B.11

Require PQR to be an isosceles triangle, with base lying on l.

It follows that M must lie at the midpoint of the base of the triangle, since (PM) is perpendicular to (QR).

M has $\lambda = 2$ and Q has $\lambda = 5$, so R must have $\lambda = -1$ for M to be the midpoint of [QR].

$\therefore$ the coordinates of R are $(3, 4, -1)$

12 a At $\mu = 3$, $r_2 = \begin{pmatrix} 5 \\ 2 \\ 6 \end{pmatrix} = \overrightarrow{OQ}$,

so Q lies on l_2.

b By inspection, the two lines share a common position vector

$\therefore P = (2, -1, 0)$

> **COMMENT**
>
> Here you could carry out the standard procedure for finding the point of intersection, but it is much easier if you spot the common position vector!

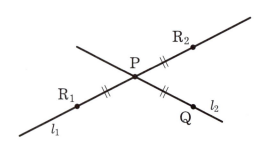

Figure 14B.12

152 Topic 14B Solving problems with lines

Require PQR to be an isosceles triangle with PR=PQ (from Figure 14B.12 there are two possible positions for R).

$$PQ = \left\| \begin{pmatrix} 3 \\ 3 \\ 6 \end{pmatrix} \right\| = 3\sqrt{6}$$

$$|d_1| = \left\| \begin{pmatrix} 1 \\ -2 \\ 2 \end{pmatrix} \right\| = 3 \Rightarrow PR = 3|\lambda|$$

$$\therefore 3|\lambda| = 3\sqrt{6}$$

$$\lambda = \pm\sqrt{6}$$

∴ the coordinates of R are

$$(2+\sqrt{6}, -1-2\sqrt{6}, 2\sqrt{6}) \text{ or}$$

$$(2-\sqrt{6}, -1+2\sqrt{6}, -2\sqrt{6})$$

Exercise 14C

3 a The vector equation of the line is

$$r = \begin{pmatrix} 1 \\ 4 \\ -1 \end{pmatrix} + \lambda \begin{pmatrix} 3 \\ -2 \\ 3 \end{pmatrix}$$

Cartesian form:

$$\frac{x-1}{3} = \frac{y-4}{-2} = \frac{z+1}{3}$$

b Unit direction vector is

$$\frac{1}{\sqrt{3^2+2^2+3^2}} \begin{pmatrix} 3 \\ -2 \\ 3 \end{pmatrix} = \frac{1}{\sqrt{22}} \begin{pmatrix} 3 \\ -2 \\ 3 \end{pmatrix}$$

4 a Rewrite the Cartesian equation as

$$\frac{x - \frac{1}{2}}{2} = \frac{y+2}{3} = \frac{z - \frac{4}{3}}{-2}$$

Vector equation:

$$r = \begin{pmatrix} \frac{1}{2} \\ -2 \\ \frac{4}{3} \end{pmatrix} + \lambda \begin{pmatrix} 2 \\ 3 \\ -2 \end{pmatrix}$$

b Intersection with x-axis occurs where $y = z = 0$:

Require λ such that

$$\begin{cases} -2 + 3\lambda = 0 \\ \frac{4}{3} - 2\lambda = 0 \end{cases}$$

This pair of simultaneous equations has consistent solution $\lambda = \frac{2}{3}$, so the line does intersect the x-axis, at $\left(\frac{11}{6}, 0, 0\right)$

c Let θ be the angle between the

direction vectors $\mathbf{a} = \begin{pmatrix} 1 \\ 0 \\ 0 \end{pmatrix}$ (for the

x-axis) and $\mathbf{d} = \begin{pmatrix} 2 \\ 3 \\ -2 \end{pmatrix}$ (for the line). Then

$$\cos\theta = \frac{\mathbf{a} \cdot \mathbf{d}}{|\mathbf{a}||\mathbf{d}|} = \frac{2}{\sqrt{17}}$$

$$\theta = \arccos\left(\frac{2}{\sqrt{17}}\right) = 61.0° \text{ (3SF)}$$

5 a Vector forms of the lines:

$$\frac{x-3}{5} = \frac{y-2}{1} = \frac{z-\frac{3}{2}}{-1} \Rightarrow r_1 = \begin{pmatrix} 3 \\ 2 \\ \frac{3}{2} \end{pmatrix} + \lambda \begin{pmatrix} 5 \\ 1 \\ -1 \end{pmatrix}$$

$$\frac{x+1}{3} = \frac{z-3}{-1}, y = 1 \Rightarrow r_2 = \begin{pmatrix} -1 \\ 1 \\ 3 \end{pmatrix} + \mu \begin{pmatrix} 3 \\ 0 \\ -1 \end{pmatrix}$$

Let θ be the angle between the

direction vectors $d_1 = \begin{pmatrix} 5 \\ 1 \\ -1 \end{pmatrix}$ and

$d_2 = \begin{pmatrix} 3 \\ 0 \\ -1 \end{pmatrix}$. Then

$\cos\theta = \dfrac{d_1 \cdot d_2}{|d_1||d_2|} = \dfrac{16}{\sqrt{27}\sqrt{10}}$

$\theta = \arccos\left(\dfrac{16}{\sqrt{270}}\right) = 13.2°$ (3SF)

b The lines intersect if there are λ and μ that satisfy

$3 + 5\lambda = -1 + 3\mu$ ...(1)

$2 + \lambda = 1$ ...(2)

$\dfrac{3}{2} - \lambda = 3 - \mu$ ...(3)

$(2) \Rightarrow \lambda = -1$

Substituting into (1):

$-2 = -1 + 3\mu \Rightarrow \mu = -\dfrac{1}{3}$

Check for consistency in (3):

LHS $= \dfrac{5}{2}$, RHS $= \dfrac{10}{3}$

There is no consistent solution, so the two lines do not intersect.

6 a Parameterised form of the first line:

$r_1 = (2 + 3v, -1 + 4v, -1 + v)$

Parameterised form of the second line:

$r_2 = (5 - \mu, -2 - 3\mu, 7 + 2\mu)$

The lines intersect if there are v and μ that satisfy the simultaneous equations

$2 + 3v = 5 - \mu$ ...(1)

$-1 + 4v = -2 - 3\mu$ ...(2)

$-1 + v = 2\mu + 7$ ...(3)

$(3) \Rightarrow v = 2\mu + 8$

Substituting into (1):

$2 + 6\mu + 24 = 5 - \mu \Rightarrow 21 = -7\mu$

Substituting into (2):

$-1 + 8\mu + 32 = -2 - 3\mu \Rightarrow 33 = -11\mu$

These give the same value, $\mu = -3$, and hence $v = 2$.

Therefore the equations are consistent for $\mu = -3$, $v = 2$ and the point of intersection is (8, 7, 1).

b By observation, at $\lambda = 1$, the line passes through (8, 7, 1).

Exercise 14E

5 a Normal vector for plane Π_1:

$n_1 = \begin{pmatrix} 3 \\ -1 \\ 1 \end{pmatrix}$

b Normal vector for plane Π_2:

$n_2 = \begin{pmatrix} 1 \\ -5 \\ 5 \end{pmatrix}$

Angle θ between the planes is the same as the angle between the normal vectors:

$\cos\theta = \dfrac{n_1 \cdot n_2}{|n_1||n_2|} = \dfrac{13}{\sqrt{11}\sqrt{51}}$

$\theta = \arccos\left(\dfrac{13}{\sqrt{561}}\right) = 57°$

(to the nearest degree)

6 The line has vector equation

$r = \begin{pmatrix} 2 \\ 1 \\ 0 \end{pmatrix} + \lambda \begin{pmatrix} 3 \\ 2 \\ 1 \end{pmatrix}$

The plane has equation $r \cdot \begin{pmatrix} 2 \\ -1 \\ -2 \end{pmatrix} = 5$

For intersection, require

$$\left(\begin{pmatrix}2\\1\\0\end{pmatrix}+\lambda\begin{pmatrix}3\\2\\1\end{pmatrix}\right)\cdot\begin{pmatrix}2\\-1\\-2\end{pmatrix}=5$$

$$2(2+3\lambda)-(1+2\lambda)-2\lambda=5$$

$$2\lambda=2$$

$$\lambda=1$$

Substituting this value of λ into the equation for the line:
the intersection point is (5, 3, 1)

7

COMMENT

There are a great many ways to approach vector problems, and a selection of methods for this question is given below. You should make sure you understand all of them, and try to identify which approach you prefer in which circumstances. Each approach will have its advantages, and if you can master multiple approaches you can choose the most efficient for any given problem.

Method 1: using the parameterised form for one line and substituting it into the Cartesian form for the other line.

For the first line,

$$x=4-3\lambda,\ y=1+3\lambda,\ z=2+\lambda$$

Substituting into the equation for the second line:

$$\frac{3-3\lambda}{4}=\frac{3+3\lambda}{3}=\frac{3+2\lambda}{4}$$

From the first and third expressions:
$3-3\lambda=3+2\lambda\Rightarrow\lambda=0$

But this is inconsistent with the second expression.

Therefore the two lines do not intersect.

COMMENT

Note that if the lines do intersect, this method gives you the value of the parameter λ for the intersection point.

Method 2: using simultaneous equations for the parameterised forms.

The second line $\dfrac{x-1}{4}=\dfrac{y+2}{3}=\dfrac{z-\frac{1}{2}}{2}$

has vector equation $\mathbf{r}=\begin{pmatrix}1\\-2\\\frac{1}{2}\end{pmatrix}+\mu\begin{pmatrix}4\\3\\2\end{pmatrix}$

An intersection of the two lines represents a solution to the simultaneous equations

$$4-3\lambda=1+4\mu \quad \ldots(1)$$

$$1+3\lambda=-2+3\mu \quad \ldots(2)$$

$$2+\lambda=\frac{1}{2}+2\mu \quad \ldots(3)$$

$$(3)\Rightarrow\lambda=2\mu-\frac{3}{2}$$

Substituting into (1):

$$4-6\mu+\frac{9}{2}=1+4\mu\Rightarrow10\mu=\frac{15}{2}$$

$$\therefore\mu=\frac{3}{4},\lambda=0$$

But this is inconsistent for (2):

$$1+0\neq-2+\frac{9}{4}$$

So there is no solution.

COMMENT

Note that if the lines do intersect, this method also gives you the value of the parameter λ (or μ) for the intersection point.

Method 3: using the vector equations of the lines and taking scalar products with a common normal.

The second line $\dfrac{x-1}{4} = \dfrac{y+2}{3} = \dfrac{z-\frac{1}{2}}{2}$

has vector equation $\boldsymbol{r} = \begin{pmatrix} 1 \\ -2 \\ \frac{1}{2} \end{pmatrix} + \mu \begin{pmatrix} 4 \\ 3 \\ 2 \end{pmatrix}$

The common normal to the two lines is

$\boldsymbol{n} = \begin{pmatrix} -3 \\ 3 \\ 1 \end{pmatrix} \times \begin{pmatrix} 4 \\ 3 \\ 2 \end{pmatrix} = \begin{pmatrix} 3 \\ 10 \\ -21 \end{pmatrix}$

For an intersection of the two lines:

$\begin{pmatrix} 4 \\ 1 \\ 2 \end{pmatrix} + \lambda \begin{pmatrix} -3 \\ 3 \\ 1 \end{pmatrix} = \begin{pmatrix} 1 \\ -2 \\ \frac{1}{2} \end{pmatrix} + \mu \begin{pmatrix} 4 \\ 3 \\ 2 \end{pmatrix}$

Taking the scalar product with $\hat{\boldsymbol{n}}$:

$\dfrac{1}{|\boldsymbol{n}|} \begin{pmatrix} 3 \\ 10 \\ -21 \end{pmatrix} \cdot \begin{pmatrix} 4 \\ 1 \\ 2 \end{pmatrix} + 0\lambda$

$= \dfrac{1}{|\boldsymbol{n}|} \begin{pmatrix} 3 \\ 10 \\ -21 \end{pmatrix} \cdot \begin{pmatrix} 1 \\ -2 \\ \frac{1}{2} \end{pmatrix} + 0\mu$

$\Rightarrow -\dfrac{20}{|\boldsymbol{n}|} = -\dfrac{27.5}{|\boldsymbol{n}|}$

This equation is invalid, therefore the two lines do not intersect.

> **COMMENT**
>
> Although this method does not give the intersection point if the lines do intersect, in the event that they do not, the discrepancy between the two sides of the invalid equation is the shortest distance between the lines, which in this case is $\dfrac{7.5}{|\boldsymbol{n}|}$. While this detail is not needed in this question, method 3 would be extremely useful for a question which did ask for the distance between the lines in a subsequent part.

Method 4: using projections – scalar products with two arbitrary vectors normal to one of the lines.

The second line $\dfrac{x-1}{4} = \dfrac{y+2}{3} = \dfrac{z-\frac{1}{2}}{2}$

has vector equation $\boldsymbol{r} = \begin{pmatrix} 1 \\ -2 \\ \frac{1}{2} \end{pmatrix} + \mu \begin{pmatrix} 4 \\ 3 \\ 2 \end{pmatrix}$

For an intersection of the two lines:

$\begin{pmatrix} 4 \\ 1 \\ 2 \end{pmatrix} + \lambda \begin{pmatrix} -3 \\ 3 \\ 1 \end{pmatrix} = \begin{pmatrix} 1 \\ -2 \\ \frac{1}{2} \end{pmatrix} + \mu \begin{pmatrix} 4 \\ 3 \\ 2 \end{pmatrix}$ …(*)

By inspection, $\begin{pmatrix} 1 \\ 0 \\ -2 \end{pmatrix}$ and $\begin{pmatrix} 3 \\ -4 \\ 0 \end{pmatrix}$ are both perpendicular to the direction vector of the second line.

Taking the scalar product of (*) with $\begin{pmatrix} 1 \\ 0 \\ -2 \end{pmatrix}$:

$0 - 5\lambda = 0 + 0\mu$

$\Rightarrow \lambda = 0$

Taking the scalar product of (*) with $\begin{pmatrix} 3 \\ -4 \\ 0 \end{pmatrix}$:

$8 - 21\lambda = 11 + 0\mu$

$\Rightarrow \lambda = -\dfrac{1}{7}$

These give inconsistent values of λ.

Therefore the lines do not intersect.

> **COMMENT**
>
> This method, like the first two, gives the intersection point if there is one. Geometrically it represents a projection of the second line onto the first in a direction perpendicular to it. If two such projections are made, the same point will be found if the lines intersect; otherwise the projections will locate different points, since two skew lines will appear to overlie each other at different places when viewed from different directions.

8 $12x - 3y + 5z = 60$

a Intersection with x-axis has $y = z = 0$:

$12x = 60 \Rightarrow x = 5$

$\therefore P(5, 0, 0)$

Intersection with y-axis has $x = z = 0$:

$-3y = 60 \Rightarrow y = -20$

$\therefore Q(0, -20, 0)$

Intersection with z-axis has $x = y = 0$:

$5z = 60 \Rightarrow z = 12$

$\therefore R(0, 0, 12)$

b $\overrightarrow{PR} = \begin{pmatrix} -5 \\ 0 \\ 12 \end{pmatrix}$, $\overrightarrow{PQ} = \begin{pmatrix} -5 \\ -20 \\ 0 \end{pmatrix}$

Area PQR $= \dfrac{1}{2}|\overrightarrow{PR} \times \overrightarrow{PQ}|$

$= \dfrac{1}{2}\left|\begin{pmatrix} 240 \\ -60 \\ 100 \end{pmatrix}\right|$

$= \left|\begin{pmatrix} 120 \\ -30 \\ 50 \end{pmatrix}\right|$

$= \sqrt{120^2 + 30^2 + 50^2}$

$= \sqrt{17\,800} = 133$ (3SF)

9 a Substituting $x = 2$, $y = 1$, $z = 6$ into the equation for the plane:

LHS $= 5(2) - 3(1) - 6 = 10 - 3 - 6 = 1 =$ RHS

So the point P lies in the plane Π.

b $\overrightarrow{PQ} = \begin{pmatrix} 5 \\ -2 \\ -4 \end{pmatrix}$

Normal to plane Π is $\mathbf{n} = \begin{pmatrix} 5 \\ -3 \\ -1 \end{pmatrix}$

Angle θ between (PQ) and Π is the complement of the angle ϕ between $\overrightarrow{PQ}$ and $\mathbf{n}$:

$\cos\phi = \dfrac{\overrightarrow{PQ} \cdot \mathbf{n}}{|\overrightarrow{PQ}||\mathbf{n}|} = \dfrac{35}{\sqrt{45}\sqrt{35}} = \sqrt{\dfrac{35}{45}}$

$\therefore \sin(\theta) = \sin(\pi - \phi) = \cos(\phi)$

$= \sqrt{\dfrac{35}{45}} = \sqrt{\dfrac{7}{9}} = \dfrac{\sqrt{7}}{3}$

c $PQ = |\overrightarrow{PQ}| = \sqrt{5^2 + 2^2 + 4^2} = \sqrt{45} = 3\sqrt{5}$

d Let R be the point on Π such that (QR) is perpendicular to Π.

Then $P\hat{R}Q = 90°$ and QR is the distance of Q from Π.

$$\therefore QR = PQ \sin\theta = \sqrt{45} \times \sqrt{\frac{35}{45}} = \sqrt{35}$$

10 a $\begin{pmatrix} 3 \\ -1 \\ 1 \end{pmatrix} \times \begin{pmatrix} 1 \\ 2 \\ -1 \end{pmatrix} = \begin{pmatrix} 1-2 \\ 1+3 \\ 6+1 \end{pmatrix} = \begin{pmatrix} -1 \\ 4 \\ 7 \end{pmatrix}$

b The normal vectors to the two planes are $\boldsymbol{n}_1 = \begin{pmatrix} 3 \\ -1 \\ 1 \end{pmatrix}$ and $\boldsymbol{n}_2 = \begin{pmatrix} 1 \\ 2 \\ -1 \end{pmatrix}$

Since $\boldsymbol{n}_1 \cdot \boldsymbol{n}_2 = 0$, the normals are perpendicular and hence the planes are perpendicular.

c Π_1 is given by $\boldsymbol{r} \cdot \begin{pmatrix} 3 \\ -1 \\ 1 \end{pmatrix} = 17$

$\begin{pmatrix} 1 \\ 1 \\ 2 \end{pmatrix} \cdot \begin{pmatrix} 3 \\ -1 \\ 1 \end{pmatrix} = 4 \neq 17$, so M does not lie in Π_1.

Π_2 is given by $\boldsymbol{r} \cdot \begin{pmatrix} 1 \\ 2 \\ -1 \end{pmatrix} = 4$

$\begin{pmatrix} 1 \\ 1 \\ 2 \end{pmatrix} \cdot \begin{pmatrix} 1 \\ 2 \\ -1 \end{pmatrix} = 1 \neq 17$, so M does not lie in Π_2.

d The vector from (a) is perpendicular to the normal vectors of both planes, so it is the direction vector of the line of intersection.

Hence the vector equation of a line parallel to this direction and passing through M is given by

$$\boldsymbol{r} = \begin{pmatrix} 1 \\ 1 \\ 2 \end{pmatrix} + \lambda \begin{pmatrix} -1 \\ 4 \\ 7 \end{pmatrix}$$

11 a $\begin{pmatrix} 1 \\ 3 \\ -2 \end{pmatrix} \times \begin{pmatrix} 3 \\ 5 \\ -1 \end{pmatrix} = \begin{pmatrix} -3+10 \\ -6+1 \\ 5-9 \end{pmatrix} = \begin{pmatrix} 7 \\ -5 \\ -4 \end{pmatrix}$

b i Substituting $x = y = z = 0$ into the equation of either plane leads to a true statement $0 = 0$, so the origin lies in both planes, and hence their intersection also contains the origin.

ii The normal vectors for the planes are

$\boldsymbol{n}_1 = \begin{pmatrix} 1 \\ 3 \\ -2 \end{pmatrix}$ and $\boldsymbol{n}_2 = \begin{pmatrix} 3 \\ 5 \\ -1 \end{pmatrix}$

The direction found in (a), which is perpendicular to both normal vectors, is the direction of the line of intersection.

It was shown in (i) that the origin lies on the line of intersection.

Therefore the vector equation of the line of intersection is

$$\boldsymbol{r} = \lambda \begin{pmatrix} 7 \\ -5 \\ -4 \end{pmatrix}$$

c Substituting $x = 7\lambda$, $y = -5\lambda$, $z = -4\lambda$ into the equation for Π_3:

$7\lambda - 5(-5\lambda) + (-4\lambda) = 8$

$28\lambda = 8$

$\lambda = \frac{2}{7}$

$\therefore$ the point of intersection of the three planes is $\left(2, -\frac{10}{7}, -\frac{8}{7}\right)$

Exercise 14F

1 From GDC, the intersection point is
$\left(\dfrac{5}{3}, \dfrac{16}{3}, -\dfrac{7}{3}\right)$

Alternatively, using algebraic elimination:

$3x + y + z = 8$ ...(1)
$-7x + 3y + z = 2$ ...(2)
$x + y + 3z = 0$ ...(3)

(1)−(3): $2x - 2z = 8$ ...(4)
3×(1)−(2): $16x + 2z = 22$ ...(5)
(4)+(5): $18x = 30 \Rightarrow x = \dfrac{5}{3}$

Then (4) $\Rightarrow z = x - 4 = -\dfrac{7}{3}$

and (1) $\Rightarrow y = 8 - z - 3x = \dfrac{16}{3}$

∴ intersection point is $\left(\dfrac{5}{3}, \dfrac{16}{3}, -\dfrac{7}{3}\right)$

2 Using Gaussian elimination:

$x = 2$ ...(1)
$x + y - z = 7$ ...(2)
$2x + y + z = 3$ ...(3)

(1) $\quad x = 2$ ...(1)
(2)−(1) $\quad y - z = 5$ ...(4)
(3)−2×(1) $\quad y + z = -1$ ...(5)

(1) $\quad x = 2$ ...(1)
(4) $\quad y - z = 5$ ...(4)
(5)−(4) $\quad 2z = -6$ ...(6)

(6) $\Rightarrow z = -3$
Then (4) $\Rightarrow y = 5 + z = 2$
∴ intersection point is $(2, 2, -3)$

3
$2x - z = 1$ ...(1)
$4x + y - z = 5$ ...(2)
$y + z = 3$ ...(3)

(1) $\quad 2x - z = 1$ ...(1)
(2)−2×(1) $\quad y + z = 3$ ...(4)
(3) $\quad y + z = 3$ ...(3)

Clearly these are consistent equations, but the system has only two independent equations, so the intersection is a line. Let $x = t$; then $z = 2t - 1$ and $y = 3 - z = 4 - 2t$.

∴ the line has equation

$\mathbf{r} = \begin{pmatrix} 0 \\ 4 \\ -1 \end{pmatrix} + t \begin{pmatrix} 1 \\ -2 \\ 2 \end{pmatrix}$

(Cartesian form $\dfrac{x}{1} = \dfrac{4-y}{2} = \dfrac{z+1}{2}$)

4
$x - 2y + z = 5$ ...(1)
$2x + y + z = 1$ ...(2)
$x + 2y - z = -2$ ...(3)

(1) $\quad x - 2y + z = 5$ ...(1)
(2)−2×(1) $\quad 5y - z = -9$ ...(4)
(3)−(1) $\quad 4y - 2z = -7$ ...(5)

(1) $\quad x - 2y + z = 5$ ...(1)
4×(4) $\quad 20y - 4z = -36$ ...(6)
5×(5) $\quad 20y - 10z = -35$ ...(7)

(1) $\quad x - 2y + z = 5$ ...(1)
(6) $\quad 20y - 4z = -36$ ...(6)
(7)−(6) $\quad -6z = 1$ ...(8)

(8) $\Rightarrow z = -\dfrac{1}{6}$

Then (4) $\Rightarrow y = \dfrac{z-9}{5} = -\dfrac{11}{6}$

and (1) $\Rightarrow x = 5 + 2y - z = \dfrac{9}{6}$

∴ intersection point is $\left(\dfrac{3}{2}, -\dfrac{11}{6}, -\dfrac{1}{6}\right)$

5 $\quad 2x - y + z = 6 \quad \ldots(1)$
$\quad 3x + y + 5z = -7 \quad \ldots(2)$
$\quad x - 3y - 3z = 8 \quad \ldots(3)$

$\quad (1) \qquad\qquad 2x - y + z = 6 \quad \ldots(1)$
$\quad 2\times(2) - 3\times(1) \qquad 5y + 7z = -32 \quad \ldots(4)$
$\quad 2\times(3) - (1) \qquad\quad -5y - 7z = 10 \quad \ldots(5)$

Clearly, (4) and (5) are inconsistent, and so this system of equations has no solution, i.e. there is no point at which the three planes all intersect.

6 a $\quad 2x + y - 2z = 0 \quad \ldots(1)$
$\quad x - 2y - z = 2 \quad \ldots(2)$
$\quad 3x + 4y - 3z = d \quad \ldots(3)$

$\quad (1) \qquad\qquad\quad 2x + y - 2z = 0 \quad \ldots(1)$
$\quad 2\times(2) - (1) \qquad -5y = 4 \quad \ldots(4)$
$\quad 2\times(3) - 3\times(1) \quad 5y = 2d \quad \ldots(5)$

For this to be a consistent set of equations, require $d = -2$

b With $d = -2$, (2) and (3) are consistent but the system has only two independent equations, so the intersection is a line:

$$y = -\dfrac{4}{5}, \quad x - \dfrac{2}{5} = z$$

In the form of a vector equation, this is

$$\mathbf{r} = \begin{pmatrix} \frac{2}{5} \\ -\frac{4}{5} \\ 0 \end{pmatrix} + \lambda \begin{pmatrix} 1 \\ 0 \\ 1 \end{pmatrix}$$

7 $\quad x - y = 4 \quad \ldots(1)$
$\quad y + z = 1 \quad \ldots(2)$
$\quad x \quad - z = d \quad \ldots(3)$

$\quad (1) \qquad\qquad x - y = 4 \qquad \ldots(1)$
$\quad (2) \qquad\qquad y + z = 1 \qquad \ldots(2)$
$\quad (3) - (1) \qquad y - z = d - 4 \quad \ldots(4)$

$\quad (1) \qquad\qquad x - y = 4 \qquad \ldots(1)$
$\quad (2) \qquad\qquad y + z = 1 \qquad \ldots(2)$
$\quad (2) - (4) \qquad 2z = 5 - d \quad \ldots(5)$

$(5) \Rightarrow z = \dfrac{5-d}{2}$

$(2) \Rightarrow y = 1 - z = \dfrac{d-3}{2}$

$(1) \Rightarrow x = 4 + y = \dfrac{d+5}{2}$

∴ intersection point is $\left(\dfrac{d+5}{2}, \dfrac{d-3}{2}, \dfrac{5-d}{2}\right)$

8 a If each plane is written in the form $\mathbf{r} \cdot \hat{\mathbf{n}}_i = k_i$, where $\hat{\mathbf{n}}_i$ is the unit normal to plane Π_i, then k_i is the perpendicular distance from the origin to the plane.

In all three cases, this distance is zero, hence the origin lies in all three planes.

The origin therefore lies in the intersection of the planes.

> **COMMENT**
>
> Coherently explaining something which seems clear can sometimes be tricky. It would be equally valid to show that (0, 0, 0) is consistent with the equation of each plane.

b $\quad x+y=0 \quad \ldots(1)$

$\quad x-4y-2z=0 \quad \ldots(2)$

$\quad \frac{1}{2}x+3y+z=0 \quad \ldots(3)$

(1)	$x+y=0$	...(1)
(1)−(2)	$5y+2z=0$	...(4)
2×(3)−(1)	$5y+2z=0$	...(5)

COMMENT

As always, if you can avoid having fractions in your answer, it will appear tidier and you will lower your chances of making arithmetical errors. In the elimination process, consider multiplying to make terms match instead of dividing (or multiplying by fractions).

Equations (4) and (5) are consistent, but the system has only two independent equations, so the intersection is a line.

Let $x=2t$; then $y=-2t$ and $z=5t$, so the equation of the line is

$$\frac{x}{2}=-\frac{y}{2}=\frac{z}{5}$$

or, in vector form,

$$r = t\begin{pmatrix} 2 \\ -2 \\ 5 \end{pmatrix}$$

The direction vector is $d = 2i - 2j + 5k$

9 a $\quad x-2y+z=7 \quad \ldots(1)$

$\quad 2x+y-3z=9 \quad \ldots(2)$

$\quad x+y-az=3 \quad \ldots(3)$

(1)	$x-2y+z=7$	...(1)
(2)−2×(1)	$5y-5z=-5$	...(4)
(3)−(1)	$3y-(1+a)z=-4$	...(5)
(1)	$x-2y+z=7$	...(1)
(2)÷5	$y-z=-1$	...(6)
(5)−3×(6)	$(2-a)z=-1$	...(7)

Equation (7) is invalid if $a=2$, so for this value the planes do not intersect.

b Taking only (1) and (6), the line of intersection of Π_1 and Π_2 can be obtained:

Let $z=t$; then

$(6) \Rightarrow y = z-1 = t-1$

$(1) \Rightarrow x = 7+2y-z = 5+t$

So the equation of the line is, in vector form,

$$r = \begin{pmatrix} 5 \\ -1 \\ 0 \end{pmatrix} + t\begin{pmatrix} 1 \\ 1 \\ 1 \end{pmatrix}$$

or, in Cartesian form, $x-5 = y+1 = z$

10 a $\quad x-y-z=-2 \quad \ldots(1)$

$\quad 2x+3y-7z=a+4 \quad \ldots(2)$

$\quad x+2y+pz=a^2 \quad \ldots(3)$

(1)	$x-y-z=-2$	...(1)
(2)−2×(1)	$5y-5z=a+8$	...(4)
(3)−(1)	$3y+(p+1)z=a^2+2$	...(5)
(1)	$x-y-z=-2$	...(1)
(4)	$5y-5z=a+8$	...(4)
5×(5)−3×(4)	$(5p+20)z=5a^2-3a-14$	...(6)

14 Lines and planes in space

Equation (6) is invalid if $p = -4$, unless $5a^2 - 3a - 14 = (5a+7)(a-2) = 0$, in which case equation (6) degenerates to $0 = 0$ and there are only two independent equations in the system, so that the intersection of the planes is a line.

$p = -4$ and $(5a+7)(a-2) = 0$

$\Rightarrow p = -4$ and $a = -\dfrac{7}{5}$ or 2

b With $p = -4$ and $a = 2$, equation (4) becomes $5y - 5z = 10 \Rightarrow y - z = 2$

Let $z = t$; then $y = 2 + z = 2 + t$

and $(1) \Rightarrow x = y + z - 2 = 2t$

$\therefore$ the equation of the line is

$$r = \begin{pmatrix} 0 \\ 2 \\ 0 \end{pmatrix} + t \begin{pmatrix} 2 \\ 1 \\ 1 \end{pmatrix}$$

or, in Cartesian form, $\dfrac{x}{2} = y - 2 = z$

Exercise 14G

> **COMMENT**
>
> As suggested in the preamble to this exercise, there are many ways to approach these problems. Each of the worked solutions below is an example only, and should not be taken as the 'best' way. You should try a variety of methods and see which of them feel most intuitive. Always remember that when asked to find an intersection, you can take two equations simultaneously – what is true for the general point r in one equation can be applied to the expression of r in the other – to find parameters. Don't be afraid to try novel approaches.

1 a A normal to the plane is $n = \begin{pmatrix} 2 \\ 2 \\ -1 \end{pmatrix}$

Line with direction n which passes through $(-3, -3, 4)$ has equation

$$r = \begin{pmatrix} -3 \\ -3 \\ 4 \end{pmatrix} + \lambda \begin{pmatrix} 2 \\ 2 \\ -1 \end{pmatrix}$$

or, in Cartesian form, $\dfrac{x+3}{2} = \dfrac{y+3}{2} = 4 - z$

b The plane has equation $r \cdot n = 11$

The intersection satisfies both this plane equation and the line equation

$$r = \begin{pmatrix} -3 \\ -3 \\ 4 \end{pmatrix} + \lambda \begin{pmatrix} 2 \\ 2 \\ -1 \end{pmatrix}$$

Taking the scalar product of the line equation with n:

$11 = -16 + 9\lambda$

$9\lambda = 27$

$\lambda = 3$

This describes the intersection point Q, which therefore has coordinates $(3, 3, 1)$

c Shortest distance from point P to plane Π will be the distance PQ.

Since Q corresponds to $\lambda = 3$,

$PQ = 3|n|$

$= 3\sqrt{2^2 + 2^2 + 1^2}$

$= 9$

2 a Normal to Π_1: $n_1 = \begin{pmatrix} 1 \\ -3 \\ 1 \end{pmatrix}$

Normal to Π_2: $n_2 = \begin{pmatrix} 3 \\ -9 \\ 3 \end{pmatrix} = 3n_1$

The normal vectors are parallel, so the planes must be parallel.

b Substituting $x = y = z = 0$ into the equation for Π_2 leads to a true statement $0 = 0$, so the point $(0, 0, 0)$ does lie in Π_2.

c $r = \lambda \begin{pmatrix} 1 \\ -3 \\ 1 \end{pmatrix}$

d

> **COMMENT**
>
> The question is leading you to establish the position of point P where the line in (c) intersects plane Π_1, and then calculate the distance OP.
>
> However, there is a faster way to answer the question: the distance between two parallel planes $\mathbf{r} \cdot \mathbf{n} = k_1$ and $\mathbf{r} \cdot \mathbf{n} = k_2$ is
>
> $\dfrac{|k_1 - k_2|}{|\mathbf{n}|}$; this is a result you can quote and use.

The distance d between $\mathbf{r} \cdot \mathbf{n}_1 = 6$ and $\mathbf{r} \cdot \mathbf{n}_1 = 0$ is equal to $\dfrac{6}{|\mathbf{n}_1|}$

$\therefore d = \dfrac{6}{\sqrt{1^2 + 3^2 + 1^2}} = \dfrac{6\sqrt{11}}{11}$

3 a $\begin{pmatrix} -1 \\ 0 \\ 2 \end{pmatrix} \times \begin{pmatrix} 0 \\ 1 \\ 3 \end{pmatrix} = \begin{pmatrix} 0-2 \\ 0+3 \\ -1-0 \end{pmatrix} = \begin{pmatrix} -2 \\ 3 \\ -1 \end{pmatrix}$

b i At the intersection,

$\begin{pmatrix} 7 \\ -3 \\ 2 \end{pmatrix} + t \begin{pmatrix} -1 \\ 0 \\ 2 \end{pmatrix} = \begin{pmatrix} 1 \\ 1 \\ 26 \end{pmatrix} + s \begin{pmatrix} 0 \\ 1 \\ 3 \end{pmatrix}$ ...(*)

Taking the scalar product of equation (*) with $\begin{pmatrix} -2 \\ 3 \\ -1 \end{pmatrix}$ gives

$-25 + 0t = -25 + 0s$

This is a valid statement, so the two lines do intersect.

ii By observation, $\begin{pmatrix} 0 \\ 1 \\ 0 \end{pmatrix}$ is a vector perpendicular to the direction vector of the first line.

Taking the scalar product of equation (*) with $\begin{pmatrix} 0 \\ 1 \\ 0 \end{pmatrix}$ gives

$-3 + 0t = 1 + s$

$\Rightarrow s = -4$

$\therefore$ the intersection is at $(1, -3, 14)$

(As a check, this lies on the first line for $t = 6$.)

c A normal to plane Π is $\mathbf{n} = \begin{pmatrix} -2 \\ 3 \\ -1 \end{pmatrix}$ and the plane contains point $(1, 1, 26)$

$\therefore$ the equation of Π is

$\mathbf{r} \cdot \begin{pmatrix} -2 \\ 3 \\ -1 \end{pmatrix} = \begin{pmatrix} 1 \\ 1 \\ 26 \end{pmatrix} \cdot \begin{pmatrix} -2 \\ 3 \\ -1 \end{pmatrix} = -25$

or, in Cartesian form,

$-2x + 3y - z = -25$ or $2x - 3y + z = 25$

4 a $\overrightarrow{AD} = \begin{pmatrix} -1 \\ 5 \\ 2 \end{pmatrix}$, $\overrightarrow{AB} = \begin{pmatrix} 1 \\ -1 \\ 3 \end{pmatrix}$, $\overrightarrow{AC} = \begin{pmatrix} 2 \\ 0 \\ 1 \end{pmatrix}$

$\overrightarrow{AD} \cdot \overrightarrow{AB} = -1 - 5 + 6 = 0 \Rightarrow \overrightarrow{AD}$ is perpendicular to $\overrightarrow{AB}$

$\overrightarrow{AD} \cdot \overrightarrow{AC} = -2 + 0 + 2 = 0 \Rightarrow \overrightarrow{AD}$ is perpendicular to $\overrightarrow{AC}$

b From (a), $\overrightarrow{AD}$ is a normal $\mathbf{n}$ to the plane containing points A, B and C.

So the equation of plane Π is given by

$$\mathbf{r} \cdot \mathbf{n} = \mathbf{a} \cdot \mathbf{n} = \begin{pmatrix} 7 \\ 0 \\ 1 \end{pmatrix} \cdot \begin{pmatrix} -1 \\ 5 \\ 2 \end{pmatrix} = -5$$

i.e. $\mathbf{r} \cdot \begin{pmatrix} -1 \\ 5 \\ 2 \end{pmatrix} = -5$

c Distance from point D to the plane is equal to the length AD, since $\overrightarrow{AD}$ is perpendicular to Π.

$$|\overrightarrow{AD}| = \left\| \begin{pmatrix} -1 \\ 5 \\ 2 \end{pmatrix} \right\| = \sqrt{1^2 + 5^2 + 2^2} = \sqrt{30}$$

d $\overrightarrow{AD_1} = -\overrightarrow{AD}$, so

$$\overrightarrow{OD_1} = \overrightarrow{OA} + \overrightarrow{AD_1}$$
$$= \overrightarrow{OA} - \overrightarrow{AD}$$
$$= \begin{pmatrix} 7 \\ 0 \\ 1 \end{pmatrix} - \begin{pmatrix} -1 \\ 5 \\ 2 \end{pmatrix}$$
$$= \begin{pmatrix} 8 \\ -5 \\ -1 \end{pmatrix}$$

∴ the coordinates of D_1 are $(8, -5, -1)$.

5 a In vector form, l_1 is given by

$$\mathbf{r} = \begin{pmatrix} 2 \\ -1 \\ 2 \end{pmatrix} + \lambda \begin{pmatrix} 3 \\ -1 \\ 1 \end{pmatrix}$$

The equation of l_2 may be rewritten in standardised Cartesian form as

$$\frac{x-5}{3} = \frac{y-1}{-1} = \frac{z+4}{1}$$

So in vector form l_2 is given by

$$\mathbf{r} = \begin{pmatrix} 5 \\ 1 \\ -4 \end{pmatrix} + \mu \begin{pmatrix} 3 \\ -1 \\ 1 \end{pmatrix}$$

The two lines have the same direction vector and so are parallel.

b For $\lambda = 4$, in l_1 the position is

$$\mathbf{r} = \begin{pmatrix} 2 \\ -1 \\ 2 \end{pmatrix} + 4 \begin{pmatrix} 3 \\ -1 \\ 1 \end{pmatrix} = \begin{pmatrix} 14 \\ -5 \\ 6 \end{pmatrix}$$

So A $(14, -5, 6)$ does lie on l_1.

c Point B lies on l_2 and so has position

vector $\begin{pmatrix} 5 \\ 1 \\ -4 \end{pmatrix} + \mu \begin{pmatrix} 3 \\ -1 \\ 1 \end{pmatrix}$ for some value of μ.

$$\therefore \overrightarrow{AB} = \begin{pmatrix} -9 + 3\mu \\ 6 - \mu \\ -10 + \mu \end{pmatrix}$$

Require that $\overrightarrow{AB} \cdot \begin{pmatrix} 3 \\ -1 \\ 1 \end{pmatrix} = 0$:

$-27 + 9\mu - 6 + \mu - 10 + \mu = 0$

$11\mu = 43$

$\mu = \dfrac{43}{11}$

∴ the coordinates of B are

$$\left(\frac{184}{11}, -\frac{32}{11}, -\frac{1}{11} \right)$$

d The distance d between l_1 and l_2 equals $|\overrightarrow{AB}|$.

$$\therefore d = \left\| \frac{1}{11} \begin{pmatrix} 30 \\ 23 \\ -67 \end{pmatrix} \right\|$$

$$= \frac{1}{11} \sqrt{30^2 + 23^2 + 67^2}$$

$$= 6.99 \text{ (3SF)}$$

6 a l_1 has general point $(1+3\lambda, -1+4\lambda, 3-3\lambda)$

Substituting into the Cartesian form of l_2:

$$\frac{13+3\lambda}{2} = 4\lambda - 1 = 20 - 3\lambda$$

$$13 + 3\lambda = 8\lambda - 2 = 40 - 6\lambda$$

$$15 = 5\lambda = 42 - 9\lambda$$

The consistent solution is $\lambda = 3$

∴ the point of intersection is $(10, 11, -6)$

> **COMMENT**
> Remember that there are many easy 'checks' for this answer; for example, substituting in the other known points on the lines, $(1, -1, 3)$ and $(-12, 0, -17)$, provides a quick check, and if the values work then you can be confident in your answer.

b Direction vector of l_1 is $\mathbf{d}_1 = \begin{pmatrix} 3 \\ 4 \\ -3 \end{pmatrix}$

Direction vector of l_2 is $\mathbf{d}_2 = \begin{pmatrix} 2 \\ 1 \\ 1 \end{pmatrix}$

$$\mathbf{d}_1 \times \mathbf{d}_2 = \begin{pmatrix} 7 \\ -9 \\ -5 \end{pmatrix}$$

∴ a vector perpendicular to both lines is $\begin{pmatrix} 7 \\ -9 \\ -5 \end{pmatrix}$

c Equation of the plane with normal $\mathbf{n}$ containing a point with position vector $\mathbf{a}$ is given by $\mathbf{r} \cdot \mathbf{n} = \mathbf{a} \cdot \mathbf{n}$:

$$\mathbf{r} \cdot \begin{pmatrix} 7 \\ -9 \\ -5 \end{pmatrix} = \begin{pmatrix} 10 \\ 11 \\ -6 \end{pmatrix} \cdot \begin{pmatrix} 7 \\ -9 \\ -5 \end{pmatrix}$$

$$\Rightarrow 7x - 9y - 5z = 1$$

7 a $\overrightarrow{AB} = \begin{pmatrix} 4 \\ -1 \\ 1 \end{pmatrix}$, $\overrightarrow{AC} = \begin{pmatrix} 2 \\ 0 \\ 3 \end{pmatrix}$

$$\Rightarrow \overrightarrow{AB} \times \overrightarrow{AC} = \begin{pmatrix} -3 \\ -10 \\ 2 \end{pmatrix}$$

b Area $ABC = \frac{1}{2}|\overrightarrow{AB} \times \overrightarrow{AC}|$

$$= \frac{1}{2}\sqrt{3^2 + 10^2 + 2^2}$$

$$= \frac{1}{2}\sqrt{113}$$

$$= 5.32 \text{ (3SF)}$$

c Π has normal $\mathbf{n} = \begin{pmatrix} -3 \\ -10 \\ 2 \end{pmatrix}$ and passes through point $A(8, 0, 4)$.

So equation of Π is given by $\mathbf{r} \cdot \mathbf{n} = \mathbf{a} \cdot \mathbf{n}$:

$$\mathbf{r} \cdot \begin{pmatrix} -3 \\ -10 \\ 2 \end{pmatrix} = \begin{pmatrix} 8 \\ 0 \\ 4 \end{pmatrix} \cdot \begin{pmatrix} -3 \\ -10 \\ 2 \end{pmatrix}$$

$$-3x - 10y + 2z = -16$$

or $3x + 10y - 2z = 16$

d Line through $D(-7, -28, 11)$ with direction $\mathbf{n}$ has equation

$$\mathbf{r} = \begin{pmatrix} -7 \\ -28 \\ 11 \end{pmatrix} + \lambda \begin{pmatrix} -3 \\ -10 \\ 2 \end{pmatrix}$$

e Substituting the general point on this line, $(-7-3\lambda, -28-10\lambda, 11+2\lambda)$, into the Cartesian equation of the plane gives

$$3(-7-3\lambda) + 10(-28-10\lambda) - 2(11+2\lambda) = 16$$
$$-21 - 9\lambda - 280 - 100\lambda - 22 - 4\lambda = 16$$
$$-113\lambda = 339$$
$$\lambda = -3$$

So the intersection point E of the line with the plane is $(2, 2, 5)$.

Distance from point D to the plane is
$$DE = |\lambda \mathbf{n}| = 3\sqrt{113} = 31.9 \text{ (3SF)}$$

f Volume of the pyramid is
$$\frac{1}{3}(\text{Area ABC}) \times DE$$

$$\text{Volume} = \frac{1}{3} \times \left(\frac{1}{2}\sqrt{113}\right) \times 3\sqrt{113}$$
$$= \frac{113}{2}$$
$$= 56.5$$

> **COMMENT**
>
> Where parameters and coordinates of intersections are 'nice' numbers (i.e. integers or simple rational values), it can be useful to keep distances in surd form when combining, and then calculations need not be subject to rounding errors or extensive use of calculator memory.

8 a Vector equation for l:
$$\mathbf{r} = \begin{pmatrix} -1 \\ 1 \\ 4 \end{pmatrix} + \lambda \begin{pmatrix} 6 \\ 1 \\ 5 \end{pmatrix}$$

b The normal vector of the plane must be perpendicular to each vector in the plane.

As A and B are both points in the plane, $\overrightarrow{AB}$ is a vector in the plane and so $\overrightarrow{AB}$ is perpendicular to $\mathbf{n}$.

Line l lies in the plane, so its direction vector $\mathbf{d}$ is a vector in the plane and therefore $\mathbf{d}$ is perpendicular to $\mathbf{n}$.

c $\overrightarrow{AB} = \begin{pmatrix} 4 \\ 2 \\ -3 \end{pmatrix}$. A vector perpendicular to both $\overrightarrow{AB}$ and $\mathbf{d}$ is

$$\overrightarrow{AB} \times \mathbf{d} = \begin{pmatrix} 4 \\ 2 \\ -3 \end{pmatrix} \times \begin{pmatrix} 6 \\ 1 \\ 5 \end{pmatrix} = \begin{pmatrix} 13 \\ -38 \\ -8 \end{pmatrix}$$

Hence $\mathbf{n}$ may equal $\begin{pmatrix} 13 \\ -38 \\ -8 \end{pmatrix}$

d Π has normal $\mathbf{n}$ and contains point A. Its equation is given by $\mathbf{r} \cdot \mathbf{n} = \mathbf{a} \cdot \mathbf{n}$:

$$\mathbf{r} \cdot \begin{pmatrix} 13 \\ -38 \\ -8 \end{pmatrix} = \begin{pmatrix} -1 \\ 1 \\ 4 \end{pmatrix} \cdot \begin{pmatrix} 13 \\ -38 \\ -8 \end{pmatrix}$$
$$\Rightarrow 13x - 38y - 8z = -83$$

9 a A normal to the plane is $\begin{pmatrix} 6 \\ -2 \\ 1 \end{pmatrix}$

$\therefore$ line l has equation $\mathbf{r} = \lambda \begin{pmatrix} 6 \\ -2 \\ 1 \end{pmatrix}$

This intersects the plane where

$6(6\lambda) - 2(-2\lambda) + 1(\lambda) = 16$

$41\lambda = 16$

$\lambda = \dfrac{16}{41}$

So the foot of the perpendicular is at

$P\left(\dfrac{96}{41}, -\dfrac{32}{41}, \dfrac{16}{41}\right)$

b The shortest distance from the plane to the origin is the distance OP:

$OP = \left|\dfrac{16}{41}\begin{pmatrix}6\\-2\\1\end{pmatrix}\right|$

$= \dfrac{16}{41}\sqrt{6^2 + 2^2 + 1^2}$

$= \dfrac{16\sqrt{41}}{41}$

10 a A normal vector to a plane

$ax + by + cz = d$ is $\begin{pmatrix}a\\b\\c\end{pmatrix}$.

$\therefore$ normal to Π_1 is $\boldsymbol{n}_1 = \begin{pmatrix}1\\0\\-1\end{pmatrix}$ and

normal to Π_2 is $\boldsymbol{n}_2 = \begin{pmatrix}-1\\0\\1\end{pmatrix}$;

since $\boldsymbol{n}_1 = -\boldsymbol{n}_2$, the normal vectors are parallel and therefore the planes are parallel.

b Line through origin that is perpendicular to the two planes:

$\boldsymbol{r} = \lambda\begin{pmatrix}1\\0\\-1\end{pmatrix}$

c General point on the line is given by $(\lambda, 0, -\lambda)$.

i Substituting the general point into the equation for Π_1 gives

$2\lambda = 4 \Rightarrow \lambda = 2$

Point of intersection is $P_1(2, 0, -2)$

ii Substituting the general point into the equation for Π_2 gives

$-2\lambda = 8 \Rightarrow \lambda = -4$

Point of intersection is $P_2(-4, 0, 4)$

d Distance between the two planes is

$P_1P_2 = \sqrt{6^2 + 0^2 + 6^2} = 6\sqrt{2}$

> **COMMENT**
>
> This could also have been determined without using the results from (b), as described in the comment associated with question 2(d) of this exercise:
>
> The distance between two parallel planes $\boldsymbol{r}\cdot\boldsymbol{n} = k_1$ and $\boldsymbol{r}\cdot\boldsymbol{n} = k_2$ is $\dfrac{|k_1 - k_2|}{|\boldsymbol{n}|}$, so the distance between Π_1 and Π_2 is
>
> $\dfrac{|4 - (-8)|}{\sqrt{2}} = 6\sqrt{2}$; remember that you would need to rewrite the equation for Π_2 as $x - z = -8$, so that both equations use the same normal vector $\boldsymbol{n}$.

Mixed examination practice 14

Short questions

1 Direction vector is $\begin{pmatrix}3\\1\\0\end{pmatrix}$, and the line passes through point $(6, 0, 1)$

$\therefore$ equation of line is $\boldsymbol{r} = \begin{pmatrix}6\\0\\1\end{pmatrix} + \lambda\begin{pmatrix}3\\1\\0\end{pmatrix}$

2 Substituting $x=3$, $y=-1$, $z=2$ into

$$\frac{x+3}{2}=\frac{y-8}{-3}=\frac{z+13}{p}:$$

$$\frac{6}{2}=\frac{-9}{-3}=\frac{15}{p}$$

$$\Rightarrow p=5$$

3 a $\mathbf{r}\cdot\mathbf{n}=\begin{pmatrix}3\\-1\\2\end{pmatrix}\cdot\begin{pmatrix}3\\1\\-1\end{pmatrix}$

$$\Rightarrow \mathbf{r}\cdot\begin{pmatrix}3\\1\\-1\end{pmatrix}=6$$

or, in Cartesian form, $3x+y-z=6$

b Substituting $(x,y,z)=(a, 2a, a-1)$ into the plane equation:

$3a+2a-(a-1)=6$

$4a=5$

$a=\dfrac{5}{4}$

4 $x-2y+z=5$ ...(1)
$2x+y+z=1$ ...(2)
$x+2y-z=-2$ ...(3)

(1) $x-2y+z=5$ ...(1)
(2)−2×(1) $5y-z=-9$ ...(4)
(3)−(1) $4y-2z=-7$ ...(5)

(1) $x-2y+z=5$ ...(1)
(4) $5y-z=-9$ ...(4)
$4\times(4)-5\times(5)$ $6z=-1$ ...(6)

$(6)\Rightarrow z=-\dfrac{1}{6}$

$(4)\Rightarrow y=\dfrac{z-9}{5}=-\dfrac{11}{6}$

$(1)\Rightarrow x=2y-z+5=\dfrac{9}{6}$

∴ the point of intersection of the three planes is $\left(\dfrac{3}{2},-\dfrac{11}{6},-\dfrac{1}{6}\right)$

5 Point A corresponds to $\lambda=0$, and point B corresponds to $\lambda=2$

$AP=3AB \Rightarrow$ P corresponds to $\lambda=6$

∴ the coordinates of P are $(11, 13, 8)$

6 $\mathbf{r}=\begin{pmatrix}-3\\0\\4\end{pmatrix}+\lambda\begin{pmatrix}2\\2\\-1\end{pmatrix}$

$|\mathbf{d}|=\sqrt{2^2+2^2+1}=3$

Point A corresponds to $\lambda=0$

So a point that is 10 units from A has $\lambda=\pm\dfrac{10}{3}$

The possible coordinates are

$\begin{pmatrix}-3\\0\\4\end{pmatrix}\pm\dfrac{10}{3}\begin{pmatrix}2\\2\\-1\end{pmatrix}=\left(\dfrac{11}{3},\dfrac{20}{3},\dfrac{2}{3}\right)$ and

$\left(-\dfrac{29}{3},-\dfrac{20}{3},\dfrac{22}{3}\right)$

7 a $\begin{pmatrix}8\\-11\\20\end{pmatrix}-\begin{pmatrix}4\\1\\12\end{pmatrix}=\begin{pmatrix}4\\-12\\8\end{pmatrix}$, so a direction

vector for the line is $\mathbf{d}=\begin{pmatrix}1\\-3\\2\end{pmatrix}$

∴ the line has equation

$\mathbf{r}=\begin{pmatrix}4\\1\\12\end{pmatrix}+\lambda\begin{pmatrix}1\\-3\\2\end{pmatrix}$

b Since point P lies on l, $\overrightarrow{OP}=\begin{pmatrix}4+\lambda\\1-3\lambda\\12+2\lambda\end{pmatrix}$

for some value of λ

168 Mixed examination practice 14

Require that $\overrightarrow{OP} \cdot \mathbf{d} = 0$:

$$\begin{pmatrix} 4+\lambda \\ 1-3\lambda \\ 12+2\lambda \end{pmatrix} \cdot \begin{pmatrix} 1 \\ -3 \\ 2 \end{pmatrix} = 0$$

$(4-3+24) + \lambda(1+9+4) = 0$

$25 + 14\lambda = 0$

$\lambda = -\dfrac{25}{14}$

∴ the coordinates of P are $\left(\dfrac{31}{14}, \dfrac{89}{14}, \dfrac{118}{14}\right)$

8 a $\mathbf{b} = \begin{pmatrix} 1 \\ -1 \\ 4 \end{pmatrix}$, $\mathbf{a} = \begin{pmatrix} 2 \\ -1 \\ 1 \end{pmatrix}$

$\mathbf{b} \times \mathbf{a} = \begin{pmatrix} (-1)(1)-(-1)(4) \\ (2)(4)-(1)(1) \\ (1)(-1)-(2)(-1) \end{pmatrix} = \begin{pmatrix} 3 \\ 7 \\ 1 \end{pmatrix}$

b $\Pi_1: \mathbf{r} \cdot \begin{pmatrix} 2 \\ -1 \\ 1 \end{pmatrix} = 5$, $\Pi_2: \mathbf{r} \cdot \begin{pmatrix} 1 \\ -1 \\ 4 \end{pmatrix} = 12$

$\begin{pmatrix} 2 \\ 2 \\ 3 \end{pmatrix} \cdot \begin{pmatrix} 2 \\ -1 \\ 1 \end{pmatrix} = 4-2+3 = 5 \Rightarrow (2, 2, 3)$ lies in plane Π_1

$\begin{pmatrix} 2 \\ 2 \\ 3 \end{pmatrix} \cdot \begin{pmatrix} 1 \\ -1 \\ 4 \end{pmatrix} = 2-2+12 = 12 \Rightarrow (2, 2, 3)$ lies in plane Π_2

c $\mathbf{a}$ and $\mathbf{b}$ are the normals to planes Π_1 and Π_2 respectively.

The line of intersection of the two planes must be perpendicular to both normals, and so lies in the direction $\mathbf{b} \times \mathbf{a}$.

Since $(2, 2, 3)$ lies in both planes, it must lie on the line of intersection.

Therefore the line of intersection has vector equation $\mathbf{r} = \begin{pmatrix} 2 \\ 2 \\ 3 \end{pmatrix} + \lambda \begin{pmatrix} 3 \\ 7 \\ 1 \end{pmatrix}$

In Cartesian form, this is

$$\dfrac{x-2}{3} = \dfrac{y-2}{7} = \dfrac{z-3}{1}$$

9

> **COMMENT**
>
> It may seem routine to solve this problem using a substitution, as shown in the first method. In fact, some thought leads to the second, faster, approach.

Method 1: substituting a parameterised point into the equation.

The general point on the line is

$\left(3+\lambda, -1+\dfrac{5\lambda}{2}, 5+k\lambda\right)$

Substituting into the equation for the plane:

$3(3+\lambda) + 2\left(-1+\dfrac{5\lambda}{2}\right) - (5+k\lambda) = 2$

$9 + 3\lambda - 2 + 5\lambda - 5 - k\lambda = 2$

$(8-k)\lambda = 0$

∴ $k = 8$ (since λ is arbitrary)

Method 2: using the normal.

Require that the direction vector of the line $\dfrac{x-3}{1} = \dfrac{y+1}{\frac{5}{2}} = \dfrac{z-5}{k}$ be perpendicular to the normal of the plane:

$$\begin{pmatrix} 1 \\ \frac{5}{2} \\ k \end{pmatrix} \cdot \begin{pmatrix} 3 \\ 2 \\ -1 \end{pmatrix} = 0$$

$3 + 5 - k = 0$

$\therefore k = 8$

10 a $u = \begin{pmatrix} 1 \\ 2 \\ 3 \end{pmatrix}$, $v = \begin{pmatrix} 2 \\ -1 \\ 2 \end{pmatrix}$

$$u \times v = \begin{pmatrix} (2)(2)-(3)(-1) \\ (3)(2)-(1)(2) \\ (1)(-1)-(2)(2) \end{pmatrix} = \begin{pmatrix} 7 \\ 4 \\ -5 \end{pmatrix}$$

b The normals to the two planes are u and v, so the direction vector of the intersection line is $d = u \times v$, since this is perpendicular to both normals.

Using $d \cdot u = d \cdot v = 0$:

$d \cdot w = d \cdot (\lambda u + \mu v)$
$ = \lambda d \cdot u + \mu d \cdot v$
$ = 0$

That is, the direction of the intersection line is perpendicular to $w = \lambda u + \mu v$ for all values of λ and μ.

11 Rewrite the Cartesian equation of the first line as $l_1 : \dfrac{x-0}{1} = \dfrac{y-3}{-2} = \dfrac{z-1}{1}$

Vector equation for $l_1 : r = \begin{pmatrix} 0 \\ 3 \\ 1 \end{pmatrix} + \lambda \begin{pmatrix} 1 \\ -2 \\ 1 \end{pmatrix}$

$l_2 : \dfrac{x-2}{3} = \dfrac{y+1}{-3} = \dfrac{z-3}{5}$

Vector equation for $l_2 : r = \begin{pmatrix} 2 \\ -1 \\ 3 \end{pmatrix} + \mu \begin{pmatrix} 3 \\ -3 \\ 5 \end{pmatrix}$

The plane containing both lines must have normal vector n perpendicular to both direction vectors.

$$\begin{pmatrix} 1 \\ -2 \\ 1 \end{pmatrix} \times \begin{pmatrix} 3 \\ -3 \\ 5 \end{pmatrix} = \begin{pmatrix} -7 \\ -2 \\ 3 \end{pmatrix}$$

Choose $n = \begin{pmatrix} 7 \\ 2 \\ -3 \end{pmatrix}$

The plane contains point $(0, 3, 1)$ in line l_1.

Equation of a plane with normal n passing through a point with position vector a has equation $r \cdot n = a \cdot n$:

$$r \cdot \begin{pmatrix} 7 \\ 2 \\ -3 \end{pmatrix} = \begin{pmatrix} 0 \\ 3 \\ 1 \end{pmatrix} \cdot \begin{pmatrix} 7 \\ 2 \\ -3 \end{pmatrix}$$

$\Rightarrow 7x + 2y - 3z = 3$

Check that l_2 does lie in this plane:

If $x = 2$, $y = -1$, $z = 3$ then
$7x + 2y - 3z = 14 - 2 - 9 = 3$

$\therefore (2, -1, 3)$ also lies in the plane
$7x + 2y - 3z = 0$. This plane therefore does indeed contain both l_1 and l_2.

> **COMMENT**
>
> The question effectively asserts that there is a plane containing the two lines. This need not be the case, of course, since a pair of lines exist together in a plane only if they are parallel or intersect; two skew lines cannot lie in the same plane.
>
> It is good practice to check that the solution is valid, either by finding the point of intersection of l_1 and l_2 or by checking that a point on l_2 actually does lie in the proposed plane, as shown.

170 Mixed examination practice 14

Long questions

1 a When $t=1$, $r_2 = \begin{pmatrix} 4 \\ 1 \\ 2 \end{pmatrix} = \overrightarrow{OA}$,

so point A lies on l_2.

b $AB = \left\| \begin{pmatrix} -4 \\ 4 \\ -1 \end{pmatrix} \right\| = \sqrt{33}$

c $d_1 = \begin{pmatrix} 2 \\ -1 \\ 3 \end{pmatrix}$, $d_2 = \begin{pmatrix} 4 \\ -4 \\ 1 \end{pmatrix}$

Angle between l_1 and l_2 is

$\theta = \arccos\left(\dfrac{d_1 \cdot d_2}{|d_1||d_2|}\right)$

$= \arccos\left(\dfrac{8+4+3}{\sqrt{4+1+9}\,\sqrt{16+16+1}}\right)$

$= \arccos\left(\dfrac{15}{\sqrt{14}\,\sqrt{33}}\right)$

$= 45.7°$ (3SF)

d

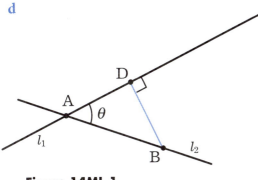

Figure 14ML.1

If D is the point on l_1 that is closest to B, then triangle ABD has

$AB = \sqrt{33}$, $A\hat{D}B = 90°$, $D\hat{A}B = \theta$

$\therefore BD = AB \sin\theta = 4.11$ (3SF)

2 a Suppose that the two lines intersect; then at the intersection point,

$\begin{pmatrix} -3 \\ 3 \\ 18 \end{pmatrix} + \lambda \begin{pmatrix} 2 \\ -1 \\ -8 \end{pmatrix} = \begin{pmatrix} 5 \\ 0 \\ 2 \end{pmatrix} + \mu \begin{pmatrix} 1 \\ 1 \\ -1 \end{pmatrix}$

$2\lambda - \mu = 8$ ...(1)
$-\lambda - \mu = -3$ ...(2)
$-8\lambda + \mu = -16$ ...(3)

(1)+(3):
$-6\lambda = -8$
$\Rightarrow \lambda = \dfrac{4}{3}$

(2)+(3):
$-9\lambda = -19$
$\Rightarrow \lambda = \dfrac{19}{9}$

These two values of λ are not the same, so there is no single value of λ that satisfies all the equations; hence there is no intersection point.

b i $\overrightarrow{PQ} = \begin{pmatrix} 5+\mu \\ \mu \\ 2-\mu \end{pmatrix} - \begin{pmatrix} -3+2\lambda \\ 3-\lambda \\ 18-8\lambda \end{pmatrix}$

$= \begin{pmatrix} \mu - 2\lambda + 8 \\ \mu + \lambda - 3 \\ -\mu + 8\lambda - 16 \end{pmatrix}$

ii (PQ) is perpendicular to l_1

$\therefore \overrightarrow{PQ} \cdot \begin{pmatrix} 2 \\ -1 \\ -8 \end{pmatrix} = 0$

$2(\mu - 2\lambda + 8) - (\mu + \lambda - 3)$
$-8(-\mu + 8\lambda - 16) = 0$
$9\mu - 69\lambda + 147 = 0$

iii (PQ) is perpendicular to l_2

$$\therefore \vec{PQ} \cdot \begin{pmatrix} 1 \\ 1 \\ -1 \end{pmatrix} = 0$$

$(\mu - 2\lambda + 8) + (\mu + \lambda - 3)$
$-(-\mu + 8\lambda - 16) = 0$
$3\mu - 9\lambda + 21 = 0$

iv $9\mu - 69\lambda = -147$ ...(1)
$3\mu - 9\lambda = -21$ ...(2)
$3 \times (2) - (1)$:
$42\lambda = 84$
$\Rightarrow \lambda = 2$

Substituting into (2):
$3\mu - 18 = -21$
$\Rightarrow \mu = -1$

$\therefore P = (1, 1, 2)$ and $Q = (4, -1, 3)$

v The shortest distance between l_1 and l_2 is PQ:

$PQ = \sqrt{3^2 + 2^2 + 1^2}$
$= \sqrt{14}$
$= 3.74$ (3SF)

3 a Normal to Π: $n = \begin{pmatrix} 1 \\ -2 \\ 1 \end{pmatrix}$

Line in direction n through point $(4, -1, 2)$ has equation

$$r = \begin{pmatrix} 4 \\ -1 \\ 2 \end{pmatrix} + \lambda \begin{pmatrix} 1 \\ -2 \\ 1 \end{pmatrix}$$

b Π has vector equation $r \cdot n = 20$

Taking the scalar product of the line equation with n:

$$r \cdot n = \begin{pmatrix} 4 \\ -1 \\ 2 \end{pmatrix} \cdot \begin{pmatrix} 1 \\ -2 \\ 1 \end{pmatrix} + \lambda \begin{pmatrix} 1 \\ -2 \\ 1 \end{pmatrix} \cdot \begin{pmatrix} 1 \\ -2 \\ 1 \end{pmatrix}$$

$20 = 8 + 6\lambda$
$\lambda = 2$

$\therefore$ intersection of l and Π is at
B $(6, -5, 4)$

c Shortest distance from point A to plane Π is

$AB = |\lambda n| = 2\sqrt{1^2 + 2^2 + 1^2} = 2\sqrt{6}$

4 a $|v_1| = \sqrt{7^2 + 10^2 + 3^2}$
$= \sqrt{158}$
$= 12.6$ km/min (3SF)

b At time t, the position vector of the second aircraft is $\begin{pmatrix} 24 + 3t \\ 66 - 8t \\ 12 - 4t \end{pmatrix}$

c The positions are the two aircraft are

$$r_1 = \begin{pmatrix} 16 + 7t \\ 30 + 10t \\ 3 + 3t \end{pmatrix}, \quad r_2 = \begin{pmatrix} 24 + 3t \\ 66 - 8t \\ 12 - 4t \end{pmatrix}$$

Separation distance

$$d = |r_2 - r_1| = \left\| \begin{pmatrix} 8 - 4t \\ 36 - 18t \\ 9 - 7t \end{pmatrix} \right\|$$

At $t = 3$, $d = \sqrt{4^2 + 18^2 + 12^2} = 22$ km

d One aircraft is vertically above the other when the first two components of the separation vector

$$r_2 - r_1 = \begin{pmatrix} 8-4t \\ 36-18t \\ 9-7t \end{pmatrix} \text{ are zero.}$$

At $t = 2$, the separation vector is $\begin{pmatrix} 0 \\ 0 \\ -5 \end{pmatrix}$,

which represents plane 1 being exactly 5 km above plane 2.

5 a $\begin{pmatrix} -1 \\ 1 \\ 3 \end{pmatrix} \times \begin{pmatrix} 2 \\ 1 \\ 1 \end{pmatrix} = \begin{pmatrix} 1-3 \\ 6+1 \\ -1-2 \end{pmatrix} = \begin{pmatrix} -2 \\ 7 \\ -3 \end{pmatrix}$

b At the intersection,

$\begin{pmatrix} 5 \\ 1 \\ 2 \end{pmatrix} + t \begin{pmatrix} -1 \\ 1 \\ 3 \end{pmatrix} = \begin{pmatrix} 5 \\ 4 \\ 9 \end{pmatrix} + s \begin{pmatrix} 2 \\ 1 \\ 1 \end{pmatrix} \quad \ldots(*)$

$\begin{pmatrix} 1 \\ 1 \\ 0 \end{pmatrix}$ is perpendicular to the direction vector of the first line.

Taking the scalar product of (*) with $\begin{pmatrix} 1 \\ 1 \\ 0 \end{pmatrix}$ gives

$6 + 0t = 9 + 3s$

$3s = -3$

$s = -1$

∴ the intersection point is (3, 3, 8), which occurs on the first line for $t = 2$.

c The vector $\begin{pmatrix} -2 \\ 7 \\ -3 \end{pmatrix}$ found in (a) is perpendicular to the direction vectors of both lines, so it is perpendicular to the plane containing the two lines.

d A plane with normal **n** passing through a point with position vector **a** has equation $r \cdot n = r \cdot a$:

$r \cdot \begin{pmatrix} -2 \\ 7 \\ -3 \end{pmatrix} = \begin{pmatrix} 3 \\ 3 \\ 8 \end{pmatrix} \cdot \begin{pmatrix} -2 \\ 7 \\ -3 \end{pmatrix}$

$\Rightarrow -2x + 7y - 3z = -9$

or $2x - 7y + 3z = 9$

6 a $3x - y + z = 2 \quad \ldots(1)$

$x + 2y - z = -1 \quad \ldots(2)$

$5x - 4y + dz = 3 \quad \ldots(3)$

(1) $\qquad 3x - y + \qquad z = 2 \quad \ldots(1)$

$3 \times (2) - (1) \qquad 7y - \qquad 4z = -5 \quad \ldots(4)$

$3 \times (3) - 5 \times (1) \qquad -7y + (3d-5)z = -1 \quad \ldots(5)$

This system of equations has no solution if (4) and (5) are incompatible, which is the case if $3d - 5 = 4$

i.e. $d = 3$

b The line of intersection must be normal to both $n_1 = \begin{pmatrix} 3 \\ -1 \\ 1 \end{pmatrix}$ and $n_2 = \begin{pmatrix} 1 \\ 2 \\ -1 \end{pmatrix}$

Take $d = \begin{pmatrix} 3 \\ -1 \\ 1 \end{pmatrix} \times \begin{pmatrix} 1 \\ 2 \\ -1 \end{pmatrix} = \begin{pmatrix} -1 \\ 4 \\ 7 \end{pmatrix}$ as the direction vector of the line.

To find a point on the line of intersection, using (4) above,

if $z = -4$ then $y = \dfrac{4z-5}{7} = -3$,

and from (1), $x = \dfrac{2+y-z}{3} = 1$

> **COMMENT**
>
> You can set z to be any value; 0 would be perfectly reasonable, but if you can look ahead for a value that leads to integer coordinates, it is often more convenient for future working and intuition.

Hence $(1, -3, -4)$ lies on the line of intersection.

The equation of the line is therefore

$$r = \begin{pmatrix} 1 \\ -3 \\ -4 \end{pmatrix} + \lambda \begin{pmatrix} -1 \\ 4 \\ 7 \end{pmatrix}$$

c i For $(0, 1, p)$ to lie on l_1, take $\lambda = 1$, and then the third component is
$$p = -4 + 7 = 3$$

ii l_2 has direction vector

$$n_3 = \begin{pmatrix} 5 \\ -4 \\ d \end{pmatrix} = \begin{pmatrix} 5 \\ -4 \\ 3 \end{pmatrix} \text{ and passes}$$

through $(0, 1, 3)$, so it has equation

$$r = \begin{pmatrix} 0 \\ 1 \\ 3 \end{pmatrix} + \mu \begin{pmatrix} 5 \\ -4 \\ 3 \end{pmatrix}$$

iii Since the three planes do not intersect and no two are parallel, they must form a triangular prism. Therefore, the distance from one plane to the line of intersection of the other two can be calculated using a normal from the plane to any point on the line.

Distance from Π_3 to l_1 is AB, where B is the intersection point of l_2 with Π_3.

Π_3 has equation $r \cdot \begin{pmatrix} 5 \\ -4 \\ 3 \end{pmatrix} = 3$

Substitute the equation of l_2 into the equation of Π_3 to find the intersection point B:

$$\left(\begin{pmatrix} 0 \\ 1 \\ 3 \end{pmatrix} + \mu \begin{pmatrix} 5 \\ -4 \\ 3 \end{pmatrix} \right) \cdot \begin{pmatrix} 5 \\ -4 \\ 3 \end{pmatrix} = 3$$

$$25\mu - 4(1 - 4\mu) + 3(3 + 3\mu) = 3$$

$$50\mu = -2$$

$$\mu = -\frac{1}{25}$$

Then the distance AB is

$$|\mu n_3| = \frac{\sqrt{50}}{25} = \frac{\sqrt{2}}{5}$$

7 a Direction vector of l_2 is the same as that for l_1: $\begin{pmatrix} 4 \\ -3 \\ 3 \end{pmatrix}$

l_2 passes through $(0, -1, 2)$

$\therefore$ equation of l_2 is $r = \begin{pmatrix} 0 \\ -1 \\ 2 \end{pmatrix} + \lambda \begin{pmatrix} 4 \\ -3 \\ 3 \end{pmatrix}$

b A general point on l_1 has position

vector $r = \begin{pmatrix} 2 \\ -1 \\ 0 \end{pmatrix} + \mu \begin{pmatrix} 4 \\ -3 \\ 3 \end{pmatrix}$

$$\therefore \overrightarrow{AB} = \begin{pmatrix} 2 + 4\mu \\ -3\mu \\ -2 + 3\mu \end{pmatrix}$$

Require $\overrightarrow{AB}$ to be perpendicular

to $\begin{pmatrix} 4 \\ -3 \\ 3 \end{pmatrix}$:

$$\begin{pmatrix} 2+4\mu \\ -3\mu \\ -2+3\mu \end{pmatrix} \cdot \begin{pmatrix} 4 \\ -3 \\ 3 \end{pmatrix} = 0$$

$8 + 16\mu + 9\mu - 6 + 9\mu = 0$

$34\mu = -2$

$\mu = -\dfrac{1}{17}$

∴ the coordinates of B are

$\left(\dfrac{30}{17}, -\dfrac{14}{17}, -\dfrac{3}{17} \right)$

c Shortest distance between two lines is the length of a vector perpendicular to both, so the distance d between l_1 and l_2 is equal to $|\overrightarrow{AB}|$:

$d = \left| \dfrac{1}{17} \begin{pmatrix} 30 \\ 3 \\ -37 \end{pmatrix} \right|$

$= \dfrac{1}{17} \sqrt{30^2 + 3^2 + 37^2}$

$= 2.81 \, (3\text{SF})$

8 a Line L has general point with position

vector $r = \begin{pmatrix} -5 \\ 1 \\ 2 \end{pmatrix} + \lambda \begin{pmatrix} 3 \\ 3 \\ -1 \end{pmatrix}$

For $\lambda = 3$, the coordinates are $(4, 10, -1)$, so point A does indeed lie on L.

b $\overrightarrow{AB} = \begin{pmatrix} -2 \\ -9 \\ 3 \end{pmatrix}$

$\Rightarrow AB = \sqrt{2^2 + 9^2 + 3^2} = \sqrt{94}$

c Let θ be the angle between (AB) and L;

then since $d = \begin{pmatrix} 3 \\ 3 \\ -1 \end{pmatrix}$ is a direction

vector for line L,

$\cos\theta = \dfrac{\overrightarrow{AB} \cdot d}{|\overrightarrow{AB}||d|}$

$= \dfrac{-36}{\sqrt{94}\sqrt{19}}$

$\theta = \arccos\left(-\dfrac{36}{\sqrt{94 \times 19}} \right) = 2.59$

So the acute angle is $\pi - 2.59 = 0.551$.

d Let C be the point on L such that (BC) is perpendicular to L.

Then BC is the shortest distance from B to L; $B\hat{C}A$ is a right angle and $B\hat{A}C = \theta$

∴ $BC = AB \sin\theta$

$= \sqrt{94} \sin(0.551)$

$= 5.08 \, (3\text{SF})$

9 a i $\begin{pmatrix} -2 \\ 1 \\ 8 \end{pmatrix} \times \begin{pmatrix} 1 \\ -3 \\ -9 \end{pmatrix} = \begin{pmatrix} 15 \\ -10 \\ 5 \end{pmatrix}$

$\Rightarrow$ take $n_1 = \begin{pmatrix} 3 \\ -2 \\ 1 \end{pmatrix}$ to be the normal to Π_1

$\begin{pmatrix} 1 \\ 2 \\ 1 \end{pmatrix} \times \begin{pmatrix} 1 \\ 1 \\ 1 \end{pmatrix} = \begin{pmatrix} 1 \\ 0 \\ -1 \end{pmatrix}$

$\Rightarrow$ take $n_2 = \begin{pmatrix} 1 \\ 0 \\ -1 \end{pmatrix}$ to be the normal to Π_2

For the intersection of Π_1 and Π_2, require that

$\begin{pmatrix} 2 \\ 1 \\ 1 \end{pmatrix} + \lambda \begin{pmatrix} -2 \\ 1 \\ 8 \end{pmatrix} + \mu \begin{pmatrix} 1 \\ -3 \\ -9 \end{pmatrix} = \begin{pmatrix} 2 \\ 0 \\ 1 \end{pmatrix} + s \begin{pmatrix} 1 \\ 2 \\ 1 \end{pmatrix} + t \begin{pmatrix} 1 \\ 1 \\ 1 \end{pmatrix}$

Taking the scalar product with n_2:

$1 - 10\lambda + 10\mu = 1 + 0s + 0t$

$\Rightarrow \lambda = \mu$

14 Lines and planes in space

ii Using the result in (i), the equation for plane Π_1 reduces to the equation for the intersection line:

$$r = \begin{pmatrix} 2 \\ 1 \\ 1 \end{pmatrix} + \lambda \begin{pmatrix} -1 \\ -2 \\ -1 \end{pmatrix}$$

b A normal to Π_3 is $n_3 = \begin{pmatrix} 3 \\ -2 \\ 1 \end{pmatrix}$, and the plane contains point $(2, 0, -1)$

∴ the equation for Π_3 is

$$r \cdot \begin{pmatrix} 3 \\ -2 \\ 1 \end{pmatrix} = \begin{pmatrix} 2 \\ 0 \\ -1 \end{pmatrix} \cdot \begin{pmatrix} 3 \\ -2 \\ 1 \end{pmatrix}$$

or $3x - 2y + z = 5$ in Cartesian form.

c The intersection of the 3 planes is the intersection of the line found in (a)(ii) and the plane in (b).

Using the vector equation of the plane, we require that

$$\left(\begin{pmatrix} 2 \\ 1 \\ 1 \end{pmatrix} + \lambda \begin{pmatrix} -1 \\ -2 \\ -1 \end{pmatrix} \right) \cdot \begin{pmatrix} 3 \\ -2 \\ 1 \end{pmatrix} = 5$$

$3(2-\lambda) - 2(1-2\lambda) + (1-\lambda) = 5$

$5 + 0\lambda = 5$

This equality is always true; there are no restrictions on values for λ, and so the intersection of the three planes is the common line

$$r = \begin{pmatrix} 2 \\ 1 \\ 1 \end{pmatrix} + \lambda \begin{pmatrix} -1 \\ -2 \\ -1 \end{pmatrix}$$

10 a $r = \begin{pmatrix} -2 \\ 4 \\ 2 \end{pmatrix} + \lambda \begin{pmatrix} 1 \\ 1 \\ 0 \end{pmatrix}$

b $\overrightarrow{AB} = \begin{pmatrix} 4 \\ -1 \\ 1 \end{pmatrix}$, and the direction of the line is $d = \begin{pmatrix} 1 \\ 1 \\ 0 \end{pmatrix}$

Let θ be the angle between the line and (AB); then

$$\cos\theta = \frac{\overrightarrow{AB} \cdot d}{|\overrightarrow{AB}||d|}$$

$$= \frac{4-1+0}{\sqrt{18}\sqrt{2}}$$

$$= \frac{3}{\sqrt{36}}$$

$$= \frac{1}{2}$$

c $AB = \sqrt{4^2 + 1^2 + 1^2} = \sqrt{18} = 3\sqrt{2}$

d $A\hat{C}B = 90°$ and, from (b), $\cos B\hat{A}C = \frac{1}{2}$

∴ $AC = AB \cos B\hat{A}C$

$= 3\sqrt{2} \times \frac{1}{2}$

$= \frac{3\sqrt{2}}{2}$

176 Mixed examination practice 14

11 a Plane Π has vector equation

$$\mathbf{r} \cdot \begin{pmatrix} 1 \\ -4 \\ 2 \end{pmatrix} = 7$$

$$\begin{pmatrix} 5 \\ 1 \\ 3 \end{pmatrix} \cdot \begin{pmatrix} 1 \\ -4 \\ 2 \end{pmatrix} = 5 - 4 + 6 = 7$$

$\Rightarrow (5, 1, 3)$ lies in Π.

b $\overrightarrow{PR} = \begin{pmatrix} -4 \\ 8 \\ -3 \end{pmatrix}$, and line (PR) passes through the point R $(5, 1, 3)$, so the equation of the line is

$$\mathbf{r} = \begin{pmatrix} 5 \\ 1 \\ 3 \end{pmatrix} + \lambda \begin{pmatrix} -4 \\ 8 \\ -3 \end{pmatrix}$$

c Normal to Π is $\mathbf{n} = \begin{pmatrix} 1 \\ -4 \\ 2 \end{pmatrix}$, so the line through P $(9, -7, 6)$ parallel to $\mathbf{n}$ is given by

$$\mathbf{r} = \begin{pmatrix} 9 \\ -7 \\ 6 \end{pmatrix} + \mu \begin{pmatrix} 1 \\ -4 \\ 2 \end{pmatrix}$$

d The intersection of the line in (c) with Π can be found by substituting the line equation into the plane equation:

$$\left(\begin{pmatrix} 9 \\ -7 \\ 6 \end{pmatrix} + \mu \begin{pmatrix} 1 \\ -4 \\ 2 \end{pmatrix} \right) \cdot \begin{pmatrix} 1 \\ -4 \\ 2 \end{pmatrix} = 7$$

$(9 + \mu) - 4(-7 - 4\mu) + 2(6 + 2\mu) = 7$

$49 + 21\mu = 7$

$\mu = -2$

$\therefore$ the coordinates of N are $(7, 1, 2)$

e Distance from P to Π is

$PN = |\mu \mathbf{n}|$

$= 2\sqrt{1^2 + 4^2 + 2^2}$

$= 2\sqrt{21}$

12 a Normal to the plane is $\mathbf{n} = \begin{pmatrix} 3 \\ -1 \\ -1 \end{pmatrix}$

So L has vector equation

$$\mathbf{r} = \begin{pmatrix} 3 \\ 1 \\ -4 \end{pmatrix} + \lambda \begin{pmatrix} 3 \\ -1 \\ -1 \end{pmatrix}$$

In Cartesian form, this is

$$\frac{x-3}{3} = \frac{y-1}{-1} = \frac{z+4}{-1}$$

b The plane has vector equation $\mathbf{r} \cdot \mathbf{n} = 1$

To find the intersection of the line and the plane, substitute the vector form of the line equation into the plane equation:

$$\left(\begin{pmatrix} 3 \\ 1 \\ -4 \end{pmatrix} + \lambda \begin{pmatrix} 3 \\ -1 \\ -1 \end{pmatrix} \right) \cdot \begin{pmatrix} 3 \\ -1 \\ -1 \end{pmatrix} = 1$$

$3(3 + 3\lambda) - (1 - \lambda) - (-4 - \lambda) = 1$

$12 + 11\lambda = 1$

$\lambda = -1$

$\therefore$ intersection point N is $(0, 2, -3)$

c On line L, point A corresponds to $\lambda = 0$, and the intersection with the plane is at $\lambda = -1$, so the image of A reflected through the plane is at $\lambda = -2$:

$A' = (-3, 3, -2)$

d $\begin{pmatrix} 1 \\ 1 \\ 1 \end{pmatrix} \cdot \begin{pmatrix} 3 \\ -1 \\ -1 \end{pmatrix} = 3 - 1 - 1 = 1$

$\Rightarrow (1, 1, 1)$ lies in Π

e Since L is perpendicular to Π, the shortest distance from point B in the plane to line L is BN, where N is the intersection point found in (b):

$$\overrightarrow{BN} = \begin{pmatrix} -1 \\ 1 \\ -4 \end{pmatrix}$$

$\therefore BN = \sqrt{1^2 + 1^2 + 4^2} = 3\sqrt{2}$

13 a $\begin{pmatrix} 2 \\ -1 \\ 1 \end{pmatrix} \times \begin{pmatrix} 3 \\ 1 \\ -1 \end{pmatrix} = \begin{pmatrix} 1-1 \\ 3+2 \\ 2+3 \end{pmatrix} = \begin{pmatrix} 0 \\ 5 \\ 5 \end{pmatrix}$

b Π_1 has normal $\mathbf{n}_1 = \begin{pmatrix} 2 \\ -1 \\ 1 \end{pmatrix}$ and passes through A $(3, 4, -2)$

$\therefore \Pi_1$ has equation $\mathbf{r} \cdot \mathbf{n}_1 = \mathbf{a} \cdot \mathbf{n}_1$
i.e. $2x - y + z = 0$

c Π_2 has equation $\mathbf{r} \cdot \begin{pmatrix} 3 \\ 1 \\ -1 \end{pmatrix} = 15$

$\begin{pmatrix} 3 \\ 4 \\ -2 \end{pmatrix} \cdot \begin{pmatrix} 3 \\ 1 \\ -1 \end{pmatrix} = 9 + 4 + 2 = 15$

$\Rightarrow$ A $(3, 4, -2)$ lies in Π_2

d The line of intersection has direction vector $\mathbf{d}$ perpendicular to the normal vectors of both planes. From (a), take $\mathbf{d} = \begin{pmatrix} 0 \\ 1 \\ 1 \end{pmatrix}$. The line passes through point A $(3, 4, -2)$, since this lies in both planes.

$\therefore$ equation of the line is

$$\mathbf{r} = \begin{pmatrix} 3 \\ 4 \\ -2 \end{pmatrix} + \lambda \begin{pmatrix} 0 \\ 1 \\ 1 \end{pmatrix}$$

e The intersection of all 3 planes is the intersection of the line in (d) with

$$\Pi_3 : \mathbf{r} \cdot \begin{pmatrix} 2 \\ 1 \\ 2 \end{pmatrix} = 12$$

To find this, substitute the line equation from (d) into the equation of Π_3:

$$\left(\begin{pmatrix} 3 \\ 4 \\ -2 \end{pmatrix} + \lambda \begin{pmatrix} 0 \\ 1 \\ 1 \end{pmatrix} \right) \cdot \begin{pmatrix} 2 \\ 1 \\ 2 \end{pmatrix} = 12$$

$2 \times 3 + (4 + \lambda) + 2(-2 + \lambda) = 12$

$6 + 3\lambda = 12$

$\lambda = 2$

$\therefore$ point of intersection is $(3, 6, 0)$

f Angle between the planes Π_1 and Π_3 is equal to the angle between their normal vectors $\mathbf{n}_1$ and $\mathbf{n}_3$:

$$\mathbf{n}_1 = \begin{pmatrix} 2 \\ -1 \\ 1 \end{pmatrix}, \quad \mathbf{n}_3 = \begin{pmatrix} 2 \\ 1 \\ 2 \end{pmatrix}$$

For the angle θ between the normals,

$\cos\theta = \dfrac{\mathbf{n}_1 \cdot \mathbf{n}_3}{|\mathbf{n}_1||\mathbf{n}_3|}$

$= \dfrac{4 - 1 + 2}{\sqrt{6}\sqrt{9}}$

$\Rightarrow \theta = \arccos\left(\dfrac{5}{\sqrt{54}}\right) = 47.1°$ (3SF)

15 Complex numbers

Exercise 15A

10 $(3+ai)(b-i) = -4i$
$3b + a + i(ab - 3) = -4i$

Comparing real and imaginary parts:
Re: $3b + a = 0 \Rightarrow a = -3b$
Im: $ab - 3 = -4 \Rightarrow ab = -1$
$\therefore -3b^2 = -1$

$\Rightarrow b = \pm\sqrt{\dfrac{1}{3}} = \pm\dfrac{1}{3}\sqrt{3}$

and $a = -3\left(\pm\dfrac{1}{3}\sqrt{3}\right) = \mp\sqrt{3}$

11 $(1+ai)(1+bi) = b + 9i - a$
$1 - ab + i(a+b) = b - a + 9i$

Comparing real and imaginary parts:
Re: $1 - ab = b - a$
$1 + a - b - ab = 0$
$(1-b)(1+a) = 0$
$a = -1$ or $b = 1$

Im: $a + b = 9$
$\therefore$ the solutions are
$(a, b) = (-1, 10)$ or $(8, 1)$

12 $iz + 2 = i - 3z$
Let $z = x + iy$; then
$2 - y + ix = -3x + (1 - 3y)i$

Comparing real and imaginary parts:
Re: $2 - y = -3x \Rightarrow y = 2 + 3x$
Im: $x = 1 - 3y$

$\therefore x = -5 - 9x$
$10x = -5$
$\Rightarrow x = -\dfrac{1}{2},\ y = \dfrac{1}{2}$

$\therefore z = \dfrac{1}{2}(-1 + i)$

13 a $(x + iy)(2 + i) = -i$
$2x - y + i(2y + x) = -i$

Comparing real and imaginary parts:
Re: $2x - y = 0 \Rightarrow y = 2x$
Im: $2y + x = -1$
$\therefore 5x = -1$
$\Rightarrow x = -\dfrac{1}{5},\ y = -\dfrac{2}{5}$

b From (a): $\dfrac{1}{5}(-1 - 2i)(2 + i) = -i$

$\therefore \dfrac{3}{5}(-1 - 2i)(2 + i) = -3i$

and hence $-\dfrac{3i}{2+i} = \dfrac{3}{5}(-1 - 2i) = -\dfrac{3}{5} - \dfrac{6}{5}i$

14 a $(a + bi)^2 = -3 - 4i$
$a^2 - b^2 + 2abi = -3 - 4i$

Comparing real and imaginary parts:
Re: $a^2 - b^2 = -3$
Im: $2ab = -4 \Rightarrow b = -\dfrac{2}{a}$

$\therefore a^2 - \dfrac{4}{a^2} = -3$

$a^4 + 3a^2 - 4 = 0$
$(a^2 + 4)(a^2 - 1) = 0$
$a^2 = 1$ (reject $a^2 = -4$)
$\Rightarrow a = \pm 1,\ b = \mp 2$

b $z^2 + i\sqrt{3}z + i = 0$

$\Rightarrow z = \dfrac{-i\sqrt{3} \pm \sqrt{-3-4i}}{2}$

$= \dfrac{-i\sqrt{3} \pm (1-2i)}{2}$ by (a)

$= \dfrac{1}{2} - \left(1 + \dfrac{\sqrt{3}}{2}\right)i$ or $-\dfrac{1}{2} + \left(1 - \dfrac{\sqrt{3}}{2}\right)i$

b

Figure 15B.10 $z = 1 + \sqrt{3}\,i$ and $w = 3\sqrt{3} - 3i$ on Argand diagram

Exercise 15B

9 a $3\operatorname{cis}\left(\dfrac{7\pi}{4}\right) = 3\cos\left(\dfrac{7\pi}{4}\right) + 3i\sin\left(\dfrac{7\pi}{4}\right)$

$= \dfrac{3\sqrt{2}}{2} - \dfrac{3\sqrt{2}}{2}i$

b $|4i - 4| = \sqrt{4^2 + 4^2}$

$= 4\sqrt{2}$

$\arg(4i - 4) = \arctan\left(\dfrac{-4}{4}\right)$

$= \dfrac{3\pi}{4}$

(select angle in the second quadrant, as $\operatorname{Re}(4i-4) < 0$ and $\operatorname{Im}(4i-4) > 0$)

$\therefore 4i - 4 = 4\sqrt{2}\operatorname{cis}\left(\dfrac{3\pi}{4}\right)$

10 a $|z| = |1 + \sqrt{3}i| = \sqrt{1^2 + 3} = 2$

$\arg(z) = \arctan\left(\dfrac{\sqrt{3}}{1}\right) = \dfrac{\pi}{3}$

(select angle in the first quadrant, as $\operatorname{Im}(z), \operatorname{Re}(z) > 0$)

$|w| = |3\sqrt{3} - 3i| = \sqrt{27 + 3^2} = 6$

$\arg(w) = \arctan\left(\dfrac{-3}{3\sqrt{3}}\right) = -\dfrac{\pi}{6}$

(select angle in the fourth quadrant, as $\operatorname{Re}(w) > 0$ and $\operatorname{Im}(w) < 0$)

c $zw = (1 + \sqrt{3}i)(3\sqrt{3} - 3i)$

$= 3\sqrt{3} - 3i + 9i + 3\sqrt{3}$

$= 6\sqrt{3} + 6i$

$|zw| = \sqrt{108 + 36} = 12$

$\arg(zw) = \arctan\left(\dfrac{6}{6\sqrt{3}}\right) = \dfrac{\pi}{6}$

(select angle in first quadrant as $\operatorname{Im}(zw), \operatorname{Re}(zw) > 0$)

Notice that $|zw| = |z||w|$ and $\arg(zw) = \arg(z) + \arg(w)$

11 $|z + w| = |z - w|$

Let $z = x + iy$ and $w = u + iv$; then the equation becomes

$|x + u + i(y + v)| = |x - u + i(y - v)|$

$(x+u)^2 + (y+v)^2 = (x-u)^2 + (y-v)^2$

$2xu + 2yv = -2xu - 2yv$

$4xu = -4yv$

$\dfrac{x}{y} = -\dfrac{v}{u}$

$\arg(z) = \arctan\left(\dfrac{y}{x}\right) + n\pi$

$\arg(w) = \arctan\left(\dfrac{v}{u}\right) + m\pi$

$= \arctan\left(-\dfrac{x}{y}\right) + m\pi$

$= -\arctan\left(\dfrac{x}{y}\right) + m\pi$

(where $n, m = 0$ or 1)

Using $\arctan\left(\dfrac{1}{a}\right) = \dfrac{\pi}{2} - \arctan(a)$ with $a = \dfrac{y}{x}$:

$\arg(z) - \arg(w)$

$= \arctan\left(\dfrac{y}{x}\right) + n\pi + \arctan\left(\dfrac{x}{y}\right) - m\pi$

$= \arctan\left(\dfrac{y}{x}\right) + \dfrac{\pi}{2} - \arctan\left(\dfrac{y}{x}\right) + (n-m)\pi$

$= \dfrac{\pi}{2} + (n-m)\pi$ where $n - m = 0, \pm 1$

$\therefore \arg(z) - \arg(w) = \pm\dfrac{\pi}{2}$

(allowing equivalence of angles that differ by 2π)

Exercise 15C

9 a $3iz + 2z^* = 3i(x+iy) + 2(x-iy)$
$= 3ix - 3y + 2x - 2iy$
$= (2x - 3y) + i(3x - 2y)$
$\text{Re}(3iz + 2z^*) = 2x - 3y$
$\text{Im}(3iz + 2z^*) = 3x - 2y$

b Compare real and imaginary parts:
$2x - 3y = 4$ …(1)
$3x - 2y = -4$ …(2)
$2 \times (1) - 3 \times (2) \Rightarrow -5x = 20$
$\therefore x = -4, \ y = -4$
and so $z = -4 - 4i$

10 $z + 4iz^* = 2 + i + 4i(2 - i) = 6 + 9i$

So the equation $x + 3iy = z + 4iz^*$ becomes
$x + 3iy = 6 + 9i$
$\therefore x = 6, \ y = 3$

11 Let $z = x + iy$ for $x, y \in \mathbb{R}$
$(z^*)^2 = (x - iy)^2$
$= x^2 - y^2 - 2ixy$
$= (x^2 - y^2 + 2ixy)^*$
$= ((x+iy)^2)^*$
$= (z^2)^*$

12 Let $z = x + iy$ for $x, y \in \mathbb{R}$; then
$z + 3z^* = x + iy + 3(x - iy) = 4x - 2yi$
So the equation becomes $4x - 2yi = i$
$\therefore x = 0, \ y = -\dfrac{1}{2}$

Hence $z = -\dfrac{1}{2}i$

13 Let $z = x + iy$ for $x, y \in \mathbb{R}$; then
$z + i - (1 - z^*) = x + iy + i - 1 + x - iy$
$= 2x - 1 + i$
Require this to equal zero, but this is not possible, since no value of y will make the imaginary part vanish.
$\therefore$ the equation has no solution.

14 a $z + z^* = 2\text{Re}(z) = 2r\cos\theta$

b $zz^* = |z|^2 = r^2$

c $\dfrac{z}{z^*} = \dfrac{z^2}{|z|^2}$
$= \dfrac{r^2 \text{cis } 2\theta}{r^2}$
$= \text{cis } 2\theta$

15 $2z^* + \dfrac{3}{iz} = \dfrac{2izz^* + 3}{iz}$

$= \dfrac{2i|z|^2 + 3}{iz}$

$= \dfrac{3 + 6i}{iz}$ (using $|z| = \sqrt{3}$)

Require this to equal $\sqrt{15}$

$\therefore 3 + 6i = \sqrt{15}\, iz$

$\Rightarrow z = \dfrac{3 + 6i}{\sqrt{15}\, i} = \dfrac{6 - 3i}{\sqrt{15}}$

Check: $|z| = \sqrt{\dfrac{6^2}{15} + \dfrac{3^2}{15}} = \sqrt{\dfrac{45}{15}} = \sqrt{3}$ as required.

COMMENT

Note that this question gives more information than strictly needed. Try to solve the equation without using the fact that $|z| = \sqrt{3}$. You should still get the same unique answer.

16 $z - \dfrac{12i}{z^*} = \dfrac{zz^* - 12i}{z^*}$

$= \dfrac{|z|^2 - 12i}{z^*}$

$= \dfrac{9 - 12i}{z^*}$ (using $|z| = 3$)

Require this to equal 5

$\therefore 9 - 12i = 5z^*$

$\Rightarrow z = \dfrac{9 + 12i}{5}$

Check: $|z| = \sqrt{\dfrac{9^2}{5^2} + \dfrac{12^2}{5^2}} = \sqrt{\dfrac{225}{25}} = 3$ as required.

COMMENT

As in question 15, try to solve this equation without using knowledge of $|z|$.

17 Let $z = x + iy$ for $x, y \in \mathbb{R}$

$z + \dfrac{1}{z} = z + \dfrac{z^*}{|z|^2}$

$\Rightarrow \mathrm{Im}\left(z + \dfrac{1}{z}\right) = y\left(1 - \dfrac{1}{x^2 + y^2}\right)$

Require this imaginary part to equal zero:

$y = 0$ or $x^2 + y^2 = 1$

$\therefore z \in \mathbb{R}$ or $|z| = 1$

18 $\dfrac{z}{z+1} = \dfrac{x + iy}{1 + x + iy}$

$= \dfrac{(x + iy)(1 + x - iy)}{(1 + x + iy)(1 + x - iy)}$

$= \dfrac{x(1 + x) + y^2 + iy}{(1 + x)^2 + y^2}$

$\therefore \mathrm{Re}\left(\dfrac{z}{1+z}\right) = \dfrac{x(1 + x) + y^2}{(1 + x)^2 + y^2}$

$\mathrm{Im}\left(\dfrac{z}{1+z}\right) = \dfrac{y}{(1 + x)^2 + y^2}$

19 $z = x + iy$ where $x = \cos\theta$, $y = \sin\theta$, so $x^2 + y^2 = 1$

$\dfrac{z - 1}{z + 1} = \dfrac{x - 1 + iy}{x + 1 + iy}$

$= \dfrac{(x - 1 + iy)(x + 1 - iy)}{(x + 1 + iy)(x + 1 - iy)}$

$= \dfrac{x^2 + y^2 - 1 + 2iy}{(x + 1)^2 + y^2}$

$= \dfrac{1 - 1 + 2iy}{x^2 + y^2 + 1 + 2x}$

$= \dfrac{2iy}{2 + 2x}$

$= \dfrac{iy}{1 + x}$

$= i\dfrac{\sin\theta}{1 + \cos\theta}$

$\therefore \mathrm{Re}\left(\dfrac{z-1}{z+1}\right) = 0$, $\mathrm{Im}\left(\dfrac{z-1}{z+1}\right) = \dfrac{\sin\theta}{1 + \cos\theta}$

Exercise 15D

5 By the Fundamental Theorem of Algebra, the cubic has 3 roots.
It has real coefficients, so complex roots occur in conjugate pairs.
$(z-3i)$ is a factor, so $(z+3i)$ is also a factor.
$\therefore z^3 - 2z^2 + 9z - 18 = (z-3i)(z+3i)(z-k)$
$= (z^2 + 9)(z-k)$
$= z^3 - kz^2 + 9z - 9k$

Comparing coefficients:
$z^3 : 1 = 1$
$z^2 : -k = -2 \Rightarrow k = 2$
$z^1 : 9 = 9$
$z^0 : -9k = -18$ is consistent with $k = 2$
So the remaining roots are $-3i$ and 2

6 By the Fundamental Theorem of Algebra, the cubic has 3 roots.
It has real coefficients, so complex roots occur in conjugate pairs.
$(z-1-2i)$ is a factor, so $(z-1+2i)$ is also a factor.
$\therefore z^3 + z^2 - z + 15$
$= (z-1-2i)(z-1+2i)(z-k)$
$= (z^2 - 2z + 5)(z-k)$
$= z^3 - (k+2)z^2 + (2k+5)z - 5k$

Comparing coefficients:
$z^3 : 1 = 1$
$z^2 : -(k+2) = 1 \Rightarrow k = -3$
$z^1 : 2k+5 = -1$ is consistent with $k = -3$
$z^0 : -5k = 15$ is consistent with $k = -3$
So the remaining roots are $1 - 2i$ and -3

7 a The cubic has real coefficients, so complex roots occur in conjugate pairs.
Therefore the third root is $2 + 3i$

b $z^3 + bz^2 + cz + d$
$= (z+2)(z-2+3i)(z-2-3i)$
$= (z+2)(z^2 - 4z + 13)$
$= z^3 - 2z^2 + 5z + 26$
$\therefore b = -2, c = 5, d = 26$

8 The quartic has real coefficients, so complex roots occur in conjugate pairs.
Therefore it has the form
$(z-3i)(z+3i)(z-5+i)(z-5-i)$
$= (z^2 + 9)(z^2 - 10z + 26)$
$= z^4 - 10z^3 + 35z^2 - 90z + 234$

9 By the Fundamental Theorem of Algebra, this quintic polynomial (of degree 5) has 5 roots.
It has real coefficients, so complex roots occur in conjugate pairs.
Therefore it must have roots $\pm 2i$, $3 \pm i$ and one other.
This final root cannot be complex, since that would require a sixth root as its conjugate.
Hence there are four complex roots and a single real root.

10 a $f(x)$ has a root at $x = 1$ and a repeated root at $x = 5$.
Since it is a quartic polynomial, by the Fundamental Theorem of Algebra it has 4 roots.
It has real coefficients, so complex roots occur in conjugate pairs. Since three roots (1, 5 and 5) are known to be real, there cannot be a complex root. The final root must therefore also be 1 or 5.
$\therefore f_1(x) = a(x-1)^2(x-5)^2$
or $f_2(x) = a(x-1)(x-5)^3$
for some real $a \neq 0$.

15 Complex numbers 183

b

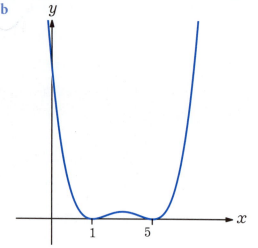

Figure 15D.10.1 $f_1(x)=(x-1)^2(x-5)^2$ is a positive quartic with two repeated roots at 1 and 5

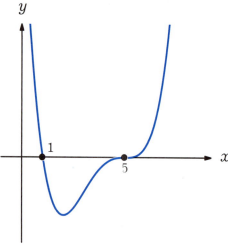

Figure 15D.10.2 $f_2(x)=(x-1)(x-5)^3$ is a positive quartic with one root at 1 and a triple root at 5

11 $f(z)=z^4+z^3+5z^2+4z+4$

a $f(2i)=16-8i-20+8i+4=0$

b The quartic has real coefficients, so its complex roots occur in conjugate pairs.

$\therefore f(z)=(z-2i)(z+2i)(z^2+az+b)$

for some $a, b \in \mathbb{R}$

$=(z^2+4)(z^2+az+b)$

$=z^4+az^3+(4+b)z^2+4az+4b$

Comparing coefficients:

$z^4 : 1=1$

$z^3 : a=1$

$z^2 : 4+b=5 \Rightarrow b=1$

$z^1 : 4a=4$ is consistent with $a=1$

$z^0 : 4b=4$ is consistent with $b=1$

$\therefore f(z)=(z-2i)(z+2i)(z^2+z+1)$

Since the roots of $z^2+z+1=0$ are $z=\dfrac{-1\pm\sqrt{3}i}{2}$, the remaining solutions are $-2i$ and $-\dfrac{1}{2}\pm\dfrac{\sqrt{3}}{2}i$

c $f(z)=(z^2+4)(z^2+z+1)$

Exercise 15E

3 a By the Fundamental Theorem of Algebra, a quartic polynomial has 4 roots.
It has real coefficients, so complex roots occur in conjugate pairs. Therefore the other two roots are $-3i$ and $3+i$

b $\dfrac{a}{1}=3i+(-3i)+(3-i)+(3+i)$

$\Rightarrow a=6$

$d=(3i)(-3i)(3-i)(3+i)=90$

4 $p^2qr+pq^2r+pqr^2=(pqr)(p+q+r)$

$=\left(-\dfrac{1}{4}\right)\left(-\dfrac{-2}{4}\right)$

$=-\dfrac{1}{8}$

5 $R_1+R_2=-\dfrac{-12}{3}=4$

$R_1R_2=\dfrac{4}{3}$

i $R=R_1+R_2=4$

ii $\dfrac{1}{R} = \dfrac{1}{R_1} + \dfrac{1}{R_2}$

$= \dfrac{R_2 + R_1}{R_1 R_2}$

$\therefore R = \dfrac{R_1 R_2}{R_2 + R_1}$

$= \dfrac{4}{\dfrac{3}{4}}$

$= \dfrac{1}{3}$

6 a $\alpha\beta\gamma = -\dfrac{-3}{3} = 1$

b Since there is no term in x^2, it follows that $\alpha + \beta + \gamma = 0$

$\therefore \gamma = -(\alpha + \beta)$

From (a), $\gamma = \dfrac{1}{\alpha\beta}$

$\therefore -(\alpha + \beta) = \dfrac{1}{\alpha\beta}$

$\Rightarrow \alpha + \beta = -\dfrac{1}{\alpha\beta}$

7 $p + q = -\dfrac{-3}{5} = \dfrac{3}{5}$

$pq = \dfrac{2}{5}$

$\therefore \dfrac{1}{p} + \dfrac{1}{q} = \dfrac{p+q}{pq} = \dfrac{\frac{3}{5}}{\frac{2}{5}} = \dfrac{3}{2}$

and $\dfrac{1}{p} \times \dfrac{1}{q} = \dfrac{1}{pq} = \dfrac{5}{2}$

Therefore, a quadratic with roots $\dfrac{1}{p}$ and $\dfrac{1}{q}$ could be $x^2 - \dfrac{3}{2}x + \dfrac{5}{2} = 0$ or, equivalently, $2x^2 - 3x + 5 = 0$

8 For the four roots $\alpha + \beta + \gamma + \delta$:

$\alpha + \beta + \gamma + \delta = -\dfrac{-9}{1} = 9$

The mean of a large sample would be representative of the mean of these 4 values: $\dfrac{9}{4}$

9 Product of the roots is
$e = p \times 2p \times 3p \times 4p = 24p^4$

Sum of the roots is
$-b = p + 2p + 3p + 4p = 10p$

$\therefore 1250 e = 1250 \times 24 p^4$

$= 30\,000 p^4$

$= 3(-b)^4$

$= 3b^4$

10 a i $(p + q + r)^2$

$= p(p + q + r) + q(p + q + r)$
$\quad + r(p + q + r)$

$= p^2 + q^2 + r^2 + 2(pq + qr + rp)$

$\Rightarrow p^2 + q^2 + r^2$

$= (p + q + r)^2 - 2(pq + qr + rp)$

ii $(pq + qr + rp)^2$

$= pq(pq + qr + rp) + qr(pq + qr + rp)$
$\quad + rp(pq + qr + rp)$

$= p^2q^2 + q^2r^2 + r^2p^2$
$\quad + 2(pqr^2 + pq^2r + p^2qr)$

$= p^2q^2 + q^2r^2 + r^2p^2$
$\quad + 2pqr(p + q + r)$

$\Rightarrow p^2q^2 + q^2r^2 + r^2p^2$

$= (pq + qr + rp)^2 - 2pqr(p + q + r)$

b i $p + q + r = -\dfrac{b}{a}$

$pqr = -\dfrac{d}{a}$

ii $a(x-p)(x-q)(x-r) = ax^3 + bx^2 + cx + d$

$a(x^3 - (p+q+r)x^2 + (pq+qr+rp)x - pqr) = ax^3 + bx^2 + cx + d$

Comparing the coefficient of x:

$c = a(pq+qr+rp)$

$\Rightarrow pq+qr+rp = \dfrac{c}{a}$

c From (b): $x_1 + x_2 + x_3 = -\dfrac{0}{2} = 0$

$x_1 x_2 + x_2 x_3 + x_3 x_1 = \dfrac{-5}{2} = -\dfrac{5}{2}$

$x_1 x_2 x_3 = -\dfrac{2}{2} = -1$

i From (a)(i):

$x_1^2 + x_2^2 + x_3^2 = (x_1 + x_2 + x_3)^2 - 2(x_1 x_2 + x_2 x_3 + x_3 x_1) = 0 - 2\left(-\dfrac{5}{2}\right) = 5$

ii From (a)(ii):

$x_1^2 x_2^2 + x_2^2 x_3^2 + x_3^2 x_1^2 = (x_1 x_2 + x_2 x_3 + x_3 x_1)^2 - 2 x_1 x_2 x_3 (x_1 + x_2 + x_3) = \left(-\dfrac{5}{2}\right)^2 - 2(-1)(0) = \dfrac{25}{4}$

$x_1^2 x_2^2 x_3^2 = (x_1 x_2 x_3)^2 = (-1)^2 = 1$

iii Using the above, a cubic equation with coefficients x_1^2, x_2^2 and x_3^2 could be

$x^3 - 5x^2 + \dfrac{25}{4}x - 1 = 0$

or, equivalently, $4x^3 - 20x^2 + 25x - 4 = 0$

Exercise 15F

3 a $z = \operatorname{cis}\left(\dfrac{\pi}{6}\right)$

By De Moivre's theorem:

$z^2 = \operatorname{cis}\left(\dfrac{\pi}{3}\right)$

$z^3 = \operatorname{cis}\left(\dfrac{\pi}{2}\right)$

$z^4 = \operatorname{cis}\left(\dfrac{2\pi}{3}\right)$

b

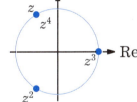

Figure 15F.3

4 a $z = \text{cis}\left(\dfrac{2\pi}{3}\right)$. So

$z^2 = \text{cis}\left(\dfrac{4\pi}{3}\right)$

$z^3 = \text{cis}(2\pi) = \text{cis}(0) = 1$

$z^4 = \text{cis}\left(\dfrac{8\pi}{3}\right) = \text{cis}\left(\dfrac{2\pi}{3}\right)$

b

Im, Re (figure with z, z^4, z^3, z^2 on unit circle)

Figure 15F.4

c Since $z^3 = 1$, $z^n = z$ for all $n = 3k+1$, $k \in \mathbb{N}$

5 a $|1+\sqrt{3}i| = \sqrt{1+3} = 2$

$\arg(1+\sqrt{3}i) = \arctan\left(\dfrac{\sqrt{3}}{1}\right) = \dfrac{\pi}{3}$

(select argument in first quadrant of Argand plane)

b From (a), $1+\sqrt{3}i = 2\,\text{cis}\left(\dfrac{\pi}{3}\right)$

By De Moivre:

$\left(1+\sqrt{3}i\right)^5 = 2^5 \text{cis}\left(\dfrac{5\pi}{3}\right) = 32\,\text{cis}\left(-\dfrac{\pi}{3}\right)$

c $\text{cis}\left(-\dfrac{\pi}{3}\right) = \dfrac{1}{2} - \dfrac{\sqrt{3}}{2}i$

$\therefore \left(1+\sqrt{3}i\right)^5 = 16 - 16\sqrt{3}\,i$

6 a $\left|-\sqrt{2}+\sqrt{2}i\right| = \sqrt{2+2} = 2$

$\arg\left(-\sqrt{2}+\sqrt{2}i\right) = \arctan\left(\dfrac{\sqrt{2}}{-\sqrt{2}}\right) = \dfrac{3\pi}{4}$

(select argument in second quadrant of Argand plane)

$\therefore -\sqrt{2}+\sqrt{2}i = 2\,\text{cis}\left(\dfrac{3\pi}{4}\right)$

b $\left(-\sqrt{2}+\sqrt{2}i\right)^6 = 2^6 \text{cis}\left(\dfrac{6 \times 3\pi}{4}\right)$

$= 64\,\text{cis}\left(\dfrac{9\pi}{2}\right)$

$= 64i$

7 a $\dfrac{1}{\text{cis}\,\theta} = \dfrac{1}{\cos\theta + i\sin\theta}$

$= \dfrac{1}{\cos\theta + i\sin\theta} \times \dfrac{\cos\theta - i\sin\theta}{\cos\theta - i\sin\theta}$

$= \dfrac{\cos\theta - i\sin\theta}{\left(\cos^2\theta + \sin^2\theta\right)}$

$= \dfrac{\cos(-\theta) + i\sin(-\theta)}{1}$

$= \text{cis}(-\theta)$

$\cos(-\theta) = \cos(2\pi - \theta)$ since $\cos x$ has period 2π;

for the same reason, $\sin(-\theta) = \sin(2\pi - \theta)$

$\therefore \dfrac{1}{\text{cis}\,\theta} = \text{cis}(-\theta) = \text{cis}(2\pi - \theta)$

15 Complex numbers 187

b $\dfrac{\text{cis}\,\theta_1}{\text{cis}\,\theta_2} = \dfrac{\cos\theta_1 + i\sin\theta_1}{\cos\theta_2 + i\sin\theta_2}$

$= \dfrac{\cos\theta_1 + i\sin\theta_1}{\cos\theta_2 + i\sin\theta_2} \times \dfrac{\cos\theta_2 - i\sin\theta_2}{\cos\theta_2 - i\sin\theta_2}$

$= \dfrac{(\cos\theta_1 \cos\theta_2 + \sin\theta_1 \sin\theta_2) + i(\sin\theta_1 \cos\theta_2 - \sin\theta_2 \cos\theta_1)}{\cos^2\theta_2 + \sin^2\theta_2}$

$= \dfrac{\cos(\theta_1 - \theta_2) + i\sin(\theta_1 - \theta_2)}{1}$

$= \text{cis}(\theta_1 - \theta_2)$

8 a $A = 4 - i$

b A rotation of $\dfrac{2\pi}{3}$ about the origin in the Argand plane is equivalent to multiplication by $\text{cis}\left(\dfrac{2\pi}{3}\right)$, and a rotation of $\dfrac{4\pi}{3}$ about the origin is equivalent to multiplication by $\text{cis}\left(\dfrac{-2\pi}{3}\right)$.

$|4 - i| = \sqrt{17}$

$\arg(4 - i) = \arctan\left(-\dfrac{1}{4}\right) = \theta$

where θ is selected to be in the fourth quadrant, so that $\cos\theta = \dfrac{4}{\sqrt{17}}$, $\sin\theta = -\dfrac{1}{\sqrt{17}}$

$\therefore 4 - i = \sqrt{17}\,\text{cis}\,\theta$

Then the other two points are represented by $\sqrt{17}\,\text{cis}\left(\theta + \dfrac{2\pi}{3}\right)$ and $\sqrt{17}\,\text{cis}\left(\theta - \dfrac{2\pi}{3}\right)$

$\sqrt{17}\,\text{cis}\left(\theta + \dfrac{2\pi}{3}\right) = \sqrt{17}\left(\cos\left(\theta + \dfrac{2\pi}{3}\right) + i\sin\left(\theta + \dfrac{2\pi}{3}\right)\right)$

$= \sqrt{17}\left(\cos\theta\cos\left(\dfrac{2\pi}{3}\right) - \sin\theta\sin\left(\dfrac{2\pi}{3}\right)\right) + i\sqrt{17}\left(\sin\theta\cos\left(\dfrac{2\pi}{3}\right) + \sin\left(\dfrac{2\pi}{3}\right)\cos\theta\right)$

$= \sqrt{17}\left(\left(\dfrac{4}{\sqrt{17}}\right)\times\left(\dfrac{-1}{2}\right) - \left(\dfrac{-1}{\sqrt{17}}\right)\left(\dfrac{\sqrt{3}}{2}\right)\right) + i\sqrt{17}\left(\left(\dfrac{-1}{\sqrt{17}}\right)\left(\dfrac{-1}{2}\right) + \left(\dfrac{\sqrt{3}}{2}\right)\left(\dfrac{4}{\sqrt{17}}\right)\right)$

$= -2 + \dfrac{\sqrt{3}}{2} + i\left(\dfrac{1}{2} + 2\sqrt{3}\right)$

Similarly,

$\sqrt{17}\,\text{cis}\left(\theta - \dfrac{2\pi}{3}\right) = -2 - \dfrac{\sqrt{3}}{2} + i\left(\dfrac{1}{2} - 2\sqrt{3}\right)$

So the vertices have coordinates $\left(-2 + \dfrac{\sqrt{3}}{2}, \dfrac{1}{2} + 2\sqrt{3}\right)$ and $\left(-2 - \dfrac{\sqrt{3}}{2}, \dfrac{1}{2} - 2\sqrt{3}\right)$

188 Topic 15F Operations in polar form

Exercise 15G

5 $\dfrac{e^{iz}+e^{-iz}}{2} = \dfrac{1}{2}\left[(\cos z + i\sin z)+(\cos(-z)+i\sin(-z))\right]$

$= \dfrac{1}{2}\left[(\cos z + i\sin z)+(\cos z - i\sin z)\right]$

$= \dfrac{1}{2}(2\cos z)$

$= \cos z$

Hence $\cos(2i) = \dfrac{e^{-2}+e^{2}}{2} = 3.76$ (3SF)

6 $5^i = e^{i(\ln 5)}$

$= \cos(\ln 5)+i\sin(\ln 5)$

$= -0.0386 + 0.999i$

7 $3^{2-i} = 9 \times 3^{-i}$

$= 9e^{-i(\ln 3)}$

$= 9(\cos(\ln 3) - i\sin(\ln 3))$

$= 9\cos(\ln 3) - 9\sin(\ln 3)i$

$= 4.09 - 8.02i$ (3SF)

8 $\cos z = 2 \Rightarrow e^{iz}+e^{-iz} = 4$

$\Rightarrow e^{2iz} - 4e^{iz}+1 = 0$

This is a quadratic equation in e^{iz}:

$e^{iz} = \dfrac{4 \pm \sqrt{4^2-4}}{2} = 2 \pm \sqrt{3}$

$\therefore iz = \ln(2 \pm \sqrt{3})$

and hence $z = -i\ln(2 \pm \sqrt{3})$

9 a $i = e^{i\frac{\pi}{2}}$

b $i^i = \left(e^{i\frac{\pi}{2}}\right)^i = e^{i\frac{\pi}{2} \times i} = e^{-\frac{\pi}{2}}$

Exercise 15H

2 $z^4 = 1 \Rightarrow z = \pm 1, \pm i$

The 4th roots of unity are

$\operatorname{cis} 0$, $\operatorname{cis} \dfrac{\pi}{2}$, $\operatorname{cis} \pi$, $\operatorname{cis}\left(-\dfrac{\pi}{2}\right)$

3 Let $z = r\operatorname{cis}\theta$; then

$(r\operatorname{cis}\theta)^3 = -8 = 2^3 \operatorname{cis} \pi$

$\therefore r^3 \operatorname{cis} 3\theta = 2^3 \operatorname{cis} \pi$

$\Rightarrow r = 2$ and $3\theta = \pi, 3\pi, 5\pi$

$\therefore r = 2$ and $\theta = \dfrac{\pi}{3}, \pi, \dfrac{5\pi}{3}$

So the solutions to $z^3 = -8$ are

$z_1 = 2\operatorname{cis}\dfrac{\pi}{3} = 1 + \sqrt{3}\,i$

$z_2 = 2\operatorname{cis}\pi = -2$

$z_3 = 2\operatorname{cis}\left(-\dfrac{\pi}{3}\right) = 1 - \sqrt{3}\,i$

4 $z^3 = \sqrt{2}(4 - 4i)$

Let $z = r\operatorname{cis}\theta$

$\left|\sqrt{2}(4 - 4i)\right| = \sqrt{32 + 32} = 8$

$\arg\left(\sqrt{2}(4-4i)\right) = \arctan\left(-\dfrac{4}{4}\right) = -\dfrac{\pi}{4}$

(choose argument in fourth quadrant of Argand plane)

$\therefore r^3 \operatorname{cis} 3\theta = 8\operatorname{cis}\left(-\dfrac{\pi}{4}\right)$

$\Rightarrow r = 2$ and $3\theta = -\dfrac{\pi}{4}, \dfrac{7\pi}{4}, \dfrac{15\pi}{4}$

$\therefore r = 2$ and $\theta = -\dfrac{\pi}{12}, \dfrac{7\pi}{12}, \dfrac{5\pi}{4}$

Using double angle formulae:

$\cos\left(\dfrac{x}{2}\right) = \sqrt{\dfrac{1+\cos x}{2}}$, $\sin\left(\dfrac{x}{2}\right) = \sqrt{\dfrac{1-\cos x}{2}}$

$\therefore \cos\left(\dfrac{\pi}{12}\right) = \sqrt{\dfrac{1+\cos\left(\dfrac{\pi}{6}\right)}{2}}$

$= \sqrt{\dfrac{1+\dfrac{\sqrt{3}}{2}}{2}} = \sqrt{\dfrac{2+\sqrt{3}}{4}} = \dfrac{1}{2}\sqrt{2+\sqrt{3}}$

and

$\sin\left(\dfrac{\pi}{12}\right) = \sqrt{\dfrac{1-\cos\left(\dfrac{\pi}{6}\right)}{2}}$

$= \sqrt{\dfrac{1-\dfrac{\sqrt{3}}{2}}{2}} = \sqrt{\dfrac{2-\sqrt{3}}{4}} = \dfrac{1}{2}\sqrt{2-\sqrt{3}}$

Therefore the solutions are

$z_1 = 2\operatorname{cis}\left(-\dfrac{\pi}{12}\right) = \sqrt{2+\sqrt{3}} - \sqrt{2-\sqrt{3}}\,i$

$z_2 = 2\operatorname{cis}\left(\dfrac{7\pi}{12}\right) = 2\operatorname{cis}\left(\dfrac{\pi}{2} + \dfrac{\pi}{12}\right)$

$= -\sqrt{2-\sqrt{3}} + \sqrt{2+\sqrt{3}}\,i$

$z_3 = 2\operatorname{cis}\left(\dfrac{5\pi}{4}\right) = 2\left(-\dfrac{1}{\sqrt{2}} - \dfrac{1}{\sqrt{2}}i\right) = -\sqrt{2} - \sqrt{2}\,i$

> **COMMENT**
>
> The sine and cosine of $\dfrac{\pi}{12}$ can be expressed in several ways using surds. As shown in Long Question 1 of the mixed practice exercise at the end of this chapter, you may also use $\cos\left(\dfrac{\pi}{12}\right) = \dfrac{\sqrt{6}+\sqrt{2}}{4}$ and $\sin\left(\dfrac{\pi}{12}\right) = \dfrac{\sqrt{6}-\sqrt{2}}{4}$.

5 $z^4 = -81i = 3^4 \text{cis}\left(-\dfrac{\pi}{2}\right)$

Let $z = r\text{cis}\,\theta$

$r^4 \text{cis}\,4\theta = 3^4 \text{cis}\left(-\dfrac{\pi}{2}\right)$

$\Rightarrow r = 3, \quad 4\theta = -\dfrac{5\pi}{2}, -\dfrac{\pi}{2}, \dfrac{3\pi}{2}, \dfrac{7\pi}{2}$

$\therefore r = 3, \quad \theta = -\dfrac{5\pi}{8}, -\dfrac{\pi}{8}, \dfrac{3\pi}{8}, \dfrac{7\pi}{8}$

Therefore the solutions are

$z_1 = 3\text{cis}\left(-\dfrac{\pi}{8}\right), z_2 = 3\text{cis}\left(\dfrac{3\pi}{8}\right), z_3 = 3\text{cis}\left(\dfrac{7\pi}{8}\right), z_4 = 3\text{cis}\left(-\dfrac{5\pi}{8}\right)$

6 a $|4 + 4\sqrt{3}i| = \sqrt{16 + 48} = 8$

$\arg(4 + 4\sqrt{3}i) = \arctan\left(\dfrac{4\sqrt{3}}{4}\right) = \dfrac{\pi}{3}$

(in first quadrant of Argand plane)

$\therefore 4 + 4\sqrt{3}i = 8\text{cis}\left(\dfrac{\pi}{3}\right)$

b Let $z = r\text{cis}\,\theta$

$r^4 \text{cis}\,4\theta = 8\text{cis}\left(\dfrac{\pi}{3}\right)$

$\Rightarrow r = 2^{\frac{3}{4}}, \quad 4\theta = \dfrac{\pi}{3}, \dfrac{7\pi}{3}, \dfrac{13\pi}{3}, \dfrac{19\pi}{3}$

$\therefore r = 2^{\frac{3}{4}}, \quad \theta = \dfrac{\pi}{12}, \dfrac{7\pi}{12}, \dfrac{13\pi}{12}, \dfrac{19\pi}{12}$

So the solutions are

$z_1 = 2^{\frac{3}{4}}\text{cis}\left(\dfrac{\pi}{12}\right), z_2 = 2^{\frac{3}{4}}\text{cis}\left(\dfrac{7\pi}{12}\right), z_3 = 2^{\frac{3}{4}}\text{cis}\left(\dfrac{13\pi}{12}\right), z_4 = 2^{\frac{3}{4}}\text{cis}\left(\dfrac{19\pi}{12}\right)$

c

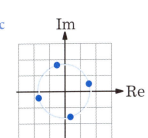

Figure 15H.6

7 **a** Let $z = r\operatorname{cis}\theta$

$r^4 \operatorname{cis} 4\theta = -16 = 2^4 \operatorname{cis}(\pi)$

$\Rightarrow r = 2, \quad 4\theta = \pi, 3\pi, 5\pi, 7\pi$

$\therefore r = 2, \quad \theta = \dfrac{\pi}{4}, \dfrac{3\pi}{4}, \dfrac{5\pi}{4}, \dfrac{7\pi}{4}$

So the solutions are

$z = \sqrt{2} + \sqrt{2}i, -\sqrt{2} + \sqrt{2}i, -\sqrt{2} - \sqrt{2}i,$
$\sqrt{2} - \sqrt{2}i$

b $z^4 + 16 = 0$ has the roots found in (a). Pairing up the conjugates:

$z^4 + 16$

$= \left[(z - \sqrt{2} - \sqrt{2}i)(z - \sqrt{2} + \sqrt{2}i)\right]$

$\quad \left[(z + \sqrt{2} - \sqrt{2}i)(z + \sqrt{2} + \sqrt{2}i)\right]$

$= (z^2 - 2\sqrt{2}z + 4)(z^2 + 2\sqrt{2}z + 4)$

8 **a** $\omega_1 = \operatorname{cis}\left(\dfrac{2\pi}{6}\right) = \operatorname{cis}\left(\dfrac{\pi}{3}\right)$

For $n = 2, 3, 4, 5$:

$\omega_n = \operatorname{cis}\left(n \times \dfrac{2\pi}{6}\right) = \operatorname{cis}\left(\dfrac{n\pi}{3}\right)$

$\therefore$ by De Moivre's theorem,

$\omega_n = \left(\operatorname{cis}\left(\dfrac{\pi}{3}\right)\right)^n = \omega_1^n$

b $1 + \omega_1 + \omega_2 + \omega_3 + \omega_4 + \omega_5$

$= 1 + \omega_1 + \omega_1^2 + \omega_1^3 + \omega_1^4 + \omega_1^5$

This is a geometric series with first term 1 and common ratio ω_1

$\therefore 1 + \omega_1 + \omega_2 + \omega_3 + \omega_4 + \omega_5 = S_6$

$= \dfrac{1 - \omega_1^6}{1 - \omega_1}$

But $\omega_1^6 = 1$ by definition, so

$1 + \omega_1 + \omega_2 + \omega_3 + \omega_4 + \omega_5 = 0$

9 $\omega = \operatorname{cis}\left(\dfrac{2\pi}{3}\right)$

$\Rightarrow \omega^2 = \operatorname{cis}\left(\dfrac{4\pi}{3}\right) = \operatorname{cis}\left(\dfrac{-2\pi}{3}\right), \omega^3 = \operatorname{cis}\left(\dfrac{6\pi}{3}\right) = 1,$

and $\omega + \omega^2 = -\dfrac{1}{2} + \dfrac{\sqrt{3}}{2}i - \dfrac{1}{2} - \dfrac{\sqrt{3}}{2}i = -1$

$\therefore (a + b\omega)(a + b\omega^2) = a^2 + b^2\omega^3 + ab(\omega + \omega^2)$

$= a^2 + b^2 - ab$

10 **a** $z^3 = -1 = \operatorname{cis}(\pi)$

Let $z = r\operatorname{cis}\theta$

$r^3 \operatorname{cis} 3\theta = \operatorname{cis}(\pi)$

$\Rightarrow r = 1, \quad 3\theta = \pi, 3\pi, 5\pi$

$\therefore r = 1, \quad \theta = \dfrac{\pi}{3}, \pi, \dfrac{5\pi}{3}$

So the cube roots of -1 are

$z = -1, \dfrac{1}{2} \pm \dfrac{\sqrt{3}}{2}i$

b $(x + 2)^3 = x^3 + 6x^2 + 12x + 8$

c $x^3 + 6x^2 + 12x + 9 = 0$

$x^3 + 6x^2 + 12x + 8 = -1$

$(z + 2)^3 = -1 \quad$ by (b)

$z + 2 = -1$ or $\dfrac{1}{2} \pm \dfrac{\sqrt{3}}{2}i \quad$ by (a)

$\therefore z = -3$ or $-\dfrac{3}{2} \pm \dfrac{\sqrt{3}}{2}i$

Exercise 15I

1 $(\cos\theta + i\sin\theta)^3$

$= \cos^3\theta + 3i\cos^2\theta\sin\theta - 3\cos\theta\sin^2\theta - i\sin^3\theta$

$\operatorname{Re}((\cos\theta + i\sin\theta)^3) = \cos^3\theta - 3\cos\theta\sin^2\theta$

$\operatorname{Im}((\cos\theta + i\sin\theta)^3) = 3\cos^2\theta\sin\theta - \sin^3\theta$

By De Moivre,

$(\cos\theta + i\sin\theta)^3 = \cos 3\theta + i\sin 3\theta$

Comparing imaginary parts:
$$\sin 3\theta = 3\cos^2\theta \sin\theta - \sin^3\theta = 3(1-\sin^2\theta)\sin\theta - \sin^3\theta = 3\sin\theta - 4\sin^3\theta$$

2 a Using the binomial theorem:
$$(\cos\theta + i\sin\theta)^4 = \cos^4\theta + 4i\cos^3\theta \sin\theta - 6\cos^2\theta \sin^2\theta - 4i\cos\theta \sin^3\theta + \sin^4\theta$$
$$\Rightarrow \mathrm{Re}\left((\cos\theta + i\sin\theta)^4\right) = \cos^4\theta - 6\cos^2\theta \sin^2\theta + \sin^4\theta$$

b By De Moivre, $(\cos\theta + i\sin\theta)^4 = \cos 4\theta + i\sin 4\theta$
$$\therefore \cos 4\theta = \cos^4\theta - 6\cos^2\theta \sin^2\theta + \sin^4\theta$$
$$= \cos^4\theta - 6\cos^2\theta(1-\cos^2\theta) + (1-\cos^2\theta)^2$$
$$= 8\cos^4\theta - 8\cos^2\theta + 1$$

3 a By De Moivre, $(\cos\theta + i\sin\theta)^n = \cos n\theta + i\sin n\theta$

Hence
$$z^n = \cos n\theta + i\sin n\theta$$
$$z^{-n} = \cos(-n\theta) + i\sin(-n\theta) = \cos n\theta - i\sin n\theta$$
$$\therefore z^n + z^{-n} = 2\cos n\theta$$

b From the above with $n = 1$,
$$2\cos\theta = z + z^{-1}$$
$$\therefore (2\cos\theta)^5 = (z + z^{-1})^5$$

Expanding using the binomial theorem:
$$32\cos^5\theta = z^5 + 5z^3 + 10z + 10z^{-1} + 5z^{-3} + z^{-5}$$
$$= (z^5 + z^{-5}) + 5(z^3 + z^{-3}) + 10(z + z^{-1})$$
$$= 2\cos 5\theta + 10\cos 3\theta + 20\cos\theta$$
$$\therefore A = 2,\ B = 10,\ C = 20$$

4 a By De Moivre, $(\cos\theta + i\sin\theta)^n = \cos n\theta + i\sin n\theta$

Hence
$$z^n = \cos n\theta + i\sin n\theta$$
$$z^{-n} = \cos(-n\theta) + i\sin(-n\theta) = \cos n\theta - i\sin n\theta$$
$$\therefore z^n - z^{-n} = 2i\sin n\theta$$

b Also, $z^n + z^{-n} = 2\cos n\theta$, so $z + z^{-1} = 2\cos\theta$

Using the binomial theorem:

$$(z+z^{-1})^6 = z^6 + 6z^4 + 15z^2 + 20 + 15z^{-2} + 6z^{-4} + z^{-6}$$

$$(z-z^{-1})^6 = z^6 - 6z^4 + 15z^2 - 20 + 15z^{-2} - 6z^{-4} + z^{-6}$$

Taking the difference of these two statements:

$$(2\cos\theta)^6 - (2i\sin\theta)^6 = 12z^4 + 40 + 12z^{-4}$$

$$64(\cos^6\theta + \sin^6\theta) = 12(z^4 + z^{-4}) + 40$$

$$= 24\cos 4\theta + 40$$

$$\cos^6\theta + \sin^6\theta = \frac{1}{64}(24\cos 4\theta + 40)$$

$$= \frac{1}{8}(3\cos 4\theta + 5)$$

5 a $(\cos\theta + i\sin\theta)^5 = \cos^5\theta + 5i\cos^4\theta\sin\theta - 10\cos^3\theta\sin^2\theta - 10i\cos^2\theta\sin^3\theta + 5\cos\theta\sin^4\theta + i\sin^5\theta$

$\text{Re}((\cos\theta + i\sin\theta)^5) = \cos^5\theta - 10\cos^3\theta\sin^2\theta + 5\cos\theta\sin^4\theta$

$\text{Im}((\cos\theta + i\sin\theta)^5) = 5\cos^4\theta\sin\theta - 10\cos^2\theta\sin^3\theta + \sin^5\theta$

b By De Moivre, $(\cos\theta + i\sin\theta)^5 = \cos 5\theta + i\sin 5\theta$

Comparing imaginary parts:

$\sin 5\theta = 5\cos^4\theta\sin\theta - 10\cos^2\theta\sin^3\theta + \sin^5\theta$

$$\therefore \frac{\sin 5\theta}{\sin\theta} = 5\cos^4\theta - 10\cos^2\theta\sin^2\theta + \sin^4\theta$$

$$= 5\cos^4\theta - 10\cos^2\theta(1-\cos^2\theta) + (1-\cos^2\theta)^2$$

$$= 16\cos^4\theta - 12\cos^2\theta + 1$$

c As $\theta \to 0$, $\cos\theta \to 1$

$$\therefore \lim_{\theta \to 0}\left(\frac{\sin 5\theta}{\sin\theta}\right) = \lim_{\theta \to 0}(16\cos^4\theta - 12\cos^2\theta + 1) = 16 - 12 + 1 = 5$$

COMMENT

There are many other ways to establish this result. The approximation that $\sin\theta \approx \theta$ for small values of θ suggests that $\lim_{\theta \to 0}\left(\frac{\sin n\theta}{\sin\theta}\right) = n$ for any n. You might want to research l'Hôpital's rule, found in the Calculus option, which would also demonstrate this result.

Mixed examination practice 15

Short questions

1 $z = 3i - \dfrac{2}{\sqrt{3}+i}$

$= 3i - \dfrac{2(\sqrt{3}-i)}{(\sqrt{3}+i)(\sqrt{3}-i)}$

$= 3i - \dfrac{2\sqrt{3}-2i}{4}$

$= -\dfrac{\sqrt{3}}{2} + \dfrac{7}{2}i$

2 $3z + w = 9 + 11i \quad \ldots(1)$
$iw - z = -8 - 2i \quad \ldots(2)$

$(1) + 3 \times (2) \Rightarrow (1+3i)w = -15 + 5i$

$\therefore w = \dfrac{-15+5i}{(1+3i)}$

$= \dfrac{(-15+5i)(1-3i)}{(1+3i)(1-3i)}$

$= \dfrac{-15+45i+5i+15}{10}$

$= 5i$

$(2) \Rightarrow z = iw + 8 + 2i = 3 + 2i$

3 A polynomial with real coefficients must have complex roots in conjugate pairs.

$\therefore$ roots are 1 and $1 \pm 2i$. Hence

$f(z) = (z-1)(z-1-2i)(z-1+2i)$
$= (z-1)(z^2 - 2z + 5)$
$= z^3 - 3z^2 + 7z - 5$

$\therefore a = -3, \ b = 7, \ c = -5$

4 Let $z = x + iy$ for $x, y \in \mathbb{R}$

Then $z^* = x - iy$

$\therefore 3z - 5z^* = -2x + 8iy$

To solve $-2x + 8iy = 4 - 3i$, compare real and imaginary parts:

Re: $-2x = 4 \Rightarrow x = -2$

Im: $8y = -3 \Rightarrow y = -\dfrac{3}{8}$

$\therefore z = -2 - \dfrac{3}{8}i$

5 $\dfrac{1}{\sqrt{3}+i} = \dfrac{\sqrt{3}-i}{4}$

$\left|\dfrac{\sqrt{3}-i}{4}\right| = \dfrac{1}{4}\sqrt{3+1} = \dfrac{1}{2}$

$\arg\left(\dfrac{\sqrt{3}-i}{4}\right) = \arctan\left(-\dfrac{1}{\sqrt{3}}\right) = -\dfrac{\pi}{6}$

(choose argument in fourth quadrant of Argand plane as $\text{Re}(z) > 0, \text{Im}(z) < 0$)

$\therefore \dfrac{1}{\sqrt{3}+i} = \dfrac{1}{2}\text{cis}\left(-\dfrac{\pi}{6}\right)$

$\Rightarrow \left(\dfrac{1}{\sqrt{3}+i}\right)^6 = \left(\dfrac{1}{2}\right)^6 \text{cis}(-\pi) = -\dfrac{1}{64}$

6 A polynomial with real coefficients must have complex roots in conjugate pairs.

$\therefore$ roots are $2 \pm 3i$ and k for some $k \in \mathbb{R}$

$z^3 + az^2 + bz - 65 = (z-k)(z-2-3i)$
$(z-2+3i)$
$= (z-k)(z^2 - 4z + 13)$
$= z^3 - (k+4)z^2$
$+ (13+4k)z - 13k$

Comparing coefficients:

$z^0: \ -13k = -65 \Rightarrow k = 5$

$z^1: \ 13 + 4k = b \Rightarrow b = 33$

$z^2: \ -k - 4 = a \Rightarrow a = -9$

$z^3: \ 1 = 1$

$\therefore a = -9, b = 33$

15 Complex numbers 195

7 $w = 1+\sqrt{3}i$, $z = 1+i$

$$\therefore \frac{w+\sqrt{2}z}{w-\sqrt{2}z} = \frac{1+\sqrt{3}i+\sqrt{2}(1+i)}{1+\sqrt{3}i-\sqrt{2}(1+i)}$$

$$= \frac{1+\sqrt{2}+i(\sqrt{3}+\sqrt{2})}{1-\sqrt{2}+i(\sqrt{3}-\sqrt{2})}$$

$$= \frac{\left(1+\sqrt{2}+i(\sqrt{3}+\sqrt{2})\right)}{\left(1-\sqrt{2}+i(\sqrt{3}-\sqrt{2})\right)} \times \frac{\left(1-\sqrt{2}-i(\sqrt{3}-\sqrt{2})\right)}{\left(1-\sqrt{2}-i(\sqrt{3}-\sqrt{2})\right)}$$

$$\Rightarrow \operatorname{Re}\left(\frac{w+\sqrt{2}z}{w-\sqrt{2}z}\right) = \frac{(1+\sqrt{2})(1-\sqrt{2})+(\sqrt{3}+\sqrt{2})(\sqrt{3}-\sqrt{2})}{(1-\sqrt{2}+i(\sqrt{3}-\sqrt{2}))(1-\sqrt{2}-i(\sqrt{3}-\sqrt{2}))}$$

$$= \frac{1-2+3-2}{(1-\sqrt{2}+i(\sqrt{3}-\sqrt{2}))(1-\sqrt{2}-i(\sqrt{3}-\sqrt{2}))}$$

$$= 0$$

8 $z = 4\operatorname{cis}\left(\frac{\pi}{4}\right)$, $w = 2\operatorname{cis}\left(\frac{\pi}{6}\right)$

$$\frac{z}{w} = \frac{4}{2}\operatorname{cis}\left(\frac{\pi}{4}-\frac{\pi}{6}\right) = 2\operatorname{cis}\left(\frac{\pi}{12}\right)$$

$$\Rightarrow \left(\frac{z}{w}\right)^6 = 2^6 \operatorname{cis}\left(\frac{\pi}{2}\right) = 64i$$

9 $\arg\left((a+i)^3\right) = \pi$

So $(a+i)^3$ is a negative real value

$(a+i)^3 = -k$ for some $k \in \mathbb{R}^+$

$(a+i)^3 = k\operatorname{cis}(\pi) = k\operatorname{cis}(-\pi) = \operatorname{cis}(3\pi)$

$$\therefore a+i = k^{\frac{1}{3}}\operatorname{cis}\left(\pm\frac{\pi}{3}\right) \text{ or } k^{\frac{1}{3}}\operatorname{cis}(\pi)$$

$$= k^{\frac{1}{3}}\left(\frac{1}{2}+\frac{\sqrt{3}}{2}i\right) \text{ or } -k^{\frac{1}{3}}$$

Only $a+i = k^{\frac{1}{3}}\left(\frac{1}{2}+\sqrt{\frac{3}{2}}i\right)$ gives a solution for $a \in \mathbb{R}^+$.

Comparing real and imaginary parts:

Re: $a = \frac{1}{2}k^{\frac{1}{3}}$

Im: $1 = k^{\frac{1}{3}} \times \dfrac{\sqrt{3}}{2} \Rightarrow k^{\frac{1}{3}} = \dfrac{2}{\sqrt{3}}$

$\therefore a = \dfrac{1}{\sqrt{3}}$

10 a $\dfrac{w+i}{w-i} = \dfrac{z+1}{z-1}$

$(w+i)(z-1) = (w-i)(z+1)$

$wz - w + iz - i = wz + w - iz - i$

$2iz = 2w$

$w = iz$

b Let $z = x + iy$ and $w = u + iv$ for $x, y, u, v \in \mathbb{R}$

Then $\text{Re}(w) = \text{Re}(iz) = -y = -\text{Im}(z)$

$\therefore \text{Im}(z) = 0 \Rightarrow \text{Re}(w) = 0$

11 $|z + 2i| = |z - 6i|$

Let $z = x + iy$ for $x, y \in \mathbb{R}$; then

$\sqrt{x^2 + (y+2)^2} = \sqrt{x^2 + (y-6)^2}$

$(y+2)^2 = (y-6)^2$

$y + 2 = \pm(y - 6)$

$y + 2 = -(y - 6)$

(reject other option as it leads to $2 = -6$, a contradiction)

$\therefore 2y = 4$

$\Rightarrow y = 2$

$\therefore \text{Im}(z) = 2$

12 $|z + 25| = 5|z + 1|$

Let $z = x + iy$ for $x, y \in \mathbb{R}$; then

$\sqrt{(x+25)^2 + y^2} = 5\sqrt{(x+1)^2 + y^2}$

$x^2 + 50x + 625 + y^2 = 25(x^2 + 2x + 1 + y^2)$

$24x^2 + 24y^2 = 600$

$x^2 + y^2 = 25$

$\sqrt{x^2 + y^2} = 5$

$\therefore |z| = 5$

13 a The sum of the root is $-a$, so

$a = -\left(\dfrac{1}{3} + \dfrac{2}{3} + 1 + 1 + 3\right) = -6$

b Require $\omega^5 = 1 = \text{cis}(0)$

$\therefore \omega = \text{cis}(0), \text{cis}\left(\dfrac{2\pi}{5}\right), \text{cis}\left(\dfrac{4\pi}{5}\right),$

$\text{cis}\left(\dfrac{6\pi}{5}\right) \text{ or } \text{cis}\left(\dfrac{8\pi}{5}\right)$

Then $\omega_1 = \text{cis}\left(\dfrac{2\pi}{5}\right)$,

$\omega_n = \omega_1^n$ for $n = 2, 3, 4,$

$\omega_1^5 = 1$

Hence

$\omega_1 + \omega_2 + \omega_3 + \omega_4 = \omega_1 + \omega_1^2 + \omega_1^3 + \omega_1^4$

$= \dfrac{\omega_1(1 - \omega_1^4)}{1 - \omega_1}$

$= \dfrac{\omega_1 - 1}{1 - \omega_1}$

$= -1$

14 a $(\alpha + \beta)^3 = \alpha^3 + 3\alpha^2\beta + 3\alpha\beta^2 + \beta^3$

$\Rightarrow \alpha^3 + \beta^3 = (\alpha + \beta)^3 - (3\alpha^2\beta + 3\alpha\beta^2)$

$= (\alpha + \beta)^3 - 3\alpha\beta(\alpha + \beta)$

b $x^2 + 7x + 2 = 0$

$\Rightarrow \alpha + \beta = -7, \alpha\beta = 2$

$\therefore \alpha^3\beta^3 = 2^3 = 8$

and $\alpha^3 + \beta^3 = (-7)^3 - 3(2)(-7) = -301$

A quadratic with roots α^3 and β^3 is

$x^2 + 301x + 8 = 0$

15 Let $w = 2+i$ and $z = 3+i$

Then $wz = 5+5i$

$\arg(w) = \arctan\left(\dfrac{1}{2}\right)$

$\arg(z) = \arctan\left(\dfrac{1}{3}\right)$

$\arg(wz) = \arctan\left(\dfrac{5}{5}\right) = \dfrac{\pi}{4}$

Since $\arg(w) + \arg(z) = \arg(wz)$, it follows that $\arctan\left(\dfrac{1}{2}\right) + \arctan\left(\dfrac{1}{3}\right) = \dfrac{\pi}{4}$

16 Let $w = \sin\theta + i(1-\cos\theta)$; then

$\arg w = \arctan\left(\dfrac{1-\cos\theta}{\sin\theta}\right) = \arctan\left(\dfrac{2\sin^2\left(\dfrac{\theta}{2}\right)}{2\sin\left(\dfrac{\theta}{2}\right)\cos\left(\dfrac{\theta}{2}\right)}\right) = \arctan\left(\tan\left(\dfrac{\theta}{2}\right)\right) = \dfrac{\theta}{2}$

$\arg z = \arg w^2 = 2\arg w$

$\Rightarrow \arg z = \theta$

17 $w = \dfrac{1}{1-z}$

Let $z = x+iy$ for $x, y \in \mathbb{R}$; then

$w = \dfrac{1}{1-x-iy}$

$= \dfrac{1-x+iy}{(1-x-iy)(1-x+iy)}$

$= \dfrac{1-x+iy}{(1-x)^2 + y^2}$

$= \dfrac{1-x+iy}{1+x^2+y^2-2x}$

$|z| = 1 \Rightarrow x^2 + y^2 = 1$

$\therefore w = \dfrac{1-x+iy}{2-2x}$

Hence $\text{Re}(w) = \dfrac{1-x}{2(1-x)} = \dfrac{1}{2}$

198 Mixed examination practice 15

18 $z = \text{cis}\,\theta$

$\Rightarrow z^2 = \cos^2\theta - \sin^2\theta + 2i\sin\theta\cos\theta$

$\therefore z^2 - 1 = -2\sin^2\theta + 2i\sin\theta\cos\theta$

$z^2 + 1 = 2\cos^2\theta + 2i\sin\theta\cos\theta$

So $\dfrac{z^2-1}{z^2+1} = \dfrac{-2\sin^2\theta + 2i\sin\theta\cos\theta}{2\cos^2\theta + 2i\sin\theta\cos\theta}$

$= \dfrac{\sin\theta}{\cos\theta} \dfrac{(-\sin\theta + i\cos\theta)}{\cos\theta + i\sin\theta}$

$= \tan\theta \dfrac{(-\sin\theta + i\cos\theta)(\cos\theta - i\sin\theta)}{(\cos\theta + i\sin\theta)(\cos\theta - i\sin\theta)}$

$= \tan\theta \dfrac{-\sin\theta\cos\theta + \cos\theta\sin\theta + i(\cos^2\theta + \sin^2\theta)}{\cos^2\theta + \sin^2\theta}$

$= i\tan\theta$

19 $\text{Im}(k) = 0 \Rightarrow k \in \mathbb{R}$

Let $z = x + iy$; then

$z^2 + 1 = 1 + x^2 - y^2 + 2ixy$

$w = \dfrac{kz}{z^2 + 1}$

$= \dfrac{kz(z^2+1)^*}{(z^2+1)(z^2+1)^*}$

$= \dfrac{k(x+iy)(1 + x^2 - y^2 - 2ixy)}{(z^2+1)(z^2+1)^*}$

$\therefore \text{Im}(w) = \dfrac{ky(1 + x^2 - y^2) - 2kx^2 y}{(z^2+1)(z^2+1)^*}$

$\text{Im}(w) = 0 \Rightarrow ky(1 + x^2 - y^2) - 2kx^2 y = 0$

$\Rightarrow ky(1 - x^2 - y^2) = 0$

If $k \neq 0$ then, given that $\text{Im}(z) = y \neq 0$,

$1 - x^2 - y^2 = 0$

$x^2 + y^2 = 1$

$\Rightarrow |z| = 1$

COMMENT

Note that there was no need to calculate or resolve the denominator of the expression for w in terms of x and y; avoid doing redundant calculations. If you know a rational expression is to equal zero, then aside from ensuring that the denominator is non-zero (which is given in this question), you need not worry about the denominator at all; simply set the numerator equal to zero.

Long questions

1 a $z_1 = \dfrac{\sqrt{6} - i\sqrt{2}}{2}$, $z_2 = 1 - i$

$|z_1| = \dfrac{1}{2}\sqrt{6+2} = \sqrt{2}$

$\arg(z_1) = \arctan\left(\dfrac{-\sqrt{2}}{\sqrt{6}}\right) = -\dfrac{\pi}{6}$

(choose argument in fourth quadrant as $\text{Re}(z) > 0, \text{Im}(z) < 0$)

$|z_2| = \sqrt{1+1} = \sqrt{2}$

$\arg(z_2) = \arctan\left(\dfrac{-1}{1}\right) = -\dfrac{\pi}{4}$

(choose argument in fourth quadrant as $\text{Re}(z) > 0, \text{Im}(z) < 0$)

$\therefore z_1 = \sqrt{2}\,\text{cis}\left(-\dfrac{\pi}{6}\right),\ z_2 = \sqrt{2}\,\text{cis}\left(-\dfrac{\pi}{4}\right)$

b $\dfrac{z_1}{z_2} = \dfrac{\sqrt{2}}{\sqrt{2}}\,\text{cis}\left(\left(-\dfrac{\pi}{6}\right) - \left(-\dfrac{\pi}{4}\right)\right) = \text{cis}\left(\dfrac{\pi}{12}\right)$

c $\dfrac{z_1}{z_2} = \dfrac{\sqrt{6} - i\sqrt{2}}{2(1-i)}$

$= \dfrac{(\sqrt{6} - i\sqrt{2})}{2(1-i)} \cdot \dfrac{(1+i)}{(1+i)}$

$= \dfrac{\sqrt{6} + \sqrt{2} + i(\sqrt{6} - \sqrt{2})}{4}$

$\therefore \cos\left(\dfrac{\pi}{12}\right) = \dfrac{1}{4}(\sqrt{6} + \sqrt{2})$,

$\sin\left(\dfrac{\pi}{12}\right) = \dfrac{1}{4}(\sqrt{6} - \sqrt{2})$

COMMENT

As illustrated in the answer to Exercise 15H question 4, there can be alternative expressions for $\sin\left(\dfrac{\pi}{12}\right)$ and $\cos\left(\dfrac{\pi}{12}\right)$ using nested surds. As an exercise, show that the formulation $\dfrac{\sqrt{6} + \sqrt{2}}{4}$ is equivalent to $\dfrac{\sqrt{2 + \sqrt{3}}}{2}$.

2 a $\left|\dfrac{\sqrt{3}}{2} - \dfrac{1}{2}i\right| = \sqrt{\dfrac{3}{4} + \dfrac{1}{4}} = 1$

$\arg\left(\dfrac{\sqrt{3}}{2} - \dfrac{1}{2}i\right) = \arctan\left(\dfrac{-\dfrac{1}{2}}{\dfrac{\sqrt{3}}{2}}\right) = -\dfrac{\pi}{6}$

(choose argument in fourth quadrant as $\text{Re}(z) > 0, \text{Im}(z) < 0$)

$\therefore \dfrac{\sqrt{3}}{2} - \dfrac{1}{2}i = \text{cis}\left(-\dfrac{\pi}{6}\right)$

b By De Moivre,

$\left(\dfrac{\sqrt{3}}{2} - \dfrac{1}{2}i\right)^9 = \left(\text{cis}\left(-\dfrac{\pi}{6}\right)\right)^9 = \text{cis}\left(-\dfrac{9\pi}{6}\right)$

$= \text{cis}\left(\dfrac{\pi}{2}\right) = i$

$\therefore c = 1$

c $\left|\dfrac{\sqrt{2}}{2}+\dfrac{\sqrt{2}}{2}i\right|=\sqrt{\dfrac{1}{2}+\dfrac{1}{2}}=1$

$\arg\left(\dfrac{\sqrt{2}}{2}+\dfrac{\sqrt{2}}{2}i\right)=\arctan\left(\dfrac{\frac{\sqrt{2}}{2}}{\frac{\sqrt{2}}{2}}\right)=\dfrac{\pi}{4}$

(choose argument in first quadrant as $\operatorname{Re}(z)>0, \operatorname{Im}(z)>0$)

$\therefore \dfrac{\sqrt{2}}{2}+\dfrac{\sqrt{2}}{2}i=\operatorname{cis}\left(\dfrac{\pi}{4}\right)$

$\left(\dfrac{\sqrt{2}}{2}+\dfrac{\sqrt{2}}{2}i\right)^n=\operatorname{cis}\left(\dfrac{n\pi}{4}\right)$

Require $\operatorname{cis}\left(-\dfrac{m\pi}{6}\right)=\operatorname{cis}\left(\dfrac{n\pi}{4}\right)$

$\therefore 2k\pi - \dfrac{m\pi}{6} = \dfrac{n\pi}{4}$ for $k \in \mathbb{Z}$

$\Rightarrow 2m+3n=24k$

Any combination of positive integers m and n for which $2m+3n$ is a multiple of 24 will fulfil the requirement.

A possible pair would be $m=6, n=4$

3 a i $(\cos\theta+i\sin\theta)^3$
$= \cos^3\theta + 3i\cos^2\theta\sin\theta$
$\quad - 3\cos\theta\sin^2\theta - i\sin^3\theta$

ii By De Moivre's theorem,
$(\cos\theta+i\sin\theta)^3 = \cos 3\theta + i\sin 3\theta$
Comparing real and imaginary parts with the expression in (i):
$\cos 3\theta = \cos^3\theta - 3\cos\theta\sin^2\theta$
$\qquad = \cos^3\theta - 3\cos\theta(1-\cos^2\theta)$
$\qquad = 4\cos^3\theta - 3\cos\theta$

$\sin 3\theta = 3\cos^2\theta\sin\theta - \sin^3\theta$
$\qquad = 3(1-\sin^2\theta)\sin\theta - \sin^3\theta$
$\qquad = 3\sin\theta - 4\sin^3\theta$

b $\dfrac{\sin 3\theta - \sin\theta}{\cos 3\theta + \cos\theta} = \dfrac{3\sin\theta - 4\sin^3\theta - \sin\theta}{4\cos^3\theta - 3\cos\theta + \cos\theta}$

$= \dfrac{2\sin\theta - 4\sin^3\theta}{4\cos^3\theta - 2\cos\theta}$

$= \dfrac{\sin\theta}{\cos\theta} \times \dfrac{1 - 2\sin^2\theta}{2\cos^2\theta - 1}$

$= \tan\theta \times \dfrac{\cos 2\theta}{\cos 2\theta}$

$= \tan\theta$

c $\sin\theta = \dfrac{1}{3} \Rightarrow \cos\theta = \sqrt{1-\left(\dfrac{1}{3}\right)^2} = \dfrac{2\sqrt{2}}{3}$

$\left(\begin{array}{l}\text{choose positive root}\\ \text{because } \theta \in \left]-\dfrac{\pi}{4}, \dfrac{\pi}{4}\right]\end{array}\right)$

Using (a):

$\tan 3\theta = \dfrac{\sin 3\theta}{\cos 3\theta}$

$= \dfrac{3\left(\dfrac{1}{3}\right) - 4\left(\dfrac{1}{3}\right)^3}{4\left(\dfrac{2\sqrt{2}}{3}\right)^3 - 3\left(\dfrac{2\sqrt{2}}{3}\right)}$

$= \dfrac{27-4}{64\sqrt{2} - 54\sqrt{2}}$

$= \dfrac{23}{10\sqrt{2}}$

$= \dfrac{23\sqrt{2}}{20}$

4 a Let $1 + \omega + \omega^2 = z$; then
$\omega z = \omega(1 + \omega + \omega^2)$
$\omega^3 = 1$ by definition, so
$\omega z = \omega + \omega^2 + 1 = z$
$\Rightarrow z(\omega - 1) = 0$
But $\omega \neq 1$, so $z = 0$

> **COMMENT**
> This tidy proof works to show that the sum of the nth roots of unity will always equal zero, without any need for identifying $\omega_k = \text{cis}\left(\dfrac{2k\pi}{n}\right)$.

b $(\omega x + \omega^2 y)(\omega^2 x + \omega y) = \omega^3 x^2 + \omega^3 y^2 + xy(\omega^2 + \omega^4)$

Using $\omega^3 = 1$ and $\omega^4 = \omega = -\omega^2 - 1$ from (a):
$(\omega x + \omega^2 y)(\omega^2 x + \omega y) = x^2 + y^2 - xy$

5 a i $x_1 + x_2 + x_3 = -\dfrac{b}{a}$

$x_1 x_2 x_3 = -\dfrac{d}{a}$

ii By the factor theorem,
$ax^3 + bx^2 + cx + d$
$= a(x - x_1)(x - x_2)(x - x_3)$
$= a\left[x^3 - (x_1 + x_2 + x_3)x^2 + (x_1 x_2 + x_2 x_3 + x_3 x_1)x - x_1 x_2 x_3\right]$

Comparing the coefficients of x:
$c = a(x_1 x_2 + x_2 x_3 + x_3 x_1)$

$\therefore x_1 x_2 + x_2 x_3 + x_3 x_1 = \dfrac{c}{a}$

b Let r be the common ratio of the geometric progression.

Then $\alpha = \dfrac{\beta}{r}$, $\gamma = \beta r$ so $\alpha \beta \gamma = \beta^3$

i From (a)(i), the product of the roots is $-\dfrac{d}{a} = -\dfrac{16}{2} = -8$

$\therefore \beta^3 = -8$
$\Rightarrow \beta = -2$

ii $\alpha + \beta + \gamma = \beta\left(\dfrac{1}{r} + 1 + r\right) = -\dfrac{b}{2}$

$\Rightarrow b = 4\left(\dfrac{1}{r} + 1 + r\right)$

$\alpha\beta + \beta\gamma + \gamma\alpha = \beta^2\left(\dfrac{1}{r} + r + 1\right) = \dfrac{c}{2}$

$\Rightarrow c = 8\left(\dfrac{1}{r} + 1 + r\right)$

$\therefore c = 2b$

6 a $z = \cos\theta + i\sin\theta$
By De Moivre's theorem,
$z^{-1} = \cos(-\theta) + i\sin(-\theta) = \cos\theta - i\sin\theta$
$\therefore z + z^{-1} = 2\cos\theta$

b Also by De Moivre's theorem,
$z^n = \cos(n\theta) + i\sin(n\theta)$
$z^{-n} = \cos(-n\theta) + i\sin(-n\theta)$
$= \cos(n\theta) - i\sin(n\theta)$
$\therefore z^n + z^{-n} = 2\cos(n\theta)$

c i $3z^4 - z^3 + 2z^2 - z + 3 = 0$
Dividing through by z^2 (clearly $z \neq 0$, so this is valid) gives
$3z^2 - z + 2 - z^{-1} + 3z^{-2} = 0$
$3(z^2 + z^{-2}) - (z + z^{-1}) + 2 = 0$

From (b):
$3(2\cos 2\theta) - 2\cos\theta + 2 = 0$
$6\cos 2\theta - 2\cos\theta + 2 = 0$

ii Using $\cos 2\theta = 2\cos^2\theta - 1$, the equation in (i) becomes
$12\cos^2\theta - 2\cos\theta - 4 = 0$
$6\cos^2\theta - \cos\theta - 2 = 0$
$\cos\theta = \dfrac{1 \pm \sqrt{1^2 + 4\times 6\times 2}}{12}$
$= \dfrac{1}{12} \pm \dfrac{7}{12}$
$= \dfrac{8}{12}$ or $-\dfrac{6}{12}$
$= \dfrac{2}{3}$ or $-\dfrac{1}{2}$

Corresponding values of $\sin\theta$ are:
$\cos\theta = \dfrac{2}{3} \Rightarrow \sin\theta = \sqrt{1 - \left(\dfrac{2}{3}\right)^2} = \pm\dfrac{\sqrt{5}}{3}$
$\cos\theta = -\dfrac{1}{2} \Rightarrow \sin\theta$
$= \sqrt{1 - \left(-\dfrac{1}{2}\right)^2} = \pm\dfrac{\sqrt{3}}{2}$
$\therefore z = \dfrac{2}{3} \pm \dfrac{\sqrt{5}}{3}i,\ -\dfrac{1}{2} \pm \dfrac{\sqrt{3}}{2}i$

7 a $\omega = e^{\frac{2i\pi}{5}}$
$\Rightarrow \omega^2 = e^{\frac{4i\pi}{5}},\ \omega^3 = e^{\frac{6i\pi}{5}},\ \omega^4 = e^{\frac{8i\pi}{5}}$

b The nth roots of unity always sum to zero (see question 4(a)),
so $\omega^1 + \omega^2 + \omega^3 + \omega^4 + 1 = 0$
Proof in this case, if needed:
$\omega^5 = e^{\frac{10i\pi}{5}} = 1$
$\omega^1 + \omega^2 + \omega^3 + \omega^4$ is a geometric series with common ratio ω
$\therefore \omega^1 + \omega^2 + \omega^3 + \omega^4 = \dfrac{\omega(1 - \omega^4)}{1 - \omega}$
$= \dfrac{\omega - \omega^5}{1 - \omega}$
$= \dfrac{\omega - 1}{1 - \omega}$
$= -1$

c $\omega = \cos\left(\dfrac{2\pi}{5}\right) + i\sin\left(\dfrac{2\pi}{5}\right)$
Using $\omega^5 = 1$:
$\omega^4 = \omega^{5-1} = \omega^{-1} = \cos\left(\dfrac{2\pi}{5}\right) - i\sin\left(\dfrac{2\pi}{5}\right)$
$\therefore \omega + \omega^4 = 2\cos\left(\dfrac{2\pi}{5}\right)$

Similarly,

$$\omega^2 = \cos\left(\frac{4\pi}{5}\right) + i\sin\left(\frac{4\pi}{5}\right)$$

$$\omega^3 = \omega^{5-2} = \omega^{-2} = \cos\left(\frac{4\pi}{5}\right) - i\sin\left(\frac{4\pi}{5}\right)$$

$$\therefore \omega^2 + \omega^3 = 2\cos\left(\frac{4\pi}{5}\right)$$

d $\omega + \omega^4 + \omega^2 + \omega^3 = -1$ from (b)

$$\therefore 2\cos\left(\frac{2\pi}{5}\right) + 2\cos\left(\frac{4\pi}{5}\right) = -1 \text{ from (c)}$$

Using $\cos 2x = 2\cos^2 x - 1$:

$$2\cos\left(\frac{2\pi}{5}\right) + 2\left(2\cos^2\left(\frac{2\pi}{5}\right) - 1\right) = -1$$

$$4\cos^2\left(\frac{2\pi}{5}\right) + 2\cos\left(\frac{2\pi}{5}\right) - 1 = 0$$

$$\therefore \cos\left(\frac{2\pi}{5}\right) = \frac{-2 \pm \sqrt{2^2 + 16}}{8}$$

$$= \frac{-1 \pm \sqrt{5}}{4}$$

The argument lies in the first quadrant, so its cosine must be positive.

Therefore $\cos\left(\frac{2\pi}{5}\right) = \frac{\sqrt{5}-1}{4}$

8 a Using binomial expansion:

$$(\cos\theta + i\sin\theta)^3 = \cos^3\theta + 3i\cos^2\theta\sin\theta$$
$$- 3\cos\theta\sin^2\theta - i\sin^3\theta$$

$$\text{Re}\left((\cos\theta + i\sin\theta)^3\right)$$
$$= \cos^3\theta - 3\cos\theta\sin^2\theta$$

$$\text{Im}\left((\cos\theta + i\sin\theta)^3\right)$$
$$= 3\cos^2\theta\sin\theta - \sin^3\theta$$

By De Moivre,
$$(\cos\theta + i\sin\theta)^3 = \cos 3\theta + i\sin 3\theta$$

Comparing real and imaginary parts:

Re: $\cos 3\theta = \cos^3\theta - 3\cos\theta\sin^2\theta$
$$= \cos^3\theta - 3\cos\theta(1 - \cos^2\theta)$$
$$= 4\cos^3\theta - 3\cos\theta$$

Im: $\sin 3\theta = 3\cos^2\theta\sin\theta - \sin^3\theta$
$$= 3(1 - \sin^2\theta)\sin\theta - \sin^3\theta$$
$$= 3\sin\theta - 4\sin^3\theta$$

b $\tan 3\theta = \dfrac{\sin 3\theta}{\cos 3\theta}$

$$= \frac{3\cos^2\theta\sin\theta - \sin^3\theta}{\cos^3\theta - 3\cos\theta\sin^2\theta}$$

Dividing through by $\cos^3\theta$ in numerator and denominator gives

$$\tan 3\theta = \frac{3\tan\theta - \tan^3\theta}{1 - 3\tan^2\theta}$$

c $x^3 - 3x^2 - 3x + 1 = 0$

$$1 - 3x^2 = 3x - x^3$$

$$\Rightarrow 1 = \frac{3x - x^3}{1 - 3x^2}$$

Let $x = \tan\theta$

From (b), it follows that $\tan 3\theta = 1$

$$\therefore 3\theta = \arctan(1) = \frac{\pi}{4} + k\pi \text{ for } k \in \mathbb{Z}$$

$$\Rightarrow \theta = \frac{\pi}{12} + \frac{k\pi}{3}$$

Hence $x = \tan\left(\dfrac{\pi}{12}\right)$ is a root of the cubic.

d Let $f(x) = x^3 - 3x^2 - 3x + 1$

$f(-1) = -1 - 3 + 3 + 1 = 0$

$\therefore$ by the factor theorem, $(x+1)$ is a factor of $f(x)$

So $f(x) = (x+1)(x^2 + bx + c)$

$$= x^3 + (1+b)x^2 + (b+c)x + c$$

204 Mixed examination practice 15

Comparing coefficients:

x^3: $1 = 1$

x^2: $1 + b = -3 \Rightarrow b = -4$

x^1: $b + c = -3 \Rightarrow c = 1$

x^0: $c = 1$ is consistent with the value found above

$\therefore f(x) = (x+1)(x^2 - 4x + 1)$

$f(x) = 0 \Rightarrow x = -1, \dfrac{4 \pm \sqrt{12}}{2}$

$\Rightarrow x = -1, 2 \pm \sqrt{3}$

e $\tan\theta$ is increasing for $\theta \in \left[0, \dfrac{\pi}{4}\right]$

$\therefore 0 = \tan(0) < \tan\left(\dfrac{\pi}{12}\right) < \tan\left(\dfrac{\pi}{4}\right) = 1$

f From (e), $\tan\left(\dfrac{\pi}{12}\right)$ is a positive value less than 1.

So of the 3 values found in (d),

$\tan\left(\dfrac{\pi}{12}\right) = 2 - \sqrt{3}$

9 a Distance between two points (x_1, y_1) and (x_2, y_2) is given by

$PQ = \sqrt{(x_1 - x_2)^2 + (y_1 - y_2)^2}$

$= |x_1 - x_2 + i(y_1 - y_2)|$

$= |x_1 + iy_1 - (x_2 + iy_2)|$

$= |z_1 - z_2|$

b i $A: a + 0i$

ii $B: b \operatorname{cis} \theta$

iii From (a):

$AB = |b\operatorname{cis}\theta - a|$

$= \sqrt{(b\cos\theta - a)^2 + b^2 \sin^2\theta}$

$= \sqrt{a^2 + b^2 - 2ab\cos\theta}$

iv

$|AB|^2 = a^2 + b^2 - 2ab\cos\theta$

$= |OA|^2 + |OB|^2 - 2|OA| \times |OB|\cos\theta$

16 Basic differentiation and its applications

Exercise 16A

3 a Sometimes true: derivative indicates the slope of the curve, not its position. For example, $y = 2x$ has constant positive gradient 2, both at point $(1, 2)$, where $y > 0$, and at point $(-1, -2)$, where $y < 0$.

b Sometimes true: as for (a). For example, $(1, -2)$ lies on the line $y = -2x$ (with negative gradient) and also on $y = x - 3$ (with positive gradient).

c Always true: $\dfrac{dy}{dx} = 0$ is a defining property of a stationary point.

d Sometimes true: there is also the possibility of a horizontal inflexion point. For example, $\dfrac{dx^3}{dx}(0) = 0$, but $(0, 0)$ is neither a local maximum nor a local minimum of the curve $y = x^3$.

e Sometimes true; for example, the function $y = -e^{-x}$ has a positive gradient throughout, but its graph is always below the x-axis.

f Sometimes true; for example, the lowest value of the function $y = \sqrt{x}$ is 0, at $x = 0$, but the gradient at $x = 0$ is not zero.

Exercise 16B

2 If $f(x) = x^2 + 1$ then $f(x+h) = (x+h)^2 + 1$.

$$\dfrac{dy}{dx} = \lim_{h \to 0} \left\{ \dfrac{(x+h)^2 + 1 - (x^2 + 1)}{h} \right\}$$

$$= \lim_{h \to 0} \left\{ \dfrac{x^2 + 2xh + h^2 + 1 - x^2 - 1}{h} \right\}$$

$$= \lim_{h \to 0} \left\{ \dfrac{2xh + h^2}{h} \right\}$$

$$= \lim_{h \to 0} \{2x + h\}$$

$$= 2x$$

3 If $f(x) = 8$ then $f(x+h) = 8$ as well.

$$\dfrac{dy}{dx} = \lim_{h \to 0} \left\{ \dfrac{8 - 8}{h} \right\}$$

$$= \lim_{h \to 0} \{0\}$$

$$= 0$$

4 $y = \dfrac{1}{x}$

$$\dfrac{dy}{dx} = \lim_{h \to 0} \left\{ \dfrac{\dfrac{1}{x+h} - \dfrac{1}{x}}{h} \right\}$$

$$= \lim_{h \to 0} \left\{ \dfrac{x - (x+h)}{h(x+h)x} \right\}$$

$$= \lim_{h \to 0} \left\{ \dfrac{-h}{hx(x+h)} \right\}$$

$$= \lim_{h \to 0} \left\{ \dfrac{-1}{x(x+h)} \right\}$$

$$= -\dfrac{1}{x^2}$$

5 $y = kf(x)$

$$\frac{dy}{dx} = \lim_{h \to 0} \left\{ \frac{kf(x+h) - kf(x)}{h} \right\}$$

$$= \lim_{h \to 0} \left\{ k \frac{f(x+h) - f(x)}{h} \right\}$$

$$= k \lim_{h \to 0} \left\{ \frac{f(x+h) - f(x)}{h} \right\}$$

$$= kf'(x)$$

6 $y = \dfrac{1}{\sqrt{x}}$

$$\frac{dy}{dx} = \lim_{h \to 0} \left\{ \frac{\frac{1}{\sqrt{x+h}} - \frac{1}{\sqrt{x}}}{h} \right\}$$

$$= \lim_{h \to 0} \left\{ \frac{\sqrt{x} - \sqrt{x+h}}{h\sqrt{x+h}\sqrt{x}} \right\}$$

$$= \lim_{h \to 0} \left\{ \frac{(\sqrt{x} - \sqrt{x+h})(\sqrt{x} + \sqrt{x+h})}{h\sqrt{x^2 + hx}(\sqrt{x} + \sqrt{x+h})} \right\}$$

$$= \lim_{h \to 0} \left\{ \frac{x - (x+h)}{h\sqrt{x^2 + hx}(\sqrt{x} + \sqrt{x+h})} \right\}$$

$$= \lim_{h \to 0} \left\{ \frac{-h}{h\sqrt{x^2 + hx}(\sqrt{x} + \sqrt{x+h})} \right\}$$

$$= \lim_{h \to 0} \left\{ \frac{-1}{\sqrt{x^2 + hx}(\sqrt{x} + \sqrt{x+h})} \right\}$$

$$= -\frac{1}{\sqrt{x^2}(\sqrt{x} + \sqrt{x})}$$

$$= -\frac{1}{2x\sqrt{x}}$$

Exercise 16D

8 $\dfrac{dy}{dx} = 3x^2 - 4x$

$\dfrac{dy}{dx} = y$

$3x^2 - 4x = x^3 - 2x^2 + 1$

$x^3 - 5x^2 + 4x + 1 = 0$

From GDC: $x = -0.199, 1.29, 3.91$ (3SF)

The points are $(-0.199, 0.913)$, $(1.29, -0.181)$, $(3.91, 30.3)$

9 The gradient is decreasing where $\dfrac{d^2 y}{dx^2} < 0$

$\dfrac{dy}{dx} = 7 - 2x - 3x^2$

$\therefore \dfrac{d^2 y}{dx^2} = -2 - 6x$

$-2 - 6x < 0$

$\Rightarrow 6x > -2$

i.e. $x > -\dfrac{1}{3}$

10 $y = \dfrac{1}{4}x^4 + x^3 - \dfrac{1}{2}x^2 - 3x + 6$

$\dfrac{dy}{dx} = x^3 + 3x^2 - x - 3$

$\dfrac{d^2 y}{dx^2} = 3x^2 + 6x - 1$

Gradient is increasing where $\dfrac{d^2 y}{dx^2} > 0$:

$3x^2 + 6x - 1 > 0$

Roots of $3x^2 + 6 - 1 = 0$ are

$x = \dfrac{-6 \pm \sqrt{6^2 + 12}}{6} = -1 \pm \dfrac{2\sqrt{3}}{3}$

A positive quadratic is greater than zero outside the roots

$\therefore x < -1 - \dfrac{2\sqrt{3}}{3}$ or $x > -1 + \dfrac{2\sqrt{3}}{3}$

11 $\dfrac{d^n}{dx^n}(x^n) = \dfrac{d^{n-1}}{dx^{n-1}}\left(\dfrac{d}{dx}(x^n)\right)$

$= \dfrac{d^{n-1}}{dx^{n-1}}(nx^{n-1})$

$= n\dfrac{d^{n-1}}{dx^{n-1}}(x^{n-1})$

$= n\dfrac{d^{n-2}}{dx^{n-2}}\left(\dfrac{d}{dx}(x^{n-1})\right)$

$= n(n-1)\dfrac{d^{n-2}}{dx^{n-2}}(x^{n-2})$

$\vdots$

$= n!$

Exercise 16E

2 $f'(x) = \cos x + 2x$

$f'\left(\dfrac{\pi}{2}\right) = \cos\dfrac{\pi}{2} + 2 \times \dfrac{\pi}{2}$

$= \pi$

3 $g'(x) = \dfrac{1}{4\cos^2 x} + 3\sin x - 3x^2$

$g'\left(\dfrac{\pi}{6}\right) = \dfrac{1}{4\cos^2\left(\dfrac{\pi}{6}\right)} + 3\sin\dfrac{\pi}{6} - 3\left(\dfrac{\pi}{6}\right)^2$

$= \dfrac{1}{4\left(\dfrac{\sqrt{3}}{2}\right)^2} + 3\times\dfrac{1}{2} - 3\times\dfrac{\pi^2}{36}$

$= \dfrac{11}{6} - \dfrac{\pi^2}{12}$

4 $h'(x) = \cos x - \sin x$

$h'(x) = 0$

$\Rightarrow \cos x - \sin x = 0$

$\tan x = 1$

$\therefore x = \dfrac{\pi}{4}, \dfrac{5\pi}{4}$

5 $\dfrac{dy}{dx} = 1 - \dfrac{2}{x^3}$

$\dfrac{1}{4\cos^2 x} - \dfrac{2}{x^3} = 1 - \dfrac{2}{x^3}$

$4\cos^2 x = 1$

$\cos x = \pm\dfrac{1}{2}$

$\therefore x = \dfrac{\pi}{3}, \dfrac{2\pi}{3}, \dfrac{4\pi}{3}, \dfrac{5\pi}{3}$

Exercise 16F

2 i $f'(x) = \dfrac{1}{2}e^x - \dfrac{7}{x}$

$f'(\ln 4) = \dfrac{1}{2}e^{\ln 4} - \dfrac{7}{\ln 4}$

$= \dfrac{1}{2}\times 4 - \dfrac{7}{\ln 4} = 2 - \dfrac{7}{\ln 4}$

ii $f'(x) = e^x - \dfrac{1}{2x}$

$f'(\ln 3) = e^{\ln 3} - \dfrac{1}{2\ln 3}$

$= 3 - \dfrac{1}{2\ln 3} = 3 - \dfrac{1}{\ln 9}$

3 $f'(x) = -6$

$-2e^x = -6$

$e^x = 3$

$x = \ln 3$

4 $g'(x) = 2$

$2x - \dfrac{12}{x} = 2$

$x^2 - x - 6 = 0$

$(x+2)(x-3) = 0$

$x = -2$ or 3

However, reject $x = -2$ as not within the domain of g.

Hence $x = 3$

> **COMMENT**
> Always check for the validity of solutions in any question containing a logarithm or square root, since the working can give rise to solution values outside the domain of the original function.

5 a i $y = 3\ln x \Rightarrow \dfrac{dy}{dx} = \dfrac{3}{x}$

ii $y = \ln 5 + \ln x \Rightarrow \dfrac{dy}{dx} = \dfrac{1}{x}$

b i $y = e^3 \times e^x \Rightarrow \dfrac{dy}{dx} = e^3 \times e^x = e^{3+x}$

ii $y = e^{-3} \times e^x \Rightarrow \dfrac{dy}{dx} = e^{-3} \times e^x = e^{x-3}$

c i $y = e^{\ln(x^2)} = x^2 \Rightarrow \dfrac{dy}{dx} = 2x$

ii $y = e^2 \times e^{\ln(x^3)} = e^2 \times x^3$

$\Rightarrow \dfrac{dy}{dx} = e^2 \times 3x^2 = 3e^2 x^2$

d i $y = \log_3 x = \dfrac{\ln x}{\ln 3} \Rightarrow \dfrac{dy}{dx} = \dfrac{1}{\ln 3} \times \dfrac{1}{x} = \dfrac{1}{x \ln 3}$

ii $y = 4\log_6 x = \dfrac{4 \ln x}{\ln 6}$

$\Rightarrow \dfrac{dy}{dx} = \dfrac{4}{\ln 6} \times \dfrac{1}{x} = \dfrac{4}{x \ln 6}$

Exercise 16G

2 $y = \sqrt{x} + 3x$

$\Rightarrow \dfrac{dy}{dx} = \dfrac{1}{2} x^{-\frac{1}{2}} + 3$

Require $\dfrac{1}{2} x^{-\frac{1}{2}} + 3 = 5$

$x^{-\frac{1}{2}} = 4$

$x = \dfrac{1}{16}$

$y\left(\dfrac{1}{16}\right) = \dfrac{1}{4} + \dfrac{3}{16} = \dfrac{7}{16}$

∴ coordinates of the point are $\left(\dfrac{1}{16}, \dfrac{7}{16}\right)$

3 $\dfrac{dy}{dx} = e^x + 1$

Gradient of $y = 3x$ is 3, so require $\dfrac{dy}{dx} = 3$:

$e^x + 1 = 3$

$\Rightarrow x = \ln 2$

$y(\ln 2) = 2 + \ln 2$

$= \ln 2e^2$

Equation of the tangent is

$y - y_1 = m(x - x_1)$

$y - \ln 2e^2 = 3(x - \ln 2)$

$y = 3x - \ln 8 + \ln 2e^2$

$y = 3x + \ln\left(\dfrac{e^2}{4}\right)$

$y = 3x + 2 - \ln 4$

4 $\dfrac{dy}{dx} = 3x^2 - 6x$

$\Rightarrow \dfrac{dy}{dx}(1) = -3$

Therefore the gradient of the normal at point $(1, -2)$ is $\dfrac{1}{3}$

Require $\dfrac{dy}{dx} = \dfrac{1}{3}$

∴ $3x^2 - 6x = \dfrac{1}{3}$

$9x^2 - 18x - 1 = 0$

$x = \dfrac{18 \pm \sqrt{18^2 + 36}}{18}$

$= 1 \pm \dfrac{\sqrt{360}}{18}$

$= 1 \pm \dfrac{\sqrt{10}}{3}$

$= 2.05, -0.0541 \,(3\text{SF})$

16 Basic differentiation and its applications   209

5 $\dfrac{dy}{dx} = 3x^2 - 6x$

$\Rightarrow \dfrac{dy}{dx}(2) = 12 - 12 = 0$

$y(2) = 8 - 12 = -4$

∴ the equation of the tangent is $y = -4$

This intersects the curve where

$x^3 - 3x^2 = -4$

$x^3 - 3x^2 + 4 = 0$

$(x-2)^2(x+1) = 0$

$x = 2$ or -1

Thus the tangent meets the curve again at $x = -1$, at the point $(-1, -4)$.

> **COMMENT**
>
> Since there is a tangent at $x = 2$, we already know that the cubic factorises with $(x-2)^2$ as a factor (repeated root at a tangent).

6 $y = (x-1)^2 = x^2 - 2x + 1$

$\Rightarrow \dfrac{dy}{dx} = 2x - 2$

Normal at the point $(a, (a-1)^2)$ has gradient $\dfrac{-1}{2a-2}$

Equation of the normal is

$y - y_1 = m(x - x_1)$

$y - (a-1)^2 = \dfrac{-1}{2a-2}(x-a)$

Require that this passes through $(0, 0)$

∴ $0 - (a-1)^2 = \dfrac{-1}{2a-2}(0-a)$

$-(a-1)^2 = \dfrac{a}{2(a-1)}$

$-2(a-1)^3 = a$

From GDC: $a = 0.410$ (3SF)

∴ the coordinates of the point are $(0.410, 0.348)$

7 $P\left(\dfrac{\pi}{6}, 1\right), Q\left(\dfrac{\pi}{4}, \sqrt{2}\right)$

a $f'(x) = 2\cos x$

$\Rightarrow f'\left(\dfrac{\pi}{6}\right) = \sqrt{3}$

b Chord PQ has gradient

$\dfrac{\sqrt{2} - 1}{\dfrac{\pi}{4} - \dfrac{\pi}{6}} = \dfrac{12(\sqrt{2}-1)}{\pi}$

Elevation of a line with gradient a is the angle $\arctan(a)$, so elevation of chord PQ is

$\arctan\left(\dfrac{12(\sqrt{2}-1)}{\pi}\right) = 57.7°$

Elevation of the tangent at P is

$\arctan(\sqrt{3}) = 60°$

The difference in elevations is the angle between the lines: $60° - 57.7° = 2.3°$

8

> **COMMENT**
>
> The question requires you to prove that the area is independent of a; this means that the end answer for the area should be an expression in which a does not appear. Calculate in the normal way, with the expectation that a will cancel out in the final part of the working.

$y = kx^{-1} \Rightarrow \dfrac{dy}{dx} = -kx^{-2}$

Tangent at the point (a, ka^{-1}) has gradient $m = -ka^{-2}$

Equation of the tangent is

$y - y_1 = m(x - x_1)$

$y - ka^{-1} = -ka^{-2}(x - a)$

$y = \dfrac{-k}{a^2}(x - 2a)$

This line intersects the y-axis at $P\left(0, \frac{2k}{a}\right)$
and intersects the x-axis at $Q(2a, 0)$

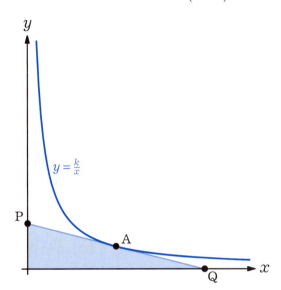

Figure 16G.8

$OP = \frac{2k}{a}$, $OQ = 2a$

$\therefore$ Area $OPQ = \frac{1}{2}(OP)(OQ)$

$= \frac{1}{2} \times \frac{2k}{a} \times 2a$

$= 2k$

Hence the area of triangle OPQ is independent of a.

9 $\frac{dy}{dx} = 3x^2 - 1$

Tangent at the point $(a, a^3 - a)$ has gradient $3a^2 - 1$

Equation of the tangent is

$y - y_1 = m(x - x_1)$

$y - (a^3 - a) = (3a^2 - 1)(x - a)$

$y = (3a^2 - 1)x - 2a^3$

This line intersects the curve where

$x^3 - x = (3a^2 - 1)x - 2a^3$

$x^3 - 3a^2 x + 2a^3 = 0$

Since the line is tangent to the curve at $x = a$, this cubic must have $(x - a)^2$ as a factor (repeated root at a tangent). Hence the cubic factorises as

$(x - a)^2 (x + 2a) = 0$

Thus the tangent intersects the curve again at $x = -2a$, as required.

Exercise 16H

2 An example for which the statement is not true:

$f(x) = x + \frac{1}{x}$ has a local minimum at $(x_1, f(x_1)) = (1, 2)$ and a local maximum at $(x_2, f(x_2)) = (-1, -2)$.

Other examples include functions for which the stationary point $(x_1, f(x_1))$ is a horizontal inflexion, not a maximum; for example, $y = (x+1)(x-2)^3$ has a negative-valued minimum for $x \in [-1, 2]$ and a horizontal inflexion at $(2, 0)$.

The statement will be true if the function has no discontinuities (although there will be some discontinuous functions for which it is also true) and no horizontal inflexions.

3 $\frac{dy}{dx} = 3x^2 + 6x - 24$

Stationary points where $\frac{dy}{dx} = 0$:

$3x^2 + 6x - 24 = 0$

$x^2 + 2x - 8 = 0$

$(x+4)(x-2) = 0$

$x = -4$ or 2

$x = -4 \Rightarrow y = (-4)^3 + 3(-4)^2 - 24(-4) + 12 = 92$

$x = 2 \Rightarrow y = 2^3 + 3(2)^2 - 24(2) + 12 = -16$

$\therefore$ stationary points are $(-4, 92)$ and $(2, -16)$

16 Basic differentiation and its applications

$\dfrac{d^2y}{dx^2} = 6x+6$

$\dfrac{d^2y}{dx^2}(-4) = -18 < 0 \Rightarrow (-4, 92)$ is a local maximum

$\dfrac{d^2y}{dx^2}(2) = 18 > 0 \Rightarrow (2, -16)$ is a local minimum

4 $\dfrac{dy}{dx} = 1 - \dfrac{1}{2}x^{-\tfrac{1}{2}} = 1 - \dfrac{1}{2\sqrt{x}}$

Stationary points where $\dfrac{dy}{dx} = 0$:

$1 - \dfrac{1}{2\sqrt{x}} = 0$

$2\sqrt{x} - 1 = 0$

$\sqrt{x} = \dfrac{1}{2}$

$x = \dfrac{1}{4}$

$x = \dfrac{1}{4} \Rightarrow y = \dfrac{1}{4} - \sqrt{\dfrac{1}{4}} = -\dfrac{1}{4}$

∴ stationary point is $\left(\dfrac{1}{4}, -\dfrac{1}{4}\right)$

$\dfrac{d^2y}{dx^2} = \dfrac{1}{4}x^{-\tfrac{3}{2}}$

$\dfrac{d^2y}{dx^2}\left(\dfrac{1}{4}\right) = \dfrac{1}{4} \times 8 > 0 \Rightarrow \left(\dfrac{1}{4}, -\dfrac{1}{4}\right)$ is a local minimum

5 $\dfrac{dy}{dx} = \cos x - 4\sin x$

Stationary points where $\dfrac{dy}{dx} = 0$:

$\cos x - 4\sin x = 0$

$4\sin x = \cos x$

$\tan x = \dfrac{1}{4}$

∴ $x = 0.245$ or 3.39 (3SF)

$x = 0.245 \Rightarrow y = \sin 0.245 + 4\cos 0.245 = 4.12$

$x = 3.39 \Rightarrow y = \sin 3.39 + 4\cos 3.39 = -4.12$

∴ stationary points are $(0.245, 4.12)$ and $(3.39, -4.12)$

$\dfrac{d^2y}{dx^2} = -\sin x - 4\cos x = -y$

$\dfrac{d^2y}{dx^2}(0.245) = -4.12 < 0 \Rightarrow (0.245, 4.12)$ is a local maximum

$\dfrac{d^2y}{dx^2}(3.39) = 4.12 > 0 \Rightarrow (3.39, -4.12)$ is a local minimum

6 $f'(x) = \dfrac{1}{x} - \dfrac{k}{x^{k+1}}$

Stationary points where $f'(x) = 0$:

$\dfrac{1}{x} - \dfrac{k}{x^{k+1}} = 0$

$\dfrac{1}{x} = \dfrac{k}{x^{k+1}}$

$x^k = k$

$x = k^{\tfrac{1}{k}}$

$x = k^{\tfrac{1}{k}} \Rightarrow y = \ln k^{\tfrac{1}{k}} + \dfrac{1}{k}$

$= \dfrac{1}{k}\ln k + \dfrac{1}{k}$

$= \dfrac{\ln k + 1}{k}$

∴ $f(x)$ has a stationary point with y-coordinate $\dfrac{\ln k + 1}{k}$

7 $f'(x) = 12x^3 - 48x^2 + 36x$

Stationary points where $f'(x) = 0$:

$12x^3 - 48x^2 + 36x = 0$

$12x(x^2 - 4x + 3) = 0$

$x(x-1)(x-3) = 0$

$x = 0, 1$ or 3

$f(0) = 3(0)^4 - 16(0)^3 + 18(0)^2 + 6 = 6$

$f(1) = 3(1)^4 - 16(1)^3 + 18(1)^2 + 6 = 11$

$f(3) = 3(3)^4 - 16(3)^3 + 18(3)^2 + 6 = -21$

∴ stationary points are $(0, 6)$, $(1, 11)$ and $(3, -21)$

$f''(x) = 36x^2 - 96x + 36$

$f''(0) = 36 > 0 \Rightarrow (0, 6)$ is a local minimum

$f''(1) = -24 < 0 \Rightarrow (1, 11)$ is a local maximum

$f''(3) = 72 > 0 \Rightarrow (3, -21)$ is another local minimum

∴ range of f is $[-21, \infty[$

> **COMMENT**
>
> Instead of using second derivative analysis, it would also be valid to use knowledge of the form of a positive quartic equation to assert that the first and third stationary points must be the local minima.

8 $f'(x) = e^x - 4$

Stationary points where $f'(x) = 0$:

$e^x - 4 = 0$

$x = \ln 4$

$f(\ln 4) = e^{\ln 4} - 4\ln 4 + 2 = 6 - 4\ln 4$

∴ stationary point is $(\ln 4, 6 - 4\ln 4)$

$f''(x) = e^x$

$f''(\ln 4) = 4 > 0 \Rightarrow (\ln 4, 6 - 4\ln 4)$ is a local minimum

∴ range of f is $[6 - \ln 4, \infty[$

9 $\dfrac{dy}{dx} = 3kx^2 + 12x$

Stationary points where $\dfrac{dy}{dx} = 0$:

$3kx^2 + 12x = 0$

$kx^2 + 4x = 0$

$x(kx + 4) = 0$

$x = 0$ or $-\dfrac{4}{k}$

$x = 0 \Rightarrow y = k(0)^3 + 6(0)^2 = 0$

$x = -\dfrac{4}{k} \Rightarrow y = k\left(-\dfrac{4}{k}\right)^3 + 6\left(-\dfrac{4}{k}\right)^2 = \dfrac{32}{k^2}$

∴ stationary points are $(0, 0)$ and $\left(-\dfrac{4}{k}, \dfrac{32}{k^2}\right)$

$\dfrac{d^2y}{dx^2} = 6kx + 12$

$\dfrac{d^2y}{dx^2}(0) = 12 > 0 \Rightarrow (0, 0)$ is a local minimum

$\dfrac{d^2y}{dx^2}\left(-\dfrac{4}{k}\right) = -12 < 0 \Rightarrow \left(-\dfrac{4}{k}, \dfrac{32}{k^2}\right)$ is a local maximum

Exercise 16I

1 $\dfrac{dy}{dx} = e^x - 2x$

$\dfrac{d^2y}{dx^2} = e^x - 2$

Points of inflexion where $\dfrac{d^2y}{dx^2} = 0$:

$e^x - 2 = 0$

$x = \ln 2$

$x = \ln 2 \Rightarrow y = e^{\ln 2} - (\ln 2)^2 = 2 - (\ln 2)^2$

∴ point of inflexion is at $(\ln 2, \ 2 - (\ln 2)^2)$

2 $\dfrac{dy}{dx} = 4x^3 - 12x + 7$

$\dfrac{d^2y}{dx^2} = 12x^2 - 12$

Points of inflexion where $\dfrac{d^2y}{dx^2} = 0$:

$12x^2 - 12 = 0$

$x^2 = 1$

$x = \pm 1$

$x = 1 \Rightarrow y = 1^4 - 6(1)^2 + 7(1) + 2 = 4$

$x = -1 \Rightarrow y = (-1)^4 - 6(-1)^2 + 7(-1) + 2 = -10$

∴ points of inflexion are at $(1, 4)$ and $(-1, -10)$

16 Basic differentiation and its applications

3 $\dfrac{dy}{dx} = \cos x$

$\dfrac{d^2y}{dx^2} = -\sin x = -y$

Points of inflexion have $\dfrac{d^2y}{dx^2} = 0$, which must therefore be on $y = 0$, the x-axis.

4 $\dfrac{dy}{dx} = -2\sin x + 1$

$\dfrac{d^2y}{dx^2} = -2\cos x$

Points of inflexion where $\dfrac{d^2y}{dx^2} = 0$:

$-2\cos x = 0$

$\cos x = 0$

$\therefore x = \dfrac{\pi}{2}, \dfrac{3\pi}{2}$

$x = \dfrac{\pi}{2} \Rightarrow y = 2\cos\left(\dfrac{\pi}{2}\right) + \dfrac{\pi}{2} = \dfrac{\pi}{2}$

$x = \dfrac{3\pi}{2} \Rightarrow y = 2\cos\left(\dfrac{3\pi}{2}\right) + \dfrac{3\pi}{2} = \dfrac{3\pi}{2}$

Verifying that these are points of inflexion:

For a small positive value δ, $\cos\left(\dfrac{\pi}{2} - \delta\right) > 0$ and so $\dfrac{d^2y}{dx^2}\left(\dfrac{\pi}{2} - \delta\right) < 0$ (gradient of curve is decreasing); similarly, $\cos\left(\dfrac{\pi}{2} + \delta\right) < 0$ and so $\dfrac{d^2y}{dx^2}\left(\dfrac{\pi}{2} + \delta\right) > 0$ (gradient of curve is increasing). Therefore $\left(\dfrac{\pi}{2}, \dfrac{\pi}{2}\right)$ is a genuine point of inflexion.

For a small positive value δ,

$\cos\left(\dfrac{3\pi}{2} - \delta\right) < 0$ and so $\dfrac{d^2y}{dx^2}\left(\dfrac{3\pi}{2} - \delta\right) > 0$

(gradient increasing); similarly,

$\cos\left(\dfrac{3\pi}{2} + \delta\right) > 0$ and so $\dfrac{d^2y}{dx^2}\left(\dfrac{3\pi}{2} + \delta\right) < 0$

(gradient decreasing). Therefore $\left(\dfrac{3\pi}{2}, \dfrac{3\pi}{2}\right)$ is a genuine point of inflexion.

> **COMMENT**
>
> Remember that just showing that the second derivative is zero is not sufficient for the point to be an inflexion. Further working is needed, either using values on either side, as above, or by showing that the first non-zero derivative after the second derivative is odd. An alternative to the justifications using δ would be:
>
> $\dfrac{d^3y}{dx^3} = 2\sin x$
>
> $\dfrac{d^3y}{dx^3}\left(\dfrac{\pi}{2}\right) = 2 \neq 0 \Rightarrow \left(\dfrac{\pi}{2}, \dfrac{\pi}{2}\right)$ is a genuine point of inflexion
>
> $\dfrac{d^3y}{dx^3}\left(\dfrac{3\pi}{2}\right) = -2 \neq 0 \Rightarrow \left(\dfrac{3\pi}{2}, \dfrac{3\pi}{2}\right)$ is a genuine point of inflexion

5 $\dfrac{dy}{dx} = 3x^2 - 2ax - b$

$\dfrac{d^2y}{dx^2} = 6x - 2a$

Points of inflexion where $\dfrac{d^2y}{dx^2} = 0$:

$6x - 2a = 0$

$x = \dfrac{a}{3}$

If this is also to be a stationary point, then require $\dfrac{dy}{dx}\left(\dfrac{a}{3}\right) = 0$:

$3\left(\dfrac{a}{3}\right)^2 - 2a\left(\dfrac{a}{3}\right) - b = 0$

$\dfrac{a^2}{3} - \dfrac{2a^2}{3} - b = 0$

$\Rightarrow b = -\dfrac{a^2}{3}$

COMMENT

Only positive cubics with a horizontal inflexion point and leading coefficient 1 have the form $y = (x-k)^3 + d$, which expands to $y = x^3 - 3kx^2 + 3k^2x - k^3 + d$. Comparing with the equation in the question gives $a = 3k$ and $b = -3k^2 = -\frac{a^2}{3}$, as required.

6 Graph shows the gradient function.

When $f'(x) = 0$ there is a stationary point.

If at a stationary point the gradient changes from negative to positive, it is a local minimum (A).

If at a stationary point the gradient changes from positive to negative, it is a local maximum (B).

When $f'(x)$ is itself at a local maximum or minimum, $f''(x) = 0$ and there is a point of inflexion (C).

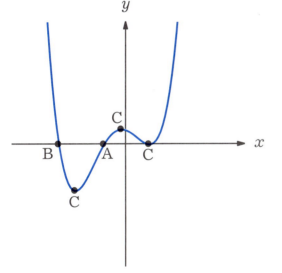

Figure 16I.6

Exercise 16J

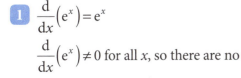

$\frac{d}{dx}(e^x) \neq 0$ for all x, so there are no

stationary points; the end points $x = 0$ and $x = 1$ must give the minimum and maximum for the interval.

$e^0 = 1$ is the minimum

$e^1 = e$ is the maximum

2 a Area $A = x(30-x) = 30x - x^2$

$\frac{dA}{dx} = 30 - 2x$

Stationary point when $\frac{dA}{dx} = 0$:

$30 - 2x = 0$

$x = 15$

Since the extreme values $x = 0$ and $x = 30$ clearly give zero area while intermediate values give a positive area, the end points do not provide a maximum value, and the stationary point is a maximum.

∴ maximum area $= 15 \times 15 = 225 \text{ m}^2$

b Perimeter $= 2x + 2(30-x) = 60$ m, a constant

3 $\frac{dy}{dx} = 3x^2 - 9$

Stationary points where $\frac{dy}{dx} = 0$:

$3x^2 - 9 = 0$

$x^2 = 3$

$x = \pm\sqrt{3}$

$x = -\sqrt{3} \Rightarrow y = (-\sqrt{3})^3 - 9(-\sqrt{3}) = 6\sqrt{3}$

$x = \sqrt{3} \Rightarrow y = (\sqrt{3})^3 - 9(\sqrt{3}) = -6\sqrt{3}$

Checking the end points:

$x = -2 \Rightarrow y = (-2)^3 - 9(-2) = 10 > -6\sqrt{3}$

$x = 5 \Rightarrow y = 5^3 - 9 \times 5 = 80 > 6\sqrt{3}$

The minimum value over the interval $[-2, 5]$ is $-6\sqrt{3} = -10.4$

The maximum value over the interval $[-2, 5]$ is 80

4 $f'(x) = e^x - 3$

Stationary points when $\dfrac{dy}{dx} = 0$:

$e^x - 3 = 0$

$x = \ln 3$

$x = \ln 3 \Rightarrow y = 3 - 3\ln 3 = -0.296$

Checking the end points:

$x = 0 \Rightarrow y = 1$

$x = 2 \Rightarrow y = e^2 - 6 = 1.39$

The minimum value over the interval $[0, 2]$ is $3 - 3\ln 3 = -0.296$

The maximum value over the interval $[0, 2]$ is $e^2 - 6 = 1.39$

5 $\dfrac{dy}{dx} = \cos x + 2$

Stationary points when $\dfrac{dy}{dx} = 0$:

$\cos x + 2 = 0$

$\cos x = -2$

No solutions so no stationary points.

Checking the end points:

$x = 0 \Rightarrow y = 0$

$x = 2\pi \Rightarrow y = 4\pi$

The minimum value over the interval $[0, 2\pi]$ is 0

The maximum value over the interval $[0, 2\pi]$ is 4π

6 $f(x) = x + \dfrac{1}{x}$ for $x > 0$

$f'(x) = 1 - \dfrac{1}{x^2}$

Stationary points where $f'(x) = 0$:

$1 - \dfrac{1}{x^2} = 0$

$x^2 = 1$

$\therefore x = 1$ (as $x > 0$)

Classify the stationary point:

$f''(x) = 2x^{-3}$

$f''(1) = 2 > 0 \Rightarrow$ local minimum

$\therefore$ minimum value of f is 2

7 Distance $D = \left(1 - w^{-\frac{1}{2}}\right) \times \dfrac{5}{w} = 5\left(w^{-1} - w^{-\frac{3}{2}}\right)$

$\dfrac{dD}{dw} = 5\left(\dfrac{3}{2}w^{-\frac{5}{2}} - w^{-2}\right)$

Stationary points when $\dfrac{dD}{dw} = 0$:

$5\left(\dfrac{3}{2}w^{-\frac{5}{2}} - w^{-2}\right) = 0$

$\dfrac{3}{2}w^{-\frac{5}{2}} = w^{-2}$

$\dfrac{3}{2} = \sqrt{w}$

$w = \dfrac{9}{4}$

Classify the stationary point:

$\dfrac{d^2D}{dw^2} = 5\left(2w^{-3} - \dfrac{15}{4}w^{-\frac{7}{2}}\right)$

$\dfrac{d^2D}{dw^2}\left(\dfrac{9}{4}\right) = -0.219 < 0 \Rightarrow$ local maximum.

So a weight of $\dfrac{9}{4} = 2.25$ will maximise the distance travelled.

8 $\dfrac{dt}{dp} = \dfrac{p}{5000} + \dfrac{1}{100} = \dfrac{p + 50}{5000}$

Stationary points when $\dfrac{dt}{dp} = 0$:

$\dfrac{p + 50}{5000} = 0$

$\Rightarrow p = -50 \notin [0, 100]$

So no stationary point in the interval.

Checking the end points:

$p = 0 \Rightarrow t = 2$ minutes, the minimum time to melt $100\,g$

$p = 100 \Rightarrow t = 4$ minutes, the maximum time to melt $100\,g$

9 a $-1 \le \cos t \le 1$

$\therefore V$ has a range of $[40, 160]$

So the minimum volume is 40 million litres.

b Water flow will equal the rate of change in volume:

$$\text{Flow} = \frac{dV}{dt} = -60\sin t$$

Maximum flow occurs when $\frac{d\text{ Flow}}{dt} = 0$:

$-60\cos t = 0$

$\cos t = 0$

$t = \frac{\pi}{2}, \frac{3\pi}{2} \quad (t \in [0,6])$

So the maximum flow in the first 6 days occurs at $t = \frac{\pi}{2}$ (1.6 days) and $\frac{3\pi}{2}$ (4.7 days)

> **COMMENT**
> Although $t = \frac{\pi}{2}$ is actually a local minimum of the 'Flow' function, it still represents a maximum flow of water: the negative sign of 'Flow' when $t = \frac{\pi}{2}$ (which makes it a minimum in the sense of being at its most negative) just means that water is flowing out at that point and not in, but the rate at which the water is flowing is exactly the same as at the local maximum when $t = \frac{3\pi}{2}$ (both are 60 million litres per day).

10 a $\frac{dF}{ds} = 4 - 2s$

Maximum F occurs when $\frac{dF}{ds} = 0$ or at an end point of the domain [0, 4.2]

$\frac{dF}{ds} = 0$

$4 - 2s = 0$

$\Rightarrow s = 2$

and $F(2) = 5$

Check end points: $F(0) = 1, F(4.2) = 0.16$

∴ maximum F occurs at $s = 2$

b Minimum C occurs when $\frac{dC}{ds} = 0$ or at an end point of the domain.

$\frac{dC}{ds} = 0$

$0.2(4 - 2s) + 0.1 = 0$

$0.9 - 0.4s = 0$

$\Rightarrow s = \frac{9}{4}$

and $C\left(\frac{9}{4}\right) = 1.5125$

Check end points:

$C(0) = 0.5, C(4.2) = 0.752$

∴ minimum C occurs when $s = 0$

c Profit P is given by

$P = F - C$

$= F - (0.3 + 0.2F + 0.1s)$

$= 0.8F - 0.3 - 0.1s$

$= 0.8(4s + 1 - s^2) - 0.3 - 0.1s$

$= -0.8s^2 + 3.1s + 0.5$

Stationary point of P occurs when $\frac{dP}{ds} = 0$:

$-1.6s + 3.1 = 0$

$s = \frac{31}{16} = 1.94 \text{ (3SF)}$

Since this value of s lies inside the domain [0, 4.2] and is the position of the vertex of a negative quadratic, it must give the global maximum of P over the domain.

> **COMMENT**
> In part (b) we can actually avoid the calculus altogether, because the minimum value for a negative quadratic over a restricted domain must lie at one of the end points; it cannot be the stationary point, since that must be a maximum. If you prefer to use this argument in an examination, be explicit about your reasoning.

16 Basic differentiation and its applications

11 a $V(0) = 4$, so 4 litres of petrol was initially in the tank.

b 30 seconds = 0.5 minutes

$V(0.5) = 41.5$, so the capacity of the tank is 41.5 litres.

c Flow $= \dfrac{dV}{dt} = 600t - 900t^2$

Maximum flow when $\dfrac{d\,\text{Flow}}{dt} = 0$:

$600 - 1800t = 0$

$t = \dfrac{1}{3}$

∴ maximum flow is at 20 seconds

12 a Total energy $E = x\left(2 - \dfrac{x}{10}\right)$

$= 2x - \dfrac{x^2}{10}$ kJ

b Maximum energy when $\dfrac{dE}{dt} = 0$:

$2 - \dfrac{2x}{10} = 0$

$x = 10$

∴ a total surface area of 10 m² provides maximum energy.

c Net energy

$N = E - 0.01x^3 = 2x - \dfrac{x^2}{10} - 0.01x^3$

Leaves produce more energy than they require when $N > 0$:

$2x - \dfrac{x^2}{10} - 0.01x^3 > 0$

$\dfrac{1}{100}(200x - 10x^2 - x^3) > 0$

$x(200 - 10x - x^2) > 0$

$x(20 + x)(10 - x) > 0$

$\Rightarrow x \in \,]0, 10[$

d Maximum net energy when $\dfrac{dN}{dt} = 0$:

$2 - \dfrac{x}{5} - 0.03x^2 = 0$

$\dfrac{1}{100}(200 - 20x - 3x^2) = 0$

$3x^2 + 20x - 200 = 0$

$x = \dfrac{-20 \pm \sqrt{20^2 + 2400}}{6}$

Require the positive solution:

$x = \dfrac{-20 + \sqrt{2800}}{6} = \dfrac{10(-1 + \sqrt{7})}{3}$

Mixed examination practice 16

Short questions

1 $\dfrac{dy}{dx} = e^x + 2\cos x$

$\dfrac{dy}{dx}\left(\dfrac{\pi}{2}\right) = e^{\frac{\pi}{2}}$

$y\left(\dfrac{\pi}{2}\right) = e^{\frac{\pi}{2}} + 2$

The equation of the tangent is

$y - y_1 = m(x - x_1)$

$y - \left(e^{\frac{\pi}{2}} + 2\right) = e^{\frac{\pi}{2}}\left(x - \dfrac{\pi}{2}\right)$

$y = 2 + e^{\frac{\pi}{2}}\left(x + 1 - \dfrac{\pi}{2}\right)$

2 $y = x^3 - 6x^2 + 12x - 8$

$\Rightarrow \dfrac{dy}{dx} = 3x^2 - 12x + 12 = 3(x - 2)^2$

At $x = 2$, $\dfrac{dy}{dx} = 0$ and $y = 0$, so the normal is a vertical line through (2, 0)

i.e. its equation is $x = 2$

3 $f(1) = 2$
$\Rightarrow 1 + b + c = 2$
$\Rightarrow b + c = 1$

$f'(x) = 2x + b$

$f'(2) = 12$
$\Rightarrow 4 + b = 12$
$\Rightarrow b = 8$
$\therefore c = 1 - 8 = -7$

4 $y = \dfrac{x^3}{6} - x^2 + x$

$\dfrac{dy}{dx} = \dfrac{x^2}{2} - 2x + 1$

$\dfrac{d^2y}{dx^2} = x - 2$

Point of inflexion occurs where $\dfrac{d^2y}{dx^2} = 0$:

$x - 2 = 0$
$\Rightarrow x = 2$

$y(2) = \dfrac{8}{6} - 4 + 2 = -\dfrac{2}{3}$

$\therefore$ coordinates of the point of inflexion are $\left(2, -\dfrac{2}{3}\right)$

5 $\dfrac{dy}{dx} = \sec^2 x - \dfrac{4}{3}$

Stationary points where $\dfrac{dy}{dx} = 0$:

$\sec^2 x - \dfrac{4}{3} = 0$

$\cos^2 x = \dfrac{3}{4}$

$\cos x = \pm \dfrac{\sqrt{3}}{2}$

$\therefore x = [2n\pi +] \dfrac{\pi}{6}, \dfrac{5\pi}{6}, \dfrac{7\pi}{6}, \dfrac{11\pi}{6}$ $(n \in \mathbb{Z})$

$\dfrac{d^2y}{dx^2} = 2\sec^2 x \tan x$

(so the sign of $\dfrac{d^2y}{dx^2}$ is determined by the sign of $\tan x$)

At $x = [2n\pi +] \dfrac{\pi}{6}, \dfrac{7\pi}{6}, \dfrac{d^2y}{dx^2} > 0 \Rightarrow$ local minima

At $x = [2n\pi +] \dfrac{5\pi}{6}, \dfrac{11\pi}{6}, \dfrac{d^2y}{dx^2} < 0 \Rightarrow$ local maxima

So there are local maxima at $x = \dfrac{5\pi}{6} + n\pi$ and local minima at $x = \dfrac{\pi}{6} + n\pi$ $(n \in \mathbb{Z})$.

6 $f(x) = ax^3 + bx^2 + cx + d$

$f(0) = 2 \Rightarrow d = 2$

$f'(x) = 3ax^2 + 2bx + c$

$f'(0) = -3 \Rightarrow c = -3$

Now $f(1) = a + b + c + d = a + b - 1$

and $f'(1) = 3a + 2b + c = 3a + 2b - 3$

$f(1) = f'(1) \Rightarrow a + b - 1 = 3a + 2b - 3$
$\Rightarrow b = 2 - 2a$

$f''(x) = 6ax + 2b = 6ax + 4 - 4a$

$f''(-1) = 6 \Rightarrow -6a + 4 - 4a = 6$
$\Rightarrow -10a = 2$
$\Rightarrow a = -\dfrac{1}{5}$

$\therefore b = 2 - 2\left(-\dfrac{1}{5}\right) = \dfrac{12}{5}$

Therefore the cubic equation is
$f(x) = -\dfrac{1}{5}x^3 + \dfrac{12}{5}x^2 - 3x + 2$

7 a Local minimum: $f'(x) = 0$ and gradient on graph of $f'(x)$ is positive (A)

b Local maximum: $f'(x) = 0$ and gradient on graph of $f'(x)$ is negative (B)

c Inflexion: turning points on graph of $f'(x)$ (C)

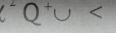

Figure 16MS.7

8 $\dfrac{dy}{dx} = 3x^2$

Tangent at (a, a^3) has gradient $3a^2$ and so has equation

$y - a^3 = 3a^2(x - a)$

$y = 3a^2 x - 2a^3$

It has y-intercept at $(0, -2a^3)$

Gradient of curve at $(-a, -a^3)$ is $3a^2$

Normal at $(-a, -a^3)$ has gradient $-\dfrac{1}{3a^2}$ and so has equation

$y - (-a^3) = -\dfrac{1}{3a^2}(x + a)$

$y = -\dfrac{1}{3a^2} x - a^3 - \dfrac{1}{3a}$

It has y-intercept at $\left(0, -a^3 - \dfrac{1}{3a}\right)$

If the y-intercepts are the same point, then

$-2a^3 = -a^3 - \dfrac{1}{3a}$

$\dfrac{1}{3a} = a^3$

$3a^4 = 1$

$a = 3^{-\frac{1}{4}} = \dfrac{1}{\sqrt[4]{3}}$

(choose positive root since $a > 0$)

Long questions

1 a Point of contact at $x = 2$

On the tangent line, $y = 24(2-1) = 24$

On the curve,

$y = a(2)^3 + b(2)^2 + 4$
$= 8a + 4b + 4$

$\therefore 8a + 4b + 4 = 24$

$\Rightarrow 2a + b = 5$

b For the curve, $\dfrac{dy}{dx} = 3ax^2 + 2bx$

At $x = 2$,

$\dfrac{dy}{dx} = 3a(2)^2 + 2b(2)$

$= 12a + 4b$

and gradient of the tangent is 24

$\therefore 12a + 4b = 24$

$\Rightarrow 3a + b = 6$

c $2a + b = 5$ …(1)

$3a + b = 6$ …(2)

$(2) - (1) \Rightarrow a = 1$

$\therefore b = 3$

d Points of intersection occur when

$x^3 + 3x^2 + 4 = 24(x - 1)$

$x^3 + 3x^2 - 24x + 28 = 0$

One solution is known to be at $x = 2$, which is a double root.

So $(x - 2)^2$ is a factor of $x^3 + 3x^2 - 24x + 28$; factorising gives

$(x - 2)^2 (x + 7) = 0$

$\therefore$ the other intersection point is

$(x, y) = (-7, -192)$

2 a i $\dfrac{dy}{dx} = 3x^2 - 2x - 1$

Stationary points (including the turning point at A) occur where $\dfrac{dy}{dx} = 0$:

$3x^2 - 2x - 1 = 0$

$(3x + 1)(x - 1) = 0$

$x = -\dfrac{1}{3}$ or 1

The point A has negative x-coordinate, so its coordinates are $\left(-\dfrac{1}{3}, \dfrac{86}{27}\right)$

ii $\dfrac{d^2y}{dx^2} = 6x - 2$

Points of inflexion occur where $\dfrac{d^2y}{dx^2} = 0$:

$6x - 2 = 0$

$\Rightarrow x = \dfrac{1}{3}$

∴ the coordinates of B are $\left(\dfrac{1}{3}, \dfrac{70}{27}\right)$

b i The line containing A and B has gradient

$m = \dfrac{\dfrac{86}{27} - \dfrac{70}{27}}{-\dfrac{1}{3} - \dfrac{1}{3}} = -\dfrac{\dfrac{16}{27}}{\dfrac{2}{3}} = -\dfrac{8}{9}$

The equation of the line is

$y - y_1 = m(x - x_1)$

$y - \dfrac{70}{27} = -\dfrac{8}{9}\left(x - \dfrac{1}{3}\right)$

$27y - 70 = -24x + 8$

$27y + 24x = 78$ or $y = -\dfrac{8}{9}x + \dfrac{78}{27}$

ii Require gradient of tangent to be $-\dfrac{8}{9}$

i.e. $\dfrac{dy}{dx} = -\dfrac{8}{9}$

$3x^2 - 2x - 1 = -\dfrac{8}{9}$

$27x^2 - 18x - 1 = 0$

$x = \dfrac{18 \pm \sqrt{324 + 108}}{54}$

$= \dfrac{18 \pm 12\sqrt{3}}{54}$

$= \dfrac{3 \pm 2\sqrt{3}}{9}$

3 a

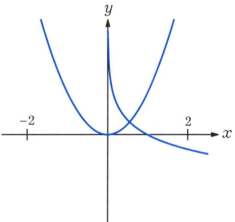

Figure 16ML.3 Graphs of $y = x^2$ and $y = -\dfrac{1}{2}\ln x$

b At intersection, $x^2 = -\dfrac{1}{2}\ln x$

From GDC: $x = 0.548$

c Let $c = 0.548$, the x-coordinate of the intersection point P.

Then point P has coordinates (c, c^2)

For the tangent to $y = x^2$:

$\dfrac{dy}{dx} = 2x \Rightarrow$ gradient at P is $2c$

∴ tangent has equation

$y - c^2 = 2c(x - c)$

Hence the point Q has coordinates $(0, -c^2)$

For the tangent to $y = -\frac{1}{2}\ln x$:

$\frac{dy}{dx} = -\frac{1}{2x} \Rightarrow$ gradient at P is $-\frac{1}{2c}$

∴ tangent has equation

$y + \frac{1}{2}\ln c = -\frac{1}{2c}(x-c)$

Hence the point R has coordinates

$\left(0, \frac{1}{2} - \frac{1}{2}\ln c\right)$

Since c is defined by $c^2 = -\frac{1}{2}\ln c$, the coordinates of R can also be expressed as $\left(0, \frac{1}{2} + c^2\right)$

The tangents are perpendicular, since the product of their gradients is

$2c \times \left(-\frac{1}{2c}\right) = -1$

∴ Area PQR $= \frac{1}{2}(PQ)(PR)$

$= \frac{1}{2}\sqrt{c^2 + 4c^4}\sqrt{c^2 + \frac{1}{4}}$

$= \frac{c}{4}(4c^2 + 1)$

$= 0.302$

d As observed in (c), the gradient at the point where $x = a$ on the $y = x^2$ curve has gradient $2a$, and the gradient on the $y = -\frac{1}{2}\ln x$ curve has gradient $-\frac{1}{2a}$.

The product of these gradients is -1, so the two tangents are always perpendicular.

4 a i $P(0) = 10 + 1 - 0 = 11$, so the initial population is 11 000.

ii 14 million = 14 000 thousand

$P = 14\,000$

$10 + e^t - 3t = 14\,000$

From GDC, $t = 9.55$ (3SF)

So after 9.55 hours the population reaches 14 million.

b i $\frac{dP}{dt} = e^t - 3$

ii 6 million = 6000 thousand

$\frac{dP}{dt} = 6000$

$e^t - 3 = 6000$

$e^t = 6003$

$t = \ln 6003$

$= 8.70$ hours (3SF)

c i $\frac{d^2P}{dt^2} = e^t$, the rate of acceleration of the population

ii $\frac{dP}{dt} = 0$

$e^t - 3 = 0$

$\Rightarrow t = \ln 3$

At $t = \ln 3$, $\frac{d^2P}{dt^2} = 3 > 0$, so this is a local minimum.

$P(\ln 3) = 10 + 3 - 3\ln 3 = 9.704$

∴ the minimum population is 9704.

17 Basic integration and its applications

Exercise 17C

4 $\int \dfrac{1+x}{\sqrt{x}}\,dx = \int x^{-\frac{1}{2}} + x^{\frac{1}{2}}\,dx$

$= 2x^{\frac{1}{2}} + \dfrac{2}{3}x^{\frac{3}{2}} + c$

$= \dfrac{2\sqrt{x}}{3}(3+x) + c$

> **COMMENT**
> After studying Section 19B, try performing this integration using a substitution $u = \sqrt{x}$.

Exercise 17E

2 $\int \dfrac{\sin x + \cos x}{2\cos x}\,dx = \dfrac{1}{2}\int \tan x + 1\,dx$

$= \dfrac{1}{2}\ln|\sec x| + \dfrac{x}{2} + c$

3

$\int \dfrac{\cos 2x}{\cos x - \sin x}\,dx = \int \dfrac{\cos^2 x - \sin^2 x}{\cos x - \sin x}\,dx$

$= \int \dfrac{(\cos x + \sin x)(\cos x - \sin x)}{\cos x - \sin x}\,dx$

$= \int \cos x + \sin x\,dx$

$= \sin x - \cos x + c$

Exercise 17F

2 a $f'(x) = \dfrac{1}{2x}$

$f(x) = \int \dfrac{1}{2x}\,dx$

$= \dfrac{1}{2}\ln x + c$

$= \ln \sqrt{x} + c$

> **COMMENT**
> An equivalent solution is $f(x) = \ln\left(k\sqrt{x}\right)$ where the unknown $k = e^c$ is restricted to a positive value. Unless the question requires it, or if doing so simplifies the appearance of the equation, there is no need to rewrite logarithm solutions in this way.

b $f(2) = 7$

$\dfrac{1}{2}\ln 2 + c = 7$

$c = 7 - \dfrac{1}{2}\ln 2$

$\therefore f(x) = \dfrac{1}{2}\ln x + 7 - \dfrac{1}{2}\ln 2$

$= 7 + \ln\sqrt{\dfrac{x}{2}}$

3 a Maximum occurs where $\dfrac{dy}{dx} = 0$:

$x^2 - 4 = 0$

$x = \pm 2$

$\dfrac{d^2 y}{dx^2} = 2x$

$\dfrac{d^2y}{dx^2}(2) = 4 > 0 \Rightarrow$ local minimum

$\dfrac{d^2y}{dx^2}(-2) = -4 < 0 \Rightarrow$ local maximum

$\therefore$ maximum point is at $x = -2$

b $y = \int x^2 - 4 \, dx$

$= \dfrac{x^3}{3} - 4x + c$

$y(0) = 2$

$\Rightarrow \dfrac{0^3}{3} - 4(0) + c = 2$

$\Rightarrow c = 2$

$\therefore y(x) = \dfrac{x^3}{3} - 4x + 2$

Hence $y(-2) = -\dfrac{8}{3} + 8 + 2 = 7\dfrac{1}{3}$

4 Gradient of normal is $x \Rightarrow$ gradient of tangent is $-\dfrac{1}{x}$

$\dfrac{dy}{dx} = -\dfrac{1}{x}$

$\Rightarrow y = \int -\dfrac{1}{x} dx$

$= -\ln|x| + c$

$y(e^2) = 3$

$\Rightarrow -2 + c = 3$

$\Rightarrow c = 5$

$\therefore y = 5 - \ln x = \ln\left(\dfrac{e^5}{x}\right) \quad (x > 0)$

Exercise 17G

3 $\int_0^\pi e^x + \sin x + 1 \, dx = \left[e^x - \cos x + x\right]_0^\pi$

$= (e^\pi - (-1) + \pi) - (1 - 1 + 0)$

$= e^\pi + 1 + \pi$

4 $\int_k^{2k} \dfrac{1}{x} dx = [\ln x]_k^{2k}$

$= \ln 2k - \ln k$

$= \ln\left(\dfrac{2k}{k}\right)$

$= \ln 2$

and this is independent of k.

5 $\int_3^9 2f(x) + 1 \, dx = 2\int_3^9 f(x) dx + \int_3^9 1 \, dx$

$= 2 \times 7 + [x]_3^9$

$= 14 + (9 - 3)$

$= 20$

6 $\int_1^a t^{\frac{1}{2}} dt = 42$

$\left[\dfrac{2}{3} t^{\frac{3}{2}}\right]_1^a = 42$

$\dfrac{2}{3} a^{\frac{3}{2}} - \dfrac{2}{3} = 42$

$a^{\frac{3}{2}} - 1 = 63$

$a = 64^{\frac{2}{3}} = 16$

Exercise 17H

2

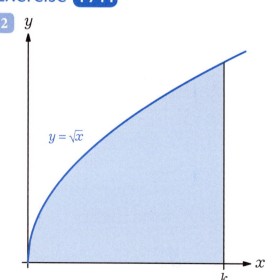

Figure 17H.2 Area enclosed by the curve $y = \sqrt{x}$, the x-axis and the line $x = k$

$$\int_0^k x^{\frac{1}{2}} \, dx = 18$$

$$\left[\frac{2}{3} x^{\frac{3}{2}}\right]_0^k = 18$$

$$\frac{2}{3} k^{\frac{3}{2}} = 18$$

$$k = 27^{\frac{2}{3}} = 9$$

3 a $\int_0^3 x^2 - 1 \, dx = \left[\frac{x^3}{3} - x\right]_0^3$

$= (9 - 3) - 0$

$= 6$

b

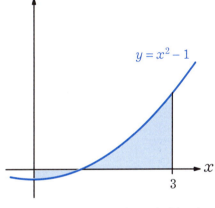

Figure 17H.3 Area bounded by the curve $y = x^2 - 1$ and the x-axis between $x = 0$ and $x = 3$

Intersections of $y = x^2 - 1$ with the x-axis occur at $x = \pm 1$.

Area $= \int_0^3 |x^2 - 1| \, dx$

$= \int_0^1 1 - x^2 \, dx + \int_1^3 x^2 - 1 \, dx$

$= \left[x - \frac{x^3}{3}\right]_0^1 + \left[\frac{x^3}{3} - x\right]_1^3$

$= \left(1 - \frac{1}{3}\right) - 0 + (9 - 3) - \left(\frac{1}{3} - 1\right)$

$= 8 - \frac{2}{3}$

$= 7\frac{1}{3}$

> **COMMENT**
> Alternatively, use a GDC to calculate the integral of the modulus function. Unless the question explicitly calls for an 'exact' solution, this is often a faster way of finding the solution.

4

> **COMMENT**
> If the area above the x-axis equals the area below it, then the net area will equal zero, i.e. the integral is zero. Using this fact is much simpler than splitting the integral into two parts and equating them.

Require that the net area equals zero:

$$\int_0^3 x^2 - kx \, dx = 0$$

$$\left[\frac{x^3}{3} - \frac{kx^2}{2}\right]_0^3 = 0$$

$$9 - \frac{9k}{2} = 0$$

$$k = 2$$

5 Intersections with the x-axis occur where

$7x - x^2 - 10 = 0$

$x^2 - 7x + 10 = 0$

$(x - 2)(x - 5) = 0$

$x = 2$ or 5

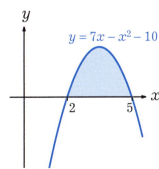

Figure 17H.5 Area enclosed by the curve $y = 7x - x^2 - 10$ and the x-axis

Hence the enclosed area is given by

$$\int_2^5 7x - x^2 - 10 \, dx = \left[\frac{7x^2}{2} - \frac{x^3}{3} - 10x\right]_2^5$$

$$= \left(\frac{175}{2} - \frac{125}{3} - 50\right) - \left(\frac{28}{2} - \frac{8}{3} - 20\right)$$

$$= \frac{-25}{6} - \left(-\frac{52}{6}\right)$$

$$= \frac{27}{6} = \frac{9}{2}$$

COMMENT

Once the limits are established, the integral could alternatively be calculated using a GDC.

Exercise 17I

2 $y = \sqrt{x} \Rightarrow x = y^2$

Area $= \int_a^{2a} y^2 \, dy$

$= \left[\frac{y^3}{3}\right]_a^{2a}$

$= \frac{1}{3}(8a^3 - a^3)$

$= \frac{7}{3}a^3$

$\therefore 504 = \frac{7}{3}a^3$

$a^3 = 216$

$\Rightarrow a = 6$

3 $y = \ln(x+1) \Rightarrow x = e^y - 1$

Area $= \int_0^2 (e^y - 1) \, dy$

$= [e^y - y]_0^2$

$= e^2 - 2 - (1 - 0)$

$= e^2 - 3$

4 $y = \sqrt{x} \Rightarrow x = y^2$

At $x = 4$, $y = 2$

At $x = a$, $y = \sqrt{a}$

Area $= \int_2^{\sqrt{a}} x \, dy$

$= \int_2^{\sqrt{a}} y^2 \, dy$

$= \left[\frac{y^3}{3}\right]_2^{\sqrt{a}}$

$= \frac{1}{3}\left(a^{\frac{3}{2}} - 8\right)$

$\therefore \frac{1}{3}\left(a^{\frac{3}{2}} - 8\right) = 39$

$a^{\frac{3}{2}} = 125$

$\Rightarrow a = 25$

5 The diagram in the question should look as follows:

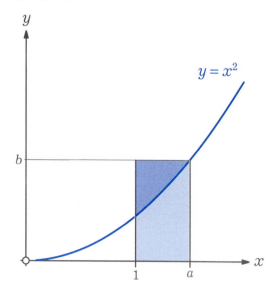

Figure 17I.5 Graph of $y = x^2$ with pale region under the curve and dark (pink) region above the curve, for x between 1 and a

$y = x^2 \Rightarrow x = \sqrt{y}$

From the equation $y = x^2$, $b = a^2$
Let B be the area below the curve and P the area above.

Then

$$B = \int_1^a x^2 \, dx$$

$$= \left[\frac{x^3}{3}\right]_1^a$$

$$= \frac{1}{3}(a^3 - 1)$$

P can be evaluated as the area extended to the y-axis, less the area of the rectangle with width 1 and height $b-1$:

$$P = \int_1^b \sqrt{y} \, dy - 1 \times (b-1)$$

$$= \left[\frac{2}{3} y^{\frac{3}{2}}\right]_1^b - (b-1)$$

$$= \frac{2}{3}\left(b^{\frac{3}{2}} - 1\right) - b + 1$$

$$= \frac{2}{3} b^{\frac{3}{2}} + \frac{1}{3} - b$$

$$P = B \Rightarrow a^3 - 1 = 2b^{\frac{3}{2}} + 1 - 3b$$

$$\Rightarrow a^3 = 2b^{\frac{3}{2}} + 2 - 3b$$

Substituting $b = a^2$:

$$a^3 = 2a^3 + 2 - 3a^2$$

$$a^3 - 3a^2 + 2 = 0$$

The trivial solution with $P = B = 0$ is equivalent to $a = 1$, which is clearly a root of this cubic. Factorising:

$$(a-1)(a^2 - 2a - 2) = 0$$

$$a = 1 \text{ or } a = \frac{2 \pm \sqrt{4+8}}{2}$$

$$\therefore a = 1, \, 1 \pm \sqrt{3}$$

With $a > 1$ as shown in the diagram, the solution is $a = 1 + \sqrt{3}$

and in this case the area B is

$$\frac{1}{3}\left[(1+\sqrt{3})^3 - 1\right] = 3 + 2\sqrt{3}$$

Exercise 17J

2 Intersections when

$$x^2 + x - 2 = x + 2$$

$$x^2 - 4 = 0$$

$$x = \pm 2$$

Enclosed area $= \int_{-2}^{2} 4 - x^2 \, dx$

$$= \left[4x - \frac{x^3}{3}\right]_{-2}^{2}$$

$$= \left(8 - \frac{8}{3}\right) - \left(-8 + \frac{8}{3}\right)$$

$$= 16 - \frac{16}{3}$$

$$= 10\frac{2}{3}$$

3 Intersections when $e^x = x^2$

From GDC: $x = -0.703$, which lies outside the interval of interest, $[0, 2]$.

At $x = 0$, $e^x > x^2$, so $e^x - x^2$ is positive over the whole of the interval of interest.

Enclosed area $= \int_0^2 e^x - x^2 \, dx$

$$= \left[e^x - \frac{x^3}{3}\right]_0^2$$

$$= \left(e^2 - \frac{8}{3}\right) - 1$$

$$= e^2 - \frac{11}{3}$$

COMMENT

Instead of checking for roots to $e^x = x^2$, this question could be answered by using a GDC to calculate $\int_0^2 |e^x - x^2| \, dx$ directly.

17 Basic integration and its applications 227

4 Intersections when $\frac{1}{x} = \sin x$

From GDC: $x = 1.11$ or 2.77 for $x \in {]}0, \pi[$

Enclosed area $= \int_{1.11}^{2.77} \sin x - \frac{1}{x} \, dx$

$= 0.462$ (3SF)

5 The y-coordinates of the intersections are:

at $x = -1$, $y = (-1)^2 = 1$

at $x = 2$, $y = 2^2 = 4$

i.e. the intersection points are $(-1, 1)$ and $(2, 4)$.

Gradient of line $= \frac{4-1}{2-(-1)} = 1$

∴ equation of the line is

$y - y_1 = m(x - x_1)$

$y - 1 = 1(x - (-1))$

$y = x + 2$

The shaded region is the area between $y = x^2$ and $y = x + 2$:

Shaded area $= \int_{-1}^{2} x + 2 - x^2 \, dx$

$= \left[\frac{x^2}{2} + 2x - \frac{x^3}{3} \right]_{-1}^{2}$

$= \left(\frac{4}{2} + 4 - \frac{8}{3} \right) - \left(\frac{1}{2} - 2 + \frac{1}{3} \right)$

$= \frac{9}{2}$

> **COMMENT**
>
> An alternative approach would be to integrate $y = x^2$ between 1 and 2, and subtract that result from the area of the trapezium. As long as your working is clearly laid out, any valid method is acceptable.

6 Intersection occurs at $x = \frac{\pi}{4}$

(by symmetry or by solving $\sin x = \cos x$, i.e. $\tan x = 1$)

Shaded area $= \int_{0}^{\frac{\pi}{4}} \sin x \, dx + \int_{\frac{\pi}{4}}^{\frac{\pi}{2}} \cos x \, dx$

$= \left[-\cos x \right]_0^{\frac{\pi}{4}} + \left[\sin x \right]_{\frac{\pi}{4}}^{\frac{\pi}{2}}$

$= \left(-\frac{1}{\sqrt{2}} + 1 \right) + \left(1 - \frac{1}{\sqrt{2}} \right)$

$= 2 - \sqrt{2}$

> **COMMENT**
>
> Since the graph is symmetrical about $x = \frac{\pi}{4}$, the area could instead be calculated as $2 \int_0^{\frac{\pi}{4}} \sin x \, dx$.

7 Intersections where $x(x-4)^2 = x^2 - 7x + 15$

From GDC: $x = 1, 3, 5$

Area enclosed $= \int_1^5 \left| x(x-4)^2 - x^2 + 7x - 15 \right| dx$

$= 8$ (from GDC)

> **COMMENT**
>
> Clearly the answer could be obtained by integrating term by term in the usual way; however, since the question is evidently intended to be answered using a GDC, this is not necessary.

8 Intersections when

$x^2 = mx$

$x(x - m) = 0$

$x = 0$ or m

Enclosed area $= \frac{32}{3}$

∴ $\int_0^m mx - x^2 \, dx = \frac{32}{3}$

$$\left[\frac{mx^2}{2} - \frac{x^3}{3}\right]_0^m = \frac{32}{3}$$

$$\frac{m^3}{2} - \frac{m^3}{3} = \frac{32}{3}$$

$$\frac{m^3}{6} = \frac{32}{3}$$

$$m^3 = 64$$

$$m = 4$$

9 The curves are $x_1 = 2 - y$ and $x_2 = y^2$

Intersections when

$$2 - y = y^2$$

$$y^2 + y - 2 = 0$$

$$(y+2)(y-1) = 0$$

$$y = -2 \text{ or } 1$$

$$\text{Area} = \int_{-2}^{1} x_1 - x_2 \, dy$$

$$= \int_{-2}^{1} 2 - y - y^2 \, dy$$

$$= \left[2y - \frac{y^2}{2} - \frac{y^3}{3}\right]_{-2}^{1}$$

$$= \left(2 - \frac{1}{2} - \frac{1}{3}\right) - \left(-4 - 2 + \frac{8}{3}\right)$$

$$= \frac{9}{2}$$

Mixed examination practice 17
Short questions

1 $f(x) = -\cos x + c$

$$f\left(\frac{\pi}{3}\right) = 0$$

$$-\frac{1}{2} + c = 0$$

$$\Rightarrow c = \frac{1}{2}$$

$$\therefore f(x) = \frac{1}{2} - \cos x$$

2 Intersections when $\ln x = e^x - e$

From GDC, $x = 1$ or 0.233

From GDC, the area between the curves is 0.201 (3SF)

> **COMMENT**
> Using techniques from Chapter 19, you can integrate the function directly and evaluate it exactly in terms of the intersection values. This method is shown below, but note that in the absence of instructions otherwise, it is appropriate (and much faster) to calculate the area using the GDC once you have found the intersections.

Let $a = 0.233$, the lower intersection value; then a satisfies $\ln a = e^a - e$.

$$\text{Area} = \int_a^1 \ln x - (e^x - e) \, dx$$

$$= \left[x \ln x - x - e^x + ex\right]_a^1$$

$$= (0 - 1 - e + e) - (a \ln a - a - e^a + ea)$$

$$= -1 - (a(e^a - e) - a - e^a + ea)$$

$$= (1 - a)e^a + a - 1$$

$$= 0.201$$

3 Intersections with the x-axis when
$$k^2 - x^2 = 0$$
$$x = \pm k$$

$$\therefore \text{Area} = \int_{-k}^{k} k^2 - x^2 \, dx$$

$$= \left[k^2 x - \frac{x^3}{3}\right]_{-k}^{k}$$

$$= \left(k^3 - \frac{k^3}{3}\right) - \left(-k^3 + \frac{k^3}{3}\right)$$

$$= \frac{4k^3}{3}$$

4 Let the blue area be B and the red area be R.

The points where the boundaries meet the curve are (a, a^n) and (b, b^n), so the total area is

$$B + R = b \times b^n - a \times a^n = b^{n+1} - a^{n+1}$$

Integrating to find the blue area under the curve:

$$B = \int_a^b x^n \, dx$$

$$= \left[\frac{x^{n+1}}{n+1} \right]_a^b$$

$$= \frac{1}{n+1}\left(b^{n+1} - a^{n+1}\right)$$

The blue area is only a quarter of the total (and the red area is three quarters of the total)

$$\therefore \frac{1}{n+1}\left(b^{n+1} - a^{n+1}\right) = \frac{1}{4}\left(b^{n+1} - a^{n+1}\right)$$

$$n+1 = 4$$

$$\Rightarrow n = 3$$

5 $\int \frac{1 + x^2\sqrt{x}}{x} \, dx = \int x^{-1} + x^{\frac{3}{2}} \, dx$

$$= \ln|x| + \frac{2x^{\frac{5}{2}}}{5} + c$$

6 a $\int_0^a x^3 - x \, dx = 0$

$$\left[\frac{x^4}{4} - \frac{x^2}{2} \right]_0^a = 0$$

$$\frac{a^2}{4}\left(a^2 - 2\right) = 0$$

$$a = 0 \text{ or } \pm\sqrt{2}$$

Since $a > 0$, $a = \sqrt{2}$

b Curve intersects the x-axis where

$$x^3 - x = 0$$

$$x(x^2 - 1) = 0$$

$$x = 0 \text{ or } \pm 1$$

Figure 17MS.6

Total area enclosed = area above x-axis + area below x-axis

Since the integral from 0 to a equals zero (defined in (a)), the area above the x-axis must equal the area below.

$$\therefore \text{Total area} = \left| \int_0^1 x^3 - x \, dx \right| + \left| \int_1^{\sqrt{2}} x^3 - x \, dx \right|$$

$$= -2\int_0^1 x^3 - x \, dx$$

$$= -2\left[\frac{x^4}{4} - \frac{x^2}{2} \right]_0^1$$

$$= \frac{1}{2}$$

7 The graphs intersect when

$$\sin x = 1 - \sin x$$

$$2\sin x = 1$$

$$\sin x = \frac{1}{2}$$

$$x = \frac{\pi}{6}, \frac{5\pi}{6} \quad (\text{for } 0 < x < \pi)$$

Difference function is

$$y_1 - y_2 = \sin x - (1 - \sin x) = 2\sin x - 1$$

Enclosed area $= \int_{\pi/6}^{5\pi/6} 2\sin x - 1 \, dx$

$$= \left[-2\cos x - x \right]_{\pi/6}^{5\pi/6}$$

$$= \left(\sqrt{3} - \frac{5\pi}{6} \right) - \left(-\sqrt{3} - \frac{\pi}{6} \right)$$

$$= 2\sqrt{3} - \frac{2\pi}{3}$$

8 a $f''(3) = 24 > 0 \Rightarrow (3, 19)$ is a local minimum

b $f'(x) = \int 6x + 6 \, dx$
$= 3x^2 + 6x + c$

Stationary point at $x = 3$
$\therefore f'(3) = 0$
$27 + 18 + c = 0$
$c = -45$

Hence $f'(x) = 3x^2 + 6x - 45$
$f(x) = \int 3x^2 + 6x - 45 \, dx$
$= x^3 + 3x^2 - 45x + d$
$f(3) = 19$
$\Rightarrow 27 + 27 - 135 + d = 19$
$\Rightarrow d = 100$
$\therefore f(x) = x^3 + 3x^2 - 45x + 100$

Long questions

1 a Intersections where
$5a^2 + 4ax - x^2 = x^2 - a^2$
$2x^2 - 4ax - 6a^2 = 0$
$x^2 - 2ax - 3a^2 = 0$
$(x - 3a)(x + a) = 0$
$x = 3a$ or $-a$

The coordinates of the points of intersection are $(-a, 0)$ and $(3a, 8a^2)$.

b Difference function is
$y_1 - y_2 = 6a^2 + 4ax - 2x^2$

Area enclosed $= \int_{-a}^{3a} 6a^2 + 4ax - 2x^2 \, dx$

$= \left[6a^2 x + 2ax^2 - \dfrac{2}{3}x^3 \right]_{-a}^{3a}$

$= \left(18a^3 + 18a^3 - 18a^3 \right)$
$- \left(-6a^3 + 2a^3 + \dfrac{2}{3}a^3 \right)$

$= \dfrac{64}{3} a^3$

c

Figure 17ML.1

Area below axis $= \left| \int_{-a}^{a} x^2 - a^2 \, dx \right|$

$= \left| \left[\dfrac{x^3}{3} - a^2 x \right]_{-a}^{a} \right|$

$= \left| \left(\dfrac{a^3}{3} - a^3 \right) - \left(-\dfrac{a^3}{3} + a^3 \right) \right|$

$= \dfrac{4}{3} a^3$

$\therefore$ Area above axis $= \dfrac{64}{3} a^3 - \dfrac{4}{3} a^3 = \dfrac{60}{3} a^3$

Fraction of the enclosed area which lies above the x-axis is $\dfrac{\left(\dfrac{60 a^3}{4} \right)}{\left(\dfrac{64 a^3}{4} \right)} = \dfrac{60}{64} = \dfrac{15}{16}$

2 a $\cos^2 \theta + \sin^2 \theta = 1$
$\Rightarrow \cos \theta = \sqrt{1 - \sin^2 \theta}$

Let $x = \sin \theta$; then $\theta = \arcsin x$ and
$\cos(\arcsin x) = \sqrt{1 - x^2}$

b $y = a \Rightarrow \sin x = a$
$\Rightarrow x = \arcsin a$
$\therefore$ the x-coordinate of P is $\arcsin a$

c Red area $= \int_0^{\arcsin a} \sin x \, dx$

$= [-\cos x]_0^{\arcsin a}$

$= -\cos(\arcsin a) - (-1)$

$= 1 - \sqrt{1-a^2}$

d Recasting the equation of the curve as $x = \arcsin y$, the blue area can be calculated directly:

Blue area $= \int_0^a \arcsin y \, dy$

$= \int_0^a \arcsin x \, dx$

(change of dummy variable does not change the value of the definite integral)

But, by subtraction,

Blue area $= a \arcsin a -$ Red area

$\therefore \int_0^a \arcsin x \, dx = a \arcsin a - 1 + \sqrt{1-a^2}$

18 Further differentiation methods

Exercise 18A

4 $y = (4x^2+1)^{-\frac{1}{2}}$

$\dfrac{dy}{dx} = -\dfrac{1}{2}(4x^2+1)^{-\frac{3}{2}} \times 8x$

$= -\dfrac{4x}{\sqrt{(4x^2+1)^3}}$

$\therefore \dfrac{dy}{dx}(\sqrt{2}) = -\dfrac{4\sqrt{2}}{\sqrt{9^3}} = -\dfrac{4\sqrt{2}}{27}$

$y(\sqrt{2}) = \dfrac{1}{\sqrt{4 \times 2 + 1}} = \dfrac{1}{3}$

Normal at $\left(\sqrt{2}, \dfrac{1}{3}\right)$ has gradient

$\dfrac{27}{4\sqrt{2}} = \dfrac{27\sqrt{2}}{8}$ and is given by

$y - y_1 = m(x - x_1)$

$y - \dfrac{1}{3} = \dfrac{27\sqrt{2}}{8}(x - \sqrt{2})$

$24y - 8 = 81\sqrt{2}x - 162$

$24y = 81\sqrt{2}x - 154$

$y = \dfrac{27\sqrt{2}}{8}x - \dfrac{77}{12}$

5 $y = e^{\sin x}$

$\Rightarrow \dfrac{dy}{dx} = e^{\sin x} \cos x$

Stationary points where $\dfrac{dy}{dx} = 0$:

$e^{\sin x} \cos x = 0$

$\cos x = 0$ (since $e^x \neq 0$)

$x = \dfrac{\pi}{2}, \dfrac{3\pi}{2}$ (for $x \in [0, 2\pi]$)

Coordinates of the stationary points are

$\left(\dfrac{\pi}{2}, e\right)$ and $\left(\dfrac{3\pi}{2}, e^{-1}\right)$

6 $f(x) = \csc^2 x = (\sin x)^{-2}$

a $f'(x) = -2(\sin x)^{-3} \cos x$

$= -2 \cot x \csc^2 x$

b $f'(x) = 2f(x)$

$-2 \cot x \csc^2 x = 2 \csc^2 x$

$\tan x = -1$

$x = -\dfrac{\pi}{4}, \dfrac{3\pi}{4}$

7 $f(x) = \ln(x^2 - 35)$

$f'(x) = (2x) \times \dfrac{1}{x^2 - 35}$

$f'(x) = 1$

$\dfrac{2x}{x^2 - 35} = 1$

$x^2 - 2x - 35 = 0$

$(x+5)(x-7) = 0$

$x = -5$ or 7

The domain of $f(x)$ is $x > \sqrt{35}$, so the only solution is $x = 7$.

8

> **COMMENT**
>
> This question was intended to appear in the next exercise (18B), as the product rule is needed to differentiate the function in part (a). Please return to it after you have worked through Section 18B or when you have completed this chapter.

a $y = (x-a)^p (x-b)^q$

$$\frac{dy}{dx} = p(x-a)^{p-1}(x-b)^q + q(x-a)^p (x-b)^{q-1}$$

$$= (x-a)^{p-1}(x-b)^{q-1}(px - pb + qx - qa)$$

Require $\frac{dy}{dx} = 0$ for $a < x < b$

$\therefore px + qx = pb + qa$

$$\Rightarrow x = \frac{pb + qa}{p+q}$$

b Positive polynomial of order 5

$p = 2 \Rightarrow$ double root at $(a, 0)$

$q = 3 \Rightarrow$ triple root at $(b, 0)$

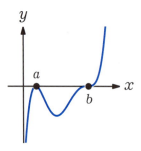

Figure 18A.8 Graph of $y = (x-a)^2 (x-b)^3$

c $y = (x-a)^p (x-b)^q$ is a positive polynomial, so the curve will finish with a positive gradient. If the larger root b is repeated an odd number of times (i.e. if q is odd), the curve will be negative in the interval $[a, b]$ and the stationary point will be a minimum. If, however, q is even, then the curve will be positive in $[a, b]$ and so the stationary point will be a maximum.

Therefore, the stationary point is a maximum if (and only if) q is even.

9 a $h = e^x + \frac{1}{e^{2x}}$ for $-1 \le x \le 2$

$h(-1) = e^{-1} + e^2$

$h(2) = e^2 + e^{-4}$

Since $e^{-1} > e^{-4}$, the post at $x = -1$ is taller.

b $\frac{dh}{dx} = e^x - 2e^{-2x}$

Stationary point occurs where $\frac{dh}{dx} = 0$:

$e^x - 2e^{-2x} = 0$

$e^x = 2e^{-2x}$

$e^{3x} = 2$

$3x = \ln 2$

$x = \frac{1}{3} \ln 2$

Classifying the stationary point:

$\frac{d^2h}{dx^2} = e^x + 4e^{-2x}$

$\frac{d^2h}{dx^2}\left(\frac{1}{3}\ln 2\right) = e^{\frac{1}{3}\ln 2} + 4e^{-\frac{2}{3}\ln 2} > 0 \Rightarrow$ local minimum

Therefore the minimum height occurs at $x = \frac{1}{3}\ln 2$

c Minimum height is

$h\left(\frac{1}{3}\ln 2\right) = h\left(\ln 2^{\frac{1}{3}}\right)$

$= 2^{\frac{1}{3}} + 2^{-\frac{2}{3}}$

$= 2^{\frac{1}{3}}\left(1 + \frac{1}{2}\right)$

$= \frac{3}{2} \times 2^{\frac{1}{3}}$

$= \frac{3\sqrt[3]{2}}{2}$

10 a $\sin 2x = \sin x$

$2 \sin x \cos x = \sin x$

$\sin x (2\cos x - 1) = 0$

$\sin x = 0$ or $\cos x = \dfrac{1}{2}$

So, for $0 \leq x \leq 2\pi$,

$x = 0, \pi, 2\pi, \dfrac{\pi}{3}, \dfrac{5\pi}{3}$

b $y = \sin 2x - \sin x$

$\Rightarrow \dfrac{dy}{dx} = 2\cos 2x - \cos x$

Stationary points where $\dfrac{dy}{dx} = 0$:

$2\cos 2x - \cos x = 0$

$2(2\cos^2 x - 1) - \cos x = 0$

$4\cos^2 x - \cos x - 2 = 0$

$\cos x = \dfrac{1 \pm \sqrt{1 + 32}}{8}$

$x = 0.568, 5.72, 2.21, 4.08$

∴ stationary points are at (0.568, 0.369), (2.21, −1.76), (4.08, 1.76), (5.72, −0.369)

c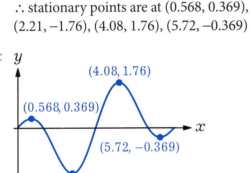

Figure 18A.10 Graph of $y = \sin 2x - \sin x$

Exercise 18B

3 Let $u = (3x^2 - x + 2)$, $v = e^{2x}$

$\dfrac{dy}{dx} = v\dfrac{du}{dx} + u\dfrac{dv}{dx}$

$= e^{2x}(6x - 1) + (3x^2 - x + 2) \times 2e^{2x}$

$= (6x^2 + 4x + 3)e^{2x}$

4 Let $u = x^2$, $v = e^{3x}$

$f'(x) = v\dfrac{du}{dx} + u\dfrac{dv}{dx}$

$= e^{3x}(2x) + x^2 \times 3e^{3x}$

$= (2x + 3x^2)e^{3x}$

Let $p = 2x + 3x^2$, $q = e^{3x}$

$f''(x) = q\dfrac{dp}{dx} + p\dfrac{dq}{dx}$

$= e^{3x}(2 + 6x) + (2x + 3x^2) \times 3e^{3x}$

$= (9x^2 + 12x + 2)e^{3x}$

5 Let $u = (2x + 1)^5$, $v = e^{-2x}$

$\dfrac{dy}{dx} = v\dfrac{du}{dx} + u\dfrac{dv}{dx}$

$= e^{-2x} \times 5(2x + 1)^4 \times 2 + (2x + 1)^5(-2e^{-2x})$

$= e^{-2x}(2x + 1)^4(10 - 4x - 2)$

$= e^{-2x}(8 - 4x)(2x + 1)^4$

Stationary points where $\dfrac{dy}{dx} = 0$:

$e^{-2x}(8 - 4x)(2x + 1)^4 = 0$

$(8 - 4x)(2x + 1)^4 = 0$

$x = 2$ or $-\dfrac{1}{2}$

6 Let $u = (3x + 1)^5$, $v = (3 - x)^3$

$\dfrac{dy}{dx} = v\dfrac{du}{dx} + u\dfrac{dv}{dx}$

$= (3 - x)^3 \times 5(3x + 1)^4 \times 3 + (3x + 1)^5$
$\times 3(3 - x)^2 \times (-1)$

$= (3 - x)^2(3x + 1)^4(15(3 - x) - 3(3x + 1))$

$= (3 - x)^2(3x + 1)^4(42 - 24x)$

$= 6(3 - x)^2(3x + 1)^4(7 - 4x)$

Stationary points where $\dfrac{dy}{dx} = 0$:

$6(3 - x)^2(3x + 1)^4(7 - 4x) = 0$

$x = 3, -\dfrac{1}{3}, \dfrac{7}{4}$

7 a Let $u = x$, $v = \sin 2x$

$$\frac{dy}{dx} = v\frac{du}{dx} + u\frac{dv}{dx}$$

$$= \sin 2x \times 1 + x \times 2\cos 2x$$

$$= \sin 2x + 2x\cos 2x$$

$$\frac{d^2y}{dx^2} = 2\cos 2x + 2\cos 2x - 4x\sin 2x$$

$$= 4\cos 2x - 4x\sin 2x$$

Inflexion points where $\frac{d^2y}{dx^2} = 0$:

$4\cos 2x - 4x\sin 2x = 0$

$\cos 2x = x\sin 2x$

b $\cos 2x = x\sin 2x$

$\Rightarrow \tan 2x = \dfrac{1}{x}$

From GDC: $x = 0.538, 1.82, 3.29, 4.81$

Inflexion points are $(0.538, 0.474)$, $(1.82, -0.877)$, $(3.29, 0.957)$, $(4.81, -0.979)$

> **COMMENT**
>
> To get 3SF accuracy on the y-coordinates, ensure that when inserting the x-coordinate into the function you use either the full x-value obtained from your GDC (saved in its memory) or several significant figures beyond the three written down in your answer.

8 Let $w = xe^x$, $u = x$, $v = e^x$

By the product rule,

$$\frac{dw}{dx} = v\frac{du}{dx} + u\frac{dv}{dx}$$

$$= e^x + xe^x$$

Then, by the chain rule, since $y = \sin(w)$,

$$\frac{dy}{dx} = \frac{dw}{dx} \times \cos(w)$$

$$= (e^x + xe^x)\cos(xe^x)$$

$$= e^x(1+x)\cos(xe^x)$$

9 a Let $u = x$, $v = \ln x$

$$f'(x) = v\frac{du}{dx} + u\frac{dv}{dx}$$

$$= \ln x \times 1 + x \times \frac{1}{x}$$

$$= \ln x + 1$$

b $\int \ln x\, dx = \int (\ln x + 1) - 1\, dx$

$$= \int \ln x + 1\, dx - \int 1\, dx$$

$$= x\ln x - x + c$$

10 Let $u = e^{-x}$, $v = \cos x$

$$\frac{dy}{dx} = v\frac{du}{dx} + u\frac{dv}{dx}$$

$$= -e^{-x}\cos x - e^{-x}\sin x$$

$$= -e^{-x}(\sin x + \cos x)$$

Stationary points where $\frac{dy}{dx} = 0$:

$-e^{-x}(\sin x + \cos x) = 0$

$\sin x + \cos x = 0$

$\sin x = -\cos x$

$\tan x = -1$

$x = \dfrac{3\pi}{4}$ (for $0 \le x \le \pi$)

$y\left(\dfrac{3\pi}{4}\right) = e^{-\frac{3\pi}{4}}\cos\left(\dfrac{3\pi}{4}\right) = -\dfrac{\sqrt{2}}{2}e^{-\frac{3\pi}{4}}$

$\therefore$ coordinates of the stationary point are

$\left(\dfrac{3\pi}{4},\; -\dfrac{\sqrt{2}}{2}e^{-\frac{3\pi}{4}}\right)$

11 Let $u = x^2$, $v = (1+x)^{\frac{1}{2}}$

$$f'(x) = v\frac{du}{dx} + u\frac{dv}{dx}$$

$$= (1+x)^{\frac{1}{2}} \times 2x + x^2 \times \frac{1}{2}(1+x)^{-\frac{1}{2}}$$

$$= \frac{x}{2(1+x)^{\frac{1}{2}}}(4(1+x) + x)$$

$$= \frac{x}{2\sqrt{1+x}}(4 + 5x)$$

$\therefore a = 4,\; b = 5$

12 a $y = x^x$

$= \left(e^{\ln x}\right)^x$

$= e^{x \ln x}$

b Let $u = x$, $v = \ln x$

By the product rule,

$\dfrac{d}{dx}(x \ln x) = v\dfrac{du}{dx} + u\dfrac{dv}{dx}$

$= \ln x \times 1 + x \times \dfrac{1}{x}$

$= \ln x + 1$

$\therefore \dfrac{dy}{dx} = (\ln x + 1)e^{x \ln x} = (\ln x + 1)x^x$

c Stationary points where $\dfrac{dy}{dx} = 0$:

$(\ln x + 1)x^x = 0$

$\ln x + 1 = 0$ (as $x^x \neq 0$)

$\ln x = -1$

$x = e^{-1}$

$\therefore$ coordinates of the stationary point are $\left(e^{-1}, e^{-e^{-1}}\right)$

Exercise 18C

2 $y\left(\dfrac{\pi}{2}\right) = \dfrac{\sin\left(\dfrac{\pi}{2}\right)}{\dfrac{\pi}{2}} = \dfrac{2}{\pi}$

Let $u = \sin x$, $v = x$

$\dfrac{dy}{dx} = \dfrac{v\dfrac{du}{dx} - u\dfrac{dv}{dx}}{v^2}$

$= \dfrac{x \cos x - \sin x}{x^2}$

$\dfrac{dy}{dx}\left(\dfrac{\pi}{2}\right) = \dfrac{-1}{\left(\dfrac{\pi}{2}\right)^2} = -\dfrac{4}{\pi^2}$

Normal at $\left(\dfrac{\pi}{2}, \dfrac{2}{\pi}\right)$ has gradient $\dfrac{\pi^2}{4}$

Equation of the normal is

$y - y_1 = m(x - x_1)$

$y - \dfrac{2}{\pi} = \dfrac{\pi^2}{4}\left(x - \dfrac{\pi}{2}\right)$

$y = \dfrac{\pi^2}{4}x - \dfrac{\pi^3}{8} + \dfrac{2}{\pi}$

$y = \dfrac{\pi^2}{4}x + \dfrac{16 - \pi^4}{8\pi}$

3 Let $u = x^2$, $v = 2x - 1$

$\dfrac{dy}{dx} = \dfrac{v\dfrac{du}{dx} - u\dfrac{dv}{dx}}{v^2}$

$= \dfrac{2x(2x-1) - 2x^2}{(2x-1)^2}$

$= \dfrac{2x^2 - 2x}{(2x-1)^2}$

$= \dfrac{2x(x-1)}{(2x-1)^2}$

Stationary points where $\dfrac{dy}{dx} = 0$:

$\dfrac{2x(x-1)}{(2x-1)^2} = 0$

$2x(x-1) = 0$

$x = 0$ or 1

$y(0) = 0$, $y(1) = \dfrac{1^2}{2(1) - 1} = 1$

$\therefore$ coordinates of the stationary points are $(0, 0)$ and $(1, 1)$

4 Let $u = x - a$, $v = x + 2$

$\dfrac{dy}{dx} = \dfrac{v\dfrac{du}{dx} - u\dfrac{dv}{dx}}{v^2}$

$= \dfrac{(x+2) - (x-a)}{(x+2)^2} = \dfrac{a+2}{(x+2)^2}$

$\dfrac{dy}{dx}(a) = \dfrac{a+2}{(a+2)^2} = \dfrac{1}{a+2}$

Require $\dfrac{dy}{dx}(a) = 1$

$$\therefore \frac{1}{a+2} = 1$$

$$a+2 = 1$$

$$a = -1$$

5 Let $u = \ln x$, $v = x$

$$\frac{dy}{dx} = \frac{v\frac{du}{dx} - u\frac{dv}{dx}}{v^2}$$

$$= \frac{\frac{x}{x} - \ln x}{x^2} = \frac{1 - \ln x}{x^2}$$

Stationary points where $\frac{dy}{dx} = 0$:

$$\frac{1 - \ln x}{x^2} = 0$$

$$\ln x = 1$$

$$x = e$$

Let $p = 1 - \ln x$, $q = x^2$

$$\frac{d^2 y}{dx^2} = \frac{q\frac{dp}{dx} - p\frac{dq}{dx}}{q^2}$$

$$= \frac{x^2\left(-\frac{1}{x}\right) - 2x(1 - \ln x)}{x^4}$$

$$= \frac{-x - 2x + 2x \ln x}{x^4}$$

$$= \frac{2\ln x - 3}{x^3}$$

$$\frac{d^2 y}{dx^2}(e) = -\frac{1}{e^3} < 0 \Rightarrow \text{local maximum}$$

$\therefore$ stationary point at $\left(e, \frac{1}{e}\right)$ is a local maximum

6 Let $u = x^2$, $v = 1 - x$

$$f'(x) = \frac{v\frac{du}{dx} - u\frac{dv}{dx}}{v^2}$$

$$= \frac{2x(1-x) + x^2}{(1-x)^2}$$

$$= \frac{2x - x^2}{(1-x)^2}$$

For the function to be increasing, require $\frac{dy}{dx} > 0$:

$$\frac{2x - x^2}{(1-x)^2} > 0$$

$$2x - x^2 > 0$$

$$x(2 - x) > 0$$

$$0 < x < 2$$

Checking for validity of the function:
$x = 1$ is not in the domain.
So $f(x)$ is increasing for $x \in \,]0, 1[\, \cup \,]1, 2[$.

7 Let $u = x^2$, $v = (x+1)^{\frac{1}{2}}$

$$\frac{dy}{dx} = \frac{v\frac{du}{dx} - u\frac{dv}{dx}}{v^2}$$

$$= \frac{2x(1+x)^{\frac{1}{2}} - \frac{x^2}{2}(x+1)^{-\frac{1}{2}}}{(x+1)}$$

$$= \frac{4x(1+x) - x^2}{2(x+1)^{\frac{3}{2}}}$$

$$= \frac{4x + 3x^2}{2(x+1)^{\frac{3}{2}}}$$

$$= \frac{x(3x + 4)}{2(x+1)^{\frac{3}{2}}}$$

$\therefore a = 3$, $b = 4$, $p = \frac{3}{2}$

8 $f(x)$ has a local maximum at $x = a$
$\Rightarrow f'(a) = 0$ and, for small $\delta > 0$,
$f'(a - \delta) > 0$ and $f'(a + \delta) < 0$

Let $y = \frac{1}{f(x)}$

By the quotient rule or chain rule,

$$\frac{dy}{dx} = -\frac{f'(x)}{(f(x))^2}$$

Then $\frac{dy}{dx}(a) = \frac{f'(a)}{(f(a))^2} = \frac{0}{(f(a))^2} = 0$,

so there is a stationary point at $\left(a, \frac{1}{f(a)}\right)$

Also, $\dfrac{dy}{dx}(a-\delta) = -\dfrac{f'(a-\delta)}{(f(a-\delta))^2} < 0$

and $\dfrac{dy}{dx}(a+\delta) = -\dfrac{f'(a+\delta)}{(f(a+\delta))^2} > 0$

Therefore $\left(a, \dfrac{1}{f(a)}\right)$ is a local minimum.

> **COMMENT**
>
> It might seem better to use the second derivative to determine the nature of the stationary point, but there are good reasons not to do this here. It is possible for a local maximum to have a zero second derivative, such as in the curve $y = -x^4$, so any proof would require contingencies for this circumstance; and in any case calculating the second derivative of $y = \dfrac{1}{f(x)}$ requires multiple uses of chain rule, product rule and quotient rule, which is unnecessarily complex.

Exercise 18D

5 a $y = \ln x \Rightarrow x = e^y$

By implicit differentiation:

$1 = e^y \dfrac{dy}{dx}$

$\therefore \dfrac{dy}{dx} = \dfrac{1}{e^y} = \dfrac{1}{x}$

b $\dfrac{d}{dx}(\log_a x) = \dfrac{d}{dx}\left(\dfrac{\ln x}{\ln a}\right)$

$= \dfrac{1}{\ln a} \times \dfrac{d}{dx}(\ln x)$

$= \dfrac{1}{x \ln a}$

c Let $u = kx$; then

$\dfrac{d}{dx}(\ln(kx)) = \dfrac{d}{dx}(\ln u)$

$= \dfrac{1}{u} \times \dfrac{du}{dx}$

$= \dfrac{1}{kx} \times k$

$= \dfrac{1}{x}$

The value of k (as long as it is positive) does not affect the derivative.

This is clear if the logarithm is rewritten using rules of logs:

$\ln(kx) = \ln k + \ln x$

That is, the value $\ln k$ represents a constant added to the logarithm, and any constant has zero derivative, so does not appear in the derivative of the function.

Let $v = x^n$. Then

$\dfrac{d}{dx}(\ln x^n) = \dfrac{d}{dx}(\ln v)$

$= \dfrac{1}{v} \times \dfrac{dv}{dx}$

$= \dfrac{1}{x^n} \times nx^{n-1}$

$= \dfrac{n}{x}$

The power n becomes a multiple of the derivative.

This is clear if the logarithm is rewritten using rules of logs:

$\ln(x^n) = n \ln x$

That is, the value n represents a constant multiplying the logarithm, so the derivative of the function is just the derivative of the logarithm multiplied by the constant.

6 Differentiating $x^2 - 3xy + y^2 + 1 = 0$ implicitly:

$$2x - 3y - 3x\frac{dy}{dx} + 2y\frac{dy}{dx} = 0$$

$$\Rightarrow \frac{dy}{dx} = \frac{3y - 2x}{2y - 3x}$$

Substitute $x = 1$, $y = 2$:

$$\frac{dy}{dx} = \frac{6-2}{4-3} = 4$$

∴ at point $(1, 2)$ the gradient is 4

7 Differentiating $4x^2 - 3xy - y^2 = 25$ implicitly:

$$8x - 3y - 3x\frac{dy}{dx} - 2y\frac{dy}{dx} = 0$$

$$\Rightarrow \frac{dy}{dx} = \frac{8x - 3y}{2y + 3x}$$

Substitute $x = 2$, $y = -3$:

$$\frac{dy}{dx} = \frac{16 + 9}{-6 + 6}$$

So at point $(2, -3)$ the gradient is infinite; that is, the tangent is vertical.

∴ the tangent at $(2, -3)$ has equation $x = 2$

8 Differentiating $x2^y = \ln y$ implicitly:

$$2^y + x\ln 2 \times 2^y \frac{dy}{dx} = \frac{1}{y}\frac{dy}{dx}$$

$$y2^y = \frac{dy}{dx}(1 - xy2^y \ln 2)$$

$$\Rightarrow \frac{dy}{dx} = \frac{y2^y}{1 - xy2^y \ln 2}$$

9 Differentiating $e^x + ye^{-x} = 2e^2$ implicitly:

$$e^x - ye^{-x} + \frac{dy}{dx}e^{-x} = 0$$

$$\Rightarrow \frac{dy}{dx} = \frac{ye^{-x} - e^x}{e^{-x}} = y - e^{2x}$$

Stationary point where $\frac{dy}{dx} = 0$:

$y = e^{2x}$

Substituting this into the equation of the curve:

$$e^x + e^{2x}e^{-x} = 2e^2$$

$$2e^x = 2e^2$$

$$\Rightarrow x = 2$$

∴ coordinates of the stationary point are $(2, e^4)$

10 a $y^2 = x^3 \Rightarrow y = \pm x^{\frac{3}{2}}$

The graph is the curve of $y_1 = x^{\frac{3}{2}}$ combined with the curve of $y_2 = -x^{\frac{3}{2}}$

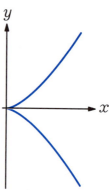

Figure 18D.10 Graph of $y = \pm x^{\frac{3}{2}}$

b Differentiating $y^2 = x^3$ implicitly:

$$2y\frac{dy}{dx} = 3x^2$$

$$\Rightarrow \frac{dy}{dx} = \frac{3x^2}{2y}$$

At $x = 4$, $y = \pm\sqrt{64} = \pm 8$ so, picking the positive root, L is tangent to the curve at $(4, 8)$

Gradient at $(4, 8)$ is $\frac{3 \times 16}{2 \times 8} = 3$

∴ L has equation $y - 8 = 3(x - 4)$, or $y = 3x - 4$

c To find intersections of L and C, substitute $y = 3x - 4$ into $y^2 = x^3$:

$$(3x - 4)^2 = x^3$$

$$9x^2 - 24x + 16 = x^3$$

$$x^3 - 9x^2 + 24x - 16 = 0$$

d One (repeated) root of the cubic equation in (c) must be $x = 4$, since L is tangent to C at this point.

$$\therefore x^3 - 9x^2 + 24x - 16 = (x-4)^2(x-a) = 0$$

Comparing the constant coefficient gives $a = 1$

So L intersects C again at $(1, -1)$

Exercise 18E

2 $y = \arccos\left(\dfrac{x}{2}\right)$

$\dfrac{dy}{dx} = \dfrac{1}{2} \times \dfrac{-1}{\sqrt{1-\left(\dfrac{x}{2}\right)^2}}$

$= -\dfrac{1}{\sqrt{4-x^2}}$

$\Rightarrow \dfrac{dy}{dx}\left(\dfrac{1}{3}\right) = -\dfrac{1}{\sqrt{4-\dfrac{1}{9}}} = -\dfrac{3}{\sqrt{35}}$

3 Using the chain rule, let $u = \dfrac{3x}{2}$

$y = \arcsin u$

$\Rightarrow \dfrac{dy}{dx} = \dfrac{du}{dx} \times \dfrac{1}{\sqrt{1-u^2}}$

$= \dfrac{3}{2} \times \dfrac{1}{\sqrt{1-\left(\dfrac{3x}{2}\right)^2}}$

$= \dfrac{3}{\sqrt{4-9x^2}}$

4 $x \arctan y = 1$

> **COMMENT**
>
> With the option of implicit differentiation, this type of problem can be approached by differentiating and then rearranging, as well as by the previous method of rearranging first. Both approaches are shown below.

Method 1: implicit differentiation followed by rearrangement and substitution.

$\arctan y + x \times \dfrac{1}{1+y^2} \dfrac{dy}{dx} = 0$

$\Rightarrow \dfrac{dy}{dx} = \dfrac{-(1+y^2)\arctan y}{x}$

$\arctan y = \dfrac{1}{x} \Rightarrow y = \tan\left(\dfrac{1}{x}\right)$

$\therefore \dfrac{dy}{dx} = -\dfrac{1+\tan^2(x^{-1})}{x^2}$

Method 2: substitution and chain rule.

$\arctan y = \dfrac{1}{x} \Rightarrow y = \tan\left(\dfrac{1}{x}\right)$

Let $u = x^{-1}$; then

$y = \tan u, \quad \dfrac{du}{dx} = -\dfrac{1}{x^2}$

$\therefore \dfrac{dy}{dx} = \dfrac{du}{dx} \times \sec^2 u$

$= -\dfrac{1}{x^2}(1+\tan^2 u)$

$= -\dfrac{1}{x^2}\left(1+\tan^2\left(\dfrac{1}{x}\right)\right)$

$= -\dfrac{1+\tan^2(x^{-1})}{x^2}$

5 a Using the product rule and the known derivative of $\arcsin x$:

$\dfrac{d}{dx}(x \arcsin x) = \arcsin x + \dfrac{x}{\sqrt{1-x^2}}$

b Integrating both sides of the result in (a) with respect to x:

$$x\arcsin x = \int \arcsin x + \frac{x}{\sqrt{1-x^2}}\,dx$$

$$\Rightarrow \int \arcsin x \,dx = x\arcsin x - \int \frac{x}{\sqrt{1-x^2}}\,dx$$

For the second integral, use the substitution $u = 1-x^2$, so that

$$\frac{du}{dx} = -2x \Rightarrow dx = -\frac{1}{2x}\,du$$

> **COMMENT**
>
> See Section 19B for the method of integration by substitution. This problem can still be solved by recognising the derivative of $\sqrt{1-x^2}$ without the need for formal substitution.

$$\therefore \int \frac{x}{\sqrt{1-x^2}}\,dx = \int \frac{x}{\sqrt{u}}\left(-\frac{1}{2x}\right)du$$

$$= -\frac{1}{2}\int u^{-\frac{1}{2}}\,du$$

$$= -u^{\frac{1}{2}} + c$$

$$= -\sqrt{1-x^2} + c$$

and hence

$$\int \arcsin x \,dx = x\arcsin x - \left(-\sqrt{1-x^2}\right) + c$$

$$= x\arcsin x + \sqrt{1-x^2} + c$$

6 $y = \arcsin(x^2)$

$$\frac{dy}{dx} = 2x \times \frac{1}{\sqrt{1-(x^2)^2}}$$

$$= 2x(1-x^4)^{-\frac{1}{2}}$$

$$\frac{d^2y}{dx^2} = 2(1-x^4)^{-\frac{1}{2}} + 2x\left(-\frac{1}{2}\right)(1-x^4)^{-\frac{3}{2}}(-4x^3)$$

$$= 2(1-x^4)^{-\frac{1}{2}} + 4x^4(1-x^4)^{-\frac{3}{2}}$$

$$= 2(1-x^4)^{-\frac{3}{2}}(1-x^4 + 2x^4)$$

$$= \frac{2(1+x^4)}{(1-x^4)^{\frac{3}{2}}}$$

At a point of inflexion, $\dfrac{d^2y}{dx^2} = 0$

$$\Rightarrow 1 + x^4 = 0$$

This equation has no real solutions, so the graph has no points of inflexion.

Mixed examination practice 18

Short questions

1 a $y = x^2 \arcsin x$

Using the product rule:

$$\frac{dy}{dx} = 2x \arcsin x + \frac{x^2}{\sqrt{1-x^2}}$$

b $xe^y = 4y^2$

By implicit differentiation:

$$e^y + xe^y \frac{dy}{dx} = 8y \frac{dy}{dx}$$

$$\Rightarrow \frac{dy}{dx} = \frac{e^y}{8y - xe^y}$$

2 $f(x) = \arccos(1 - x^2)$

Using the chain rule, let $u = 1 - x^2$

$$f'(x) = \frac{du}{dx} \times \frac{df}{du}$$

$$= -2x \times \left(-\frac{1}{\sqrt{1-u^2}}\right)$$

$$= \frac{2x}{\sqrt{1-(1-x^2)^2}}$$

$$= \frac{2x}{\sqrt{2x^2 - x^4}}$$

$$= \frac{2x}{|x|\sqrt{2-x^2}}$$

> **COMMENT**
> It would be a simple error to cancel the x^2 factor inside the square root with the x in the numerator, but this ignores the fact that $\frac{x}{\sqrt{x^2}}$ equals 1 only for positive x, and in fact is undefined for $x = 0$ and equals -1 for negative x. The undefined gradient $\frac{0}{0}$ at $x = 0$ represents the fact that the gradient curve is discontinuous at that point; in other words, the curve of $f(x)$ has a corner at $x = 0$.

> **COMMENT**
> An alternative approach to this question would be to rearrange first and then use implicit differentiation:
>
> $y = \arccos(1 - x^2)$
>
> $\Rightarrow \cos y = 1 - x^2$
>
> Differentiating implicitly gives
>
> $-\sin y \frac{dy}{dx} = -2x$
>
> $\Rightarrow \frac{dy}{dx} = \frac{2x}{\sin y}$
>
> $\sin y = \sqrt{1 - \cos^2 y}$
>
> $\qquad = \sqrt{1 - (1-x^2)^2}$
>
> $\qquad = \sqrt{2x^2 - x^4}$
>
> $\qquad = |x|\sqrt{2 - x^2}$
>
> $\therefore \frac{dy}{dx} = \frac{2x}{|x|\sqrt{2-x^2}}$

3 $y = (4 - x^2)^{-1}$

$\Rightarrow \frac{dy}{dx} = 2x(4 - x^2)^{-2}$

$\frac{dy}{dx}\left(\frac{1}{2}\right) = \frac{1}{\left(4 - \frac{1}{4}\right)^2} = \frac{16}{225}$

4 Differentiating $4x^2 + xy^2 - 3y^3 = 56$ implicitly:

$8x + y^2 + 2xy \frac{dy}{dx} - 9y^2 \frac{dy}{dx} = 0$

$\frac{dy}{dx}(9y^2 - 2xy) = 8x + y^2$

$\Rightarrow \frac{dy}{dx} = \frac{8x + y^2}{9y^2 - 2xy}$

At $(-5, 2)$, the gradient is $\frac{-40 + 4}{36 + 20} = \frac{-9}{14}$

So the normal has gradient $\frac{14}{9}$

The line through $(-5, 2)$ with gradient $\dfrac{14}{9}$ has equation

$(y-2) = \dfrac{14}{9}(x+5)$

$y = \dfrac{14}{9}x + \dfrac{88}{9}$

or $14x - 9y + 88 = 0$

5 $y = \arctan(x^2)$

$\dfrac{dy}{dx} = \dfrac{d}{dx}(x^2) \times \dfrac{1}{1+(x^2)^2}$

$= \dfrac{2x}{1+x^4}$

$= 2x(1+x^4)^{-1}$

$\dfrac{d^2y}{dx^2} = 2(1+x^4)^{-1} + 2x(-1)(1+x^4)^{-2}(4x^3)$

$= 2(1+x^4)^{-1} - 8x^4(1+x^4)^{-2}$

$= (1+x^4)^{-2}(2(1+x^4) - 8x^4)$

$= \dfrac{2 - 6x^4}{(1+x^4)^2}$

6 Differentiating $4\sin x \cos y + \sec^2 y = 5$ implicitly:

$4\cos x \cos y - 4\sin x \sin y \dfrac{dy}{dx} + 2\sec^2 y \tan y \dfrac{dy}{dx} = 0$

$\dfrac{dy}{dx}(4\sin x \sin y - 2\sec^2 y \tan y) = 4\cos x \cos y$

$\Rightarrow \dfrac{dy}{dx} = \dfrac{2\cos x}{2\sin x \tan y - \sec^3 y \tan y}$

At $x = \dfrac{\pi}{6}, y = \dfrac{\pi}{3}$:

$\dfrac{dy}{dx} = \dfrac{2 \times \dfrac{\sqrt{3}}{2}}{2 \times \dfrac{1}{2} \times \sqrt{3} - 2^3 \times \sqrt{3}} = \dfrac{\sqrt{3}}{\sqrt{3} - 8\sqrt{3}} = -\dfrac{1}{7}$

7 $y = xe^{-kx}$

$\dfrac{dy}{dx} = e^{-kx} - kxe^{-kx} = (1-kx)e^{-kx}$

$\dfrac{dy}{dx}\left(\dfrac{2}{5}\right) = 0$

$\Rightarrow \left(1 - \dfrac{2k}{5}\right)e^{-\frac{2k}{5}} = 0$

$1 - \dfrac{2k}{5} = 0$

$\therefore k = \dfrac{5}{2}$

8 a $f(x) = a(b+e^{-cx})^{-1}$

Using the chain rule:

$f'(x) = -a(b+e^{-cx})^{-2}(-ce^{-cx}) = ace^{-cx}(b+e^{-cx})^{-2}$

Using the product rule with $u = ace^{-cx}$, $v = (b+e^{-cx})^{-2}$ and the chain rule again:

$f''(x) = v\dfrac{du}{dx} + u\dfrac{dv}{dx}$

$= -ac^2e^{-cx}(b+e^{-cx})^{-2} + ace^{-cx}$

$\left(-2(b+e^{-cx})^{-3}\right)(-ce^{-cx})$

$= -ac^2e^{-cx}(b+e^{-cx})^{-2} + 2ac^2e^{-2cx}(b+e^{-cx})^{-3}$

$= \dfrac{ac^2e^{-cx}}{(b+e^{-cx})^3}\left(-(b+e^{-cx}) + 2e^{-cx}\right)$

$= \dfrac{ac^2e^{-cx}}{(b+e^{-cx})^3}(e^{-cx} - b)$

b $f''(x) = 0$

$\dfrac{ac^2e^{-cx}}{(b+e^{-cx})^3}(e^{-cx} - b) = 0$

$e^{-cx} - b = 0$

$e^{-cx} = b$

$x = -\dfrac{\ln b}{c}$

$f\left(-\dfrac{\ln b}{c}\right) = \dfrac{a}{b + e^{\ln b}} = \dfrac{a}{2b}$

$\therefore$ the coordinates of the point where $f''(x) = 0$ are $\left(-\dfrac{\ln b}{c}, \dfrac{a}{2b}\right)$

Mixed examination practice 18

c At a point of inflexion, $f''(x) = 0$ and $f''(x)$ changes sign.
For a small value $\delta > 0$:

$$ac^2 e^{-c(x\pm\delta)} > 0, \quad b + e^{-c(x\pm\delta)} > 0$$

and $e^{-c(x-\delta)} - b > 0 > e^{-c(x+\delta)} - b$

$\therefore f''(x-\delta) > 0 > f''(x+\delta)$

Hence this is a true point of inflexion.

9 Differentiating $(y-2)^2 e^x = 4x$ implicitly:

$$(y-2)^2 e^x + 2(y-2)e^x \frac{dy}{dx} = 4$$

$$\Rightarrow \frac{dy}{dx} = \frac{4 - (y-2)^2 e^x}{2(y-2)e^x}$$

At a stationary point, gradient equals zero
$\therefore 4 - (y-2)^2 e^x = 0$
$\Rightarrow (y-2)^2 e^x = 4$

Comparing with the original equation of the curve gives

$4x = 4$

$\therefore x = 1$

Substituting $x = 1$ into the equation of the curve:

$(y-2)^2 e = 4$

$\Rightarrow y = 2 \pm \sqrt{4e^{-1}}$

$\therefore$ coordinates of the stationary points are $\left(1, 2\left(1 \pm e^{-0.5}\right)\right)$

Long questions

1 a Vertical asymptote where denominator equals zero: $x = \frac{1}{2}$

b By the quotient rule,

$$\frac{dy}{dx} = \frac{2x(1-2x) - x^2(-2)}{(1-2x)^2}$$

$$= \frac{-2x^2 + 2x}{(1-2x)^2}$$

$$= \frac{2x(1-x)}{(1-2x)^2}$$

Stationary points where $\frac{dy}{dx} = 0$:

$$\frac{2x(1-x)}{(1-2x)^2} = 0$$

$2x(1-x) = 0$

$x = 0$ or 1

$\therefore$ the stationary points are $(0, 0)$ and $(1, -1)$

c $\frac{dy}{dx} = \frac{2x - 2x^2}{(1-2x)^2}$

By the quotient rule,

$$\frac{d^2y}{dx^2} = \frac{(2-4x)(1-2x)^2 + 4(2x-2x^2)(1-2x)}{(1-2x)^4}$$

$$= \frac{(2-4x)(1-2x) + 4(2x-2x^2)}{(1-2x)^3}$$

$$= \frac{2}{(1-2x)^3}$$

$\frac{d^2y}{dx^2}(0) = 2 > 0 \Rightarrow$ local minimum

$\frac{d^2y}{dx^2}(1) = -2 < 0 \Rightarrow$ local maximum

So $(0, 0)$ is a local minimum and $(1, -1)$ is a local maximum.

d

Figure 18ML.1 Graph of $y = \frac{x^2}{1-2x}$

18 Further differentiation methods 245

2 a i By the quotient rule,

$$f'(x) = \frac{2x \times 2^x - x^2 \times 2^x \ln 2}{(2^x)^2}$$

$$= \frac{2x - x^2 \ln 2}{2^x}$$

ii By the quotient rule,

$$f''(x) = \frac{(2 - 2x \ln 2)2^x - (2x - x^2 \ln 2)2^x \ln 2}{(2^x)^2}$$

$$= \frac{2 - 4x \ln 2 + x^2 (\ln 2)^2}{2^x}$$

b i $f'(x) = 0$

$$\frac{2x - x^2 \ln 2}{2^x} = 0$$

$2x - x^2 \ln 2 = 0$

$x(2 - x \ln 2) = 0$

$\therefore x = 0$ or $\dfrac{2}{\ln 2}$

Since $x > 0$, the only solution is

$$x = \frac{2}{\ln 2}$$

ii

$$f''\left(\frac{2}{\ln 2}\right) = \frac{2 - 4\left(\frac{2}{\ln 2}\right)\ln 2 + \left(\frac{4}{(\ln 2)^2}\right)(\ln 2)^2}{2^{\frac{2}{\ln 2}}}$$

$$= \frac{2 - 8 + 4}{2^{\frac{2}{\ln 2}}} < 0$$

so $x = \dfrac{2}{\ln 2}$ gives a local maximum of $f(x)$

c Points of inflexion where $f''(x) = 0$:

$2 - 4x \ln 2 + x^2 (\ln 2)^2 = 0$

This is a quadratic in $x \ln 2$.
Let $u = x \ln 2$; then

$u^2 - 4u + 2 = 0$

$$\Rightarrow u = \frac{4 \pm \sqrt{16 - 8}}{2} = 2 \pm \sqrt{2}$$

$$\therefore x = \frac{2 \pm \sqrt{2}}{\ln 2}$$

3 $f(x) = \arccos\sqrt{1 - 9x^2}$ for $0 < x < \dfrac{1}{3}$

a Let $u = (1 - 9x^2)^{\frac{1}{2}}$; then

$$\frac{du}{dx} = -9x(1 - 9x^2)^{-\frac{1}{2}}$$

$$f'(x) = \frac{du}{dx} \times \frac{d}{du}(\arccos u)$$

$$= -9x(1 - 9x^2)^{-\frac{1}{2}} \times \left(-\frac{1}{\sqrt{1 - u^2}}\right)$$

$$= -9x(1 - 9x^2)^{-\frac{1}{2}} \times \left(-\frac{1}{\sqrt{1 - (1 - 9x^2)}}\right)$$

$$= -9x(1 - 9x^2)^{-\frac{1}{2}} \times \left(-\frac{1}{3x}\right)$$

$$= \frac{3}{\sqrt{1 - 9x^2}}$$

b $f'(x) = 3(1 - 9x^2)^{-\frac{1}{2}}$

$$\Rightarrow f''(x) = 3\left(-\frac{1}{2}\right)(1 - 9x^2)^{-\frac{3}{2}}(-18x)$$

$$= \frac{27x}{\left(\sqrt{1 - 9x^2}\right)^3}$$

For $x \in \left]0, \dfrac{1}{3}\right[$, the numerator and denominator are both positive

$\therefore f''(x) > 0$ for $x \in \left]0, \dfrac{1}{3}\right[$

c $g(x) = \arccos(kx)$

$$\Rightarrow g'(x) = -\frac{k}{\sqrt{1 - k^2 x^2}}$$

Require that $g'(x) = -pf'(x)$

$$\therefore -\frac{k}{\sqrt{1 - k^2 x^2}} = -\frac{3p}{\sqrt{1 - 9x^2}}$$

Comparing gives $k^2 = 9$ from the denominator and $k = 3p$ from the numerator.

So $k = 3, p = 1$ or $k = -3, p = -1$

246 Mixed examination practice 18

4 $x^2 - xy + y^2 = 12$

a By implicit differentiation:

$$2x - y - x\frac{dy}{dx} + 2y\frac{dy}{dx} = 0 \quad \ldots(*)$$

$$\Rightarrow \frac{dy}{dx} = \frac{2x-y}{x-2y}$$

Stationary point where $\frac{dy}{dx} = 0$:

$2x - y = 0$

$\Rightarrow y = 2x$

Substituting into the original equation:

$x^2 - 2x^2 + 4x^2 = 12$

$3x^2 = 12$

$\Rightarrow x = \pm 2$

∴ the stationary points are (2, 4) and (−2, −4)

b Applying implicit differentiation to (∗):

$$2 - \frac{dy}{dx} - \frac{dy}{dx} - x\frac{d^2y}{dx^2} + 2\left(\frac{dy}{dx}\right)^2 + 2y\frac{d^2y}{dx^2} = 0$$

$$\Rightarrow (x-2y)\frac{d^2y}{dx^2} = 2 - 2\frac{dy}{dx} + 2\left(\frac{dy}{dx}\right)^2$$

At stationary points, $\frac{dy}{dx} = 0$

$$\therefore (x-2y)\frac{d^2y}{dx^2} = 2$$

c At (2, 4):

$(2-8)\frac{d^2y}{dx^2} = 2 \Rightarrow \frac{d^2y}{dx^2} < 0$

∴ (2, 4) is a local maximum

At (−2, −4):

$(-2+8)\frac{d^2y}{dx^2} = 2 \Rightarrow \frac{d^2y}{dx^2} > 0$

∴ (−2, −4) is a local minimum

5 a Domain of arcsec x is the range of sec x (with domain restricted to $[0, \pi]$):

$]-\infty, -1] \cup [1, \infty[$

b The graph of arcsec x is the graph of sec x reflected through $y = x$.

Domain is $]-\infty, -1] \cup [1, \infty[$; range is $[0, \pi]$

Figure 18ML.5 Graph of $y = \text{arcsec } x$

c $\frac{d}{dx}(\sec x) = \frac{d}{dx}(\cos x)^{-1}$

Using the chain rule, let $u = \cos x$

$$\frac{d}{dx}(\sec x) = \frac{d}{dx}(u^{-1})$$

$$= \frac{du}{dx} \times (-u^{-2})$$

$$= (-\sin x) \times \left(-\frac{1}{\cos^2 x}\right)$$

$$= \frac{\sin x}{\cos^2 x}$$

$$= \sec x \tan x$$

d $y = \text{arcsec } x \Rightarrow \sec y = x$

Differentiating $\sec y = x$ implicitly with respect to x:

$\sec y \tan y \frac{dy}{dx} = 1$

$\Rightarrow \frac{dy}{dx} = \frac{1}{\sec y \tan y}$

$\tan^2 \theta + 1 = \sec^2 \theta \Rightarrow \tan \theta = \pm\sqrt{\sec^2 \theta - 1}$

From Figure 18ML.5, the graph of $y = \text{arcsec } x$ clearly has positive gradient for all values of x in its domain, so choose the positive root.

$$\therefore \frac{dy}{dx} = \frac{1}{\sec y \sqrt{\sec^2 y - 1}}$$

$$= \frac{1}{x\sqrt{x^2 - 1}}$$

19 Further integration methods

Exercise 19A

5 Both are correct: their answers differ only in the (unknown) constant.

Marina has $f(x) = \frac{1}{3}\ln|x| + c$

Jack has
$g(x) = \frac{1}{3}\ln|3x| + c$

$= \frac{1}{3}(\ln 3 + \ln|x|) + c$

$= \frac{1}{3}\ln|x| + \frac{1}{3}\ln 3 + c$

$= \frac{1}{3}\ln|x| + d$

i.e. the constants are related by $d = \frac{1}{3}\ln 3 + c$

6 $\int_{a^2}^{a}(1-x)^{-1}\,dx = 0.4$

$[-\ln(1-x)]_{a^2}^{a} = 0.4$

$\ln(1-a^2) - \ln(1-a) = 0.4$

$\ln\left(\frac{1-a^2}{1-a}\right) = 0.4$

$\ln(1+a) = 0.4$

$1 + a = e^{0.4}$

$a = e^{0.4} - 1$

$= 0.492$ (3SF)

> **COMMENT**
> Note that because $0 < a < 1$, a^2 is the lower limit and a is the upper limit.

Exercise 19B

6 Let $u = x^2 + x - 1$; then

$\frac{du}{dx} = 2x + 1$

$dx = \frac{du}{(2x+1)}$

$\int_0^2 (2x+1)e^{x^2+x-1}\,dx = \int_{x=0}^{x=2}(2x+1)e^u \frac{du}{(2x+1)}$

$= \int_{x=0}^{x=2} e^u\,du$

$= \left[e^u\right]_{x=0}^{x=2}$

$= \left[e^{x^2+x-1}\right]_0^2$

$= e^5 - e^{-1}$

7 Let $u = x^2 - 1$; then

$\frac{du}{dx} = 2x$

$dx = \frac{du}{2x}$

$\int_2^5 \frac{2x}{x^2-1}\,dx = \int_{x=2}^{x=5} \frac{2x}{u} \frac{du}{2x}$

$= \int_{x=2}^{x=5} u^{-1}\,du$

$= [\ln u]_{x=2}^{x=5}$

$= \left[\ln(x^2-1)\right]_2^5$

$= \ln 24 - \ln 3$

$= \ln\left(\frac{24}{3}\right)$

$= \ln 8$

248 Topic 19B Integration by substitution

8 $u = x - 2 \Rightarrow du = dx$

$$\int \frac{x}{\sqrt{x-2}} \, dx = \int \frac{u+2}{\sqrt{u}} \, du$$

$$= \int u^{\frac{1}{2}} + 2u^{-\frac{1}{2}} \, du$$

$$= \frac{2}{3} u^{\frac{3}{2}} + 4u^{\frac{1}{2}} + c$$

$$= \left(\frac{2}{3} u + 4 \right) u^{\frac{1}{2}} + c$$

$$= \left(\frac{2}{3}(x-2) + 4 \right) \sqrt{x-2} + c$$

$$= \frac{2}{3}(x+4)\sqrt{x-2} + c$$

9 a Let $f(x) = x^3 - 1$

$f(1) = 1^3 - 1 = 0$

so by the factor theorem, $(x-1)$ is a factor of $f(x)$

b $\dfrac{2x^2 - x - 1}{x^3 - 1} = \dfrac{(2x^2 - 2x) + (x-1)}{x^3 - 1}$

$$= \frac{(2x+1)(x-1)}{(x-1)(x^2+x+1)}$$

$$= \frac{2x+1}{x^2+x+1}$$

Integrating by substitution, let $u = x^2 + x + 1$; then

$\dfrac{du}{dx} = (2x+1) \Rightarrow dx = \dfrac{1}{2x+1} du$

$$\int \frac{2x^2 - x - 1}{x^3 - 1} \, dx = \int \frac{2x+1}{x^2 + x + 1} \, dx$$

$$= \int \frac{2x+1}{u} \times \frac{1}{2x+1} \, du$$

$$= \int \frac{1}{u} \, du$$

$$= \ln|u| + c$$

$$= \ln|x^2 + x + 1| + c$$

> **COMMENT**
>
> If you can recognise that the numerator of the integrand is the derivative of the denominator, you can move directly to the solution being the natural logarithm of the modulus of the denominator, as highlighted in Key Point 19.4; there is no need for formal substitution.

10 Let $u = \ln x$; then

$\dfrac{du}{dx} = \dfrac{1}{x} \Rightarrow dx = x \, du$

$$\int \frac{\sec^2(\ln(x^2))}{2x} \, dx = \int \frac{\sec^2(2u)}{2x} \times x \, du$$

$$= \frac{1}{2} \int \sec^2 2u \, du$$

$$= \frac{1}{2} \times \frac{1}{2} \tan(2u) + c$$

$$= \frac{1}{4} \tan(\ln(x^2)) + c$$

11 Using substitution, let $u = \sin x$; then

$\dfrac{du}{dx} = \cos x \Rightarrow dx = \dfrac{1}{\cos x} du$

$$\int \frac{\cos x}{\sin^5 x} \, dx = \int \frac{\cos x}{u^5} \times \frac{1}{\cos x} \, du$$

$$= \int u^{-5} \, du$$

$$= -\frac{1}{4} u^{-4} + c$$

$$= -\frac{1}{4 \sin^4 x} + c$$

19 Further integration methods

12 Let $u = x^2 - 3x + 3$; then

$$\frac{du}{dx} = 2x - 3 \Rightarrow dx = \frac{du}{(2x-3)}$$

$$\int_1^3 \frac{(2x-3)\sqrt{x^2-3x+3}}{x^2-3x+3} dx = \int_{x=1}^{x=3} \frac{(2x-3)\sqrt{u}}{u} \frac{du}{(2x-3)}$$

$$= \int_{x=1}^{x=3} u^{-\frac{1}{2}} du$$

$$= \left[2u^{\frac{1}{2}} \right]_{x=1}^{x=3}$$

$$= \left[2\sqrt{x^2-3x+3} \right]_1^3$$

$$= 2\sqrt{3} - 2$$

Exercise 19C

5 All are correct: as in Exercise 15A question 5, the difference lies in the unknown constant.

$$A(x) = \frac{1}{2}\sin^2 x + c$$

$$B(x) = -\frac{1}{2}\cos^2 x + c$$

$$= -\frac{1}{2}(1 - \sin^2 x) + c$$

$$= \frac{1}{2}\sin^2 x + c - \frac{1}{2}$$

$$= \frac{1}{2}\sin^2 x + d$$

which is the same as $A(x)$ with $d = c - \frac{1}{2}$

$$C(x) = -\frac{1}{4}\cos 2x + c$$

$$= -\frac{1}{4}(1 - 2\sin^2 x) + c$$

$$= \frac{1}{2}\sin^2 x + c - \frac{1}{4}$$

$$= \frac{1}{2}\sin^2 x + k$$

which is the same as $A(x)$ with $k = c - \frac{1}{4}$

6 $\int \sin^2\left(\dfrac{x}{3}\right) dx = \dfrac{1}{2}\int 1-\cos\left(\dfrac{2x}{3}\right) dx$

$\phantom{\int \sin^2\left(\dfrac{x}{3}\right) dx} = \dfrac{1}{2}x - \dfrac{3}{4}\sin\left(\dfrac{2x}{3}\right)+c$

7 a $\tan^3 x = \tan x \times \tan^2 x$

$ = \tan x\left(\sec^2 x - 1\right)$

$ = \tan x \sec^2 x - \tan x$

b $\int \tan^3 x \, dx = \int \tan x \sec^2 x - \tan x \, dx$

$ = \dfrac{1}{2}\tan^2 x + \ln|\cos x| + c$

8 $\int_0^{\pi/12} \tan^2 kx \, dx = \int_0^{\pi/12} \sec^2 kx - 1 \, dx$

$\phantom{\int_0^{\pi/12} \tan^2 kx \, dx} = \left[\dfrac{1}{k}\tan kx - x\right]_0^{\pi/12}$

$\phantom{\int_0^{\pi/12} \tan^2 kx \, dx} = \dfrac{1}{k}\tan\left(\dfrac{k\pi}{12}\right) - \dfrac{\pi}{12}$

$\dfrac{1}{k}\tan\left(\dfrac{k\pi}{12}\right) - \dfrac{\pi}{12} = \dfrac{4-\pi}{12}$

$\dfrac{1}{k}\tan\left(\dfrac{k\pi}{12}\right) = \dfrac{4}{12}$

$\Rightarrow k = \pm 3$

9 a $\cos(A+B) = \cos A \cos B - \sin A \sin B$

Let $A = B = x$; then

$\cos(2x) = \cos^2 x - \sin^2 x$

$ = \cos^2 x - \left(1-\cos^2 x\right)$

$ = 2\cos^2 x - 1$

b $\int \cos 2x \sin x \, dx = \int \left(2\cos^2 x - 1\right)\sin x \, dx$

$ = \int 2\cos^2 x \sin x - \sin x \, dx$

$ = -\dfrac{2}{3}\cos^3 x + \cos x + c$

19 Further integration methods

10 a
$$\sin^3\theta = \sin\theta \times \sin^2\theta$$
$$= \sin\theta(1-\cos^2\theta)$$
$$= \sin\theta - \sin\theta\cos^2\theta$$

b
$$\int_0^{3\pi} \sin^3\left(\frac{x}{3}\right) dx = \int_0^{3\pi} \sin\left(\frac{x}{3}\right) - \sin\left(\frac{x}{3}\right)\cos^2\left(\frac{x}{3}\right) dx$$
$$= \left[-3\cos\left(\frac{x}{3}\right) + \cos^3\left(\frac{x}{3}\right)\right]_0^{3\pi}$$
$$= (3-1)-(-3+1)$$
$$= 4$$

11 a i
$$\frac{1}{z} = \frac{z^*}{|z|^2}$$
$$= \frac{\cos\theta - i\sin\theta}{\cos^2\theta + \sin^2\theta}$$
$$= \cos\theta - i\sin\theta$$

ii By De Moivre's theorem, $(\cos\theta + i\sin\theta)^n = \cos n\theta + i\sin n\theta$
$$\therefore (\cos\theta + i\sin\theta)^{-n} = \cos(-n\theta) + i\sin(-n\theta) = \cos n\theta - i\sin n\theta$$
$$z^n - z^{-n} = (\cos n\theta + i\sin n\theta) - (\cos n\theta - i\sin n\theta) = 2i\sin(n\theta)$$

b i $(z-z^{-1})^5 = z^5 - 5z^3 + 10z - 10z^{-1} + 5z^{-3} - z^{-5}$

ii By (a)(ii) with $n = 1$,
$$z - \frac{1}{z} = 2i\sin\theta$$
$$\therefore (2i\sin\theta)^5 = \left(z - \frac{1}{z}\right)^5 = (z^5 - z^{-5}) - 5(z^3 - z^{-3}) + 10(z - z^{-1})$$

$32i\sin^5\theta = 2i\sin 5\theta - 10i\sin 3\theta + 20i\sin\theta$ by (a)(ii) with $n = 1,3,5$

$16\sin^5\theta = \sin 5\theta - 5\sin 3\theta + 10\sin\theta$

$\Rightarrow a = 1,\ b = -5,\ c = 10$

c Using the result in (b)(ii) with $\theta = 2x$:
$$\int \sin^5 2x\, dx = \frac{1}{16}\int (\sin 10x - 5\sin 6x + 10\sin 2x)\, dx$$
$$= \frac{1}{16}\left(-\frac{1}{10}\cos 10x + \frac{5}{6}\cos 6x - 5\cos 2x\right) + c$$
$$= -\frac{1}{160}\cos 10x + \frac{5}{96}\cos 6x - \frac{5}{16}\cos 2x + c$$

252 Topic 19C Using trigonometric identities in integraion

COMMENT

By the way that the question is structured, clearly you are supposed to use the identity obtained in part (b) to calculate the integral in (c). But an alternative and valid approach would involve rewriting $\sin^5 2x$ using the identity $\sin^2 2x = 1 - \cos^2 2x$ and integrating that way; this will result in a solution of a different form, as follows:

$$\int \sin^5 2x \, dx = \int \sin 2x \left(1 - \cos^2 2x\right)^2 dx$$

$$= \int \sin 2x \left(1 - 2\cos^2 2x + \cos^4 2x\right) dx$$

$$= \int \sin 2x - 2\cos^2 2x \sin 2x + \cos^4 2x \sin 2x \, dx$$

$$= -\frac{1}{2}\cos 2x + \frac{1}{3}\cos^3 2x - \frac{1}{10}\cos^5 2x + c$$

COMMENT

Using the methods from Chapter 15, you should be able to show that these two answers are equivalent: express $\cos^3 2x$ and $\cos^5 2x$ in terms of $\cos 2x$, $\cos 6x$ and $\cos 10x$.

Exercise 19D

3
$$\int_0^{\sqrt{3}/2} \frac{3}{1+4x^2} \, dx = \left[\frac{3}{2}\arctan(2x)\right]_0^{\sqrt{3}/2}$$

$$= \frac{3}{2}\arctan\left(\sqrt{3}\right) - 0$$

$$= \frac{3}{2} \times \frac{\pi}{3}$$

$$= \frac{\pi}{2}$$

COMMENT

By this stage you may feel confident enough to jump straight to the integrated form without using rearrangement or substitution. If you want to take things more slowly, recognise that the overall form of the integral will produce arctan in the result, and substitute $u = 2x$ as a preliminary step.

4 a $2x^2 + 4x + 11 = 2(x^2 + 2) + 11$
$= 2((x+1)^2 - 1) + 11$
$= 2(x+1)^2 + 9$

b $\int \dfrac{3}{2x^2 + 4x + 11} \, dx = \int \dfrac{3}{9 + 2(x+1)^2} \, dx$

Let $u = \sqrt{2}(x+1)$; then

$\dfrac{du}{dx} = \sqrt{2} \Rightarrow dx = \dfrac{1}{\sqrt{2}} du$

$\therefore \int \dfrac{3}{9 + 2(x+1)^2} \, dx = \int \dfrac{3}{9 + u^2} \times \dfrac{1}{\sqrt{2}} \, du$

$= \dfrac{3}{\sqrt{2}} \times \dfrac{1}{3} \arctan\left(\dfrac{u}{3}\right) + c$

$= \dfrac{1}{\sqrt{2}} \arctan\left(\dfrac{\sqrt{2}}{3}(x+1)\right) + c$

5 a $1 + 6x - 3x^2 = 1 - 3(x^2 - 2x)$
$= 1 - 3((x-1)^2 - 1)$
$= 4 - 3(x-1)^2$
$= 2^2 - 3(x-1)^2$

b $\int_1^2 \dfrac{1}{\sqrt{1 + 6x - 3x^2}} \, dx = \int_1^2 \dfrac{1}{\sqrt{2^2 - 3(x-1)^2}} \, dx$

Let $u = \sqrt{3}(x-1)$; then

$\dfrac{du}{dx} = \sqrt{3} \Rightarrow dx = \dfrac{1}{\sqrt{3}} du$

$\therefore \int_1^2 \dfrac{1}{\sqrt{1 + 6x - 3x^2}} \, dx = \int_{x=1}^{x=2} \dfrac{1}{\sqrt{2^2 - u^2}} \times \dfrac{1}{\sqrt{3}} \, du$

$= \left[\dfrac{1}{\sqrt{3}} \times \arcsin\left(\dfrac{u}{2}\right)\right]_{x=1}^{x=2} = \left[\dfrac{1}{\sqrt{3}} \arcsin\left(\dfrac{\sqrt{3}}{2}(x-1)\right)\right]_1^2$

$= \dfrac{1}{\sqrt{3}}\left(\dfrac{\pi}{3} - 0\right) = \dfrac{\pi}{3\sqrt{3}}$

$= \dfrac{\sqrt{3}\pi}{9}$

6 $4x^2 - 24x + 61 = 4(x^2 - 6) + 61$
$$= 4((x-3)^2 - 9) + 61$$
$$= 4(x-3)^2 + 25$$

Let $u = 2(x-3)$; then

$$\frac{du}{dx} = 2 \Rightarrow dx = \frac{1}{2}du$$

$$\therefore \int_3^{5.5} \frac{10}{4x^2 - 24x + 61} dx = \int_3^{5.5} \frac{10}{4(x-3)^2 + 25} dx$$

$$= \int_{x=3}^{x=5.5} \frac{10}{u^2 + 25} \times \frac{1}{2} du$$

$$= \left[\frac{10}{2} \times \frac{1}{5} \arctan\left(\frac{u}{5}\right) \right]_{x=3}^{x=5.5}$$

$$= \left[\arctan\left(\frac{2(x-3)}{5}\right) \right]_3^{5.5}$$

$$= \arctan(1) - \arctan(0)$$

$$= \frac{\pi}{4}$$

7 Let $u = e^x$; then

$$\frac{du}{dx} = e^x = u \Rightarrow dx = \frac{1}{u} du$$

$$\frac{1}{2} \ln 3 = \ln \sqrt{3}$$

$$\therefore \int_0^{\frac{1}{2}\ln 3} \frac{1}{e^x + e^{-x}} dx = \int_{x=0}^{x=\ln\sqrt{3}} \frac{1}{u + u^{-1}} \times \frac{1}{u} du$$

$$= \int_{x=0}^{x=\ln\sqrt{3}} \frac{1}{u^2 + 1} du$$

$$= \left[\arctan u \right]_{x=0}^{x=\ln\sqrt{3}}$$

$$= \left[\arctan(e^x) \right]_0^{\ln\sqrt{3}}$$

$$= \arctan(\sqrt{3}) - \arctan(1)$$

$$= \frac{\pi}{3} - \frac{\pi}{4}$$

$$= \frac{\pi}{12}$$

Exercise 19E

4 a $\dfrac{1}{x-2} - \dfrac{1}{x+3} = \dfrac{(x+3)-(x-2)}{(x-2)(x+3)}$

$= \dfrac{5}{x^2+x-6}$

b $\displaystyle\int \dfrac{5}{x^2+x-6}\,dx = \int \dfrac{1}{x-2} - \dfrac{1}{x+3}\,dx$

$= \ln|x-2| - \ln|x+3| + c$

$= \ln\left|\dfrac{x-2}{x+3}\right| + c$

5 $\displaystyle\int_0^2 \dfrac{4}{x^2+4}\,dx = 4\left[\dfrac{1}{2}\arctan\left(\dfrac{x}{2}\right)\right]_0^2$

$= 2(\arctan 1 - \arctan 0)$

$= \dfrac{\pi}{2}$

6 a $\dfrac{1}{2-x} + \dfrac{2}{1+x} = \dfrac{(1+x)+2(2-x)}{(2-x)(1+x)}$

$= \dfrac{5-x}{2+x-x^2}$

b $\displaystyle\int_0^1 \dfrac{5-x}{2+x-x^2}\,dx = \int_0^1 \dfrac{1}{2-x} + \dfrac{2}{1+x}\,dx$

$= \left[-\ln|2-x| + 2\ln|1+x|\right]_0^1$

$= (-\ln 1 + 2\ln 2) - (-\ln 2 + 2\ln 1)$

$= 3\ln 2$

$= \ln 2^3$

$\therefore k = 8$

7 Let $u = 1 - x^2$; then

$$\frac{du}{dx} = -2x \Rightarrow dx = -\frac{1}{2x} du$$

$$\int \frac{4x+5}{\sqrt{1-x^2}} dx = \int \frac{4x}{\sqrt{1-x^2}} + \frac{5}{\sqrt{1-x^2}} dx$$

$$= \int \frac{4x}{\sqrt{u}} \times \left(-\frac{1}{2x}\right) du + 5 \arcsin x + c$$

$$= \int -2u^{-\frac{1}{2}} du + 5 \arcsin x + c$$

$$= -4u^{\frac{1}{2}} + 5 \arcsin x + c$$

$$= -4\sqrt{1-x^2} + 5 \arcsin x + c$$

8 a $2x^2 - 8x + 17 = 2(x^2 - 4x) + 17$

$$= 2((x-2)^2 - 4) + 17$$

$$= 2(x-2)^2 + 9$$

b Derivative of the denominator is $4x - 8 = 2(2x - 4)$

$$\frac{2x+8}{2x^2-8x+17} = \frac{2x-4}{2x^2-8x+17} + \frac{12}{2x^2-8x+17}$$

Let $u = \sqrt{2}(x-2)$; then

$$\frac{du}{dx} = \sqrt{2} \Rightarrow dx = \frac{1}{\sqrt{2}} du$$

$$\int \frac{2x+8}{2x^2-8x+17} dx = \int \frac{2x-4}{2x^2-8x+17} + \frac{12}{2x^2-8x+17} dx$$

$$= \frac{1}{2} \int \frac{4x-8}{2x^2-8x+17} dx + \int \frac{12}{2(x-2)^2+9} dx$$

$$= \frac{1}{2} \ln|2x^2-8x+17| + \int \frac{12}{u^2+9} \times \frac{1}{\sqrt{2}} du + c$$

$$= \frac{1}{2} \ln|2x^2-8x+17| + \frac{12}{\sqrt{2}} \times \frac{1}{3} \arctan\left(\frac{u}{3}\right) + c$$

$$= \frac{1}{2} \ln|2x^2-8x+17| + 2\sqrt{2} \arctan\left(\frac{\sqrt{2}}{3}(x-2)\right) + c$$

> **COMMENT**
> The answer in the book appears different, but you can check that it is equivalent to the form found above.

19 Further integration methods 257

Exercise 19F

4 In the abstract, if you include an unknown constant c when integrating $\dfrac{dv}{dx}$, you get the following:

$$\int u\dfrac{dv}{dx}\,dx = u(v+c) - \int \dfrac{du}{dx}(v+c)\,dx + k$$

$$= uv + cu - \int \dfrac{du}{dx}v\,dx - \int \dfrac{du}{dx}c\,dx + k$$

$$= uv + cu - \int \dfrac{du}{dx}v\,dx - c\int \dfrac{du}{dx}\,dx + k$$

$$= uv + cu - \int \dfrac{du}{dx}v\,dx - cu + k$$

$$= uv - \int \dfrac{du}{dx}v\,dx + k$$

As seen, the terms containing c cancel out in the final answer, leaving only the standard unknown constant (called k here).

5 Let $u = 2x$, $\dfrac{dv}{dx} = e^{-3x}$

$u = 2x \Rightarrow \dfrac{du}{dx} = 2$

$\dfrac{dv}{dx} = e^{-3x} \Rightarrow v = -\dfrac{1}{3}e^{-3x}$

$\int u\dfrac{dv}{dx}\,dx = uv - \int v\dfrac{du}{dx}\,dx$

$\therefore \int 2xe^{-3x}\,dx = -\dfrac{2}{3}xe^{-3x} + \int \dfrac{2}{3}e^{-3x}\,dx$

$= -\dfrac{2}{3}xe^{-3x} - \dfrac{2}{9}e^{-3x} + c$

$= -\dfrac{2}{9}e^{-3x}(3x+1) + c$

6 Let $u = \ln x$, $\dfrac{dv}{dx} = x^5$

$u = \ln x \Rightarrow \dfrac{du}{dx} = \dfrac{1}{x}$

$\dfrac{dv}{dx} = x^5 \Rightarrow v = \dfrac{1}{6}x^6$

$\int u\dfrac{dv}{dx}\,dx = uv - \int v\dfrac{du}{dx}\,dx$

$\therefore \int_1^e x^5 \ln x\,dx = \left[\dfrac{1}{6}x^6 \ln x\right]_1^e - \int_1^e \dfrac{1}{6}x^5\,dx$

$= \left[\dfrac{1}{6}x^6 \ln x - \dfrac{1}{36}x^6\right]_1^e$

$= \left(\dfrac{e^6}{6} - \dfrac{e^6}{36}\right) - \left(0 - \dfrac{1}{36}\right)$

$= \dfrac{5e^6 + 1}{36}$

7 a Let $u = \cos x$; then

$\dfrac{du}{dx} = -\sin x \Rightarrow dx = -\dfrac{du}{\sin x}$

$\int \tan x\,dx = \int \dfrac{\sin x}{\cos x}\,dx$

$= \int \dfrac{\sin x}{u} \times \dfrac{-1}{\sin x}\,du$

$= \int -\dfrac{1}{u}\,du$

$= -\ln|u| + c$

$= \ln\left|\dfrac{1}{u}\right| + c$

$= \ln|\sec x| + c$

b Let $u = x$, $\dfrac{dv}{dx} = \sec^2 x$

$u = x \Rightarrow \dfrac{du}{dx} = 1$

$\dfrac{dv}{dx} = \sec^2 x \Rightarrow v = \tan x$

$\int u\dfrac{dv}{dx}\,dx = uv - \int v\dfrac{du}{dx}\,dx$

$\therefore \int x\sec^2 x\,dx = x\tan x - \int \tan x\,dx$

$= x\tan x - \ln|\sec x| + c$

8 Let $u = \arccos x, \dfrac{dv}{dx} = 1$

$u = \arccos x \Rightarrow \dfrac{du}{dx} = -\dfrac{1}{\sqrt{1-x^2}}$

$\dfrac{dv}{dx} = 1 \Rightarrow v = x$

$\int u \dfrac{dv}{dx} dx = uv - \int v \dfrac{du}{dx} dx$

$\therefore \int_0^k \arccos x \, dx = \left[x \arccos x \right]_0^k + \int_0^k \dfrac{x}{\sqrt{1-x^2}} dx$

$= \left[x \arccos x - \sqrt{1-x^2} \right]_0^k$

$= \left(k \arccos k - \sqrt{1-k^2} \right) - (0 - 1)$

$= k \arccos k - \sqrt{1-k^2} + 1$

Require that $k \arccos k - \sqrt{1-k^2} + 1 = 0.5$
From GDC: $k = 0.360$ (3SF)

> **COMMENT**
>
> Take care in this sort of circumstance: the usual generalised variables u and v would cause confusion here, since u is already involved in the substitution. Simply choose a different letter and note that the integral is now being calculated with respect to u instead of x!

9 Let $u = \sqrt{x+1}$; then

$\dfrac{du}{dx} = \dfrac{1}{2}(x+1)^{-\frac{1}{2}} \Rightarrow dx = 2\sqrt{x+1}\, du = 2u\, du$

$\int_{-1}^3 \dfrac{1}{2} e^{\sqrt{x+1}}\, dx = \int_{x=-1}^{x=3} \dfrac{1}{2} e^u \times 2u\, du$

$= \int_{x=-1}^{x=3} u e^u\, du$

Using integration by parts, let $w = u, \dfrac{dv}{du} = e^u$

$w = u \Rightarrow \dfrac{dw}{du} = 1$

$\dfrac{dv}{du} = e^u \Rightarrow v = e^u$

$\int w \dfrac{dv}{du}\, du = wv - \int v \dfrac{dw}{du}\, du$

$\therefore \int_{x=-1}^{x=3} u e^u\, du = \left[u e^u \right]_{x=-1}^{x=3} - \int_{x=-1}^{x=3} e^u\, du$

$= \left[u e^u - e^u \right]_{x=-1}^{x=3}$

$= \left[\sqrt{x+1}\, e^{\sqrt{x+1}} - e^{\sqrt{x+1}} \right]_{-1}^{3}$

$= \left(\sqrt{4}\, e^{\sqrt{4}} - e^{\sqrt{4}} \right) - \left(0 - e^0 \right)$

$= e^2 + 1$

Mixed examination practice 19

Short questions

1 $\int_0^\pi \cos^2(3x)\, dx = \int_0^\pi \dfrac{1}{2}(1 + \cos(6x))\, dx$

$= \left[\dfrac{x}{2} + \dfrac{1}{12} \sin(6x) \right]_0^\pi$

$= \dfrac{\pi}{2}$

2 Let $u = x, \dfrac{dv}{dx} = \cos(2x)$

$u = x \Rightarrow \dfrac{du}{dx} = 1$

$\dfrac{dv}{dx} = \cos(2x) \Rightarrow v = \dfrac{1}{2} \sin(2x)$

$\int u \dfrac{dv}{dx}\, dx = uv - \int v \dfrac{du}{dx}\, dx$

$\therefore \int x \cos(2x)\, dx = \dfrac{1}{2} x \sin(2x) - \int \dfrac{1}{2} \sin(2x)\, dx$

$= \dfrac{1}{2} x \sin(2x) + \dfrac{1}{4} \cos(2x) + c$

3 $\int_0^m \dfrac{dx}{3x+1} = 1$

$\left[\dfrac{1}{3}\ln|3x+1|\right]_0^m = 1$

$\dfrac{1}{3}\ln|3m+1| = 1$

$3m+1 = \pm e^3$

$m = \dfrac{-1 \pm e^3}{3}$

$= 6.36$ or -7.03

Reject the negative value, since that would take the integral across $x = -\dfrac{1}{3}$, where the integrand is not defined.

$\therefore m = \dfrac{e^3 - 1}{3} = 6.36$ (3SF)

4 $\int_0^{\pi/12} \dfrac{1}{\cos^2 x}\,dx = \int_0^{\pi/12} \sec^2(4x)\,dx$

$= \left[\dfrac{1}{4}\tan(4x)\right]_0^{\pi/12}$

$= \dfrac{1}{4}\tan\left(\dfrac{\pi}{3}\right) - 0$

$= \dfrac{\sqrt{3}}{4}$

5 a $\int \dfrac{1}{1-3x}\,dx = -\dfrac{1}{3}\ln|1-3x| + c$

b $\int \dfrac{1}{(2x+3)^2}\,dx = \int (2x+3)^{-2}\,dx$

$= -\dfrac{1}{2}(2x+3)^{-1} + c$

6 Let $u = \ln x, \dfrac{dv}{dx} = 1$

$u = \ln x \Rightarrow \dfrac{du}{dx} = \dfrac{1}{x}$

$\dfrac{dv}{dx} = 1 \Rightarrow v = x$

$\int u\dfrac{dv}{dx}\,dx = uv - \int v\dfrac{du}{dx}\,dx$

$\therefore \int \ln x\,dx = x\ln x - \int 1\,dx$

$= x\ln x - x + c$

7 a $\dfrac{e^{-4x} + 3e^{-2x}}{e^{-4x} - 9} = \dfrac{e^{-2x}(e^{-2x} + 3)}{(e^{-2x} + 3)(e^{-2x} - 3)}$

$= \dfrac{e^{-2x}}{e^{-2x} - 3}$

$= \dfrac{1}{1 - 3e^{2x}}$

b

COMMENT

Although the simplest form of (a) has a numerator of 1, the more useful form for integration is the one (in the second line) that contains an exponential in both numerator and denominator, so that the numerator remains a multiple of the derivative of the denominator.

$\int \dfrac{e^{-4x} + 3e^{-2x}}{e^{-4x} - 9}\,dx = \int \dfrac{e^{-2x}}{e^{-2x} - 3}\,dx$

$= -\dfrac{1}{2}\ln|e^{-2x} - 3| + c$

$= \ln\left(\dfrac{1}{\sqrt{|e^{-2x} - 3|}}\right) + c$

8 Derivative of $x^2 + 4$ is $2x$

$\int \dfrac{6x+4}{x^2+4}\,dx = 3\int \dfrac{2x}{x^2+4}\,dx + 4\int \dfrac{1}{x^2+4}\,dx$

$= 3\ln|x^2+4| + 4 \times \dfrac{1}{2}\arctan\left(\dfrac{x}{2}\right) + c$

$= 3\ln(x^2+4) + 2\arctan\left(\dfrac{x}{2}\right) + c$

9 a $\dfrac{2}{x-1} - \dfrac{1}{x+2} = \dfrac{2(x+2)-(x-1)}{(x-1)(x+2)}$

$= \dfrac{x+5}{(x-1)(x+2)}$

b $\displaystyle\int_5^7 \dfrac{x+5}{(x-1)(x+2)}\,dx = \int_5^7 \dfrac{2}{x-1} - \dfrac{1}{x+2}\,dx$

$= \left[2\ln|x-1| - \ln|x+2|\right]_5^7$

$= (2\ln 6 - \ln 9) - (2\ln 4 - \ln 7)$

$= \ln 36 - \ln 9 - \ln 16 + \ln 7$

$= \ln\left(\dfrac{36 \times 7}{9 \times 16}\right)$

$= \ln\left(\dfrac{7}{4}\right)$

10 Let $u = \ln x$; then

$\dfrac{du}{dx} = \dfrac{1}{x} \Rightarrow dx = x\,du$

$\therefore \displaystyle\int \dfrac{1}{x\ln x}\,dx = \int \dfrac{1}{xu} \times x\,du$

$= \displaystyle\int \dfrac{1}{u}\,du$

$= \ln|u| + c$

$= \ln|\ln x| + c$

11 Let $u = \dfrac{1}{2}x - 1$; then $x = 2u + 2$ and

$\dfrac{du}{dx} = \dfrac{1}{2} \Rightarrow dx = 2\,du$

$\therefore \displaystyle\int \dfrac{x}{\sqrt{\tfrac{1}{2}x-1}}\,dx = \int \dfrac{2u+2}{\sqrt{u}} \times 2\,du$

$= \displaystyle\int 4u^{\tfrac{1}{2}} + 4u^{-\tfrac{1}{2}}\,du$

$= \dfrac{8}{3}u^{\tfrac{3}{2}} + 8u^{\tfrac{1}{2}} + c$

$= \dfrac{8}{3}\left(\dfrac{1}{2}x - 1\right)^{\tfrac{3}{2}} + 8\left(\dfrac{1}{2}x - 1\right)^{\tfrac{1}{2}} + c$

12 $\displaystyle\int_2^5 \dfrac{x-1}{x+2}\,dx = \int_2^5 \dfrac{x+2-3}{x+2}\,dx$

$= \displaystyle\int_2^5 1 - \dfrac{3}{x+2}\,dx$

$= \left[x - 3\ln|x+2|\right]_2^5$

$= (5 - 3\ln 7) - (2 - 3\ln 4)$

$= 3 + 3\ln\left(\dfrac{4}{7}\right)$

13 Let $u = \arctan x$, $\dfrac{dv}{dx} = 1$

$u = \arctan x \Rightarrow \dfrac{du}{dx} = \dfrac{1}{1+x^2}$

$\dfrac{dv}{dx} = 1 \Rightarrow v = x$

$\displaystyle\int u\dfrac{dv}{dx}\,dx = uv - \int v\dfrac{du}{dx}\,dx$

$\therefore \displaystyle\int \arctan x\,dx = x\arctan x - \int \dfrac{x}{1+x^2}\,dx$

$= x\arctan x - \dfrac{1}{2}\ln|1+x^2| + c$

14 $\dfrac{2}{1-x^2} = \dfrac{1+x+1-x}{(1+x)(1-x)}$

$= \dfrac{1}{1-x} + \dfrac{1}{1+x}$

$\displaystyle\int_{-a}^{a} \dfrac{2}{1-x^2}\,dx = \int_{-a}^{a} \dfrac{1}{1-x} + \dfrac{1}{1+x}\,dx$

$= \left[-\ln|1-x| + \ln|1+x|\right]_{-a}^{a}$

$= \left[\ln\left|\dfrac{1+x}{1-x}\right|\right]_{-a}^{a}$

$= \ln\left|\dfrac{1+a}{1-a}\right| - \ln\left|\dfrac{1-a}{1+a}\right|$

$= 2\ln\left|\dfrac{1+a}{1-a}\right|$

19 Further integration methods 261

Require that $2\ln\left|\dfrac{1+a}{1-a}\right| = 1$

$\therefore \left|\dfrac{1+a}{1-a}\right| = e^{\frac{1}{2}}$

The integral $2\ln\left|\dfrac{1+a}{1-a}\right|$ does not converge as $a \to \pm 1$, so we should not consider $|a| \geq 1$.

$\therefore \left|\dfrac{1+a}{1-a}\right| = \dfrac{1+a}{1-a} = e^{\frac{1}{2}}$

$1 + a = e^{\frac{1}{2}}(1-a)$

$1 + a = e^{\frac{1}{2}} - e^{\frac{1}{2}}a$

$\Rightarrow a = \dfrac{e^{\frac{1}{2}} - 1}{e^{\frac{1}{2}} + 1} = 0.245 \text{ (3SF)}$

> **COMMENT**
>
> Some functions can be integrated even to a value at which the function itself is not defined. For example, $y = \dfrac{1}{\sqrt{|x|}}$ is undefined at $x = 0$, but its integral over the interval $[0, b]$ is nonetheless well-defined. Such an 'improper' integral is calculated by determining the integral across the interval $[a, b]$ and then taking the limit as $a \to 0$.
> The situation in question 14 is different, however, because the limit as $a \to 1$ (or $a \to -1$) for $2\ln\left|\dfrac{1+a}{1-a}\right|$ is not finite. Therefore we cannot consider values of $|a|$ equal to or greater than 1.

Long questions

1 a $\dfrac{A}{x+2} + \dfrac{1-Bx}{x^2+1} = \dfrac{A(x^2+1) + (1-Bx)(x+2)}{(x+2)(x^2+1)}$

Require that $\dfrac{A(x^2+1) + (1-Bx)(x+2)}{(x+2)(x^2+1)} = \dfrac{4-3x}{(x+2)(x^2+1)}$

so $(A-B)x^2 + (1-2B)x + A + 2 = 4 - 3x$

Comparing coefficients:

x^2: $A - B = 0 \Rightarrow A = B$

x^1: $1 - 2B = -3 \Rightarrow B = 2$

x^0: $A + 2 = 4$ is consistent with $A = B = 2$

$\therefore \dfrac{4-3x}{(x+2)(x^2+1)} = \dfrac{2}{x+2} + \dfrac{1-2x}{x^2+1}$

b $\displaystyle\int \dfrac{4-3x}{(x+2)(x^2+1)}\, dx = \int \dfrac{2}{x+2} + \dfrac{1-2x}{x^2+1}\, dx$

$\displaystyle = 2\ln|x+2| + \int \dfrac{1}{x^2+1} - \dfrac{2x}{x^2+1}\, dx$

$= 2\ln|x+2| + \arctan x - \ln|x^2+1| + c$

c $\dfrac{d}{dx}\left(\sqrt{1-x^2}\right) = \dfrac{1}{2}(1-x^2)^{-\frac{1}{2}} \times (-2x) = -\dfrac{x}{\sqrt{1-x^2}}$

$\displaystyle\int_0^{\sqrt{3}/2} \dfrac{4-3x}{\sqrt{1-x^2}}\, dx = \int_0^{\sqrt{3}/2} \dfrac{4}{\sqrt{1-x^2}} - \dfrac{3x}{\sqrt{1-x^2}}\, dx$

$\displaystyle = \left[4\arcsin x + 3\sqrt{1-x^2}\right]_0^{\sqrt{3}/2}$

$= \left(4 \times \dfrac{\pi}{3} + \dfrac{3}{2}\right) - (0+3)$

$= \dfrac{4\pi}{3} - \dfrac{3}{2}$

2 a $\displaystyle I+J = \int \dfrac{\sin x}{\sin x + \cos x}\, dx + \int \dfrac{\cos x}{\sin x + \cos x}\, dx$

$\displaystyle = \int \dfrac{\sin x + \cos x}{\sin x + \cos x}\, dx$

$\displaystyle = \int 1\, dx$

$= x + c_1$

b $\displaystyle J-I = \int \dfrac{\cos x - \sin x}{\sin x + \cos x}\, dx$

Let $u = \sin x + \cos x$; then

$\dfrac{du}{dx} = \cos x - \sin x$

$dx = \dfrac{du}{\cos x - \sin x}$

$\displaystyle J - I = \int \dfrac{\cos x - \sin x}{u} \times \dfrac{du}{\cos x - \sin x}$

$\displaystyle = \int \dfrac{1}{u}\, du$

$= \ln|u| + c_2$

$= \ln|\sin x + \cos x| + c_2$

19 Further integration methods

c $\displaystyle \int \frac{\sin x}{\sin x + \cos x} dx = I$

$$= \frac{1}{2}((I+J)-(J-I))$$

$$= \frac{1}{2}(x + c_1 - \ln|\sin x + \cos x| - c_2)$$

$$= \frac{1}{2}(x - \ln|\sin x + \cos x|) + c$$

3 a $t = \tan\left(\dfrac{x}{2}\right)$

$\Rightarrow \dfrac{dt}{dx} = \dfrac{1}{2}\sec^2\left(\dfrac{x}{2}\right)$

$\phantom{\Rightarrow \dfrac{dt}{dx}} = \dfrac{1}{2}\left(1 + \tan^2\left(\dfrac{x}{2}\right)\right)$

$\phantom{\Rightarrow \dfrac{dt}{dx}} = \dfrac{1}{2}(1 + t^2)$

b i $\sin(2\theta) = 2\sin\theta\cos\theta$

$ = 2\dfrac{\sin\theta}{\cos\theta} \times \cos^2\theta$

$ = \dfrac{2\tan\theta}{\sec^2\theta}$

ii Take $\theta = \dfrac{x}{2}$ and recall that $t = \tan\left(\dfrac{x}{2}\right)$ from (a)

$\sec^2\theta = 1 + \tan^2\theta = 1 + t^2$

$\therefore \sin x = \sin(2\theta) = \dfrac{2\tan\theta}{\sec^2\theta} = \dfrac{2t}{1+t^2}$

c Using $t = \tan\left(\dfrac{x}{2}\right)$, $\dfrac{dt}{dx} = \dfrac{1}{2}(1+t^2) \Rightarrow dx = \dfrac{2}{1+t^2}\,dt$:

$$\int_0^{\pi/2} \dfrac{1}{1+\sin x}\,dx = \int_{x=0}^{x=\pi/2} \dfrac{1}{1+\dfrac{2t}{1+t^2}} \times \dfrac{2}{1+t^2}\,dt$$

$$= \int_{x=0}^{x=\pi/2} \dfrac{2}{1+t^2+2t}\,dt$$

$$= \int_{x=0}^{x=\pi/2} \dfrac{2}{(1+t)^2}\,dt$$

$$= \left[-2(1+t)^{-1}\right]_{x=0}^{x=\pi/2}$$

$$= \left[-\dfrac{2}{1+\tan\left(\dfrac{x}{2}\right)}\right]_0^{\pi/2}$$

$$= -\dfrac{2}{2} - \left(-\dfrac{2}{1}\right)$$

$$= 1$$

4 a By De Moivre's theorem, $z^n = \cos(n\theta) + i\sin(n\theta)$

$z^{-n} = \cos(-n\theta) + i\sin(-n\theta) = \cos(n\theta) - i\sin(n\theta)$

$\therefore z^n + z^{-n} = 2\cos(n\theta)$

b $(z+z^{-1})^4 = z^4 + 4z^2 + 6 + 4z^{-2} + z^{-4}$

$(2\cos\theta)^4 = (z^4 + z^{-4}) + 4(z^2 + z^{-2}) + 6$

$16\cos^4\theta = 2\cos(4\theta) + 8\cos(2\theta) + 6$

$\cos^4\theta = \dfrac{1}{8}(\cos(4\theta) + 4\cos(2\theta) + 3)$

c i $g(a) = \int_0^a \cos^4\theta\,d\theta = \dfrac{1}{8}\int_0^a \cos(4\theta) + 4\cos(2\theta) + 3\,d\theta$

$= \dfrac{1}{8}\left[\dfrac{1}{4}\sin(4\theta) + 2\sin(2\theta) + 3\theta\right]_0^a$

$= \dfrac{1}{32}\sin(4a) + \dfrac{1}{4}\sin(2a) + \dfrac{3}{8}a$

ii $g(a) = 1 \Rightarrow \dfrac{1}{32}\sin(4a) + \dfrac{1}{4}\sin(2a) + \dfrac{3}{8}a = 1$

$\Rightarrow a = 2.96$ (3SF, from GDC)

20 Further applications of calculus

Exercise 20A

4 Let:

r = radius (cm)

A = area (cm^2)

t = time (seconds)

$\dfrac{dr}{dt} = 1.5$

$\dfrac{dA}{dt} = \dfrac{dA}{dr} \times \dfrac{dr}{dt}$

$A = \pi r^2 \Rightarrow \dfrac{dA}{dr} = 2\pi r$

$\therefore \dfrac{dA}{dt} = 2\pi r \times 1.5 = 3\pi r$

When $r = 12$, $\dfrac{dA}{dt} = 36\pi = 113$ (3SF)

So when the radius is 12 cm, the rate of change of the area is 113 cm^2 s^{-1}.

5 Let:

l = side length (cm)

A = area (cm^2)

t = time (seconds)

$\dfrac{dA}{dt} = 50$

$\dfrac{dA}{dt} = \dfrac{dA}{dl} \times \dfrac{dl}{dt} \Rightarrow \dfrac{dl}{dt} = \dfrac{dA}{dt} \div \dfrac{dA}{dl}$

$A = l^2 \Rightarrow \dfrac{dA}{dl} = 2l$

$\therefore \dfrac{dl}{dt} = 50 \div 2l = \dfrac{25}{l}$

$l = 12.5 \Rightarrow \dfrac{dl}{dt} = \dfrac{25}{12.5} = 2$

So when the side length is 12.5 cm, the rate of increase of the side length is 2 cm s^{-1}.

6 Let:

r = radius (cm)

h = height (cm)

A = surface area (cm^2)

t = time (seconds)

$A = 2\pi r^2 + 2\pi rh$

$\Rightarrow \dfrac{dA}{dt} = 4\pi r \dfrac{dr}{dt} + 2\pi h \dfrac{dr}{dt} + 2\pi r \dfrac{dh}{dt}$

$\dfrac{dA}{dt} - 2\pi r \dfrac{dh}{dt} = \dfrac{dr}{dt}(4\pi r + 2\pi h)$

$\dfrac{dr}{dt} = \dfrac{1}{4\pi r + 2\pi h}\left(\dfrac{dA}{dt} - 2\pi r \dfrac{dh}{dt}\right)$

Substituting

$r = 4$, $h = 1$, $\dfrac{dA}{dt} = 20\pi$ and $\dfrac{dh}{dt} = -2$:

$\dfrac{dr}{dt} = \dfrac{1}{16\pi + 2\pi}(20\pi + 16\pi) = 2$

So when $\dfrac{dA}{dt} = 20\pi$, $r = 4$, $h = 1$,

and $\dfrac{dh}{dt} = -2$ the rate of change of the radius is 2 cm s^{-1}.

7 Let:

r = radius (cm)

V = volume (cm³)

t = time (seconds)

$\dfrac{dV}{dt} = 500$

$V = \dfrac{4}{3}\pi r^3$

$\Rightarrow \dfrac{dV}{dt} = 4\pi r^2 \dfrac{dr}{dt}$

$\therefore r^2 = \dfrac{1}{4\pi}\dfrac{dV}{dt} \div \dfrac{dr}{dt}$

$\dfrac{dr}{dt} = 0.5 \Rightarrow r = \sqrt{\dfrac{500}{2\pi}} = 8.92$ (3SF)

So when $\dfrac{dr}{dt} = \dfrac{1}{2}$, the radius is 8.92 cm.

8 If the lighthouse is at L and the ship is at A, then the distance d (km) between them is $d = |\overrightarrow{LA}| = \left|\begin{pmatrix} 5+16t \\ 7+12t \end{pmatrix}\right|$

$= \sqrt{(5+16t)^2 + (7+12t)^2}$

$= \sqrt{74 + 328t + 400t^2}$

$\dfrac{dd}{dt} = \dfrac{1}{2} \times \dfrac{800t + 328}{\sqrt{74 + 328t + 400t^2}}$

When $t = 0$, $\dfrac{dd}{dt} = \dfrac{164}{\sqrt{74}} = 19.1$ (3SF)

$\therefore$ the distance is increasing at 19.1 km h⁻¹

Exercise 20B

5 a $v = \int a \, dt$

$= at + c$

But $v(0) = u$

$\therefore u = 0 + c$

$\Rightarrow c = u$

$\therefore v = u + at$

b $s = \int v \, dt$

$= \int u + at \, dt$

$= ut + \dfrac{at^2}{2} + k$

But $s(0) = 0$ (by definition of displacement s being the displacement from the initial position at $t = 0$)

$\therefore 0 = 0 + 0 + k$

$\Rightarrow k = 0$

$\therefore s = ut + \dfrac{1}{2}at^2$

c $v^2 = (u + at)^2$ from (a)

$= u^2 + 2uat + a^2t^2$

$= u^2 + 2a\left(ut + \dfrac{1}{2}at^2\right)$

$= u^2 + 2as$ from (b)

6 a By the quotient rule,

$a = \dfrac{dv}{dt} = \dfrac{(t^2+1) \times 1 - t \times 2t}{(t^2+1)^2} = \dfrac{1-t^2}{(t^2+1)^2}$

b $s(5) = \int_0^5 v \, dt = \int_0^5 \dfrac{t}{t^2+1} \, dt$

Let $w = t^2 + 1$; then

$\dfrac{dw}{dt} = 2t \Rightarrow dt = \dfrac{dw}{2t}$

$\therefore s(5) = \int_{t=0}^{t=5} \dfrac{t}{w} \dfrac{dw}{2t}$

$= \dfrac{1}{2}\int_{t=0}^{t=5} \dfrac{1}{w} dw$

$= \dfrac{1}{2}[\ln w]_{t=0}^{t=5}$

$= \dfrac{1}{2}\left[\ln(t^2+1)\right]_0^5$

$= \dfrac{1}{2}(\ln 26 - \ln 1)$

$= \ln\sqrt{26}$

$= 1.63$ (3SF)

20 Further applications of calculus

7 Distance travelled in the first 2 seconds is

$$x(2) = \int_0^2 |v|\, dt$$

In $[0, 2]$, $v = 0$ when

$12 - 9.8t = 0$

$$t = \frac{12}{9.8} = \frac{60}{49}$$

so velocity graph goes negative (below the t-axis) for $t > \dfrac{60}{49}$

$$\therefore x(2) = \int_0^{60/49} v\, dt - \int_{60/49}^2 v\, dt$$

$$= \int_0^{60/49} 12 - 9.8t\, dt - \int_{60/49}^2 12 - 9.8t\, dt$$

$$= 10.3\,\text{m}\ (3\,\text{SF, from GDC})$$

8 a $s(6) = \int_0^6 v\, dt$

$$= \int_0^6 5\cos\left(\frac{t}{3}\right) dt$$

$$= \left[15\sin\left(\frac{t}{3}\right)\right]_0^6$$

$$= 15\sin(2)$$

$$= 13.6\,\text{m}\ (3\,\text{SF})$$

b $x(6) = \int_0^6 |v|\, dt$

In $[0, 6]$, $v = 0$ when

$$5\cos\left(\frac{t}{3}\right) = 0$$

$$t = \frac{3\pi}{2}$$

$$\therefore x(6) = \int_0^{3\pi/2} v\, dt - \int_{3\pi/2}^6 v\, dt$$

$$= s(6) - 2\int_{3\pi/2}^6 5\cos\left(\frac{t}{3}\right) dt$$

$$= s(6) - 2\left[15\sin\left(\frac{t}{3}\right)\right]_{3\pi/2}^6$$

$$= 13.6 - 30\left(\sin 2 - \sin\frac{\pi}{2}\right)$$

$$= 16.4\,\text{m}\ (3\,\text{SF})$$

COMMENT

Notice the importance of finding where $v = 0$; at such points the velocity graph crosses the t-axis and so the integration needs to be separated at these points (as the 'area' will be negative below the t-axis).

9 $v = \dfrac{ds}{dt} = -t^2 + 3t + 4$

$$\Rightarrow \frac{dv}{dt} = -2t + 3$$

Stationary value of v when $\dfrac{dv}{dt} = 0$:

$-2t + 3 = 0$

$$\Rightarrow t = \frac{3}{2}$$

$$v\left(\frac{3}{2}\right) = -\left(\frac{3}{2}\right)^2 + 3\left(\frac{3}{2}\right) + 4 = \frac{25}{4}$$

Check values at end points:

$$v(0) = 4 < v\left(\frac{3}{2}\right)$$

$$v(5) = -6 < v\left(\frac{3}{2}\right)$$

$\therefore$ maximum velocity is $v(1.5) = 6.25$

10 $v = \ln(s + 2)$

$$\Rightarrow \frac{dv}{dt} = \frac{1}{s+2}\frac{ds}{dt}$$

$$\frac{dv}{dt} = a,\ \frac{ds}{dt} = v$$

$$\therefore a = \frac{v}{s+2}$$

When $v = 4$, $\ln(s+2) = 4 \Rightarrow s + 2 = e^4$

$\therefore a = 4e^{-4} = 0.0733\ (3\,\text{SF})$

11 $v = \dfrac{10(s-2)}{s^2+4}$

a Maximum velocity occurs when $\dfrac{dv}{dt} = 0$:

$\dfrac{dv}{dt} = \dfrac{10(s^2+4) - 2s \times 10(s-2)}{(s^2+4)^2}$

$= \dfrac{40 + 40s - 10s^2}{(s^2+4)^2}$

$\dfrac{dv}{dt} = 0 \Rightarrow 40 + 40s - 10s^2 = 0$

$\Rightarrow s^2 - 4s - 4 = 0$

$\Rightarrow s = \dfrac{4 \pm \sqrt{16+16}}{2} = 2 \pm 2\sqrt{2}$

The negative value of s gives a negative v, so the maximum velocity is

$v|_{s=2+\sqrt{2}} = \dfrac{10(2+2\sqrt{2}-2)}{(2+2\sqrt{2})^2 + 4} = 1.04\,\text{ms}^{-1}$ (3SF)

b $a = \dfrac{dv}{dt} = \dfrac{40 + 40s - 10s^2}{(s^2+4)^2}$

$\therefore \dfrac{dv}{dt}\bigg|_{s=3} = \dfrac{40 + 120 - 90}{13^2} = 0.414\,\text{ms}^{-2}$ (3SF)

> **COMMENT**
>
> The notation of a vertical bar with a value condition on the lower right is often used when simpler notation might be ambiguous. In this problem, v is expressed as a function of s, so $v(b)$ should mean v evaluated at $s = b$; however, we commonly use $v(b)$ to mean v evaluated at time $t = b$, and this is so standard that in other situations a more explicit notation is useful for clarity.

Exercise 20C

3 $V = \pi \int_1^{2e} y^2\,dx$

$= \pi \int_1^{2e} (\ln x)^2\,dx$

$= 19.0$ (from GDC)

4 $V = \pi \int_0^{\pi/2} y^2\,dx$

$= \pi \int_0^{\pi/2} \sin x\,dx$

$= \pi\left[-\cos x\right]_0^{\pi/2}$

$= \pi(0 - (-1)) = \pi$

5 $y = \ln(x^2)$

$x = 1 \Rightarrow y = 0$

$x = e^2 \Rightarrow y = 4$

$V = \pi \int_0^4 x^2\,dy$

$= \pi \int_0^4 e^y\,dy$

$= \pi \left[e^y\right]_0^4$

$= \pi(e^4 - 1)$

6 a i Line with gradient $-\dfrac{h}{r}$ through point $(0, h)$ has equation

$y = -\dfrac{h}{r}x + h$

ii For points on the line, $x = r - \dfrac{r}{h}y$

$V = \pi \int_0^h x^2\,dy$

$= \pi \int_0^h \left(r - \dfrac{r}{h}y\right)^2 dy$

$= \pi \int_0^h r^2 - \dfrac{2r^2}{h}y + \dfrac{r^2}{h^2}y^2\,dy$

$= \pi \left[r^2 y - \dfrac{r^2}{h}y^2 + \dfrac{r^2}{3h^2}y^3\right]_0^h$

$= \pi \left(r^2 h - r^2 h + \dfrac{r^2 h}{3}\right) - 0$

$= \dfrac{\pi r^2 h}{3}$

b $x^2 + y^2 = r^2 \Rightarrow y^2 = r^2 - x^2$

$$V = \pi \int_{-r}^{r} y^2 \, dx$$

$$= \pi \int_{-r}^{r} r^2 - x^2 \, dx$$

$$= \pi \left[r^2 x - \frac{x^3}{3} \right]_{-r}^{r}$$

$$= \pi \left(r^3 - \frac{r^3}{3} \right) - \pi \left(-r^3 + \frac{r^3}{3} \right)$$

$$= \frac{4\pi r^3}{3}$$

7 $y = 2\cos x \Rightarrow x = \arccos\left(\frac{y}{2}\right)$

When $x = 0$, $y = 2$

$$V = \pi \int_0^2 x^2 \, dy$$

$$= \pi \int_0^2 \arccos^2\left(\frac{y}{2}\right) dy$$

$$= 7.17 \text{ (3SF, from GDC)}$$

8 $y = \sqrt{x} \Rightarrow x = y^2$

$$V = \frac{\pi}{2} \int_0^3 x^2 \, dy$$

$$= \frac{\pi}{2} \int_0^3 y^4 \, dy$$

$$= \frac{\pi}{2} \left[\frac{y^5}{5} \right]_0^3$$

$$= \frac{\pi}{2} \times \frac{243}{5}$$

$$= 76.3 \text{ (3SF)}$$

9 $V = \pi \int_1^a y^2 \, dx$

$$= \pi \int_1^a \frac{9}{x} \, dx$$

$$= \pi \left[9 \ln x \right]_1^a$$

$$= 9\pi \ln a$$

$$\therefore \pi \ln a^9 = \pi \ln\left(\frac{64}{27}\right)$$

$$a^9 = \frac{64}{27}$$

$$a^3 = \sqrt[3]{\frac{4}{3}}$$

$$a = \sqrt[3]{\frac{4}{3}}$$

10 $y = e^{2x} - 1 \Rightarrow x = \frac{1}{2}\ln(y+1)$

$$V = \frac{\pi}{2} \int_0^3 x^2 \, dy$$

$$= \frac{\pi}{8} \int_0^3 (\ln(y+1))^2 \, dy$$

$$= 1.02 \text{ (3SF, from GDC)}$$

11 a Intersections where

$$4\sqrt{x} = x + 3$$

$$16x = (x+3)^2$$

$$x^2 - 10x + 9 = 0$$

$$(x-1)(x-9) = 0$$

$$x = 1 \text{ or } 9$$

$\therefore$ coordinates of the intersections are (1, 4) and (9, 12)

b $y_1 = 4\sqrt{x_1} \Rightarrow x_1 = \dfrac{y_1^2}{16}$

$y_2 = x_2 + 3 \Rightarrow x_2 = y_2 - 3$

Between the intersection points, x_2 is to the right of x_1

$V = \pi \displaystyle\int_4^{12} (x_2)^2 - (x_1)^2 \, dy$

$= \pi \displaystyle\int_4^{12} (y-3)^2 - \left(\dfrac{y^2}{16}\right)^2 dy$

$= \pi \displaystyle\int_4^{12} y^2 - 6y + 9 - \dfrac{y^4}{256} \, dy$

$= \pi \left[\dfrac{y^3}{3} - 3y^2 + 9y - \dfrac{y^5}{1280}\right]_4^{12}$

$= \pi \left(576 - 432 + 108 - \dfrac{972}{5}\right)$

$- \pi \left(\dfrac{64}{3} - 48 + 36 - \dfrac{4}{5}\right)$

$= \pi \left(\dfrac{288}{5} - \dfrac{128}{15}\right)$

$= \dfrac{736}{15} \pi = 154 \text{ (3SF)}$

12 a Intersections where

$x^2 + 3 = 4x + 3$

$x^2 - 4x = 0$

$x(x-4) = 0$

$x = 0$ or 4

Substituting into $y = 4x + 3$:

$x = 0 \Rightarrow y = 3$

$x = 4 \Rightarrow y = 4 \times 4 + 3 = 19$

∴ coordinates of the intersections are $(0, 3)$ and $(4, 19)$

b Volume of revolution of the enclosed region is the difference in the volumes of revolution from the two curves:

$V = \pi \displaystyle\int_0^4 (4x+3)^2 - (x^2+3)^2 \, dx$

$= \pi \displaystyle\int_0^4 16x^2 + 24x + 9 - x^4 - 6x^2 - 9 \, dx$

$= \pi \displaystyle\int_0^4 -x^4 + 10x^2 + 24x \, dx$

$= \pi \left[-\dfrac{x^5}{5} + \dfrac{10x^3}{3} + 12x^2\right]_0^4$

$= \pi \left(-\dfrac{1024}{5} + \dfrac{640}{3} + 192\right)$

$= \dfrac{3008\pi}{15}$

$= 630 \text{ (3SF)}$

13 The volume of revolution is the sum of the volumes of revolution generated by the two graphs.

$y_1 = \ln x_1 \Rightarrow x_1 = e^{y_1}$

$y_2 = -\dfrac{1}{e} x_2 + 2 \Rightarrow x_2 = e(2 - y_2)$

$V = \pi \displaystyle\int_0^1 (x_1)^2 \, dy + \pi \displaystyle\int_1^2 (x_2)^2 \, dy$

$= \pi \displaystyle\int_0^1 (e^y)^2 \, dy + \pi \displaystyle\int_1^2 (e(2-y))^2 \, dy$

$= \pi \displaystyle\int_0^1 e^{2y} \, dy + \pi e^2 \displaystyle\int_1^2 (4 - 4y + y^2) \, dy$

$= \dfrac{\pi}{2} \left[e^{2y}\right]_0^1 + \pi e^2 \left[4y - 2y^2 + \dfrac{y^3}{3}\right]_1^2$

$= \dfrac{\pi}{2}(e^2 - 1) + \pi e^2 \left(8 - 8 + \dfrac{8}{3} - 4 + 2 - \dfrac{1}{3}\right)$

$= \dfrac{5}{6}\pi e^2 - \dfrac{\pi}{2}$

20 Further applications of calculus

Exercise 20D

3 a The base of the box is a square with side length $12-2x$, and the box sides have height x

$$\therefore V = x(12-2x)^2$$

b Let $u = x$, $v = (12-2x)^2$

$$\frac{dV}{dx} = v\frac{du}{dx} + u\frac{dv}{dx}$$
$$= (12-2x)^2 - 4x(12-2x)$$
$$= (12-2x)(12-2x-4x)$$
$$= (12-2x)(12-6x)$$
$$= 12(6-x)(2-x)$$

Stationary values for V occur when $\frac{dV}{dx} = 0$:

$$12(6-x)(2-x) = 0$$
$$\Rightarrow x = 6 \text{ or } 2$$

Clearly $x = 6$ represents a minimum, corresponding to zero volume, so $x = 2$ must represent the maximum.

Check using the second derivative:

$$\frac{d^2V}{dx^2} = -12(6-x) - 12(2-x)$$
$$= -12(8-2x)$$

$\frac{d^2V}{dx^2}(2) = -48 < 0$, so volume is indeed maximal at $x = 2$.

4 Let each side of the base have length x, and let the height of the box be h.

The surface area is $S = x^2 + 4xh$

The volume is $V = x^2 h$

$V = 32$

$x^2 h = 32$

$$\Rightarrow h = \frac{32}{x^2}$$

Substituting into the expression for S:

$$S = x^2 + 4x\left(\frac{32}{x^2}\right)$$
$$= x^2 + \frac{128}{x}$$

$$\frac{dS}{dx} = 2x - \frac{128}{x^2}$$

Stationary values of S occur when $\frac{dS}{dx} = 0$:

$$2x - \frac{128}{x^2} = 0$$
$$2x = \frac{128}{x^2}$$
$$x^3 = 64$$
$$x = 4$$

Check that this is a local minimum:

$$\frac{d^2S}{dx^2} = 2 + \frac{256}{x^3}$$

$\frac{d^2S}{dx^2}(4) = 6 > 0 \Rightarrow$ local minimum

$$\therefore S(4) = 4^2 + \frac{128}{4} = 16 + 32 = 48 \text{ cm}^2 \text{ is the}$$
minimum surface area.

5 Let the vertex A have coordinates $(x, 0)$ where $x > 0$. Then:

Area of rectangle $= 2x(4-x^2)$
$$= 8x - 2x^3$$

$$\frac{d \text{ Area}}{dx} = 8 - 6x^2$$

Stationary value when $\frac{d \text{ Area}}{dx} = 0$:

$$8 - 6x^2 = 0$$
$$x^2 = \frac{4}{3}$$
$$x = \frac{2}{\sqrt{3}} = \frac{2\sqrt{3}}{3}$$

It is clear that this is a maximum rather than a minimum, from the context.

Showing this rigorously:

$\dfrac{d^2 \text{Area}}{dx^2} = -12x$

$\dfrac{d^2 \text{Area}}{dx^2}\left(\dfrac{2}{\sqrt{3}}\right) < 0 \Rightarrow$ local maximum

Hence the coordinates of A that give the maximum possible area are $\left(\dfrac{2\sqrt{3}}{3}, 0\right)$

6 a i By symmetry, the coordinates of B are $(\pi - x, 0)$

  ii Area $= (\pi - 2x)\sin x$

 b $\dfrac{d\,\text{Area}}{dx} = (\pi - 2x)\cos x - 2\sin x$

Area has a stationary value when $\dfrac{d\,\text{Area}}{dx} = 0$:

$(\pi - 2x)\cos x - 2\sin x = 0$

$2\sin x = (\pi - 2x)\cos x$

$2\tan x = \pi - 2x$

 c From GDC: $x = 0.710$

Hence maximum area is
$(\pi - 0.710)\sin(0.710) = 1.12 \,(3\text{SF})$

7 Let the cylindrical can have radius r and height h.

Volume $V = \pi r^2 h$

Surface area $S = 2\pi r^2 + 2\pi rh$

$\therefore S = 450$

$2\pi r^2 + 2\pi rh = 450$

$2\pi r(r + h) = 450$

$\Rightarrow h = \dfrac{450}{2\pi r} - r$

Substituting into the expression for volume:

$V = \pi r^2 \left(\dfrac{450}{2\pi r} - r\right)$

$= 225r - \pi r^3$

$\therefore \dfrac{dV}{dr} = 225 - 3\pi r^2$

Stationary values of V occur when $\dfrac{dV}{dr} = 0$:

$225 - 3\pi r^2 = 0$

$3\pi r^2 = 225$

$\therefore r = \sqrt{\dfrac{75}{\pi}}$

$\dfrac{d^2 V}{dr^2} = -6\pi r$

$\dfrac{d^2 V}{dr^2}\left(\sqrt{\dfrac{75}{\pi}}\right) < 0 \Rightarrow$ local maximum

So largest possible capacity is

$V\left(\sqrt{\dfrac{75}{\pi}}\right) = \sqrt{\dfrac{75}{\pi}}(225 - 75) = 733\,\text{cm}^3\,(3\text{SF})$

8 Let each side of the base have length x, and let the height be h.

Volume $V = x^2 h$

Surface area $S = 2x^2 + 4xh$

$\therefore S = 450$

$\Rightarrow 2x^2 + 4xh = 450$

$4x\left(\dfrac{x}{2} + h\right) = 450$

$h = \dfrac{450}{4x} - \dfrac{x}{2}$

Substituting into the expression for volume:

$V = x^2\left(\dfrac{450}{4x} - \dfrac{x}{2}\right)$

$= \dfrac{1}{2}(225x - x^3)$

$\dfrac{dV}{dx} = \dfrac{1}{2}(225 - 3x^2)$

Stationary values of V occur when $\dfrac{dV}{dx} = 0$:

$\dfrac{1}{2}(225 - 3x^2) = 0$

$3x^2 = 225$

$\therefore x = \sqrt{75}$

$\dfrac{d^2 V}{dx^2} = -3x$

$\dfrac{d^2 V}{dx^2}(\sqrt{75}) < 0 \Rightarrow$ local maximum

20 Further applications of calculus   273

So largest possible capacity is

$$V(\sqrt{75}) = \frac{\sqrt{75}}{2}(225-75)$$
$$= 75\sqrt{75}$$
$$= 650\,\text{cm}^3 \text{ (3SF)}$$

9 $x + y = 6$
$\Rightarrow y = 6 - x$
$S = x^2 + y^2$
$= x^2 + (6-x)^2$
$= 2x^2 - 12x + 36$
$\dfrac{dS}{dx} = 4x - 12$

Stationary values of S when $\dfrac{dS}{dx} = 0$:
$4x - 12 = 0$
$\Rightarrow x = 3$
$\dfrac{d^2S}{dx^2} = 4 > 0 \Rightarrow$ local minimum

a So $x = 3$, $y = 3$ gives the minimum value of the sum of squares.

b End-point values of 0 and 6 produce the maximum value of the sum of squares.

10 a Curved surface area of the cone is given by $S = \pi r \sqrt{r^2 + h^2}$

Volume $V = 81\pi$
$\Rightarrow \dfrac{1}{3}\pi r^2 h = 81\pi$
$\Rightarrow h = \dfrac{243}{r^2}$

Substituting into the expression for surface area:

$$S = \pi r \sqrt{r^2 + \dfrac{243^2}{r^4}}$$
$$= \dfrac{\pi r}{r^2}\sqrt{r^6 + 243^2}$$
$$= \dfrac{\pi}{r}\sqrt{r^6 + 243^2}$$

b $S = \pi(r^4 + 243^2 r^{-2})^{\frac{1}{2}}$
$\Rightarrow \dfrac{dS}{dr} = \dfrac{\pi}{2}(4r^3 - 2 \times 243^2 r^{-3})(r^4 + 243^2 r^{-2})^{-\frac{1}{2}}$

Stationary values of S when $\dfrac{dS}{dr} = 0$:

$$\dfrac{\pi(4r^3 - 2 \times 243^2 r^{-3})}{2\sqrt{r^4 + 243^2 r^{-2}}} = 0$$

$4r^3 - 2 \times 243^2 r^{-3} = 0$

$r^6 = \dfrac{243^2}{2}$

$r = 5.56 \text{ (3SF)}$

$\therefore h = \dfrac{243}{r^2} = 7.86$

This pair of r and h values clearly produces a minimum value for the surface area rather than a maximum, since the surface area can be made arbitrarily large by taking sufficiently small or large values of r.

11 a The triangle has side lengths b, $10 - \dfrac{b}{2}$, $10 - \dfrac{b}{2}$

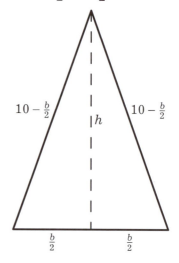

Figure 20D.11

By Pythagoras' Theorem,

$$h = \sqrt{\left(10 - \dfrac{b}{2}\right)^2 - \left(\dfrac{b}{2}\right)^2}$$
$$= \sqrt{100 - 10b}$$

Area of triangle is

$$A = \frac{bh}{2} = \frac{b}{2}\sqrt{100-10b}$$

b $A = \frac{b}{2}(100-10b)^{\frac{1}{2}}$

$$\frac{dA}{db} = \frac{1}{2}(100-10b)^{\frac{1}{2}} + \frac{b}{4}(100-10b)^{-\frac{1}{2}} \times (-10)$$

$$= \frac{1}{2\sqrt{100-10b}}(100-10b-5b)$$

$$= \frac{1}{2\sqrt{100-10b}}(100-15b)$$

Stationary value of A when $\frac{dA}{db} = 0$:

$$\frac{1}{2\sqrt{100-10b}}(100-15b) = 0$$

$100 - 15b = 0$

$15b = 100$

$b = \frac{20}{3}$

That is, the base length is one third of the perimeter of the isosceles triangle, so the triangle is equilateral.

This clearly gives a maximum value rather than a minimum, since the area can be made arbitrarily small by taking b small enough or close enough to 10.

12 Let the two numbers be x and y.

$x^2 + y^2 = a$

$\Rightarrow y = \sqrt{a-x^2}$

The product P is given by

$$P = xy = x\sqrt{a-x^2} = x(a-x^2)^{\frac{1}{2}}$$

$$\frac{dP}{dx} = (a-x^2)^{\frac{1}{2}} - x^2(a-x^2)^{-\frac{1}{2}}$$

$$= \frac{1}{\sqrt{a-x^2}}(a-x^2-x^2)$$

$$= \frac{a-2x^2}{\sqrt{a-x^2}}$$

Stationary values of P when $\frac{dP}{dx} = 0$:

$$\frac{a-2x^2}{\sqrt{a-x^2}} = 0$$

$a - 2x^2 = 0$

$2x^2 = a$

$$\therefore x = \sqrt{\frac{a}{2}} \text{ (since } x > 0)$$

$y^2 = a - x^2$

$$= a - \frac{a}{2} = \frac{a}{2}$$

$$y = \sqrt{\frac{a}{2}} \text{ (since } y > 0)$$

Hence $x = y$ for the stationary point of P.

$$\frac{d^2P}{dx^2} = \frac{-4x(a-x^2)^{\frac{1}{2}} + x(a-2x^2)(a-x^2)^{-\frac{1}{2}}}{a-x^2}$$

$$= \frac{-4x(a-x^2) + x(a-2x^2)}{(a-x^2)^{\frac{3}{2}}}$$

$$\frac{d^2P}{dx^2}\left(\sqrt{\frac{a}{2}}\right) = \frac{-4\sqrt{\frac{a}{2}}\left(\frac{a}{2}\right)}{\left(\frac{a}{2}\right)^{\frac{3}{2}}} = -4 < 0 \Rightarrow \text{local maximum}$$

Hence P is at a maximum when $x = y$.

13

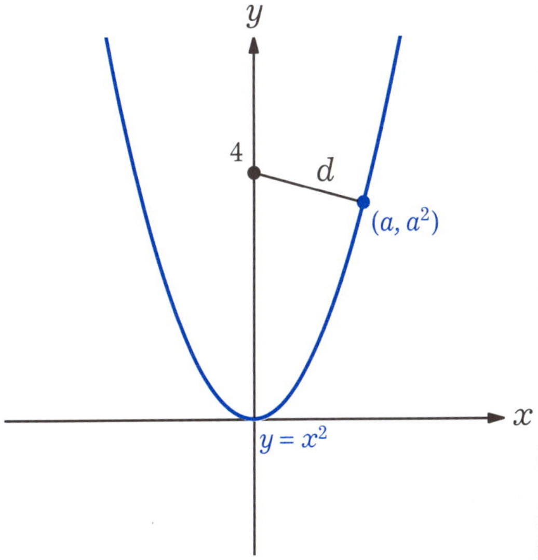

Figure 20D.13

The distance d from point (a, a^2) to $(0, 4)$ is given by

$d^2 = a^2 + (4-a^2)^2$

$= 16 - 7a^2 + a^4$

$\dfrac{d}{da}d^2 = -14a + 4a^3$

Stationary value of d^2 when $\dfrac{d}{da}d^2 = 0$:

$-14a + 4a^3 = 0$

$2a(2a^2 - 7) = 0$

$2a = 0$ or $a^2 = \dfrac{7}{2}$

$\therefore a = 0$ or $a = \sqrt{\dfrac{7}{2}}$ (since $a \geq 0$)

$\dfrac{d^2}{da^2}d^2 = -14 + 12a^2$

$\dfrac{d^2}{da^2}d^2(0) = -14 < 0 \Rightarrow a = 0$ represents a local maximum for d^2

$\dfrac{d^2}{da^2}d^2\left(\sqrt{\dfrac{7}{2}}\right) = 28 > 0 \Rightarrow a = \sqrt{\dfrac{7}{2}}$ represents a local minimum for d^2

$\therefore$ the point closest to $(0, 4)$ on the curve for $x \geq 0$ is $\left(\sqrt{\dfrac{7}{2}}, \dfrac{7}{2}\right)$

Mixed examination practice 20

Short questions

1 $y = ax - x^2 = x(a-x)$

Roots at $x = 0$ and $x = a$

$V = \pi \int_0^a y^2 \, dx$

$= \pi \int_0^a (ax - x^2)^2 \, dx$

$= \pi \int_0^a (a^2 x^2 - 2ax^3 + x^4) \, dx$

$= \pi \left[\dfrac{a^2 x^3}{3} - \dfrac{ax^4}{2} + \dfrac{x^5}{5} \right]_0^a$

$= \pi a^5 \left(\dfrac{1}{3} - \dfrac{1}{2} + \dfrac{1}{5} \right)$

$= \dfrac{\pi a^5}{30}$

2 $v = t^3 - 6t^2 + 8t$

a $s(5) = \int_0^5 v \, dt$

$= \left[\dfrac{t^4}{4} - 2t^3 + 4t^2 \right]_0^5$

$= \dfrac{625}{4} - 250 + 100$

$= \dfrac{25}{4} = 6.25 \text{ m}$

b $x(5) = \int_0^5 |v| \, dt$

Determine the boundaries of positive and negative velocity:

$v = 0 \Rightarrow t(t-2)(t-4) = 0$

$v \geq 0$ in $[0, 2] \cup [4, \infty[$ and $v \leq 0$ in $]-\infty, 0] \cup [2, 4]$

$\therefore x(5) = \int_0^2 v \, dt + \int_4^5 v \, dt - \int_2^4 v \, dt$

$= \int_0^5 v \, dt - 2\int_2^4 v \, dt$

$= \left[\dfrac{t^4}{4} - 2t^3 + 4t^2 \right]_0^5 - 2\left[\dfrac{t^4}{4} - 2t^3 + 4t^2 \right]_2^4$

$= \dfrac{25}{4} - 2((64 - 128 + 64) - (4 - 16 + 16))$

$= \dfrac{25}{4} + 8$

$= 14.25 \text{ m}$

3 Let the two positive numbers be x and y.

$x^2 + y^2 = 32 \Rightarrow y = \sqrt{32-x^2}$

Sum of the values $S = x + y = x + \sqrt{32-x^2}$

Maximum S at a stationary point on the curve of S:

$\dfrac{dS}{dx} = 0$

$1 - \dfrac{x}{\sqrt{32-x^2}} = 0$

$x = \sqrt{32-x^2}$

$x^2 = 32 - x^2$

$x^2 = 16$

$\therefore x = y = 4$

(choose positive root as $x, y > 0$)

There is only one stationary point, and since $S|_{x=y=4} = 8 > S|_{x=0, y=\sqrt{32}} = \dfrac{8}{\sqrt{2}}$, it must be a maximum.

4 Let:

r = radius (cm)

A = area (cm^2)

t = time (seconds)

Rate of increase of radius is inversely proportional to square root of radius:

$\dfrac{dr}{dt} = \dfrac{k}{\sqrt{r}}$

Initial values are $r|_{t=0} = 4$, $\dfrac{dr}{dt}\bigg|_{t=0} = 2$

$\therefore 2 = \dfrac{k}{\sqrt{4}}$

$k = 4$

so $\dfrac{dr}{dt} = \dfrac{4}{\sqrt{r}}$

$A = \pi r^2$

$\Rightarrow \dfrac{dA}{dt} = 2\pi r \dfrac{dr}{dt} = 8\pi \sqrt{r}$

$\dfrac{dA}{dt} = 115$

$\Rightarrow 8\pi \sqrt{r} = 115$

$\Rightarrow r = \left(\dfrac{115}{8\pi}\right)^2 = 20.9\,\text{cm}\,(3\text{SF})$

5 $s = 3e^{-t}\sin t$

a $v = \dfrac{ds}{dt}$

$= 3e^{-t}\cos t - 3e^{-t}\sin t$

$= 3e^{-t}(\cos t - \sin t)$

$\therefore v(3) = -0.169$ from GDC

b

$v = 3e^{-t}(\cos t - \sin t)$

Figure 20MS.5

6 a $DE = AD = 100 - h$

Area $DBCE = \dfrac{1}{2}h \times (100 + 100 - h)$

$= 100h - \dfrac{h^2}{2}$

b $\dfrac{d}{dt}(\text{Area } DBCE) = (100 - h)\dfrac{dh}{dt} = -18$

$\Rightarrow \dfrac{dh}{dt} = \dfrac{18}{h-100}$

c $h = 100 - k\sqrt{t}$

$\Rightarrow \dfrac{dh}{dt} = -\dfrac{1}{2}\dfrac{k}{\sqrt{t}}$

But, from (b),

$$\frac{dh}{dt} = \frac{18}{h-100} = \frac{18}{-k\sqrt{t}}$$

$$\therefore \frac{18}{-k\sqrt{t}} = -\frac{k}{2\sqrt{t}}$$

$$k^2 = 36$$

$$k = 6$$

(select positive root since $h \leq 100$)

> **COMMENT**
>
> Alternatively, you could solve the differential equation directly using a rearrangement, as shown below.

$$\frac{dh}{dt} = \frac{18}{h-100} \Rightarrow \frac{dt}{dh} = \frac{h-100}{18}$$

Integrating with respect to h:

$$t = \frac{1}{18}\left(\frac{h^2}{2} - 100h\right) + c$$

$$= \frac{1}{36}\left(h^2 - 200h\right) + c$$

Completing the square:

$$t = \frac{1}{36}\left((100-h)^2 - 100^2\right) + c$$

(Note that we complete the square as $(100-h)^2$ instead of $(h-100)^2$ because $0 \leq h \leq 100$.)

Substituting the initial condition $h = 100$ when $t = 0$:

$$c = \frac{100^2}{36}$$

$$\therefore t = \frac{1}{36}(100-h)^2$$

$$36t = (100-h)^2$$

$$100 - h = 6\sqrt{t}$$

$$h = 100 - 6\sqrt{t}$$

$$\therefore k = 6$$

7 Let P be the position of the plane and let A be the point that is 3 km directly above the observer O. Let $AP = x$. We know $AO = 3$, $O\hat{P}A = \theta$.

By trigonometry, $\dfrac{3}{x} = \tan\theta \Rightarrow x = 3\cot\theta$

$$\frac{dx}{dt} = -3\csc^2\theta \frac{d\theta}{dt}$$

When $\theta = \dfrac{\pi}{3}$ and $\dfrac{d\theta}{dt} = \dfrac{1}{60}$,

$$\frac{dx}{dt} = -3\left(\frac{2}{\sqrt{3}}\right)^2 \times \frac{1}{60} = -\frac{1}{15}\,\text{km s}^{-1}$$

So the plane is approaching the point above the observer at $\dfrac{1}{15}$ km s^{-1}, or

$$\frac{1}{15} \times 3600 = 240\,\text{km h}^{-1}$$

8 a $P(d, k)$ lies on the curve $x = y^2$, so $k = \sqrt{d}$

The coordinates of P can also be written as (k^2, k)

$$\therefore SP = \sqrt{(k^2-1)^2 + k^2}$$

and hence $r = \dfrac{d}{\sqrt{(k^2-1)^2 + k^2}}$

$$= \frac{k^2}{\sqrt{k^4 - k^2 + 1}}$$

b Maximum value of r will occur at the same value of k as the maximum of r^2

$$r^2 = \frac{k^4}{k^4 - k^2 + 1}$$

$$\Rightarrow \frac{dr^2}{dk} = \frac{4k^3(k^4-k^2+1) - k^4(4k^3-2k)}{(k^4-k^2+1)^2}$$

$$= \frac{-2k^5 + 4k^3}{(k^4-k^2+1)^2}$$

$$\frac{dr^2}{dk} = 0 \Rightarrow 4k^3 - 2k^5 = 0$$

$$\Rightarrow 2k^3(2-k^2) = 0$$

$$\Rightarrow k = 0 \text{ or } k = \sqrt{2} \text{ (since } k \geq 0\text{)}$$

$k = 0$ corresponds to minimum $r = 0$

$k = \sqrt{2}$ corresponds to maximum

$$r = \frac{2}{\sqrt{4-2+1}} = \frac{2}{\sqrt{3}} = \frac{2\sqrt{3}}{3}$$

9 $\dfrac{dv}{dt} = \dfrac{v^2 + 4}{2v}$

> **COMMENT**
>
> There are several approaches available. Here are a few suggestions.
>
> Method 1: Use the fact that $\dfrac{dv}{dt} = 1 \div \left(\dfrac{dt}{dv}\right)$ to rearrange the equation.
>
> Method 2: Try the proposed solution and show that it satisfies the initial condition and differential equation. This may seem like a cheat, but since there can only be one function which will satisfy a differential equation together with a point condition, it is sufficient.
>
> Method 3: Intuit a substitution – the solution gives a hint as to a suitable one.

<u>Method 1</u>: rearrange the derivative and change the variable of interest.

$$\frac{dt}{dv} = \frac{2v}{v^2 + 4}$$

Integrating with respect to v:

$$t = \ln|v^2 + 4| + c$$

$v^2 + 4$ is always positive, so the modulus signs can be discarded.

$v(0) = 3 \Rightarrow 0 = \ln(13) + c$

$\Rightarrow c = -\ln(13)$

$$\therefore t = \ln\left(\frac{v^2 + 4}{13}\right)$$

$13e^t = v^2 + 4$

$v^2 = 13e^t - 4$

<u>Method 2</u>: validate the solution.

Try $v^2 = 13e^t - 4$

Then $v = \sqrt{13e^t - 4}$

$$\frac{dv}{dt} = \frac{1}{2} \times \frac{13e^t}{\sqrt{13e^t - 4}}$$

$$= \frac{1}{2} \times \frac{v^2 + 4}{v}$$

$$= \frac{v^2 + 4}{2v}$$

If $v^2 = 13e^t - 4$ then
$v(0) = \sqrt{13e^0 - 4} = \sqrt{9} = 3$

Thus $v^2 = 13e^t - 4$ satisfies both the first-order differential equation and the initial condition; therefore, by a uniqueness theorem for first-order differential equations, this must be the solution.

<u>Method 3</u>: substitution.

Try the substitution $v^2 + 4 = e^u$

Implicit differentiation gives

$$2v\frac{dv}{dt} = e^u \frac{du}{dt}$$

Comparing this with the differential equation $\dfrac{dv}{dt} = \dfrac{v^2 + 4}{2v}$:

$$e^u \frac{du}{dt} = v^2 + 4 = e^u$$

$$\therefore \frac{du}{dt} = 1$$

$\Rightarrow u = t + c$

$e^u = e^{t+c} = e^c \times e^t$

$v^2 + 4 = e^c \times e^t$

$v(0) = 3 \Rightarrow 9 + 4 = e^c$

$\therefore v^2 + 4 = 13e^t$

and so $v^2 = 13e^t - 4$

Long questions

1 a i Perimeter of square $= 4x$

ii Perimeter of circle $= 2\pi y$

b $4x + 2\pi y = 8$

$$\Rightarrow x = \frac{8 - 2\pi y}{4} = 2 - \frac{\pi y}{2}$$

c Total area $A = x^2 + \pi y^2$

$$= \left(2 - \frac{\pi y}{2}\right)^2 + \pi y^2$$

$$= 4 - 2\pi y + \frac{\pi^2 y^2}{4} + \pi y^2$$

$$= 4 - 2\pi y + \frac{\pi}{4}(\pi + 4) y^2$$

d The formula for total area is a positive quadratic, so its minimum value will lie at the stationary point:

$$\frac{dA}{dy} = -2\pi + \frac{\pi}{2}(\pi + 4) y = 0$$

$$\frac{y}{2}(\pi + 4) = 2$$

$$\Rightarrow y = \frac{4}{\pi + 4}$$

Let c be the percentage of the wire that is used for the circle; then

$$c = \frac{2\pi y}{8} \times 100$$

$$= 25\pi y$$

$$= 25\pi \times \frac{4}{\pi + 4}$$

$$= 44.0 \text{ (3SF)}$$

∴ 44.0% of the wire is used for the circle.

2 $y_1 = x^2 - 8x + 12 = (x - 6)(x - 2)$

$y_2 = 12 + x - x^2 = -(x + 3)(x - 4)$

$0 \leq x \leq 5$

a Intersections occur where $y_1 = y_2$:

$x^2 - 8x + 12 = 12 + x - x^2$

$2x^2 - 9x = 0$

$x(2x - 9) = 0$

$x = 0$ or $x = \frac{9}{2}$

∴ the coordinates of the intersection points are $(0, 12)$ and $\left(\frac{9}{2}, -\frac{15}{4}\right)$

b Vertical distance between the curves is $|y_1 - y_2|$

$y_1 - y_2 = 2x^2 - 9x = x(2x - 9)$

The vertex of this quadratic lies halfway between the roots, at $x = \frac{9}{4}$

$$|y_1 - y_2|\big|_{x=9/4} = \frac{81}{16}$$

The boundaries are at $x = 0$, where there is an intersection, and $x = 5$.

$$|y_1 - y_2|\big|_{x=5} = 5 < \frac{81}{16}$$

Hence the greatest vertical distance between the curves is $\frac{81}{16}$

c

> **COMMENT**
>
> The full and correct solution is given below. Students may be forgiven for presenting the standard simple answer $V = \pi \int_0^{9/2} y_2^2 - y_1^2 \, dx$, which results in a value $V = 787$. In fact, because the enclosed region crosses the x-axis, this does not give the correct volume, and the solution requires consideration of each of four regions as the area rotates. Because of this issue, and the large amount of repetitive detail work required, it is unlikely that a question of this sort would be found in a real examination paper.

i The curves and the area they enclose are shown in the diagram below.

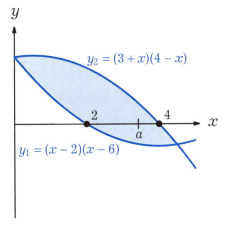

Figure 20ML.2

The volume of revolution is complicated; let a be the value such that $y_1(a) = -y_2(a)$.

Within the interval of intersection $[0, 4.5]$, y_1 has a root at 2 and y_2 has a root at 4.

Therefore the volume of revolution consists of four parts:

$V = V_1 + V_2 + V_3 + V_4$, where

$V_1 = \pi \int_0^2 y_2^2 - y_1^2 \, dx$

$V_2 = \pi \int_2^a y_2^2 \, dx$

$V_3 = \pi \int_a^4 y_1^2 \, dx$

$V_4 = \pi \int_4^{4.5} y_1^2 - y_2^2 \, dx$

$y_1^2 = (x^2 - 8x + 12)^2$
$= x^4 - 16x^3 + 24x^2 + 64x^2 - 192x + 144$
$= x^4 - 16x^3 + 88x^2 - 192x + 144$

$y_2^2 = (12 + x - x^2)^2$
$= x^4 - 16x^3 + 88x^2 - 192x + 144$
$= x^4 - 2x^3 - 23x^2 + 24x + 144$

$\therefore y_2^2 - y_1^2 = 14x^3 - 111x^2 + 216x$

Solving for a:

$y_1(a) = a^2 - 8a + 12$

$y_2(a) = 12 + a - a^2$

$y_1(a) = -y_2(a) \Rightarrow a^2 - 8a + 12 = a^2 - a - 12$
$\Rightarrow 7a = 24$
$\Rightarrow a = \dfrac{24}{7}$

ii Substituting the expressions for y_1^2, y_2^2 and a into the formulae for the volume parts and using GDC to calculate the integrals:

$V_1 = \pi \int_0^2 14x^3 - 111x^2 + 216x \, dx$

$= \pi \left[\dfrac{7}{2}x^4 - 37x^3 + 108x^2 \right]_0^2$

$= 192\pi$

$V_2 = \pi \int_2^{24/7} x^4 - 2x^3 - 23x^2 + 24x + 144 \, dx$

$= \pi \left[\dfrac{x^5}{5} - \dfrac{x^4}{2} - \dfrac{23x^3}{3} + 12x^2 + 144x \right]_2^{24/7}$

$= \dfrac{3\,952\,000}{50\,421} \pi$

$V_3 = \pi \int_{24/7}^4 x^4 - 16x^3 + 88x^2 - 192x + 144 \, dx$

$= \pi \left[\dfrac{x^5}{5} - 4x^4 + \dfrac{88x^3}{3} - 96x^2 + 144x \right]_{24/7}^4$

$= \dfrac{2\,182\,592}{252\,105} \pi$

$V_4 = \pi \int_4^{9/2} -14x^3 + 111x^2 - 216x \, dx$

$= \pi \left[-\dfrac{7}{2}x^4 + 37x^3 - 108x^2 \right]_4^{9/2}$

$= \dfrac{173}{32} \pi$

$V = V_1 + V_2 + V_3 + V_4 = 894$
(from GDC, to nearest integer)

> **COMMENT**
>
> For completeness, shown below is what would be the standard working for such a question if the enclosed area did **not** cross the axis about which it was to be rotated!

i

$$V = \pi \int_0^{9/2} y_2^2 - y_1^2 \, dx$$

$$= \pi \int_0^{9/2} (12 + x - x^2)^2 - (x^2 - 8x + 12)^2 \, dx$$

$$= \pi \int_0^{9/2} (144 + 24x - 24x^2 + x^2 - 2x^3 + x^4)$$
$$- (x^4 - 16x^3 + 24x^2 + 64x^2 - 192x + 144) \, dx$$

$$= \pi \int_0^{9/2} 216x - 111x^2 + 14x^3 \, dx$$

ii $V = \pi \left[108x^2 - 37x^3 + \dfrac{7}{2}x^4 \right]_0^{9/2}$

$$= \dfrac{8019\pi}{32}$$

$$= 787 \text{ (from GDC, to nearest integer)}$$

3 a Let ϕ_b be the elevation from the viewer's eyes to the base of the painting. Then

$$\phi_b = \arctan\left(\dfrac{2 - 1.5}{x}\right) = \arctan\left(\dfrac{0.5}{x}\right)$$

Let ϕ_t be the elevation from the viewer's eyes to the top of the painting. Then

$$\phi_t = \arctan\left(\dfrac{2.5}{x}\right)$$

$$\theta = \phi_t - \phi_b$$

$$= \arctan\left(\dfrac{2.5}{x}\right) - \arctan\left(\dfrac{0.5}{x}\right)$$

b To maximise θ, set its derivative with respect to x equal to zero.

$$\dfrac{d\theta}{dx} = \left(-\dfrac{2.5}{x^2}\right)\dfrac{1}{1 + \left(\dfrac{2.5}{x}\right)^2} - \left(-\dfrac{0.5}{x^2}\right)\dfrac{1}{1 + \left(\dfrac{0.5}{x}\right)^2}$$

$$= \dfrac{0.5}{x^2 + 0.25} - \dfrac{2.5}{x^2 + 6.25}$$

$$= \dfrac{0.5(x^2 + 6.25) - 2.5(x^2 + 0.25)}{(x^2 + 0.25)(x^2 + 6.25)}$$

$$= \dfrac{-2x^2 + 2.5}{(x^2 + 0.25)(x^2 + 6.25)}$$

$$\dfrac{d\theta}{dx} = 0 \Rightarrow x^2 = 1.25 = \dfrac{5}{4}$$

$$\Rightarrow x = \dfrac{\sqrt{5}}{2}$$

That this is a maximum is clear from the context:

Setting $x = 0$ reduces θ to zero, and as $x \to \infty$, $\theta \to 0$.

θ is bounded (it cannot be greater than π, given the context).

Hence there must be a maximum value of θ for some $x > 0$, and since there is only one stationary point, this must be the maximum.

4 a Integrating by parts, set

$$u = \ln x \Rightarrow \dfrac{du}{dx} = \dfrac{1}{x}$$

$$\dfrac{dv}{dx} = 1 \Rightarrow v = x$$

$$\int u \dfrac{dv}{dx} \, dx = uv - \int v \dfrac{du}{dx} \, dx$$

$$\therefore \int 1 \times \ln x \, dx = x \ln x - \int \dfrac{x}{x} \, dx$$

$$= x \ln x - x + c$$

b $v_1(t) = 3\ln(t+1)$, $s_1(0) = 0$

i $a_1 = \dfrac{dv_1}{dt} = \dfrac{3}{t+1}$

$a_1(5) = \dfrac{1}{2} \text{ms}^{-2}$

ii $s_1 = \int v_1(t)\,dt$

$= \int 3\ln(t+1)\,dt$

Substitute $u = t+1 \Rightarrow dt = du$

$s_1 = \int 3\ln u\, du$

$= 3(u\ln u - u) + c$

$= 3(t+1)(\ln(t+1) - 1) + c$

$s_1(0) = 0 \Rightarrow 3(-1) + c = 0$

$\Rightarrow c = 3$

$\therefore s_1(t) = 3(t+1)\ln(t+1) - 3t$

iii $v_1(0) = 0$, $v_1 > 0$ for $t > 0$

So the distance travelled is the same as the displacement.

Distance travelled in the first 5 seconds is

$s_1(5) = 18\ln 6 - 15 = 17.3\,\text{m}$ (3 SF)

c $v_2 = 8 - t$, $s_2(0) = 0$

i $v_1(t)$ is increasing and $v_2(t)$ is decreasing.
If $v_2(t) \geq v_1(t)$ for $0 \leq t \leq a$, then $v_1(a) = v_2(a)$:

$3\ln(a+1) = 8 - a$

From GDC: $a = 3.49$

ii Speed of the second object is $|v_2|$.
Since $v_2(t)$ is linear, this is maximal at one end of the domain considered.

$|v_2(0)| = 8$

$|v_2(20)| = 12$

Greatest speed in the first 20 seconds is $12\,\text{m s}^{-1}$.

iii $s_2 = \int v_2(t)\,dt$

$= 8t - \dfrac{t^2}{2} + c$

$= \dfrac{1}{2}t(16 - t) + c$

$s_2(0) = 0 \Rightarrow c = 0$

$\therefore s_2(t) = \dfrac{1}{2}t(16 - t)$

$v_2(8) = 0$ and $v_2 < 0$ for $t > 8$.

$\therefore$ distance travelled by the second object is

$x_2(t) = \begin{cases} s_2(t) & 0 \leq t \leq 8 \\ 2s_2(8) - s_2(t) & t \geq 8 \end{cases}$

Recall that distance travelled by the first object is

$x_1(t) = s_1(t) = 3(t+1)\ln(t+1) - 3t$

for all $t \geq 0$

Case 1: equal distance in the first 8 seconds

$s_1(t) = x_2(t)$ for $0 \leq t \leq 8$

$\Rightarrow 3(t+1)\ln(t+1) - 3t = \dfrac{1}{2}t(16 - t)$

From GDC: $t = 7.47$

Case 2: equal distance at a time $t > 8$

$s_1(t) = x_2(t)$ for $t > 8$

$\Rightarrow 3(t+1)\ln(t+1) - 3t = 64 - \dfrac{1}{2}t(16 - t)$

From GDC: $t = 25.4$

The objects have each travelled no distance (and are at the same point) at $t = 0\,\text{s}$.

The objects have travelled the same distance (and are at the same point) at $t = 7.47\,\text{s}$.

The objects have travelled the same distance (and are at different points) at $t = 25.4\,\text{s}$.

20 Further applications of calculus

5 a Let $AB = x$. From the given information:

$x(0) = 20$, $\dfrac{dx}{dt} = 3$

So $x = 3t + c$

and $x(0) = 20 \Rightarrow c = 20$

$\therefore x = 3t + 20$

b From the given information:

$h(0) = 30$, $\dfrac{dh}{dt} = -2$

So $h = -2t + k$

and $h(0) = 30 \Rightarrow k = 30$

$\therefore h = 30 - 2t$

c Area $ABC = \dfrac{1}{2}xh$

d $\dfrac{d\,\text{Area}}{dt} = \dfrac{1}{2}x\dfrac{dh}{dt} + \dfrac{1}{2}h\dfrac{dx}{dt}$

$= \dfrac{1}{2}(3t+20)(-2) + \dfrac{1}{2}(30-2t)(3)$

$= 25 - 6t$

d $x = h = 26 \Rightarrow t = 2$

$\left.\dfrac{d\,\text{Area}}{dt}\right|_{t=2} = 25 - 12 = 13\,\text{cm}^2\,\text{s}^{-1}$

6 a Set:

$u = (\ln x)^2 \Rightarrow \dfrac{du}{dx} = \dfrac{2\ln x}{x}$

$\dfrac{dv}{dx} = 1 \Rightarrow v = x$

$\int u\dfrac{dv}{dx}dx = uv - \int v\dfrac{du}{dx}dx$

$\therefore \int 1 \times (\ln x)^2\,dx = x(\ln x)^2 - \int \dfrac{2x}{x}\ln x\,dx$

$= x(\ln x)^2 - 2\int \ln x\,dx$

Now set:

$w = \ln x \Rightarrow \dfrac{dw}{dx} = \dfrac{1}{x}$

$\dfrac{dv}{dx} = 1 \Rightarrow v = x$

$\int w\dfrac{dv}{dx}dx = wv - \int v\dfrac{dw}{dx}dx$

$\therefore \int 1 \times \ln x\,dx = x\ln x - \int 1\,dx$

$= x\ln x - x + c$

Hence

$\int (\ln x)^2\,dx = x(\ln x)^2 - 2(x\ln x - x) + c$

$= x\left((\ln x)^2 - 2\ln x + 2\right) + c$

b Upper right vertex of the shaded rectangle is $(1, e)$; the y-intercept of the curve is $(0, 1)$.

$V_1 = \pi \int_0^1 y^2\,dx$

$= \pi \int_0^1 e^{2x}\,dx$

$= \pi \left[\dfrac{1}{2}e^{2x}\right]_0^1$

$= \dfrac{\pi}{2}(e^2 - 1)$

$V_2 = \pi \int_1^e x^2\,dy$

$= \pi \int_1^e (\ln y)^2\,dy$

$= \pi\left[y(\ln y)^2 - 2y\ln y + 2y\right]_1^e$

$= \pi(e - 2e + 2e) - \pi(0 - 0 + 2)$

$= \pi(e - 2)$

$\therefore \dfrac{V_1}{V_2} = \dfrac{\dfrac{\pi}{2}(e^2-1)}{\pi(e-2)} = \dfrac{e^2-1}{2(e-2)}$

7 $v_A = -\dfrac{1}{2}t^2 + 3t + \dfrac{3}{2}$, $v_B = e^{0.2t}$ for $0 \leq t \leq 9$

a v_A is a negative quadratic, so has a single stationary point which is a maximum.

$\dfrac{dv_A}{dt} = -t + 3 = 0 \Rightarrow t = 3$

$\therefore$ maximum value of v_A is $v_A(3) = 6$

b $a_B = \dfrac{dv_B}{dt} = 0.2e^{0.2t}$

$\Rightarrow a_B(4) = 0.2e^{0.8} = 0.445$

c $s_A = \int v_A \, dt$

$= \int -\dfrac{1}{2}t^2 + 3t + \dfrac{3}{2} \, dt$

$= -\dfrac{1}{6}t^3 + \dfrac{3}{2}t^2 + \dfrac{3}{2}t + c$

$s_A(0) = 0 \Rightarrow c = 0$

$\therefore s_A = -\dfrac{1}{6}t^3 + \dfrac{3}{2}t^2 + \dfrac{3}{2}t$

$s_B = \int v_B \, dt$

$= \int e^{0.2t} \, dt$

$= 5e^{0.2t} + k$

$s_B(0) = 5 \Rightarrow k = 0$

$\therefore s_B = 5e^{0.2t}$

d i

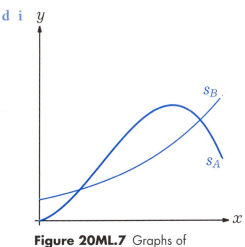

Figure 20ML.7 Graphs of
$s_A = -\dfrac{1}{6}t^3 + \dfrac{3}{2}t^2 + \dfrac{3}{2}t$ and $s_B = 5e^{0.2t}$

ii Solving $s_A = s_B$ on the GDC:
$t = 1.95, 7.81$

8 a By Pythagoras' Theorem, the walked distance is

$d = \sqrt{(10-x)^2 + 4^2} = \sqrt{16 + (10-x)^2}$

The time it takes to cycle x km at $10 \, \text{km h}^{-1}$ is $\dfrac{x}{10}$ hours.

The time it takes to walk d km at $5 \, \text{km h}^{-1}$ is $\dfrac{d}{5}$ hours.

$\therefore$ total time T is given by

$T = \dfrac{x}{10} + \dfrac{d}{5} = \dfrac{x}{10} + \dfrac{1}{5}\sqrt{16 + (10-x)^2}$

b i Minimal T will occur either when $\dfrac{dT}{dx} = 0$ or at the boundary values

$x = 0$ or $x = 10$

$T(0) = \dfrac{1}{5}\sqrt{116} = 2.15$ hours

$T(10) = 1.8$ hours

$\dfrac{dT}{dx} = \dfrac{1}{10} + \dfrac{1}{5} \times \dfrac{1}{2} \times \dfrac{-2(10-x)}{\sqrt{16+(10-x)^2}}$

$= \dfrac{1}{10}\left(1 - \dfrac{2(10-x)}{\sqrt{16+(10-x)^2}}\right)$

$\dfrac{dT}{dx} = 0 \Rightarrow 1 - 2\dfrac{(10-x)}{\sqrt{16+(10-x)^2}} = 0$

$\Rightarrow 2(10-x) = \sqrt{16+(10-x)^2}$

$\Rightarrow 4(10-x)^2 = 16 + (10-x)^2$

$\Rightarrow 3(10-x)^2 = 16$

$\Rightarrow 10 - x = \pm\sqrt{\dfrac{16}{3}}$

$\Rightarrow x = 10 - \dfrac{4}{\sqrt{3}}$

(choose this root because $x \leq 10$)

$T\left(10 - \dfrac{4}{\sqrt{3}}\right) = 1 - \dfrac{4}{10\sqrt{3}} + \dfrac{1}{5}\sqrt{16 + \dfrac{16}{3}}$

$= 1 - \dfrac{4}{10\sqrt{3}} + \dfrac{8}{5\sqrt{3}}$

$= 1 + \dfrac{2\sqrt{3}}{5} = 1.69$ hours

Therefore the minimal T occurs at the stationary point, when $3(10-x)^2 = 16$.

ii John should cycle

$x = 10 - \dfrac{4}{\sqrt{3}} = 7.69$ km

9 a i $\dfrac{9}{AX} = \sin\theta \Rightarrow AX = 9\csc\theta$

$\dfrac{\sqrt{3}}{XB} = \cos\theta \Rightarrow XB = \sqrt{3}\sec\theta$

ii $AB = AX + XB = 9\csc\theta + \sqrt{3}\sec\theta$

Minimum AB occurs when its derivative with respect to θ is zero:

$\dfrac{dAB}{d\theta} = -9\csc\theta\cot\theta + \sqrt{3}\sec\theta\tan\theta = 0$

$9\dfrac{\cos\theta}{\sin^2\theta} = \sqrt{3}\dfrac{\sin\theta}{\cos^2\theta}$

$3\sqrt{3} = \tan^3\theta$

$\tan\theta = \sqrt{3}$

$\Rightarrow \theta = \dfrac{\pi}{3}$

$AB|_{\theta=\pi/3} = 9 \times \dfrac{2}{\sqrt{3}} + \sqrt{3} \times 2$

$= 8\sqrt{3} = 13.9\,\text{m}$

b The longest possible ladder (assuming it is always carried horizontally) that can fit around the corner is 13.9 metres long, since it must be able to pass the tightest position, determined in part (a).

21 Summarising data

Exercise 21A

1
a Discrete – takes integer values only

b Discrete – takes values from a finite list only

c Continuous – takes values in an interval

d Continuous – takes values in an interval

e Discrete – takes integer values only

f Discrete – takes integer values only

2
a Restricted, self-selected population sampled; only shoppers and shop workers will be interviewed.

b Restricted, self-selected population sampled; only non-truants will be in class to be interviewed.

c Restricted, self-selected population sampled; only internet users will be interviewed.

d Cannot sample two people from the same household, so it is not true that the selection of each individual is independent of the selection of others in the same household. Also, people in small households will have a greater chance of being selected than those in large households.

Exercise 21B

3
a There are 7 data values.

The median is 8

Lower data set 5, 5, 7 has median 5, so $Q_1 = 5$

Upper data set 9, x, 13 has median x, so $Q_3 = x$

$IQR = Q_3 - Q_1 = 7$

$x - 5 = 7$

$x = 12$

b From GDC, for the data set 5, 5, 7, 9, 12, 13: $\sigma = 2.92$

4 $\mu = \dfrac{2+5+9+x+y}{5} = 5$

$2+5+9+x+y = 25$

$\Rightarrow y = 9 - x$

$s_n^2 = \dfrac{2^2+5^2+9^2+x^2+y^2}{5} - 5^2 = 6$

$2^2+5^2+9^2+x^2+y^2 = 5(6+5^2) = 155$

$\Rightarrow x^2 + y^2 = 45$

$\therefore x^2 + (9-x)^2 = 45$

$2x^2 - 18x + 36 = 0$

$x^2 - 9x + 18 = 0$

$(x-3)(x-6) = 0$

$x = 3$ or 6

$\therefore y = 6$ or 3

So $xy = 18$

Note: This result can also be obtained by observing that the product of the roots of $x^2 - 9x + 18 = 0$ is 18.

5 Let x_i be the score in test i. Then $x_6 = 32$

$$\frac{1}{5}\sum_{i=1}^{5} x_i = 23$$

$$\sum_{i=1}^{5} x_i = 5 \times 23 = 115$$

$$\sum_{i=1}^{6} x_i = 115 + 32 = 147$$

$$\frac{1}{5}\sum_{i=1}^{5} x_i^2 - 23^2 = 4^2$$

$$\sum_{i=1}^{5} x_i^2 = 5(4^2 + 23^2) = 2725$$

$$\sum_{i=1}^{6} x_i^2 = 2725 + 32^2 = 3749$$

$$\therefore s_n = \sqrt{\frac{3749}{6} - \left(\frac{147}{6}\right)^2} = 4.96$$

6 $\frac{1}{15}\sum_{i=1}^{15} x_i = 600$

$$\Rightarrow \sum_{i=1}^{15} x_i = 15 \times 600 = 9000$$

$$\frac{1}{16}\sum_{i=1}^{16} x_i = 600.25$$

$$\Rightarrow \sum_{i=1}^{16} x_i = 600.25 \times 16 = 9604$$

$$\therefore x_{16} = 9604 - 9000 = 604$$

$$\frac{1}{15}\sum_{i=1}^{15} x_i^2 - 600^2 = 12^2$$

$$\sum_{i=1}^{15} x_i^2 = 15(12^2 + 600^2) = 5\,402\,160$$

$$\sum_{i=1}^{16} x_i^2 = 5\,402\,160 + 604^2 = 5\,766\,976$$

$$\therefore s_n = \sqrt{\frac{5\,766\,976}{16} - 600.25^2} = 11.7$$

7 The minimum variance is zero (if all the values are equal). This is the case when

$$\frac{1}{20}\sum_{i=1}^{20} x_i^2 - \left(\frac{1542}{20}\right)^2 = 0$$

$$\therefore \sum_{i=1}^{20} x_i^2 = \frac{1542^2}{20} = 118\,888.2$$

8 a Range $= x_{\max} - x_{\min}$

$x_{\max} \geq x_i$ for any i, by definition of the maximum.

$x_{\min} \leq \bar{x}$, since

$$\bar{x} = \frac{1}{n}\sum x_i \geq \frac{1}{n}\sum x_{\min} = x_{\min}$$

Hence $-x_{\min} \geq -\bar{x}$

and so $x_{\max} - x_{\min} \geq x_i + (-\bar{x})$

i.e. $x_{\max} - x_{\min} \geq x_i - \bar{x}$

b Both the range and s_n are non-negative, so squaring and taking square roots through inequalities is valid in the working below.

$$s_n^2 = \frac{1}{n}\sum_i (x_i - \bar{x})^2$$

$$s_n^2 \leq \frac{1}{n}\sum_i \text{range}^2 \quad \text{by (a)}$$

$$s_n^2 \leq \frac{1}{n}(n \times \text{range}^2)$$

$$s_n^2 \leq \text{range}^2$$

$$s_n \leq \text{range}$$

Exercise 21C

5 a There are 50 boxes, so the median will be the mean of the 25th and 26th boxes.

Median $= \frac{1+2}{2} = 1.5$ broken eggs

b Mean $= \dfrac{(17\times 0)+(1\times 8)+(2\times 7)+(3\times 7)+(4\times 6)+(5\times 5)+(6\times 0)}{50}$

$= 1.84$ broken eggs

6 $\bar{x} = \dfrac{20\times 12 + 40q + 8p}{12+q+8} = 32$

$240 + 40q + 8p = 640 + 32q$

$8(p+q) = 400$

$p+q = 50$

$q = 50 - p$

$s_n^2 = \dfrac{20^2 \times 12 + 40^2 \times q + p^2 \times 8}{12+q+8} - 32^2 = 136$

$4800 + 1600q + 8p^2 = 1160(20+q)$

$440q + 8p^2 = 18\,400$

$\therefore 440(50-p) + 8p^2 = 18\,400$

$8p^2 - 440p + 3600 = 0$

$p^2 - 55p + 450 = 0$

$(p-45)(p-10) = 0$

$\Rightarrow p = 45,\ q = 5\ \text{ or }\ p = 10,\ q = 40$

Mixed examination practice 21

Short questions

1 a $\bar{x} = \dfrac{1}{12}\sum_{i=1}^{12} x_i = \dfrac{49}{12} = 4.08$ minutes

b $s_n = \sqrt{\dfrac{1}{12}\sum_{i=1}^{12} x_i^2 - (\bar{x})^2} = \sqrt{\dfrac{305.7}{12} - \left(\dfrac{49}{12}\right)^2} = 2.97$ minutes

2 a Total frequency is $22 + 18 + x + y = 50$

So $x = 10 - y$

$\bar{\lambda} = \dfrac{620\times 22 + 660 \times 18 + 700x + 740y}{50} = 653.6$

$\Rightarrow 700x + 740y = 7160$

$\therefore 7000 - 700y + 740y = 7160$

$40y = 160$

$\Rightarrow y = 4,\ x = 6$

b Estimate of variance:

$$s_n^2 \approx \frac{620^2 \times 22 + 660^2 \times 18 + 700^2 \times 6 + 740^2 \times 4}{50} - 653.6^2 = 1367.04$$

∴ estimate of variance is $1370\,\text{nm}^2$ (3SF)

c The specific data values are not known, so the estimate was obtained by assuming that the data values in each group fall at the midpoint of the group interval.

3 $\dfrac{1}{15}\sum_{i=1}^{15} t_i = 0.2$

$\Rightarrow \sum_{i=1}^{15} t_i = 15 \times 0.2 = 3$

$t_{16} = 0.16$

$\therefore \sum_{i=1}^{16} t_i = 3 + 0.16 = 3.16$

Hence $\bar{t} = \dfrac{1}{16}\sum_{i=1}^{16} t_i = \dfrac{3.16}{16} = 0.1975$

The new mean is 0.1975 seconds.

$\dfrac{1}{15}\sum_{i=1}^{15} t_i^2 - 0.2^2 = 0.0025$

$\Rightarrow \sum_{i=1}^{15} t_i^2 = 15(0.0025 + 0.04) = 0.6375$

So $\sum_{i=1}^{16} t_i^2 = 0.6375 + 0.16^2 = 0.6631$

$\therefore s_n = \sqrt{\dfrac{0.6631}{16} - 0.1975^2} = 0.0494$ (3SF)

The new standard deviation is 0.0494 seconds.

4 Let the two data items be x and y, with $x \leq y$.

range $= y - x$

$s_n^2 = k = \dfrac{x^2 + y^2}{2} - \left(\dfrac{x+y}{2}\right)^2$

$= \dfrac{1}{4}(2x^2 + 2y^2 - x^2 - 2xy - y^2)$

$= \dfrac{1}{4}(x^2 - 2xy + y^2)$

$= \dfrac{1}{4}(y - x)^2$

$\therefore k = \dfrac{1}{4} \times \text{range}^2$

$\Rightarrow \text{range} = 2\sqrt{k}$

Long questions

1 a The median will be at cumulative frequency 15, which corresponds to 11 cm.

b From the cumulative frequency diagram:

TABLE 21ML.1

Height (h)	Frequency
$0 < h \leq 5$	4
$5 < h \leq 10$	$13 - 4 = 9$
$10 < h \leq 15$	$21 - 13 = 8$
$15 < h \leq 20$	$26 - 21 = 5$
$20 < h \leq 25$	$30 - 26 = 4$

c Estimate of the mean uses midpoint values of the intervals:

Estimate of mean

$$= \frac{(2.5 \times 4)+(7.5 \times 9)+(12.5 \times 8)+(17.5 \times 5)+(22.5 \times 4)}{30} = \frac{355}{30} = 11.8 \text{ cm (3SF)}$$

2 a The data is recorded to the nearest cm. The first bar therefore represents actual lengths in the interval $[39.5, 49.5[$, since all such lengths will round to integers 40 through to 49. The histogram is correctly plotted to illustrate the distribution of the actual values, rather than the recorded data.

b Reading off values from the histogram:

TABLE 21ML.2

Length (cm)	Frequency
40–49	12
50–59	18
60–69	6
70–79	0
80–89	1

c To obtain estimates, use the central value in each group interval, e.g. centre of $[39.5, 49.5[$ is 44.5

Estimate of mean:

$$\bar{x} \approx \frac{44.5 \times 12 + 54.5 \times 18 + 64.5 \times 6 + 74.5 \times 0 + 84.5 \times 1}{12 + 18 + 6 + 0 + 1}$$

$$= 53.7 \text{ (3SF)}$$

Estimate of variance:

$$S_n \approx \sqrt{\frac{44.5^2 \times 12 + 54.5^2 \times 18 + 64.5^2 \times 6 + 74.5^2 \times 0 + 84.5^2 \times 1}{12+18+6+0+1} - \bar{x}^2} = 8.50 \text{ (3SF)}$$

d The following are some reasons that the mean arm length for the whole population may be different from the estimated mean from this sample of 37 students:

- The classroom is likely to contain children in a particular age bracket, while the population will be balanced across all ages of children, with corresponding differences in growth.
- If there are geographic or social variations in growth across the country, the classroom is likely to be biased according to its region or the socioeconomic status of its catchment.
- Even if both of the above (and equivalent) reasons can be discounted, natural variation makes it very unlikely that a sample of 37 would exactly mirror the population.

3 a Mean $= \dfrac{1 \times a + 2 \times b}{a+b} = \dfrac{a+2b}{a+b} = 1 + \dfrac{b}{a+b}$

$$s_n^2 = \frac{1^2 \times a + 2^2 \times b}{a+b} - \left(\frac{a+2b}{a+b}\right)^2 = \frac{(a+4b)(a+b)-(a+2b)^2}{(a+b)^2}$$

$$= \frac{a^2+5ab+4b^2-(a^2+4ab+4b^2)}{(a+b)^2} = \frac{ab}{(a+b)^2}$$

b If the mean equals the variance, then

$$\frac{a+2b}{a+b} = \frac{ab}{(a+b)^2}$$

$(a+b)(a+2b) = ab$

$a^2 + 3ab + 4b^2 = ab$

$a^2 + 2ab + 4b^2 = 0$

Solving for a in terms of b using the quadratic formula:

$$a = \frac{-2b \pm \sqrt{(2b)^2 - 4 \times 4b^2}}{2}$$

$= -b \pm \sqrt{-3b^2}$

∴ there are no real values of a and b that satisfy the condition for the mean to equal the variance.

c $a = 3b$

$\Rightarrow$ mean $= \dfrac{5b}{4b} = 1.25$

$$S_n = \sqrt{\frac{3b^2}{(4b)^2}} = \frac{\sqrt{3b}}{4b} = \frac{\sqrt{3}}{4} = 0.433$$

22 Probability

Exercise 22A

4 For a fair six-sided die, P(Odd) = 0.5 and P(Prime) = 0.5

So P(Odd) + P(Prime) = 1

But Odd is not the complement of Prime (3 and 5 are in both, 4 and 6 are in neither).

7 Draw out the table of the event space:

TABLE 22A.7

Die B \ Die A	1	2	3	4	5	6
1	1	1	1	1	1	1
2	1	2	1	2	1	2
3	1	1	3	1	1	3
4	1	2	1	4	1	2
5	1	1	1	1	5	1
6	1	2	3	2	1	6

$$P(\text{HCF}=1) = \frac{23}{36}$$

So expected number of '1' scores in 180 throws is

$$180 \times \frac{23}{36} = 115$$

8 Cases for score less than 6 (i.e. 5 or less):

Total 3: (1,1,1)

Total 4: (1,1,2), (1,2,1), (2,1,1)

Total 5: (1,1,3), (1,2,2), (1,3,1), (2,1,2), (2,2,1), (3,1,1)

This gives 10 cases out of the total $6^3 = 216$ possible outcomes.

$$\therefore P(\text{score} < 6) = \frac{10}{216} = \frac{5}{108}$$

Exercise 22B

4

> **COMMENT**
> A Venn diagram is sufficient working; the calculations needed to populate the diagram are given first but do not need to be explicitly shown in an examination answer. Below the diagram a stand-alone algebraic method is given, which could be used instead of a diagram.

$$P(S \cap L') = P(S) - P(S \cap L)$$
$$= 60\% - 50\%$$
$$= 10\%$$

$$P(L \cap S') = P(L) - P(L \cap S)$$
$$= 85\% - 50\%$$
$$= 35\%$$

$$P(L' \cap S') = 1 - P(L) - P(S \cap L')$$
$$= 100\% - 60\% - 35\%$$
$$= 5\%$$

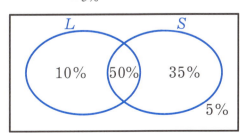

Figure 22B.4

$$P(L' \cap S') = P\big((L \cup S)'\big)$$
$$= 1 - (P(L) + P(S) - P(S \cap L))$$
$$= 1 - P(L) - P(S) + P(S \cap L)$$
$$= 100\% - 85\% - 60\% + 50\%$$
$$= 5\%$$

5

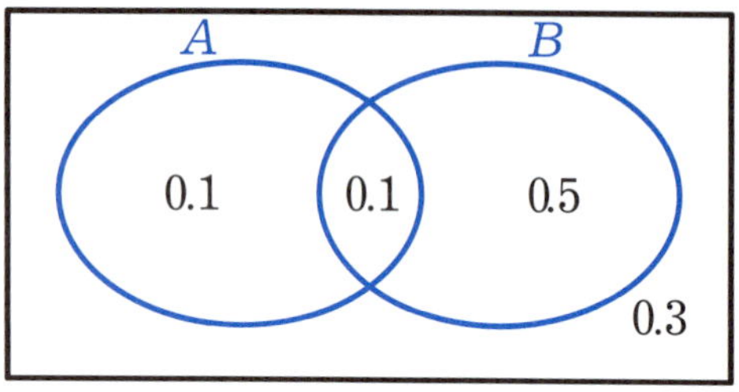

Figure 22B.5

$$P(B \cap A') = P(A \cup B) - P(A) = 0.7 - 0.2 = 0.5$$

$$P(B') = 1 - P(B) = 1 - (P(A \cap B) + P(A' \cap B)) = 1 - (0.1 + 0.5) = 0.4$$

6

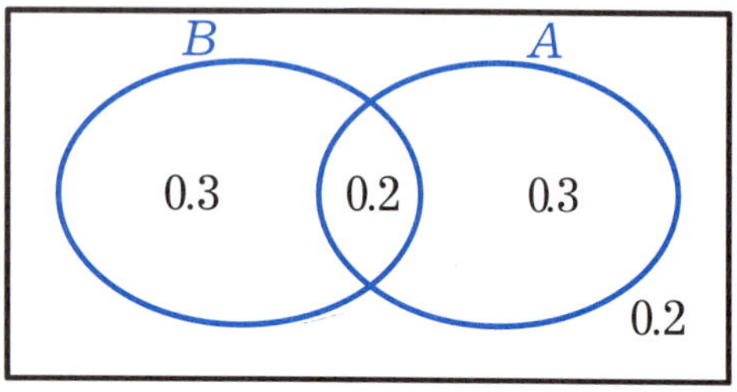

Figure 22B.6

$$P\big((A \cup B)'\big) = P(A' \cap B') = 0.2$$

$$P(A \cap B') = P(B') - P(A' \cap B') = 1 - P(B) - P(A' \cap B') = 1 - 0.5 - 0.2 = 0.3$$

7 a $\dfrac{1000}{6} = 166.7$

so there are 166 multiples of 6 among the first 1000 numbers.

$\therefore P(\text{multiple of } 6) = \dfrac{166}{1000} = 0.166$

b The numbers which are multiples of both 6 and 8 are multiples of $\text{LCM}(6, 8) = 24$

$\dfrac{1000}{24} = 41.7$

so there are 41 multiples of 24 among the first 1000 numbers

$\therefore P(\text{multiple of } 24) = \dfrac{41}{1000} = 0.041$

Exercise 22C

> **COMMENT**
>
> In all these questions a tree diagram, populated with relevant values, is sufficient preliminary working. The tree diagrams have been given in these answers, together with full algebraic working such as would be needed for an answer if a tree diagram were not drawn.

2

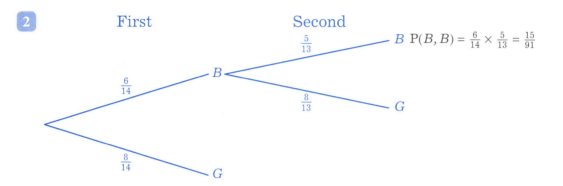

Figure 22C.2

B = boy; G = girl

$$P(B,B) = \frac{6}{14} \times \frac{5}{13} = \frac{15}{91} = 0.165 \text{ (3SF)}$$

3

> **COMMENT**
>
> In a question like this, there is no need to draw a 'full' tree, since only a few results are of interest. Thus, after blue in the first draw, all that is of interest is whether the second ball is green or not, so G / G' branches are sufficient – there is no need for three branches $B/G/R$. Similarly, after G in the first draw only B / B' branches are needed, and a first draw of R need not be detailed further.

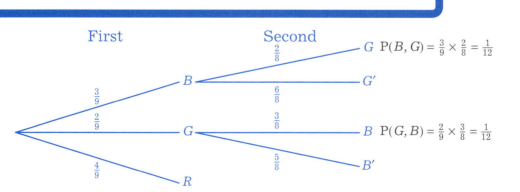

Figure 22C.3

B = blue; G = green; R = red

$P(\text{Blue and Green}) = P(B,G) + P(G,B) = \dfrac{3}{9} \times \dfrac{2}{8} + \dfrac{2}{9} \times \dfrac{3}{8} = \dfrac{1}{12} + \dfrac{1}{12} = \dfrac{1}{6}$

4

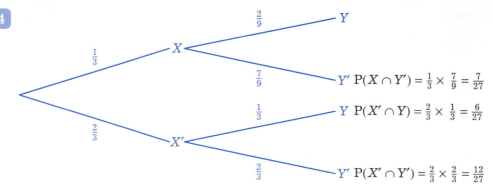

Figure 22C.4

a $P(Y') = P(X \cap Y') + P(X' \cap Y') = P(Y'|X)P(X) + P(Y'|X')P(X') = \dfrac{7}{9} \times \dfrac{1}{3} + \dfrac{2}{3} \times \dfrac{2}{3} = \dfrac{19}{27}$

b $P(X' \cap Y') = P(Y') + P(Y \cap X') = P(Y') + P(Y|X')P(X') = \dfrac{19}{27} + \dfrac{1}{3} \times \dfrac{2}{3} = \dfrac{25}{27}$

Alternatively:

$P(X' \cup Y') = 1 - P(X \cap Y) = 1 - P(Y|X)P(X) = 1 - \dfrac{2}{9} \times \dfrac{1}{3} = \dfrac{25}{27}$

> **COMMENT**
>
> Normally this alternative of calculating the complement to the union would be the faster calculation, but in this question we can with equal ease harness the answer to part (a)

5

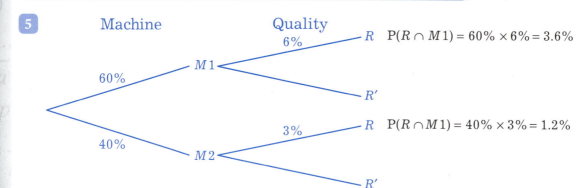

Figure 22C.5

$M1$ = machine 1; $M2$ = machine 2; R = rejected

$P(\text{Reject}) = P(R \cap M1) + P(R \cap M2) = P(R|M1)P(M1) + P(R|M2)P(M2)$
$\quad = 60\% \times 6\% + 40\% \times 3\% = 3.6\% + 1.2\% = 4.8\%$

6

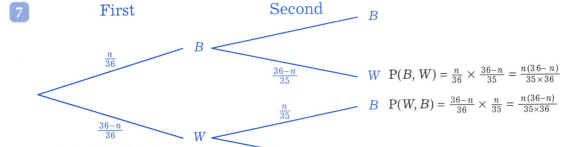

Figure 22C.6

$P(\text{Win}) = P(\text{Win}|\text{Higher opp})P(\text{Higher opp}) + P(\text{Win}|\text{Lower opp})P(\text{Lower opp})$

$= 30\% \times \dfrac{2}{11} + 70\% \times \dfrac{9}{11}$

$= \dfrac{69}{110} = 0.627 \ (3\text{SF})$

7

```
         First              Second
                               ─────── B
              n/36
           ┌─── B
           │         36-n/35
           │         ─────── W   P(B, W) = n/36 × (36-n)/35 = n(36-n)/(35×36)
           │         n/35
           │         ─────── B   P(W, B) = (36-n)/36 × n/35 = n(36-n)/(35×36)
    36-n/36
           └─── W
                     ─────── W
```

Figure 22C.7

$B = $ black; $W = $ white

Let the number of black disks be n.

If $P(\text{Same}) = P(\text{Different})$ then $P(\text{Same}) = \dfrac{1}{2}$

$P(\text{Same}) = P(B, B) + P(W, W)$

$\therefore \dfrac{n}{36} \times \dfrac{n-1}{35} + \dfrac{36-n}{36} \times \dfrac{35-n}{35} = \dfrac{1}{2}$

$n(n-1) + (36-n)(35-n) = 630$

$2n^2 - 72n + 630 = 0$

$n^2 - 36n + 315 = 0$

$(n-15)(n-21) = 0$

$n = 15$ or 21

Exercise 22D

3 a For independent events A and B,
$P(A \cap B) = P(A) \times P(B)$

$P(A \cup B) = P(A) + P(B) - P(A \cap B)$

$\therefore 0.72 = 0.6 + P(B) - 0.6P(B)$

$0.4P(B) = 0.12$

$\Rightarrow P(B) = 0.3$

b $P(A \cap B') + P(A' \cap B) = P(A \cup B) - P(A \cap B)$

$= 0.72 - 0.6 \times 0.3$

$= 0.54$

4 a $P(T \cap S) = P(T) \times P(S)$

$= 92\% \times 68\%$

$= 62.56\%$

b $P(T' \cap S') = P(T') \times P(S')$

$= 8\% \times 32\%$

$= 2.56\%$

c The event 'at least one working' is the complement of 'neither working':

$P(T \cup S) = 1 - P(T' \cap S')$

$= 97.44\%$

5 a $P(\text{all different}) = 1 \times \dfrac{7}{8} \times \dfrac{6}{8} \times \dfrac{5}{8}$

$= \dfrac{105}{256} = 0.410$

> **COMMENT**
> This calculation assumes that there is an effectively unlimited number of each type of toy, which is implicit in the way the question is phrased. If there were limited numbers, the probabilities would change to reflect this – after all, in the most extreme case, if there were only 8 packets of crisps in the world and each had a different toy, David would have complete certainty that he would get a different toy each time.

b Using the complement: let p be the probability that he fails to get any gyroscopes or yo-yos; then

$p = \left(\dfrac{6}{8}\right)^4 = \dfrac{81}{256} = 0.316$

$\therefore P(\text{at least one gyroscope or yo-yo}) = 1 - \dfrac{81}{256}$

$= \dfrac{175}{256}$

$= 0.684$

6 For independent events A and B,

$P(B) = \dfrac{P(A \cap B)}{P(A)}$

$P(A \cup B) = P(A) + P(B) - P(A \cap B)$

$P(A) = P(A \cap B) + P(A \cap B')$

$= 0.3 + 0.6 = 0.9$

$\therefore P(B) = \dfrac{0.3}{0.9} = \dfrac{1}{3}$

Hence $P(A \cup B) = 0.9 + \dfrac{1}{3} - 0.3$

$= \dfrac{28}{30} = \dfrac{14}{15}$

$= 0.933$

Exercise 22E

1 Total number of possible orderings: $4! = 24$

$\therefore P(\text{STAR, RATS or ARTS}) = \dfrac{3}{24} = \dfrac{1}{8}$

2 Total possible choices of 4 letters from 7:

$\binom{7}{4} = 35$

Possible choices containing 'P' (choose the other 3 letters from the remaining 6):

$\binom{6}{3} = 20$

$\therefore P(\text{contain P}) = \dfrac{20}{35} = \dfrac{4}{7}$

3 a Total possible choices: $\binom{18}{11} = 31824$

Possible choices including Captain and Vice Captain: $\binom{16}{9} = 11440$

$P(\text{Capt and VC}) = \dfrac{11440}{31824}$

$= \dfrac{55}{153}$

$= 0.359 \, (3\text{SF})$

b Total possible choices (choose 1 of 2 goalkeepers, then 10 from the remaining 16): $2 \times \binom{16}{10} = 16016$

Possible choices including Captain and Vice Captain: $2 \times \binom{14}{8} = 6006$

$P(\text{Capt and VC}) = \dfrac{6006}{16016} = \dfrac{3}{8} = 0.375$

4 a Total possible choices: $\binom{9}{5} = 126$

Choices that include history students only: $\binom{6}{5} = 6$

$P(5 \text{ history students}) = \dfrac{6}{126} = \dfrac{1}{21}$

$= 0.0476 \, (3\text{SF})$

b If all three philosophy students are chosen, select the remaining 2 from the 6 history students: $\binom{6}{2} = 15$

$P(3 \text{ Phil, } 2 \text{ Hist}) = \dfrac{15}{126} = \dfrac{5}{42} = 0.119 \, (3\text{SF})$

5 a Total arrangements: $6! = 720$

Arrangements with the brothers at the ends: $2 \times 4! = 48$

$P(\text{brothers at ends}) = \dfrac{48}{720} = \dfrac{1}{15}$

b For arrangements with the brothers adjacent, treat the brothers as a single unit, then multiply by the number of internal arrangements:

$5! \times 2 = 240$

$P(\text{brothers adjacent}) = \dfrac{240}{720} = \dfrac{1}{3}$

6 Total arrangements: $20!$

a For arrangements with all the 8 forwards together, treat them as a single unit (so 13 units altogether), then multiply by the number of internal arrangements:

$13! \times 8!$

$\therefore P(\text{forwards all together})$

$= \dfrac{13! \times 8!}{20!}$

$= 0.000103 \, (3\text{SF})$

b To count the arrangements with no forwards adjacent:

Line up the 8 forwards ($8!$ arrangements).

There are 7 spaces between forwards which must be filled to avoid forwards being adjacent, so choose 7 players from the remaining 12 to fill those spaces

$\Rightarrow {}^{12}P_7 = \dfrac{12!}{5!}$ arrangements.

Have each of these 7 take hold of the forward to his right; treat these as 8 units set in order.

There are a further 5 units to mix in. Consider the end arrangement as consisting of 13 ordered boxes. Into each of 5 of these boxes, one of the 5 still-unattached non-forwards will be placed.

There are ${}^{13}P_5 = \dfrac{13!}{8!}$ ways to arrange this.

The remaining 8 boxes will be filled with the pre-ordered forward units; since they are already ordered, there is only one way to do this.

22 Probability

In summary, the number of ways to arrange the team so that no forwards are together is

$$8! \times \frac{12!}{5!} \times \frac{13!}{8!} = \frac{12! \times 13!}{5!}$$

$$\therefore P(\text{all forwards apart}) = \frac{12! \times 13!}{5! \times 20!}$$

$$= 0.0102 \ (3SF)$$

Exercise 22F

3 a i $P(A \cap B) = P(A) + P(B) - P(A \cup B)$
$= 0.6 + 0.2 - 0.7$
$= 0.1$

ii $P(A) \times P(B) = 0.12 \neq P(A \cap B)$

$\therefore A$ and B are not independent.

b $P(B|A) = \dfrac{P(A \cap B)}{P(A)}$

$= \dfrac{0.1}{0.6}$

$= \dfrac{1}{6}$

4 a $P(A \cap B) = P(B|A) \times P(A)$

$= \dfrac{2}{3} \times \dfrac{1}{2}$

$= \dfrac{1}{3}$

b $P(B) = P(A \cup B) - P(A) + P(A \cap B)$

$= \dfrac{4}{5} - \dfrac{2}{3} + \dfrac{1}{3}$

$= \dfrac{7}{15}$

c $P(A) \times P(B) = \dfrac{2}{3} \times \dfrac{7}{15} = \dfrac{14}{45} \neq P(A \cap B)$

$\therefore A$ and B are not independent

5 $P(A \cup B) = P(A) + P(B) - P(A \cap B)$
$= P(A) + P(B) - P(A|B)P(B)$

$\therefore \dfrac{4}{5} = \dfrac{2}{3} + P(B) - \dfrac{1}{5} P(B)$

$\dfrac{4}{5} P(B) = \dfrac{4}{5} - \dfrac{2}{3} = \dfrac{2}{15}$

$P(B) = \dfrac{1}{6}$

Exercise 22G

> **COMMENT**
> For the questions in this section, the focus should be on completing the diagram, after which subsequent values can be read off with little or no working shown. Preliminary calculations for completing the diagram are given in each question to explain how the diagram is filled in, but usually you would not need to show these calculations in an examination.

1 a F = football; B = badminton
$n(\text{F only}) = n(\text{F}) - n(\text{F} \cap \text{B}) = 34 - 5 = 29$
$n(\text{B only}) = n(\text{B}) - n(\text{F} \cap \text{B}) = 18 - 5 = 13$
$n(\text{F}' \cap \text{B}') = 145 - 29 - 13 - 5 = 98$

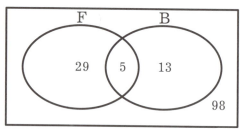

Figure 22G.1

b From diagram: $n(\text{F}' \cap \text{B}') = 98$

c $P(B) = \dfrac{18}{145}$

d $P(B|F) = \dfrac{P(B \cap F)}{P(F)}$

$= \dfrac{n(B \cap F)}{n(F)}$

$= \dfrac{5}{34}$

2 a M = mathematics; E = economics

$n(M \cup E) = 145 - 72 = 73$

$n(M \cap E) = n(M) + n(E) - n(M \cup E)$

$= 58 + 47 - 73$

$= 32$

$n(M \text{ only}) = n(M) - n(M \cap E)$

$= 58 - 32$

$= 26$

$n(E \text{ only}) = n(E) - n(M \cap E)$

$= 47 - 32$

$= 15$

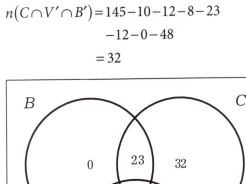

Figure 22G.2

b From diagram: $n(M \cap E) = 32$

c $P(M \cap E | M) = \dfrac{P(M \cap E)}{P(M)}$

$= \dfrac{n(M \cap E)}{n(M)}$

$= \dfrac{32}{58}$

$= \dfrac{16}{29}$

3 a B = spaghetti bolognese;
C = chilli con carne;
V = vegetable curry

$n(B \cap C \cap V') = n(B \cap C) - n(B \cap C \cap V)$

$= 35 - 12$

$= 23$

$n(B \cap V \cap C') = n(B \cap V) - n(B \cap C \cap V)$

$= 20 - 12$

$= 8$

$n(B' \cap C \cap V) = n(C \cap V) - n(B \cap C \cap V)$

$= 24 - 12$

$= 12$

$n(B \cap C' \cap V') = n(B) - n(B \cap C)$
$\qquad - n(B \cap V \cap C')$

$= 43 - 35 - 8$

$= 0$

$n(V \cap B' \cap C') = n(V) - n(V \cap B)$
$\qquad - n(V \cap C \cap B')$

$= 80 - 20 - 12$

$= 48$

$n(C \cap V' \cap B') = 145 - 10 - 12 - 8 - 23$
$\qquad - 12 - 0 - 48$

$= 32$

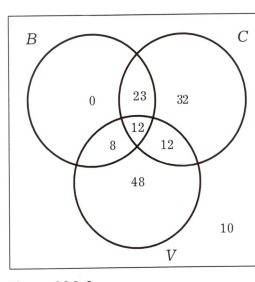

Figure 22G.3

b From diagram: $n(B \cap C' \cap V') = 0$

c $n(C) = 32 + 23 + 12 + 12 = 79$

d $P(1 \text{ meal only}) = \dfrac{0 + 32 + 48}{145}$

$= \dfrac{80}{145}$

$= \dfrac{16}{29}$

e $P(V \mid 1 \text{ meal only}) = \dfrac{n(V \text{ only})}{n(1 \text{ meal only})}$

$= \dfrac{48}{80}$

$= \dfrac{3}{5}$

f $P(\text{at least 2 meals}) = 1 - P(0 \text{ meals}) - P(1 \text{ meal})$

$= 1 - \dfrac{n(0 \text{ meals})}{145} - \dfrac{3}{5}$

$= 1 - \dfrac{10}{145} - \dfrac{80}{145}$

$= \dfrac{11}{29}$

4 a B = blue eyes; D = dark hair

$P(B \cap D') = P(B) - P(B \cap D) = 0.4 - 0.2 = 0.2$

$P(B' \cap D) = P(D) - P(B \cap D) = 0.7 - 0.2 = 0.5$

$P(B' \cap D') = 1 - 0.2 - 0.2 - 0.5 = 0.1$

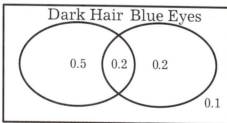

Figure 22G.4

b From diagram: $P(B' \cap D') = 0.1$

c $P(B \mid D) = \dfrac{P(B \cap D)}{P(D)}$

$= \dfrac{0.2}{0.7}$

$= \dfrac{2}{7}$

d $P(B \mid D') = \dfrac{P(B \cap D')}{P(D')}$

$= \dfrac{0.2}{1 - 0.7}$

$= \dfrac{2}{3}$

e From (c) and (d), $P(B \mid D) \neq P(B \mid D')$, so blue eyes and dark hair are not independent characteristics.

5 C = cold; R = raining

$P(C \cap R) = P(C) + P(R) - P(C \cup R)$

$= P(C) + P(R) - \left[1 - P\left((C \cup R)'\right) \right]$

$= 0.6 + 0.45 - (1 - 0.25)$

$= 0.3$

b $P(C \cap R') = P(C) - P(C \cap R)$

$= 0.6 - 0.3 = 0.3$

$P(R \cap C') = P(R) - P(C \cap R)$

$= 0.45 - 0.3 = 0.15$

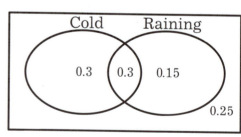

Figure 22G.5

c $P(C' \mid R) = \dfrac{P(R \cap C')}{P(R)}$

$= \dfrac{0.15}{0.45}$

$= \dfrac{1}{3}$

d $P(R|C') = \dfrac{P(R \cap C')}{P(C')}$

$= \dfrac{0.15}{1-0.6}$

$= \dfrac{3}{8}$

e $P(R \cap C) = 0.3$

$P(R) \times P(C) = 0.6 \times 0.45 = 0.27$

$P(R \cap C) \neq P(R) \times P(C)$, so the two events are not independent.

Exercise 22H

> **COMMENT**
>
> A tree diagram is supplied with each worked solution, together with full algebraic calculations; you should choose which approach you prefer for a given question. If you use a diagram, you may not need to show as much separate algebraic working to validate your result, unless answering a 'show that' question. However, filling in detail on a tree diagram can also take time; avoid calculating unnecessary values when populating a tree diagram.

1

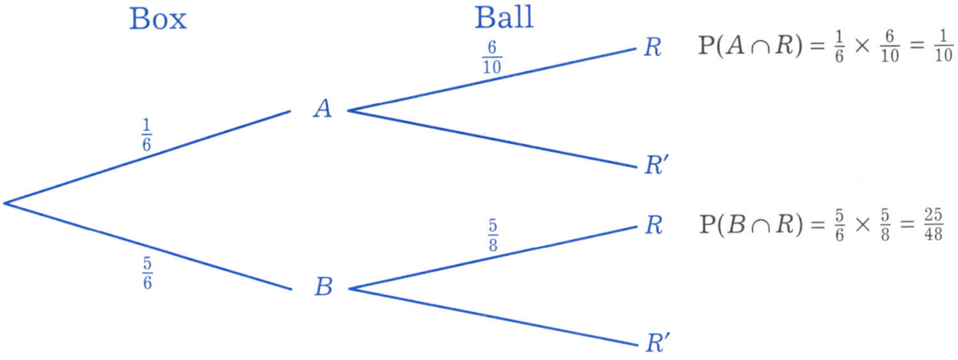

Box   Ball

$A \to R$: $P(A \cap R) = \dfrac{1}{6} \times \dfrac{6}{10} = \dfrac{1}{10}$ (branches $\dfrac{1}{6}$, $\dfrac{6}{10}$, R')

$B \to R$: $P(B \cap R) = \dfrac{5}{6} \times \dfrac{5}{8} = \dfrac{25}{48}$ (branches $\dfrac{5}{6}$, $\dfrac{5}{8}$, R')

Figure 22H.1

$A =$ box A; $B =$ box B; $R =$ red

a $P(R) = P(R|A) \times P(A) + P(R|B) \times P(B)$

$= \dfrac{6}{10} \times \dfrac{1}{6} + \dfrac{5}{8} \times \dfrac{5}{6}$

$= \dfrac{1}{10} + \dfrac{25}{48}$

$= \dfrac{149}{240}$

$= 0.621$

b $P(B|R) = \dfrac{P(B \cap R)}{P(R)} = \dfrac{P(R|B)P(B)}{P(R)} = \dfrac{\frac{5}{8} \times \frac{5}{6}}{\frac{149}{240}} = \dfrac{125}{149} = 0.839$

2

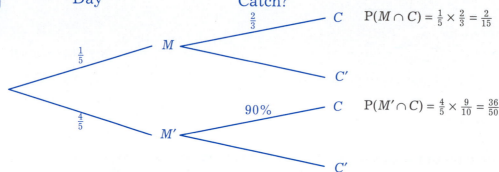

Figure 22H.2

C = Robert catches the train; M = Monday

a $P(C) = P(C|M) \times P(M) + P(C|M') \times P(M') = \dfrac{2}{3} \times \dfrac{1}{5} + \dfrac{9}{10} \times \dfrac{4}{5} = \dfrac{2}{15} + \dfrac{18}{25} = \dfrac{64}{75} = 0.853$

b $P(M|C) = \dfrac{P(M \cap C)}{P(C)} = \dfrac{\frac{2}{15}}{\frac{64}{75}} = \dfrac{5}{32} = 0.156$

3

Bag 1   Bag 2

$\frac{3}{8}$ R $\overset{\frac{7}{10}}{\longrightarrow}$ R $P(R, R) = \frac{3}{8} \times \frac{7}{10} = \frac{21}{80}$

$\phantom{\frac{3}{8} R}\overset{\frac{3}{10}}{\longrightarrow}$ B $P(R, B) = \frac{3}{8} \times \frac{3}{10} = \frac{9}{80}$

$\frac{5}{8}$ B $\overset{\frac{7}{10}}{\longrightarrow}$ R $P(B, R) = \frac{5}{8} \times \frac{7}{10} = \frac{35}{80}$

$\phantom{\frac{5}{8} B}\overset{\frac{3}{10}}{\longrightarrow}$ B $P(B, B) = \frac{5}{8} \times \frac{3}{10} = \frac{15}{80}$

Figure 22H.3

R = red; B = blue

a $P(\text{same colour}) = P(R,R) + P(B,B) = \dfrac{6}{16} \times \dfrac{7}{10} + \dfrac{10}{16} \times \dfrac{3}{10} = \dfrac{21}{80} + \dfrac{15}{80} = \dfrac{9}{20} = 0.45$

b $P(R1|\text{different}) = \dfrac{P(R1 \cap \text{different})}{P(\text{different})} = \dfrac{P(R1 \cap B2)}{P(R1 \cap B2) + P(B1 \cap R2)}$

$= \dfrac{\dfrac{6}{16} \times \dfrac{3}{10}}{\dfrac{6}{16} \times \dfrac{3}{10} + \dfrac{10}{16} \times \dfrac{7}{10}} = \dfrac{\dfrac{9}{80}}{\dfrac{9}{80} + \dfrac{35}{80}}$

$= \dfrac{9}{44} = 0.205$

4

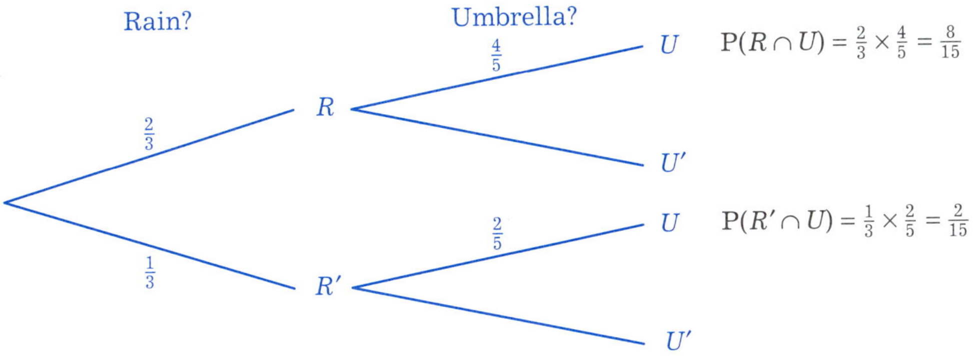

Rain? Umbrella?

- R (prob $\tfrac{2}{3}$) → U (prob $\tfrac{4}{5}$): $P(R \cap U) = \tfrac{2}{3} \times \tfrac{4}{5} = \tfrac{8}{15}$
- R → U'
- R' (prob $\tfrac{1}{3}$) → U (prob $\tfrac{2}{5}$): $P(R' \cap U) = \tfrac{1}{3} \times \tfrac{2}{5} = \tfrac{2}{15}$
- R' → U'

Figure 22H.4

$R =$ raining; $U =$ bring umbrella

a $P(U) = P(U \cap R) + P(U \cap R') = P(U|R)P(R) + P(U|R')P(R') = \dfrac{4}{5} \times \dfrac{2}{3} + \dfrac{2}{5} \times \dfrac{1}{3} = \dfrac{10}{15} = \dfrac{2}{3}$

b $P(R|U) = \dfrac{P(R \cap U)}{P(U)} = \dfrac{P(U|R)P(R)}{P(U)} = \dfrac{\tfrac{8}{15}}{\tfrac{2}{3}} = \dfrac{4}{5}$

5

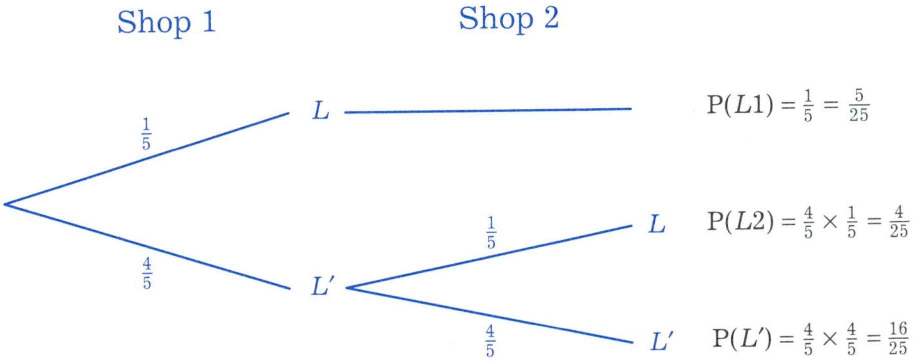

Shop 1   Shop 2

- L (prob $\tfrac{1}{5}$): $P(L1) = \tfrac{1}{5} = \tfrac{5}{25}$
- L' (prob $\tfrac{4}{5}$) → L (prob $\tfrac{1}{5}$): $P(L2) = \tfrac{4}{5} \times \tfrac{1}{5} = \tfrac{4}{25}$
- L' → L' (prob $\tfrac{4}{5}$): $P(L') = \tfrac{4}{5} \times \tfrac{4}{5} = \tfrac{16}{25}$

Figure 22H.5

$L =$ leaves umbrella in shop

$P(L2|L) = \dfrac{P(L2)}{P(L1) + P(L2)} = \dfrac{\tfrac{4}{5} \times \tfrac{1}{5}}{\tfrac{1}{5} + \tfrac{4}{5} \times \tfrac{1}{5}} = \dfrac{4}{9}$

6

Airline — Lost luggage

Tree diagram: $\frac{40}{65}$ to π, then $\frac{1}{10}$ to L and $\frac{9}{10}$ to L'; $\frac{25}{65}$ to λ, then $\frac{1}{4}$ to L and $\frac{3}{4}$ to L'.

$P(\pi \cap L) = \frac{40}{65} \times \frac{1}{10} = \frac{4}{65} = \frac{80}{1300}$

$P(\lambda \cap L) = \frac{25}{65} \times \frac{1}{4} = \frac{25}{260} = \frac{125}{1300}$

Figure 22H.6

L = lost luggage; π = Pi Air; λ = Lambda Air
Assuming that each flight has the same number of passengers,

$$P(\pi|L) = \frac{P(L|\pi)P(\pi)}{P(L|\pi)P(\pi)+P(L|\lambda)P(\lambda)} = \frac{\frac{1}{10} \times \frac{40}{65}}{\frac{1}{10} \times \frac{40}{65}+\frac{1}{4} \times \frac{25}{65}} = \frac{\frac{80}{1300}}{\frac{205}{1300}} = \frac{16}{41} = 0.390 \, (3\,\text{SF})$$

7

Raining — Late

Tree diagram: $\frac{1}{4}$ to R, then $\frac{2}{3}$ to L and $\frac{1}{3}$ to L'; $\frac{3}{4}$ to R', then $\frac{1}{5}$ to L and $\frac{4}{5}$ to L'.

$P(R \cap L) = \frac{1}{4} \times \frac{2}{3} = \frac{2}{12} = \frac{10}{60}$

$P(R' \cap L) = \frac{3}{4} \times \frac{1}{5} = \frac{3}{20} = \frac{9}{60}$

Figure 22H.7

R = raining; L = late

$$P(R|L) = \frac{P(L|R)P(R)}{P(L|R)P(R)+P(L|R')P(R')} = \frac{\frac{2}{3} \times \frac{1}{4}}{\frac{2}{3} \times \frac{1}{4}+\frac{1}{5} \times \frac{3}{4}} = \frac{\frac{10}{60}}{\frac{19}{60}} = \frac{10}{19} = 0.526 \, (3\,\text{SF})$$

8

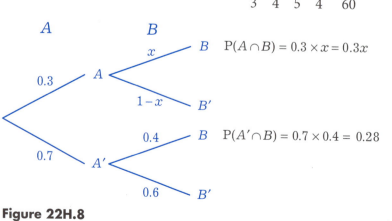

$P(A \cap B) = 0.3 \times x = 0.3x$

$P(A' \cap B) = 0.7 \times 0.4 = 0.28$

Figure 22H.8

Let $\mathrm{P}(B|A) = x$

$\mathrm{P}(B) = \mathrm{P}(B|A) \times \mathrm{P}(A) + \mathrm{P}(B|A') \times \mathrm{P}(A') = x \times 0.3 + 0.4 \times 0.7 = 0.28 + 0.3x$

$$\mathrm{P}(A|B) = \frac{\mathrm{P}(B|A) \times \mathrm{P}(A)}{\mathrm{P}(B)} = \frac{0.3x}{0.3x + 0.28} = \frac{3}{17}$$

$\therefore 5.1x = 3(0.3x + 0.28)$

$4.2x = 0.84$

$$\Rightarrow x = \frac{0.84}{4.2} = 0.2$$

9

First game
Tournament
50%
W
30%
D
20%
L
60%
T
P(W ∩ T) = 50% × 60% = 30%
40%
T'
50%
T
P(D ∩ T) = 30% × 50% = 15%
50%
T'
10%
T
P(L ∩ T) = 20% × 10% = 2%
90%
T'

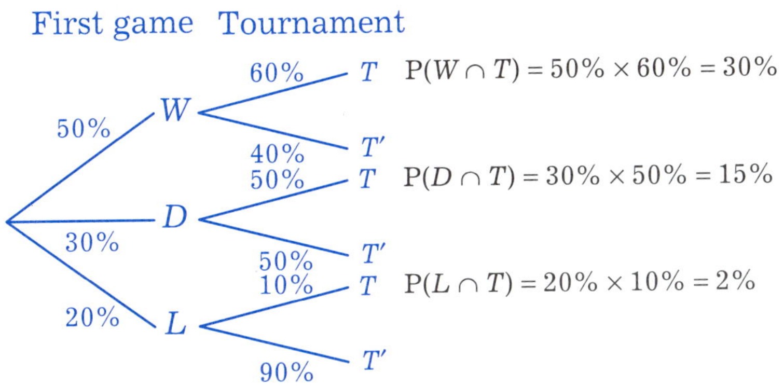

Figure 22H.9

W, L and D are the events of Lisa winning, losing and drawing the first match, respectively. T is the event of Lisa winning the tournament.

$$\mathrm{P}(D|T) = \frac{\mathrm{P}(T|D)\mathrm{P}(D)}{\mathrm{P}(T|W)\mathrm{P}(W) + \mathrm{P}(T|D)\mathrm{P}(D) + \mathrm{P}(T|L)\mathrm{P}(L)}$$

$$= \frac{50\% \times 30\%}{60\% \times 50\% + 50\% \times 30\% + 10\% \times 20\%}$$

$$= \frac{15}{47}$$

$= 31.9\%$ (3SF)

10

Transport
Late
3/12
W
4/12
B
5/12
C
50%
L
P(W ∩ L) = 3/12 × 50% = 3/24
50%
L'
25%
L
P(B ∩ L) = 4/12 × 25% = 2/24
75%
L'
10%
L
P(C ∩ L) = 5/12 × 10% = 1/24
90%
L'

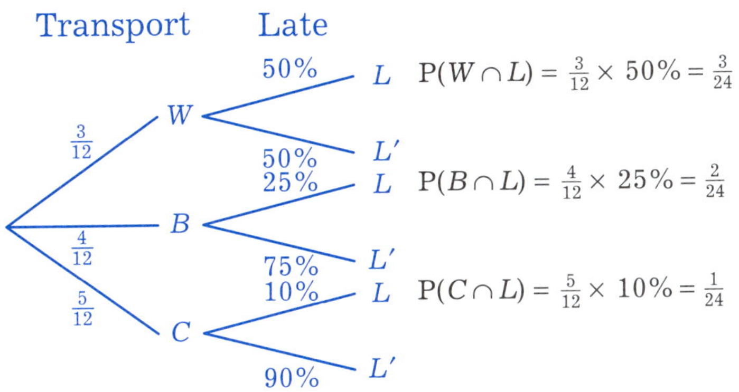

Figure 22H.10

W, B and C are the events of Omar walking, taking the bus and cycling to school, respectively. L is the event of Omar being late.

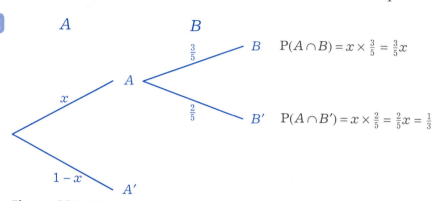

$$P(B|L) = \frac{P(L|B)P(B)}{P(L|W)P(W) + P(L|B)P(B) + P(L|C)P(C)} = \frac{25\% \times \frac{1}{3}}{50\% \times \frac{1}{4} + 25\% \times \frac{1}{3} + 10\% \times \frac{5}{12}} = \frac{1}{3}$$

11

Tree diagram:
- A with probability x, then B with $\frac{3}{5}$: $P(A \cap B) = x \times \frac{3}{5} = \frac{3}{5}x$
- A with probability x, then B' with $\frac{2}{5}$: $P(A \cap B') = x \times \frac{2}{5} = \frac{2}{5}x = \frac{1}{3}$
- A' with probability $1-x$

Figure 22H.11

Let $P(A) = x$

$P(B' \cap A) = P(B'|A)P(A)$

$$\frac{1}{3} = \frac{2}{5}x$$

$$\Rightarrow x = \frac{1}{3} \div \frac{2}{5} = \frac{5}{6}$$

$$\therefore P(A \cap B) = P(B|A)P(A) = \frac{3}{5} \times \frac{5}{6} = \frac{1}{2}$$

12 Disease state   Test positive

Tree diagram:
- D: 0.0003
  - T: 0.95, $P(D \cap T) = 0.0003 \times 0.95 = 0.000285$
  - T': 0.05
- D': 0.9997
  - T: 0.01, $P(D' \cap T) = 0.9997 \times 0.01 = 0.009997$
  - T': 0.99

Figure 22H.12

D is the event that a patient has the disease;

T is the event that the patient tested positive for the disease.

$$P(D|T) = \frac{P(T|D)P(D)}{P(T|D)P(D) + P(T|D')P(D')} = \frac{0.95 \times 0.0003}{0.95 \times 0.0003 + 0.01 \times 0.9997} = 0.0277$$

> **COMMENT**
> This is a very real problem in medical tests. For a condition that is very rare, a test regime has to be incredibly accurate to be reliable. As shown here, despite a seemingly very predictive test, the probability of a positive test result being a false positive is much more likely than it being a true positive, because the incidence of the disease (0.03%) is so much lower than the probability of an incorrect test result from a healthy patient (1%).

13

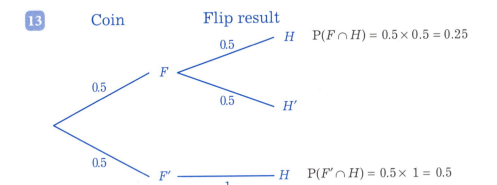

Figure 22H.13

F is the event of picking the fair coin;

H is the event of flipping Heads.

$$P(F|H) = \frac{P(H|F)P(F)}{P(H|F)P(F) + P(H|F')P(F')}$$

$$= \frac{0.5 \times 0.5}{0.5 \times 0.5 + 1 \times 0.5}$$

$$= \frac{1}{3}$$

Mixed examination practice 22

Short questions

1 $P(\text{same colour}) = P(R,R) + P(B,B) + P(W,W)$

$$= \frac{6}{18} \times \frac{5}{17} + \frac{4}{18} \times \frac{3}{17} + \frac{8}{18} \times \frac{7}{17}$$

$$= \frac{98}{306}$$

$$= \frac{49}{153}$$

$$= 0.320 \text{ (3SF)}$$

2 S = first language Spanish; A = Argentine

$$P(S|A) = \frac{P(S \cap A)}{P(A)} = \frac{n(S \cap A)}{n(A)} = \frac{12}{12+3} = \frac{4}{5}$$

3

> **COMMENT**
> A tree diagram is the clearest and fastest way to show working in this case, but the full algebraic working is also given below.

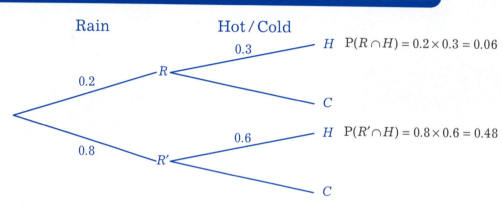

Figure 22MS.3

R = rain; H = hot (> 25°C); C = cold (≤ 25°C)

$$P(R|H) = \frac{P(R \cap H)}{P(H)} = \frac{0.2 \times 0.3}{0.2 \times 0.3 + 0.8 \times 0.6} = \frac{0.06}{0.54} = \frac{1}{9}$$

4 a There are $5! = 120$ possible arrangements in total.

$$\therefore P(\text{CHART}) = \frac{1}{120}$$

b To count the number of arrangements containing the sequence HAT, treat HAT as one unit and arrange 3 units (C, HAT and R) in $3! = 6$ ways:

$$P(\text{contains HAT}) = \frac{6}{120} = \frac{1}{20}$$

5 $P(A'|B) = \dfrac{P(A' \cap B)}{P(B)} = \dfrac{P(A') - P(A' \cap B')}{P(B)}$

$P(A') = 1 - P(A) = 1 - \dfrac{1}{14} = \dfrac{1}{15}$

$P(A' \cap B') = P\big((A \cup B)'\big) = 0$

$\therefore \dfrac{1}{5} = \dfrac{\frac{1}{15} - 0}{P(B)}$

$\Rightarrow P(B) = \dfrac{1}{3}$

Long questions

1 a R = red sweet

$$P(R,R \mid \text{Large}) = \frac{8}{20} \times \frac{7}{19} = \frac{14}{95}$$

b $P(R,R \mid \text{Small}) = \frac{4}{4+n} \times \frac{3}{3+n}$

$$\therefore \frac{12}{(4+n)(3+n)} = \frac{2}{15}$$

$$90 = (4+n)(3+n)$$

$$n^2 + 7n - 78 = 0$$

$$(n-6)(n+13) = 0$$

$$\Rightarrow n = 6$$

(as $n = -13$ is not a valid solution in this context)

c $P(\text{Large}) = \frac{1}{3}$ and $P(\text{Small}) = \frac{2}{3}$

From (a), $P(R,R \mid \text{Large}) = \frac{14}{95}$

From (b), $P(R,R \mid \text{Small}) = \frac{4}{10} \times \frac{3}{9} = \frac{2}{15}$

$P(R,R) = P(R,R \mid \text{Large}) \times P(\text{Large}) + P(R,R \mid \text{Small}) \times P(\text{Small})$

$$= \frac{14}{95} \times \frac{1}{3} + \frac{2}{15} \times \frac{2}{3}$$

$$= \frac{118}{855}$$

$$= 0.138$$

d $P(\text{Large} \mid R,R) = \dfrac{P(\text{Large} \cap \{R,R\})}{P(R,R)}$

$$= \frac{\frac{42}{855}}{\frac{118}{855}}$$

$$= \frac{42}{118}$$

$$= \frac{21}{59}$$

$$= 0.356$$

2 a Range of $P(X)$ is $[0, 1]$

b $P(A) - P(A \cap B) = P(A) - P(B \mid A)P(A)$

$$= P(A)(1 - P(B \mid A))$$

c i $P(A \cup B) = P(A) + P(B) - P(A \cap B)$

Hence $P(A \cup B) - P(A \cap B) = P(A) - P(A \cap B) + P(B) - P(A \cap B)$

From (b), $P(A) - P(A \cap B) = P(A)(1 - P(B|A))$

and similarly, $P(B) - P(A \cap B) = P(B)(1 - P(A|B))$

$\therefore P(A \cup B) - P(A \cap B) = P(A)(1 - P(B|A)) + P(B)(1 - P(A|B))$

ii The right-hand side of the equation in (i) is the sum of two products of non-negative values,

so $P(A \cup B) - P(A \cap B) \geq 0$

and hence $P(A \cup B) \geq P(A \cap B)$

3 a B = badminton; F = football

$P(B \cap F') = P(B) - P(B \cap F) = 0.3 - x$

$P(F \cap B') = 1 - P(B) - P(F' \cap B') = 1 - 0.3 - 0.5 = 0.2$

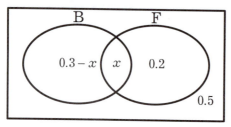

Figure 22ML.3.1

b $P(F \cap B') = 0.2$

c $P(B|F) = \dfrac{P(B \cap F)}{P(F)}$

$\therefore 0.5 = \dfrac{x}{0.2 + x}$

$0.5x + 0.1 = x$

$\Rightarrow x = P(B \cap F) = 0.2$

d

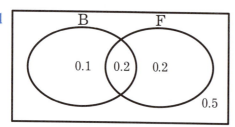

Figure 22ML.3.2

$P(B \cap F') = 0.1$

e $P(\text{B only} | \text{one sport}) = \dfrac{P(\text{B only})}{P(\text{one sport})} = \dfrac{0.1}{0.1 + 0.2} = \dfrac{1}{3}$

23 Discrete probability distributions

Exercise 23A

3 The sum of the probabilities must equal 1

$$\therefore \frac{1}{3} + \frac{1}{4} + k + \frac{1}{5} = 1$$

$$\Rightarrow k = \frac{13}{60}$$

After two rolls the total score is 4. The cases are:

$$P(1,3) = \frac{1}{3} \times k = \frac{13}{180}$$

$$P(2,2) = \frac{1}{4} \times \frac{1}{4} = \frac{1}{16}$$

$$P(3,1) = k \times \frac{1}{3} = \frac{13}{180}$$

$$\therefore P(X_1 + X_2 = 4) = \frac{13}{180} + \frac{1}{16} + \frac{13}{180}$$

$$= \frac{149}{720}$$

$$= 0.207$$

Exercise 23B

2 a $\sum_x P(X = x) = 1$

$$\Rightarrow k(3 + 4 + 5 + 6 + 7) = 1$$

$$\therefore k = \frac{1}{25} = 0.04$$

b $E(X) = \sum_x x \, P(X = x)$

$$= k(2 \times 3 + 3 \times 4 + 4 \times 5 + 5 \times 6 + 6 \times 7)$$

$$= \frac{110}{25}$$

$$= 4.4$$

3 $E(V) = \sum_v v \, P(V = v)$

$$= 1 \times 0.2 + 2 \times 0.3 + 5 \times 0.1 + 8 \times 0.1 + k \times 0.3$$

$$= 2.1 + 0.3k$$

$$\therefore 2.1 + 0.3k = 6.1$$

$$\Rightarrow k = \frac{4}{0.3} = \frac{40}{3}$$

Median is the value m for which $P(V < m) = 0.5$

From the distribution, this could be any value in the interval $[2, 5]$, so the median is defined as the midpoint of this interval: median $= 3.5$

4 a $\sum_x P(X = x) = 1$

$$\Rightarrow k(3 + 4 + 5 + 6) = 1$$

$$\therefore k = \frac{1}{18}$$

b $E(X) = \sum_x x \, P(X = x)$

$$= 0 \times 3k + 1 \times 4k + 2 \times 5k + 3 \times 6k$$

$$= 32k$$

$$= \frac{16}{9}$$

5 a $\sum_x P(X=x) = 1$

$\Rightarrow 3k + 4k + 3k = 1$

$\therefore k = \dfrac{1}{10}$

b $E(X) = \sum_x x\, P(X=x)$

$= 1 \times 3k + 2 \times 4k + 3 \times 3k$

$= 20k$

$= 2$

6 $E(X) = \sum_x x\, P(X=x)$

$= (1+1+2+2+2+5) \times \dfrac{1}{6}$

$= \dfrac{13}{6}$

$\mathrm{Var}(X) = \sum_x x^2\, P(X=x) - (E(X))^2$

$= (1^2 + 1^2 + 2^2 + 2^2 + 2^2 + 5^2) \times \dfrac{1}{6} - \left(\dfrac{13}{6}\right)^2$

$= \dfrac{39}{6} - \dfrac{169}{36}$

$= \dfrac{65}{36}$

7 a $\sum_x P(X=x) = 1$

$\Rightarrow 0.1 + p + q + 0.2 = 1$

$p + q = 0.7$

$p = 0.7 - q$

$E(X) = \sum_x x\, P(X=x) = 1.5$

$\Rightarrow 0 \times 0.1 + 1 \times p + 2q + 3 \times 0.2 = 1.5$

$p + 2q = 0.9$

$\therefore 0.7 - q + 2q = 0.9$

$q = 0.2$

and so $p = 0.5$

b $\mathrm{Var}(X) = \sum_x x^2\, P(X=x) - (E(X))^2$

$= 0 \times 0.1 + 1 \times 0.5 + 4 \times 0.2 + 9 \times 0.2 - 1.5^2$

$= 3.1 - 2.25$

$= 0.85$

8 Let X be the number of counters the player receives on a roll of the die.

To get an expected profit of 3.25 counters per roll when the player pays 5 counters per roll, require $E(X) = \sum_x x\, P(X=x) = 8.25$

i.e. $4 \times \dfrac{1}{2} + 5 \times \dfrac{1}{4} + 15 \times \dfrac{1}{5} + n \times \dfrac{1}{20} = \dfrac{33}{4}$

$\dfrac{n}{20} = 2$

$n = 40$

9 a Let a be the number that the player chooses.

The player's profit is equal to his winnings minus n.

$P(\text{Profit} = 3n) = P(a,a,a)$

$= \left(\dfrac{1}{4}\right)^3$

$= \dfrac{1}{64}$

$P(\text{Profit} = 2n) = P(a,a,a') + P(a,a',a) + P(a',a,a)$

$= 3 \times \left(\dfrac{3}{4}\right) \times \left(\dfrac{1}{4}\right)^2$

$= \dfrac{9}{64}$

$P(\text{Profit} = 1-n) = P(a,a',a') + P(a',a,a') + P(a',a',a)$

$= 3 \times \left(\dfrac{3}{4}\right)^2 \times \left(\dfrac{1}{4}\right)$

$= \dfrac{27}{64}$

$P(\text{Profit} = -n) = 1 - \dfrac{1}{64} - \dfrac{9}{64} - \dfrac{27}{64} = \dfrac{27}{64}$

TABLE 23B.9

Player's profit ($)	Probability
$-n$	$\frac{27}{64}$
$1-n$	$\frac{27}{64}$
$2n$	$\frac{9}{64}$
$3n$	$\frac{1}{64}$

b Let X be the player's profit. From the table in (a),

$$E(X) = \sum_x x\, P(X=x)$$

$$= \frac{1}{64}(-27n + 27 - 27n + 18n + 3n)$$

$$= \frac{27 - 33n}{64}$$

For the organiser to make a profit, require the player's profit to be negative:

$$E(X) < 0$$

$$27 - 33n < 0$$

$$\Rightarrow n > 0.818$$

∴ the minimum entrance fee is 82 cents.

Exercise 23C

5 a In a true binomial distribution, the probability of each 'success' is independent of the other results, as in the case of throwing a die multiple times. However, within the school population there is a fixed number of students who travel by bus, so the experiment is equivalent to drawing counters without replacement: each student chosen for the sample has a probability of bus travel which does depend on the previous students selected. Nevertheless, because the total number of students in the school is large and the base probability of 15% is not too close to 0% or 100%, the probabilities will not change very much, and so a binomial distribution will be a good approximation.

b Let X be the number of students in the sample travelling by bus.

We approximate $X \sim B(20, 0.15)$

$P(X=5) = 0.103$ (from GDC)

6 a $X \sim B\left(4050, \dfrac{2}{3}\right)$

$$\Rightarrow E(X) = 4050 \times \frac{2}{3} = 2700$$

b $SD(X) = \sqrt{Var(X)}$

$$= \sqrt{4050 \times \frac{2}{3} \times \frac{1}{3}}$$

$$= 30$$

7 Let X be the number of questions Sheila answers correctly.

$$X \sim B\left(8, \frac{1}{4}\right)$$

a From GDC: $P(X=5) = 0.0231$

b $E(X) = 8 \times \dfrac{1}{4} = 2$

c $Var(X) = 8 \times \dfrac{1}{4} \times \dfrac{3}{4} = 1.5$

$SD(X) = \sqrt{Var(X)}$

$$= \sqrt{1.5}$$

$$= 1.22$$

d From GDC: $P(X \geq 5) = 0.0273$

8 Let X be the number of people the doctor sees who have the virus.

$X \sim B(80, 0.008)$

a From GDC: $P(X=2) = 0.108$

b From GDC: $P(X \geq 3) = 0.0267$

c It is assumed that patients do or do not have the virus independently of each other; in reality, if the prevalence nationwide is 0.8%, it is likely that there will be geographical pockets which have higher and lower rates than this, since a cold virus is contagious. Therefore, if the doctor sees one patient with the virus, it may be supposed that the virus is prevalent in the locality, so the probability of seeing other patients with the virus will be higher than 0.8%.

> **COMMENT**
>
> It is also assumed that the country is large enough that the doctor will not be seeing a significant fraction of the whole population; an island nation of a few hundred, for example, would not allow for a binomial model to be used with a sample of 80, for the same reasons as outlined in Q5(a). However, given the context of the question, this would not be the answer the examiner would be looking for!

9 a $E(Y) = 12 \times 0.4 = 4.8$

b $P(Y = 4) = 0.213$

$P(Y = 5) = 0.227$

$P(Y = 6) = 0.177$

The mode is 5.

10 Let X be the number of sixes in 4 throws:

$$X \sim B\left(4, \frac{1}{6}\right)$$

$\Rightarrow P(X = 3) = 0.0154$

Let Y be the number of fives or sixes in 6 throws: $Y \sim B\left(6, \frac{1}{3}\right)$

$\Rightarrow P(Y = 5) = 0.0165$

So rolling 5 fives or sixes in 6 throws is the more probable event.

11 The question has been amended to state that Ava and Sven play n games (not x games).

$X \sim B(n, 0.4)$

a $P(X = 2) = \binom{n}{2}(0.4)^2 (0.6)^{n-2}$

$= \frac{n(n-1)}{2} \times \frac{2^2}{5^2} \times \frac{3^{n-2}}{5^{n-2}}$

$= \frac{n(n-1)}{2} \times \frac{2^2 \times 3^{n-2}}{5^n}$

$= \frac{2n(n-1) \times 3^{n-2}}{5^n}$

$= \frac{2n(n-1)}{9}\left(\frac{3}{5}\right)^n$

b $\frac{2n(n-1)}{9}\left(\frac{3}{5}\right)^n = 0.121$

$\Rightarrow n = 10$ (from GDC)

12 $X \sim B(n, p)$

$E(X) = np = 19.5$

$\text{Var}(X) = np(1-p) = 6.825$

$\therefore (1-p) = \frac{6.825}{19.5} = 0.35$

$\Rightarrow p = 0.65$

$\therefore n = \frac{19.5}{0.65} = 30$

13 Let X be the number of sixes rolled in 12 throws.

$X \sim B(12, p)$

$P(X = 2) = \binom{12}{2} p^2 (1-p)^{10}$

$= 0.283$

From GDC: $p = 0.14$ or 0.20 (2DP)

14 $X \sim B(n, p)$

$E(X) = np = 12$

$Var(X) = np(1-p) = 2^2$

$\therefore (1-p) = \dfrac{4}{12} = \dfrac{1}{3}$

$\Rightarrow p = \dfrac{2}{3}$

$\therefore n = 12 \div \dfrac{2}{3} = 18$

15 $X \sim B(4, p)$

$P(X = 3) = \binom{4}{3} p^3 (1-p) = 0.3087$

$\Rightarrow p^3 - p^4 = 0.077175$

From GDC, $p = 0.560$ or 0.891

16 $X \sim B(4, p)$

$P(X = 2) = \binom{4}{2} p^2 (1-p)^2 = \dfrac{96}{625}$

$(p(1-p))^2 = \dfrac{96}{625} \div 6 = \dfrac{16}{625}$

$p(1-p) = \dfrac{4}{25}$

$\therefore p = \dfrac{1}{5}$ or $\dfrac{4}{5}$

Exercise 23D

> **COMMENT**
>
> The Poisson distribution scales, so that if $X \sim Po(\lambda)$ is the number of events in a space of one unit then the number of events in a space of n units will have a $Po(n\lambda)$ distribution.
>
> For this reason, it can be convenient to use a subscript to indicate the number of units to which the distribution pertains. So if $X_1 \sim Po(\lambda)$ then $X_n \sim Po(n\lambda)$ and the working is clear and intuitive.

4 Let X be the number of shooting stars seen in an hour; then

$X \sim Po(12)$

$P(X > 20) = 1 - Po(X \le 20)$

$\qquad = 1 - 0.9884$

$\qquad = 0.0116$

5 Let X_n be the number of white blood cells in n high power fields.

$X_n \sim Po(4n)$

a $X_1 \sim Po(4)$

$P(X_1 = 7) = 0.0595$

b $X_6 \sim Po(24)$

$P(X_6 = 28) = 0.0548$

6 Let X_n be the number of seeds falling on an area of $n\,cm^2$.

$2\,m^2 = 20000\,cm^2$

$X_{20000} \sim Po(50000)$

$\Rightarrow X_1 \sim Po\left(\dfrac{50000}{20000}\right) = Po(2.5)$

a $X_1 \sim Po(2.5)$

$\Rightarrow E(X_1) = 2.5$

b $P(X_1 = 0) = e^{-2.5} = 0.0821$

7 Let X_n be the number of flaws in n metres of wire.

$X_1 \sim Po(1.8)$

a $P(X_1 = 1) = 1.8 e^{-1.8} = 0.298$

b $X_2 \sim Po(3.6)$

$P(X_2 \ge 1) = 1 - P(X_2 = 0)$

$\qquad = 1 - e^{-3.6}$

$\qquad = 1 - 0.0273$

$\qquad = 0.973$

8 $X \sim \text{Po}(5)$

a From GDC: $P(X \leq 5) = 0.616$

b $P(3 < X \leq 5) = P(X \leq 5) - P(X \leq 3)$
$= 0.616 - 0.265$
$= 0.351$

c $P(X \neq 4) = 1 - P(X = 4)$
$= 1 - 0.175$
$= 0.825$

d $P(3 < X \leq 5 \mid X \leq 5)$
$= \dfrac{P(\{3 < X \leq 5\} \cap \{X \leq 5\})}{P(X \leq 5)}$
$= \dfrac{P(3 < X \leq 5)}{P(X \leq 5)}$
$= \dfrac{0.351}{0.616}$
$= 0.570$

9 Let X_n be the number of people arriving over the course of n minutes.

$X_{60} \sim \text{Po}(14)$

a $X_{15} \sim \text{Po}\left(\dfrac{14}{4}\right) = \text{Po}(3.5)$

$P(X_{15} = 4) = 0.189$

b $P(X_{60} > 12 \mid X_{60} < 15)$
$= \dfrac{P(12 < X_{60} < 15)}{P(X_{60} < 15)}$
$= \dfrac{P(X_{60} \leq 14) - P(X_{60} \leq 12)}{P(X_{60} \leq 14)}$
$= \dfrac{0.570 - 0.358}{0.570}$
$= 0.372$

10 Let X be the number of eagles observed in the forest in one day.

$X \sim \text{Po}(1.4)$

a $P(X > 3) = 1 - P(X \leq 3)$
$= 1 - 0.9463$
$= 0.0537$

b $P(X = 2 \mid X \geq 1) = \dfrac{P(X \geq 1 \cap X = 2)}{P(X \geq 1)}$
$= \dfrac{P(X = 2)}{P(X \geq 1)}$
$= \dfrac{P(X = 2)}{1 - P(X = 0)}$
$= \dfrac{0.242}{1 - 0.247}$
$= 0.321$

11 $X \sim \text{Po}(m)$

a $P(X \geq 1) = 1 - P(X = 0)$
$= 1 - e^{-m} = 0.4$
$\Rightarrow e^{-m} = 0.6$
$\therefore m = -\ln(0.6) = 0.511$ is the mean of the distribution.

b $P(X > 2) = 1 - P(X \leq 1)$
$= 1 - 0.9065$
$= 0.0935$

12 $X \sim \text{Po}(m)$

a $P(X = 3) = \dfrac{m^3}{3!} e^{-m}$

$P(X < 3) = \left(\dfrac{m^2}{2!} + \dfrac{m}{1!} + \dfrac{1}{0!}\right) e^{-m}$

$\therefore \dfrac{m^3}{6} = \dfrac{m^2}{2} + m + 1$

$\Rightarrow m^3 - 3m^2 - 6m - 6 = 0$

From GDC: $m = 4.5914$ (4DP)

b $P(2 \leq X < 4) = P(X \leq 3) - P(X \leq 1)$
$= 0.327 - 0.0567$
$= 0.270$

13 $X \sim \text{Po}(m)$

$P(X > 2) = 1 - P(X \leq 2)$

$= 1 - \left(\dfrac{m^2}{2} + m + 1\right) e^{-m} = 0.3$

$\Rightarrow \left(\dfrac{m^2}{2} + m + 1\right) e^{-m} = 0.7$

From GDC, $m = 1.9138$

∴ $P(X < 2) = P(X \leq 1) = 0.430$

14 $D \sim \text{Po}(6)$, $W \sim \text{Po}(42)$

a $P(D = 6) = 0.161$

$P(W = 42) = 0.0614$

b $(P(D = 6))^7 = (0.161)^7 = 2.76 \times 10^{-6}$

c Receiving exactly 6 emails each day is only one of many possible ways to receive 42 in total over the week, and so accounts for only a small fraction of the probability. For example, receiving 5 one day, 7 another day and 6 on each of the other five days has probability $^7P_2 \times P(D = 5) \times P(D = 7) \times (P(D = 6))^5 = 0.0000993$ which is 36 times the probability of receiving exactly 6 every day.

15 Let X be the number of mistakes the teacher makes in marking one piece of homework.

$X \sim \text{Po}(1.6)$

a $P(X \geq 2) = 1 - P(X \leq 1)$

$= 1 - 0.525$

$= 0.475$

b $P(X = k) = \dfrac{1.6^k}{k!} e^{-1.6}$

$= \dfrac{1.6}{k} P(X = k-1)$ for $k \geq 1$

The ratio $\dfrac{1.6}{k}$ is greater than 1 until $k > 1.6$, so the probabilities $P(X = k)$ increase until $k = 1$ and then decrease.

∴ the most likely result is $X = 1$.

c Let Y be the number of pupils in a class of 12 who have at least one error.

$Y \sim B(12, P(X \geq 1))$

> **COMMENT**
> Be alert for questions which use the result of one distribution to provide the probability for another distribution, most often binomial. Always clearly define your new distribution, using a different letter, to keep your working clear.

$P(X \geq 1) = 1 - P(X = 0)$

$= 1 - 0.202$

$= 0.798$

∴ $Y \sim B(12, 0.798)$

$P(Y < 6) = P(Y \leq 5) = 0.00413$

16 Let X be the number of requests received for one day.

$X \sim \text{Po}(1.3)$

a $P(X = 0) = e^{-1.3} = 0.273$

b $P(X > 2) = 1 - P(X \leq 2)$

$= 1 - 0.857$

$= 0.143$

c Let Y be the number of limousines in use each day.

TABLE 23D.16

y	P(Y = y)
0	$P(X = 0) = 0.273$
1	$P(X = 1) = 0.354$
2	$P(X \geq 2) = 0.373$

Let p be the probability that a limousine is used on a particular day.

$p = P(Y = 2) + \dfrac{1}{2} P(Y = 1) = 0.550$

The expected number of days in use out of 365 is $365p = 200.9 \approx 201$

23 Discrete probability distributions

17 Let X be the number of copies requested each week.

$X \sim \text{Po}(3.2)$

a $P(X > 4) = 1 - P(X \leq 4)$
$= 1 - 0.781$
$= 0.219$

b Let Y be the number of books sold each week.

TABLE 23D.17

y	P(Y = y)
0	$P(X = 0) = 0.0408$
1	$P(X = 1) = 0.130$
2	$P(X = 2) = 0.209$
3	$P(X = 3) = 0.223$
4	$P(X \geq 4) = 0.397$

$P(X \geq 4) = 1 - P(X \leq 3)$
$= 1 - 0.603$
$= 0.397$

From the table, the most likely number sold each week is 4.

c $E(Y) = 0 \times 0.0408 + 1 \times 0.130 + 2 \times 0.209 +$
$3 \times 0.223 + 4 \times 0.397$
$= 2.81$

d Let n be the least number ordered by the shop each week.

Require that $P(X \leq n) > 98\%$

From calculator:

$P(X \leq 5) = 89.5\%$

$P(X \leq 6) = 95.5\%$

$P(X \leq 7) = 98.3\%$

Therefore the shop should order 7 copies each week in order to satisfy demand at least 98% of the time.

18 $X \sim \text{Po}(\lambda)$

$P(X = 2) = \dfrac{\lambda^2}{2} e^{-\lambda}$

$P(X = 0) + P(X = 1) = (1 + \lambda)e^{-\lambda}$

If these probabilities are equal, then

$\dfrac{\lambda^2}{2} = 1 + \lambda$

$\lambda^2 - 2\lambda - 2 = 0$

$\lambda = \dfrac{2 \pm \sqrt{4 + 8}}{2} = 1 \pm \sqrt{3}$

Since $\lambda > 0$, it follows that $\lambda = 1 + \sqrt{3}$ is the only solution.

19 $Y \sim \text{Po}(\lambda)$

a $P(Y = y) = \dfrac{\lambda^y}{y!} e^{-\lambda}$

$\Rightarrow P(Y = y + 2) = \dfrac{\lambda^{y+2}}{(y+2)!} e^{-\lambda}$

$= \dfrac{\lambda^2 \lambda^y}{(y+2)(y+1)y!} e^{-\lambda}$

$= \dfrac{\lambda^2}{(y+2)(y+1)} P(Y = y)$

b With $\lambda = 6\sqrt{2}$,

$P(Y = y + 2) = P(Y = y)$

$\Rightarrow \dfrac{72}{(y+2)(y+1)} P(Y = y) = P(Y = y)$

$\dfrac{72}{(y+2)(y+1)} = 1$

$y^2 + 3y - 70 = 0$

$(y + 10)(y - 7) = 0$

$\therefore y = 7$ (reject negative solution)

Mixed examination practice 23

Short questions

> **COMMENT**
>
> It is always wise to define your random variable clearly at the start of a question, especially in situations such as Q2 where you may want to consider the loss or profit explicitly rather than just the outcome of a die roll, or Q6 where you will use one distribution result to inform a different distribution.
>
> Defining your variable at the start of working makes the calculations clear to the reader, whether that is the examiner or yourself, when checking your answer.

1 Let X be the number of defective bottles in a sample of 20.

$X \sim B(20, 0.015)$

$P(X \geq 1) = 1 - P(X = 0)$
$= 1 - 0.985^{20}$
$= 26.1\%$

2 Let X be the expected loss on each play; then

$X = 10 - (\text{value on die})$

$E(X) = \sum_x x \, P(X = x)$

$= 10 - \left(1 \times \dfrac{1}{2} + 5 \times \dfrac{1}{5} + 10 \times \dfrac{1}{5} + \dfrac{N}{10}\right)$

$= \dfrac{13}{2} - \dfrac{N}{10}$

Require $E(X) = 1$

$\therefore \dfrac{13}{2} - \dfrac{N}{10} = 1$

$\dfrac{N}{10} = \dfrac{11}{2}$

$\Rightarrow N = 55$

3 $X \sim B(12, 0.4)$

a $E(X) = 12 \times 0.4 = 4.8$ balls

b $\text{Var}(X) = 12 \times 0.4 \times 0.6 = 2.88$

$P(X \leq 2.88) = P(X \leq 2)$
$= 0.0834$

4 Let X be the number of calls answered in a day.

$X \sim \text{Po}(35)$

a $P(X > 40) = 1 - P(X \leq 40)$
$= 1 - 0.825$
$= 0.175$

b $P(X > 35) = 1 - P(X \leq 35)$
$= 1 - 0.545$
$= 0.455$

The probability of more than 35 calls every day for 5 days is

$(P(X > 35))^5 = 0.455^5 = 0.0195$

5 Let X be the number of times Robyn hits the target in 8 attempts.

$X \sim B(8, 0.6)$

a From GDC: $P(X = 4) = 0.232$

b $P(\text{Fails to qualify}) = P(X \leq 6)$
$= 0.894$ (from GDC)

c $P(\text{Miss, Miss, Hit}) = 0.4 \times 0.4 \times 0.6$
$= 0.096$

6 Let X be the number of rainy days in August in Bangalore.

Assuming independence between days, $X \sim \text{Po}(11)$

a From GDC:
$P(X < 7) = P(X \leq 6) = 0.0786$

b Let Y be the number of years out of 10 in which there are fewer than 7 rainy days in August.

Again, assuming independence between years, $Y \sim B(10, P(X<7))$

$\therefore P(Y=5) = 0.000502$

7 Let X be the number of defective bulbs in a pack of 6.

$X \sim B(6, 0.005)$

a $P(X \geq 1) = 1 - P(X=0)$

$= 1 - (0.995)^6$

$= 0.0296$

b Let Y be the number of packs in a sample of 20 that contain at least one defective bulb.

$Y \sim B(20, P(X \geq 1))$

$\therefore P(Y > 4) = 0.000244 = 0.0244\%$

(from GDC)

8 a $X \sim Po(m)$

$P(X=0) = e^{-m} = 0.305$

$\Rightarrow m = -\ln(0.305) = 1.19$

b $Y \sim Po(k)$

$P(Y=1) = ke^{-k} = 0.2$

From GDC: $k = 0.259$ or 2.54

c $W \sim Po(\lambda)$

$P(W = w+1) = P(W=w)$

$\dfrac{\lambda^{w+1}}{(w+1)!}e^{-\lambda} = \dfrac{\lambda^w}{w!}e^{-\lambda}$

$\dfrac{\lambda \times \lambda^w}{(w+1)w!}e^{-\lambda} = \dfrac{\lambda^w}{w!}e^{-\lambda}$

$\dfrac{\lambda}{w+1} = 1$

$\lambda = w+1$

$\Rightarrow w = \lambda - 1$

9 Let X be the number of sixes from n rolls; then

$X \sim B\left(n, \dfrac{1}{6}\right)$

a $P(X \leq 2) = P(X=0) + P(X=1) + P(X=2)$

$= \binom{n}{0}\left(\dfrac{5}{6}\right)^n + \binom{n}{1}\left(\dfrac{5}{6}\right)^{n-1}\left(\dfrac{1}{6}\right)$

$+ \binom{n}{2}\left(\dfrac{5}{6}\right)^{n-2}\left(\dfrac{1}{6}\right)^2$

$= \dfrac{5^{n-2}}{6^n}\left(5^2 + 5n + \dfrac{n(n-1)}{2}\right)$

$= \dfrac{5^{n-2}}{2 \times 6^n}\left(50 + 9n + n^2\right)$

Require that $\dfrac{5^{n-2}}{2 \times 6^n}(50 + 9n + n^2) = 0.532$

From GDC, $n = 15$

b $X \sim B\left(15, \dfrac{1}{6}\right)$

$\Rightarrow P(X=2) = 0.273$

10 'More than five rolls needed to roll two sixes' is equivalent to 'Fewer than two sixes in five rolls'.

Let X be the number of sixes in five rolls.

$X \sim B\left(5, \dfrac{1}{6}\right)$

$\therefore P(X \leq 1) = 0.804$

Long questions

1 Let Y be the number of yellow ribbons in the sample of 10.

$Y \sim B\left(10, \dfrac{1}{4}\right)$

a $E(Y) = 10 \times \dfrac{1}{4} = 2.5$

b $P(Y=6) = 0.0162$ (from GDC)

c $P(Y \geq 2) = 0.756$ (from GDC)

d Expect the mode to be close to the mean for a binomial distribution.

From GDC:

$P(Y=1) = 18.8\%$
$P(Y=2) = 28.2\%$
$P(Y=3) = 25.0\%$
$P(Y=4) = 14.6\%$

From the above, the mode is 2.

e Have assumed that $P(\text{yellow}) = 0.25$ is constant.

> **COMMENT**
>
> In using a binomial distribution, we assume that each choice is independent of the previous one – that is, the probability of drawing yellow is the same each time. Since we are told that the bag contains a very large number of ribbons, this is approximately true – P(yellow) does not change much, because even after removing several ribbons, the proportion of the remaining ribbons which are yellow stays approximately one-quarter.

2 Let X_n be the number of eruptions in n hours.

$X_{24} \sim \text{Po}(20)$

a $X_1 \sim \text{Po}\left(\dfrac{20}{24}\right)$

$P(X_1 = 1) = 0.362$

b $X_{24} \sim \text{Po}(20)$

$\therefore P(X_{24} > 22) = 1 - P(X \leq 22)$

$= 1 - 0.721$

$= 0.279$

c $X_{0.5} \sim \text{Po}\left(\dfrac{20}{48}\right)$

$P(X_{0.5} = 0) = 0.659$

d 'First eruption of the day between 3 a.m. and 4 a.m.' is equivalent to 'No eruptions in 3 hours and then at least one eruption in one hour'

$X_3 \sim \text{Po}\left(\dfrac{20}{8}\right)$, $X_1 \sim \text{Po}\left(\dfrac{20}{24}\right)$

$P(X_3 = 0) \times P(X_1 > 0) = P(X_3 = 0) \times (1 - P(X_1 = 0))$

$= 0.0821 \times (1 - 0.435)$

$= 0.0464$

e $X_{7 \times 24} \sim \text{Po}(140)$

Expected volume of water in a week

$= 12\,000 \times \text{E}(X_{7 \times 24})$

$= 12\,000 \times 140$

$= 1.68 \times 10^6$ litres

f Probability of at least one eruption in an hour:

$P(X_1 > 0) = 1 - P(X_1 = 0)$

$= 1 - 0.435 = 0.565$

Let Y be the number of hours in the 8-hour period in which there is at least one eruption.

$Y \sim \text{B}(8, 0.565)$

$P(Y \geq 6) = 1 - P(Y \leq 5)$

$= 1 - 0.753$

$= 0.247$

g $P(X_1 = 1 \mid X_1 \geq 1) = \dfrac{P(X_1 = 1 \cap X_1 \geq 1)}{P(X_1 \geq 1)}$

$= \dfrac{P(X_1 = 1)}{P(X_1 \geq 1)}$

$= \dfrac{P(X_1 = 1)}{1 - P(X_1 = 0)}$

$= \dfrac{0.362}{0.565}$

$= 0.641$

3 Let X be the number of students who forget to do homework.

a $X \sim \text{B}(12, 0.05)$

$P(X \geq 1) = 1 - P(X = 0)$

$= 1 - 0.540$

$= 0.460$

b $X \sim B(n, 0.05)$

$P(X \geq 1) = 1 - P(X = 0)$
$= 1 - 0.95^n$

c Require $1 - 0.95^n \geq 0.8$

$0.95^n \leq 0.2$

$n \geq \dfrac{\log 0.2}{\log 0.95}$

$\Rightarrow n \geq 31.4$

The smallest number of students under this requirement is 32.

> **COMMENT**
>
> Remember that $\log 0.95 < 0$, so the inequality is reversed when dividing through by it.

4 Let A be the event that Anna throws a six and B be the event that Brigid throws a six.

Let A_n be the event that Anna wins on her nth throw and B_n be the event that Brigid wins on her nth throw.

a i Brigid wins on her first throw if A' then B:

$P(B_1) = P(A') \times P(B)$
$= \dfrac{5}{6} \times \dfrac{1}{6}$
$= \dfrac{5}{36}$

ii Anna wins on her second throw if A' then B' then A:

$P(A_2) = P(A') \times P(B') \times P(A)$
$= \left(\dfrac{5}{6}\right)^2 \times \dfrac{1}{6}$
$= \dfrac{25}{216}$

iii Anna wins on her nth throw if A' then B' is repeated $(n-1)$ times and followed by A:

$P(A_n) = \left(P(A') \times P(B')\right)^{n-1} \times P(A)$

$= \left(\dfrac{5}{6}\right)^{2(n-1)} \times \dfrac{1}{6}$

$= \dfrac{5^{2n-2}}{6^{2n-1}}$

b For Anna to win, either Anna wins immediately, or Anna fails then Brigid fails and then the game effectively starts again.

$\therefore p = \left(\dfrac{1}{6}\right) + \left(\dfrac{5}{6}\right)^2 \times p$

$\Rightarrow p = \dfrac{1}{6} + \dfrac{25}{36} p$

c $\dfrac{11}{36} p = \dfrac{1}{6}$

$\Rightarrow p = \dfrac{6}{11}$

$\therefore P(B) = 1 - p = \dfrac{5}{11}$

d Let X be the number of times out of six that Anna wins.

$X \sim B\left(6, \dfrac{6}{11}\right)$

$P(X > 3) = 1 - P(X \leq 3)$
$= 1 - 0.568$
$= 0.432$

5 Let X_n be the number of accidents in n weeks.

$X_1 \sim Po(2)$

$\Rightarrow X_n \sim Po(2n)$

a i $X_4 \sim Po(8)$

$P(X_4 \geq 8) = 1 - P(X_4 \leq 7)$
$= 1 - 0.453$
$= 0.547$

ii Let Y be the number of four-week periods out of 13 in which at least eight serious accidents occur.

$Y \sim B(13, P(X_4 \geq 8))$

$P(Y > 9) = 1 - P(Y \leq 9)$
$= 1 - 0.9106$
$= 0.0894$

b $P(X_n \geq 1) > 0.99$

$\Rightarrow P(X_n = 0) < 0.01$

$X_n \sim \text{Po}(2n)$

$\therefore e^{-2n} < 0.01$

$-2n < \ln(0.01)$

$n > \dfrac{1}{2}\ln(100)$

$n > 2.30$

So the least such n is 3.

6

TABLE 23ML.6.1

Score		Die 1					
		1	2	3	4	5	6
Die 2	1	2	3	4	5	6	7
	2	3	4	5	6	7	8
	3	4	5	6	7	8	9
	4	5	6	7	8	9	10
	5	6	7	8	9	10	11
	6	7	8	9	10	11	12

Let A be Aleric's score.

TABLE 23ML.6.2

a	2	3	4	5	6	7	8	9	10	11	12
$P(A = a)$	$\dfrac{1}{36}$	$\dfrac{2}{36}$	$\dfrac{3}{36}$	$\dfrac{4}{36}$	$\dfrac{5}{36}$	$\dfrac{6}{36}$	$\dfrac{5}{36}$	$\dfrac{4}{36}$	$\dfrac{3}{36}$	$\dfrac{2}{36}$	$\dfrac{1}{36}$

a i $P(A=9) = \dfrac{4}{36} = \dfrac{1}{9}$

ii Let B be Bala's score.

The scores of Aleric and Bala are independent events with the same distribution.

$$\therefore P(A=9 \cap B=9) = \left(\dfrac{1}{9}\right)^2 = \dfrac{1}{81}$$

b i $P(A=B) = \sum_x (P(X=x))^2$

$$= \dfrac{1}{36^2}\left(1^2 + 2^2 + 3^2 + 4^2 + 5^2 + 6^2 + 5^2 + 4^2 + 3^2 + 2^2 + 1^2\right)$$

$$= \dfrac{146}{36^2}$$

$$= \dfrac{73}{648}$$

$$= 0.113$$

ii By symmetry, $P(A>B) = P(B>A)$

$$\therefore P(A>B) = \dfrac{1}{2}(1 - P(A=B))$$

$$= 0.444$$

c i The number shown on each die has the same uniform distribution.

$P(X \leq x) = P(\text{Roll} \leq x \text{ four times})$

$$= (P(\text{Roll} \leq x))^4$$

$$= \left(\dfrac{x}{6}\right)^4$$

ii $P(X=x) = P(X \leq x) - P(X \leq x-1)$

$$= \dfrac{x^4}{6^4} - \dfrac{(x-1)^4}{6^4}$$

$$= \dfrac{x^4 - (x-1)^4}{6^4}$$

TABLE 23ML.6.3

x	1	2	3	4	5	6
$P(X=x)$	$\dfrac{1}{1296}$	$\dfrac{15}{1296}$	$\dfrac{65}{1296}$	$\dfrac{175}{1296}$	$\dfrac{369}{1296}$	$\dfrac{671}{1296}$

iii $E(X) = \sum_x x\, P(X=x)$

$= \dfrac{1}{1296}(1\times 1 + 2\times 15 + 3\times 65 + 4\times 175 + 5\times 369 + 6\times 671)$

$= \dfrac{6797}{1296}$

$= 5.24$

7 Let A be the number of mistakes Adele makes in a letter and B be the number of mistakes Bozena makes in a letter.

$A \sim \text{Po}(2.5), \quad B \sim \text{Po}(4.1)$

a i $P(A=3) = 0.214$

ii $P(B=3) = 0.190$

b i $P(3\text{ mistakes}) = P(3\text{ mistakes}\,|\,\text{Adele})P(\text{Adele}) + P(3\text{ mistakes}\,|\,\text{Bozena})P(\text{Bozena})$

$= 0.214 \times 80\% + 0.190 \times 20\%$

$= 0.209$

ii $P(\text{Adele}\,|\,3\text{ mistakes}) = \dfrac{P(3\text{ mistakes}\,|\,\text{Adele})P(\text{Adele})}{P(3\text{ mistakes})}$

$= \dfrac{0.214 \times 80\%}{0.209}$

$= 0.818$

c

TABLE 23ML.7 Cases for a total of three mistakes

n	P(A = n)	P(B = 3 − n)	P(A = n) × P(B = 3 − n)
0	0.0821	0.190	0.0156
1	0.205	0.139	0.0286
2	0.257	0.0679	0.0174
3	0.214	0.0166	0.00354
Total			0.0652

$P(A > B\,|\,A+B=3) = \dfrac{0.0174 + 0.00354}{0.0652}$

$= 0.322$

24 Continuous distributions

Exercise 24A

4 a $\int_{-\infty}^{\infty} f(g)\,dg = \int_0^{\pi} kg^2\,dg$

$$= \left[\frac{kg^3}{3}\right]_0^{\pi}$$

$$= \frac{k\pi^3}{3}$$

$\frac{k\pi^3}{3} = 1$

$\Rightarrow k = \frac{3}{\pi^3} = 0.0968$ (3SF)

b $P\left(G < \frac{\pi}{3}\right) = \int_0^{\pi/3} \frac{3}{\pi^3} g^2\,dg$

$$= \left[\frac{g^3}{\pi^3}\right]_0^{\pi/3}$$

$$= \frac{1}{27}$$

Expected number out of 10 000 is

$10000 \times \frac{1}{27} = 370$ (to the nearest integer)

5 $P(Y > 2) = \int_2^{\infty} 3e^{-3y}\,dy$

$$= \left[-e^{-3y}\right]_2^{\infty}$$

$$= 0 - (-e^{-6})$$

$$= e^{-6}$$

6 IQR $= b - a$ where

$\int_{-\infty}^{a} f(x)\,dx = \int_{b}^{\infty} f(x)\,dx = \frac{1}{4}$

$\int_{-\infty}^{a} f(x)\,dx = \int_0^{a} \sec^2 x\,dx$

$= [\tan x]_0^{a}$

$= \tan a = \frac{1}{4}$

$\Rightarrow a = \arctan\left(\frac{1}{4}\right)$

$\int_{b}^{\infty} f(x)\,dx = \int_{b}^{\pi/4} \sec^2 x\,dx$

$= [\tan x]_b^{\pi/4}$

$= 1 - \tan b = \frac{1}{4}$

$\Rightarrow b = \arctan\left(\frac{3}{4}\right)$

$\therefore$ IQR $= \arctan\left(\frac{3}{4}\right) - \arctan\left(\frac{1}{4}\right) = 0.399$

> **COMMENT**
> Note that you could express this value exactly, and are expected to be able to do so in a non-calculator question:
>
> $\tan(\text{IQR}) = \tan\left(\arctan\left(\frac{3}{4}\right) - \arctan\left(\frac{1}{4}\right)\right)$
>
> $= \dfrac{\frac{3}{4} - \frac{1}{4}}{1 + \frac{3}{4} \times \frac{1}{4}}$
>
> $= \dfrac{8}{19}$
>
> $\therefore$ IQR $= \arctan\left(\dfrac{8}{19}\right)$

7 a Assuming $b \geq 1$ and $b^2 \leq e$,

$$P(b < X < b^2) = \int_b^{b^2} \frac{1}{x} dx = k$$

$$[\ln x]_b^{b^2} = k$$

$$\ln b^2 - \ln b = k$$

$$\ln b = k$$

$$\therefore b = e^k$$

b Assuming that $2 - a \geq 1$ and $2 + a \leq e$

$$P(2-a < X \leq 2+a) = \int_{2-a}^{2+a} \frac{1}{x} dx = k$$

$$[\ln x]_{2-a}^{2+a} = k$$

$$\ln(2+a) - \ln(2-a) = k$$

$$\ln\left(\frac{2+a}{2-a}\right) = k$$

$$\frac{2+a}{2-a} = e^k$$

$$2 + a = (2-a)e^k$$

$$a(e^k + 1) = 2e^k - 2$$

$$\therefore a = 2 \frac{e^k - 1}{e^k + 1}$$

8 $\int_{-\infty}^{\infty} f(x) dx = \int_k^{2k} e^x dx$

$$= [e^x]_k^{2k}$$

$$= e^{2k} - e^k = 1$$

$$\therefore e^{2k} - e^k - 1 = 0$$

$$\Rightarrow e^k = \frac{1 \pm \sqrt{1^2 + 4}}{2} = \frac{1 + \sqrt{5}}{2}$$

(choose positive root since $e^k > 0$)

$$P\left(X > \frac{3k}{2}\right) = \int_{3k/2}^{2k} e^x dx$$

$$= [e^x]_{3k/2}^{2k}$$

$$= e^{2k} - e^{3k/2}$$

$$= \left(\frac{1+\sqrt{5}}{2}\right)^2 - \left(\frac{1+\sqrt{5}}{2}\right)^{3/2}$$

$$= 0.560$$

Exercise 24B

3 a $E(X) = \int_{-\infty}^{\infty} xf(x) dx$

$$= \int_0^2 x \times \frac{3}{20}(4x^2 - x^3) dx$$

$$= \frac{3}{20} \int_0^2 4x^3 - x^4 dx$$

$$= \frac{3}{20} \left[x^4 - \frac{x^5}{5}\right]_0^2$$

$$= \frac{3}{20}\left(16 - \frac{32}{5}\right)$$

$$= 1.44$$

b For a local maximum of

$y = \frac{3}{20}(4x^2 - x^3)$, $\frac{dy}{dx} = 0$:

$$\frac{3x}{20}(8 - 3x) = 0$$

$$\therefore x = 0 \text{ or } \frac{8}{3}$$

$y = \frac{3}{20}(4x^2 - x^3)$ is a negative cubic, so (the double root) $x = 0$ is a minimum and $x = \frac{8}{3}$ is a maximum. However, the latter is outside the interval $]0, 2]$ where the pdf $f(x)$ is non-zero.

Therefore the maximum of $f(x)$ occurs at $x = 2$; that is, the mode is 2.

4 a $1 = \int_{-\infty}^{\infty} f(b) db$

$$= \int_3^{10} ab^2 db$$

$$= \left[\frac{ab^3}{3}\right]_3^{10}$$

$$= \frac{1000a}{3} - \frac{27a}{3}$$

$$= \frac{973}{3} a$$

$$\therefore a = \frac{3}{973}$$

b $E(B) = \int_{-\infty}^{\infty} bf(b)\,db$

$= \int_{3}^{10} ab^3\,db$

$= \left[\dfrac{ab^4}{4}\right]_{3}^{10}$

$= \left(\dfrac{10\,000}{4} - \dfrac{81}{4}\right)a$

$= \dfrac{9919}{4} \times \dfrac{3}{973}$

$= \dfrac{29\,757}{3892}$

$= 7.65\ (3\text{SF})$

> **COMMENT**
>
> We can assert that $\lim_{y \to \infty}\left(ye^{-ky}\right) = 0$, since the ratio $\dfrac{e^{kx}}{x}$ tends to infinity as $x \to \infty$.
>
> The formal proof can be accomplished using any one of several methods, some of which (such as l'Hôpital's rule) are covered in the Calculus option, and in another context you may be expected to offer such a proof. In this question that particular detail is not being explicitly tested, so you need not offer any further reasoning.

5 a For f to be a pdf, require $f(y) \geq 0$ for all y and that the integral across all real y equals 1.

For $k > 0$, it is true that $ke^{-ky} > 0$ for all real y, since $e^x > 0$ for all values of x.

$\int_{-\infty}^{\infty} f(y)\,dy = \int_{0}^{\infty} ke^{-ky}\,dy$

$= \left[-e^{-ky}\right]_{0}^{\infty}$

$= 0 - (-1)$

$= 1$

$\therefore f(y)$ fulfils the criteria to be a pdf.

b $E(Y) = \int_{-\infty}^{\infty} yf(y)\,dy = \int_{0}^{\infty} kye^{-ky}\,dy$

Integrating by parts, set:

$u = y \Rightarrow \dfrac{du}{dy} = 1$

$\dfrac{dv}{dy} = ke^{-ky} \Rightarrow v = -e^{-ky}$

$\int u\dfrac{dv}{dy}\,dy = uv - \int v\dfrac{du}{dy}\,dy$

$\therefore E(Y) = \left[-ye^{-ky}\right]_{0}^{\infty} + \int_{0}^{\infty} e^{-ky}\,dy$

$= \left[-ye^{-ky} - \dfrac{1}{k}e^{-ky}\right]_{0}^{\infty}$

$= (0 - 0) - \left(0 - \dfrac{1}{k}\right) = \dfrac{1}{k}$

6 a $1 = \int_{-\infty}^{\infty} f(y)\,dy$

$= \int_{-k}^{k} ay^2\,dy$

$= \left[\dfrac{ay^3}{3}\right]_{-k}^{k}$

$= \dfrac{ak^3}{3} - \left(-\dfrac{ak^3}{3}\right)$

$= \dfrac{2}{3}ak^3$

$\therefore a = \dfrac{3}{2k^3}$

b $\text{Var}(Y) = \int_{-\infty}^{\infty} y^2 f(y)\,dy - \left(\int_{-\infty}^{\infty} yf(y)\,dy\right)^2$

$= \int_{-k}^{k} ay^4\,dy - \left(\int_{-k}^{k} ay^3\,dy\right)^2$

$= \left[\dfrac{ay^5}{5}\right]_{-k}^{k} - \left(\left[\dfrac{ay^4}{4}\right]_{-k}^{k}\right)^2$

$= \dfrac{2ak^5}{5} - 0^2$

$= \dfrac{2k^5}{5} \times \dfrac{3}{2k^3}$

$= \dfrac{3k^2}{5}$

$$\frac{3k^2}{5} = 5 \Rightarrow k^2 = \frac{25}{3}$$

$$\Rightarrow k = \frac{5}{\sqrt{3}}$$

(Pick positive root only, since $-k < y < k$ makes no sense as an interval for $k < 0$.)

7 First, note that since we are told $f(x)$ is a pdf, we know that

$$\int_{-\infty}^{\infty} f(x)\,dx = \int_{-\infty}^{\infty} \frac{1}{\sqrt{2\pi}} e^{-\frac{x^2}{2}}\,dx = 1$$

> **COMMENT**
> Your teacher may be concerned that the above is not rigorous, but at the level of mathematical knowledge required by the International Baccalaureate it is perfectly adequate. Properly, you should actually calculate one of the integrals, which you can do using a substitution $t = \frac{u^2}{2}$, and show that it is finite (the integral result is $\frac{1}{\sqrt{2\pi}}$).

Calculating $E(X)$:

$$E(X) = \int_{-\infty}^{\infty} x f(x)\,dx$$

$$= \int_{-\infty}^{\infty} \frac{x}{\sqrt{2\pi}} e^{-\frac{x^2}{2}}\,dx$$

$$= \int_{0}^{\infty} \frac{x}{\sqrt{2\pi}} e^{-\frac{x^2}{2}}\,dx + \int_{-\infty}^{0} \frac{x}{\sqrt{2\pi}} e^{-\frac{x^2}{2}}\,dx$$

Perform two substitutions. In the first integral, take $x = u$, which just replaces the letter x with the letter u. In the second integral, take $x = -u$, so that $dx = -du$. Then

$$E(X) = \int_{0}^{\infty} \frac{u}{\sqrt{2\pi}} e^{-\frac{u^2}{2}}\,du + \int_{x=-\infty}^{x=0} \frac{-u}{\sqrt{2\pi}} e^{-\frac{u^2}{2}}\,du$$

$$= \int_{0}^{\infty} \frac{u}{\sqrt{2\pi}} e^{-\frac{u^2}{2}}\,du + \int_{u=\infty}^{u=0} \frac{u}{\sqrt{2\pi}} e^{-\frac{u^2}{2}}\,du$$

$$= \int_{0}^{\infty} \frac{u}{\sqrt{2\pi}} e^{-\frac{u^2}{2}}\,du - \int_{0}^{\infty} \frac{u}{\sqrt{2\pi}} e^{-\frac{u^2}{2}}\,du$$

(since reversing the limits of an integral reverses its sign)

$$= 0$$

Alternatively, this result can be argued directly:
By the symmetry of the function about $x = 0$, it follows that $E(X) = 0$.

$$\mathrm{Var}(X) = E(X^2) - (E(X))^2$$

$$= \int_{-\infty}^{\infty} \frac{x^2}{\sqrt{2\pi}} e^{-\frac{x^2}{2}}\,dx - 0$$

Integrating by parts, set:

$$u = x \Rightarrow \frac{du}{dx} = 1$$

$$\frac{dv}{dx} = \frac{x}{\sqrt{2\pi}} e^{-\frac{x^2}{2}} \Rightarrow v = \int \frac{x}{\sqrt{2\pi}} e^{-\frac{x^2}{2}}\,dx$$

$$= -\frac{1}{\sqrt{2\pi}} e^{-\frac{x^2}{2}}$$

$$\int u \frac{dv}{dx}\,dx = uv - \int v \frac{du}{dx}\,dx$$

$$\therefore \int_{-\infty}^{\infty} \frac{x^2}{\sqrt{2\pi}} e^{-\frac{x^2}{2}}\,dx = \left[-\frac{x}{\sqrt{2\pi}} e^{-\frac{x^2}{2}} \right]_{-\infty}^{\infty}$$

$$+ \int_{-\infty}^{\infty} \frac{1}{\sqrt{2\pi}} e^{-\frac{x^2}{2}}\,dx$$

$$= 0 + \int_{-\infty}^{\infty} f(x)\,dx$$

$$= 1$$

> **COMMENT**
> As in question 5 of this exercise, you need to assert that x multiplied by a negative exponential of x tends to zero as x tends to infinity.

Exercise 24C

4 Let X be the life of a battery of this brand; then $X \sim N(16, 5^2)$, $x = 10.2$

  a $\dfrac{\mu - x}{\sigma} = \dfrac{16 - 10.2}{5} = 1.16$ standard deviations below the mean

  b $P(X < 10.2) = P(Z < -1.16)$
  $= 0.123$ (from GDC)

5 Let X be the length of one of Ali's jumps; then $X \sim N(5.2, 0.7^2)$

  a $P(5 < X < 5.5) = P(X < 5.5) - P(X < 5)$
  $= 0.666 - 0.388$
  $= 0.278$ (from GDC)

  b i $P(X \geq 6) = 0.127$ (from GDC)

  ii $P(\text{Qualify}) = 1 - P(\text{Fail three times})$
  $= 1 - (P(X < 6))^3$
  $= 0.334$ (from GDC)

6 Let X be the weight of a cat of this breed; then $X \sim N(16, 4^2)$
  $P(X > 13) = 0.773$ (from GDC)
  $\therefore$ expected number in a sample of 2000 is $0.773 \times 2000 = 1547$

7 $D \sim N(250, 20^2)$

  a $P(265 < D < 280) = P(D < 280) - P(D < 265)$
  $= 0.933 - 0.773$
  $= 0.160$ (from GDC)

  b $P(D > 265 \mid D < 280) = \dfrac{P(265 < D < 280)}{P(D < 280)}$
  $= \dfrac{0.160}{0.933}$
  $= 0.171$

  c $P(D < 242 \cup D > 256)$
  $= 1 - P(242 \leq D \leq 256)$
  $= 1 - P(D \leq 256) + P(D < 242)$
  $= 1 - 0.618 + 0.345$
  $= 0.727$

8 $Q \sim N(4, 160)$

  a $P(|Q| > 5) = 1 - P(-5 \leq Q \leq 5)$
  $= 1 - P(Q \leq 5) + P(Q < -5)$
  $= 1 - 0.532 + 0.238$
  $= 0.707$ (from GDC)

  b $P(Q > 5 \mid |Q| > 5) = \dfrac{P(Q > 5 \cap |Q| > 5)}{P(|Q| > 5)}$
  $= \dfrac{P(Q > 5)}{P(|Q| > 5)}$
  $= \dfrac{1 - 0.532}{0.707}$
  $= 0.663$ (from GDC)

9 Let X be the weight of an apple; then $X \sim N(150, 25^2)$

  a $P(120 < X < 170) = P(X < 170) - P(X < 120)$
  $= 0.788 - 0.115$
  $= 0.673$ (from GDC)

  b Let Y be the number of medium apples in a bag of 10; then $Y \sim B(10, 0.673)$
  From GDC: $P(Y \geq 8) = 0.314$

10 Let X be the wingspan of a pigeon; then $X \sim N(60, 6^2)$

  a From GDC: $P(X > 50) = 0.952$

  b $P(X > 55 \mid X > 50) = \dfrac{P(X > 55 \cap X > 50)}{P(X > 50)}$
  $= \dfrac{P(X > 55)}{P(X > 50)}$
  $= \dfrac{0.798}{0.952}$
  $= 0.838$ (from GDC)

11 Let X be the width of a grain of sand; then $X \sim N(2, 0.5^2)$

a From GDC: $P(X > 1.5) = 0.841$

b
$$P(X > 1.5 \mid X < 2.5) = \frac{P(1.5 < X < 2.5)}{P(X < 2.5)}$$
$$= \frac{P(X < 2.5) - P(X < 1.5)}{P(X < 2.5)}$$
$$= \frac{0.841 - 0.159}{0.841}$$
$$= 0.811 \text{ (from GDC)}$$

12 Let X be the amount of paracetamol in a tablet; then $X \sim N(500, 160^2)$

$P(X < 300) = 0.106$ (from GDC)

Let Y be the number of people in the sample of 20 who get a less than effective dose.

$Y \sim B(20, 0.106)$
$\Rightarrow P(Y \geq 2) = 0.640$ (from GDC)

13 $X \sim N(7\sigma, \sigma^2)$

$$P(X < 5\sigma) = P\left(Z < \frac{5\sigma - 7\sigma}{\sigma}\right)$$
$$= P(Z < -2) = 0.0228$$

14 $X \sim N(\mu, \sigma^2)$

$P(X \leq x) = k$

By symmetry about μ,
$P(X < \mu + a) = P(X > \mu - a)$

$P(X \leq 2\mu - x) = P(X \leq \mu + (\mu - x))$
$\qquad = P(X \geq \mu - (\mu - x))$
  by the symmetry argument above
$\qquad = P(X \geq x) = 1 - k$

Exercise 24D

4 Let X be the score in an IQ test; then $X \sim N(100, 20^2)$

$P(X > x) = 2\%$
$\Rightarrow P(X \leq x) = 0.98$
$\Rightarrow x = 141$ (from GDC)

5 Let X be the mass of a rabbit; then $X \sim N(2.6, 1.2^2)$

$P(X \geq x) = 20\%$
$\Rightarrow P(X < x) = 0.8$
$\Rightarrow x = 3.61$ kg (from GDC)

6 Let X be the diameter of a bolt; then $X \sim N(\mu, 0.02^2)$

$P(X > 2) = 6\%$
$\Rightarrow P(X \leq 2) = 0.94$
$\Rightarrow \frac{2 - \mu}{0.02} = \Phi^{-1}(0.94) = 1.55$ (from GDC)
$\Rightarrow \mu = 2 - 1.55 \times 0.02 = 1.97$ cm

7 a Let G be the diameter of a grain of sand from Playa Gauss.
$G \sim N(\mu_G, \sigma_G^2)$

$P(G < 1) = 0.3$
$\Rightarrow \frac{1 - \mu_G}{\sigma_G} = \Phi^{-1}(0.3) = -0.524$
$\Rightarrow \mu_G = 1 + 0.524\sigma_G \quad \ldots(1)$

$P(G < 2) = 0.9$
$\Rightarrow \frac{2 - \mu_G}{\sigma_G} = \Phi^{-1}(0.9) = 1.28$
$\Rightarrow \mu_G = 2 - 1.28\sigma_G \quad \ldots(2)$

Substituting (2) into (1):
$2 - 1.28\sigma_G = 1 + 0.524\sigma_G$
$1.81\sigma_G = 1$
$\Rightarrow \sigma_G = 0.554$ mm

$\therefore \mu_G = 1.29$ mm

b Let F be the diameter of a grain of sand from Playa Fermat.
$F \sim N(\mu_F, \sigma_F^2)$

$P(F < 2) = 0.8$

$\Rightarrow \dfrac{2 - \mu_F}{\sigma_F} = \Phi^{-1}(0.8) = 0.842$

$\Rightarrow \mu_F = 2 - 0.842\sigma_F$ ...(1)

$P(F < 1) = 80\% \times 40\% = 0.32$

$\Rightarrow \dfrac{1 - \mu_F}{\sigma_F} = \Phi^{-1}(0.32) = -0.468$

$\Rightarrow \mu_F = 1 + 0.468\sigma_F$ ...(2)

Substituting (2) into (1):

$1 + 0.468\sigma_F = 2 - 0.842\sigma_F$

$1.31\sigma_F = 1$

$\Rightarrow \sigma_F = 0.764$ mm

$\therefore \mu_F = 1.36$ mm

8 Let X be the voltage of a battery.
$X \sim N(9.2 - t, 0.8^2)$

$P(X < 7) = 0.1$

$\Rightarrow \dfrac{7 - (9.2 - t)}{0.8} = \Phi^{-1}(0.1) = -1.28$

$t - 2.2 = -1.28 \times 0.8$

$\therefore t = 1.17$

Estimated time of use of the batteries is 1.17 hours, or 70.5 minutes.

9 Let X be the time a student takes to complete the test; then $X \sim N(32, 6^2)$

a From GDC: $P(X < 35) = 0.691$

b $P(X < t) = 0.9$
$\Rightarrow t = 39.7$ minutes (39 minutes and 41 seconds)

c $P(X < 30) = 0.369$
Let Y be the number of the 8 students who completed the test in less than 30 minutes.
$Y \sim B(8, 0.369)$
$\Rightarrow P(Y = 2) = 0.240$

10 If $X \sim N(\mu, \sigma^2)$, then (from GDC)

$P(\mu - 3\sigma < X < \mu + 3\sigma) = \Phi(3) - \Phi(-3)$
$= 0.99865 - 0.00135$
$= 0.9973$
$= 99.73\%$

COMMENT

This assumes that the range is centred on the mean, which, while not necessarily the case, is a valid approximation.

11 Let X be the measured temperature.
$X \sim N(\mu, \sigma^2)$

$P(X < \mu - 4) = 0.36$

$\Rightarrow \dfrac{(\mu - 4) - \mu}{\sigma} = \Phi^{-1}(0.36) = -0.358$

$-4 = -0.358 \times \sigma$

$\Rightarrow \sigma = \dfrac{4}{0.358} = 11.2$

$\therefore$ standard deviation is $11.2°C$.

12 a Normal distribution is symmetrical about the mean, so median = mean,
i.e. $\dfrac{\text{median}}{\text{mean}} = 1$

b $X \sim N(\mu, \sigma^2)$
$\Phi^{-1}(0.75) = 0.674$
$\Phi^{-1}(0.25) = -0.674$
$\therefore \text{IQR} = 0.674\sigma - (-0.674\sigma) = 1.35\sigma$

Hence $\dfrac{\sigma}{\text{IQR}} = \dfrac{1}{1.35} = 0.741$

13 $Z \sim N(0,1)$

Let $a = \Phi^{-1}(x)$, so that $P(Z < a) = x$

Then $P(Z > a) = 1 - x$

But $P(Z > a) = P(Z < -a)$, by the symmetry about 0 as shown in the diagram:

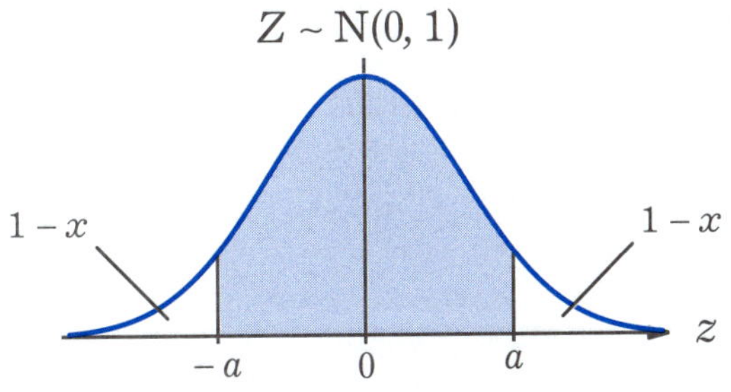

Figure 24D.13 Symmetry of standard normal distribution

$\therefore P(Z < -a) = 1 - x$

$\Rightarrow \Phi^{-1}(1-x) = -a$

So $\Phi^{-1}(x) + \Phi^{-1}(1-x) = a + (-a) = 0$

14 Let X be the breaking force for a chain link (kN).

$X \sim N(20, \sigma^2)$

a Let p be the probability of one link breaking under a force of 18 kN, so $p = P(X < 18)$

Then the probability of a 4-link chain breaking is $1 - (1-p)^4$

$1 - (1-p)^4 = 0.3$

$(1-p)^4 = 0.7$

$1 - p = 0.915$

$p = 0.0853$

b $P(X < 18) = \Phi\left(\dfrac{18 - 20}{\sigma}\right) = 0.0853$

$\Rightarrow -\dfrac{2}{\sigma} = \Phi^{-1}(0.0853) = -1.37$

$\therefore \sigma = 1.46$ kN

15 If U is a uniform continuous distribution over [0, 1], then it has pdf

$f(u) = \begin{cases} 1 & u \in [0,1] \\ 0 & \text{otherwise} \end{cases}$

so that $P(U < u) = u$ for $0 \leq u \leq 1$.

To transform U to a normal $X \sim N(\mu, \sigma^2)$, take $X = \mu + \sigma \Phi^{-1}(U)$; then

$P(X < x) = P(\mu + \sigma \Phi^{-1}(U) < x)$

$= P\left(\Phi^{-1}(U) < \dfrac{x - \mu}{\sigma}\right)$

$= P\left(U < \Phi\left(\dfrac{x - \mu}{\sigma}\right)\right)$

$= \Phi\left(\dfrac{x - \mu}{\sigma}\right)$

Mixed examination practice 24

Short questions

1 a $1 = \displaystyle\int_{-\infty}^{\infty} f(x) \, dx$

$= \displaystyle\int_0^1 k - 2x \, dx$

$= \left[kx - x^2\right]_0^1$

$= k - 1$

$\therefore k = 2$

b

$\text{Var}(X) = \displaystyle\int_{-\infty}^{\infty} x^2 f(x) \, dx - \left(\displaystyle\int_{-\infty}^{\infty} x f(x) \, dx\right)^2$

$= \displaystyle\int_0^1 2x^2 - 2x^3 \, dx - \left(\displaystyle\int_0^1 2x - 2x^2 \, dx\right)^2$

$= \left[\dfrac{2}{3}x^3 - \dfrac{1}{2}x^4\right]_0^1 - \left(\left[x^2 - \dfrac{2}{3}x^3\right]_0^1\right)^2$

$= \left(\dfrac{2}{3} - \dfrac{1}{2}\right) - \left(1 - \dfrac{2}{3}\right)^2$

$= \dfrac{1}{6} - \dfrac{1}{9}$

$= \dfrac{1}{18} = 0.0556 \text{ (3SF)}$

2 Let X be the test score of a student.
$X \sim N(62, 12^2)$

a From GDC: $P(X > 80) = 0.0668 = 6.68\%$

b $P(X \geq x) = 50\%$
$\Rightarrow x = 62$
(since the normal distribution is symmetrical about the mean)
∴ the lowest score achieved by a student in the top 50% is 62.

3 Let X be an estimate of the angle (in degrees).
$X \sim N(\mu, \sigma^2)$

$P(X < 25) = \dfrac{16}{200} = 0.08$

$\Rightarrow \dfrac{25 - \mu}{\sigma} = \Phi^{-1}(0.08) = -1.405$

$\Rightarrow \mu = 25 + 1.405\sigma$ ...(1)

$P(X > 35) = \dfrac{42}{200} = 0.21$

$\Rightarrow P(X \leq 35) = 0.79$

$\Rightarrow \dfrac{35 - \mu}{\sigma} = \Phi^{-1}(0.79) = 0.806$

$\Rightarrow \mu = 35 - 0.806\sigma$ ...(2)

Substituting (2) into (1):
$35 - 0.806\sigma = 25 + 1.405\sigma$
$2.21\sigma = 10$
∴ $\sigma = 4.52°$
and hence $\mu = 31.4°$

4 $1 = \displaystyle\int_{-\infty}^{\infty} f(x)\,dx$

$= \displaystyle\int_1^5 ax + b\,dx$

$= \left[\dfrac{ax^2}{2} + bx\right]_1^5$

$= \left(\dfrac{25}{2}a + 5b\right) - \left(\dfrac{a}{2} + b\right)$

$= 12a + 4b$

$\Rightarrow b = \dfrac{1 - 12a}{4}$

$E(X) = \displaystyle\int_{-\infty}^{\infty} xf(x)\,dx$

$= \displaystyle\int_1^5 ax^2 + bx\,dx$

$= \left[\dfrac{ax^3}{3} + \dfrac{bx^2}{2}\right]_1^5$

$= \left(\dfrac{125}{3}a + \dfrac{25}{2}b\right) - \left(\dfrac{a}{3} + \dfrac{b}{2}\right)$

$= \dfrac{124}{3}a + 12b$

$= \dfrac{124}{3}a + 3(1 - 12a)$

$= 3 + \dfrac{16}{3}a$

$E(X) = 3.5 \Rightarrow \dfrac{16}{3}a = \dfrac{1}{2}$

∴ $a = \dfrac{3}{32}$ and $b = \dfrac{1}{4} - 3a = -\dfrac{1}{32}$

5 Let X be the height of a dog of this breed; then $X \sim N(0.7, 0.05)$
From GDC: $P(X > 0.75) = 0.412$

> **COMMENT**
> Remember to use the full value from your calculator in further calculation, rather than the 3SF value you may write in working.

Let Y be the number of dogs in a sample of 6 with height greater than 0.75 m; then
$Y \sim B(6, 0.412)$

> **COMMENT**
> Always use a different letter for each variable to keep your working clear.

From GDC: $P(Y = 4) = 0.149$

6 $Z \sim N(0, 1)$ is symmetrical about the mean 0.

$$\therefore P(Z < z) = \Phi(z) = P(Z > -z)$$
$$= 1 - P(Z < -z)$$
$$\Rightarrow P(Z < -z) = 1 - \Phi(z) \quad \ldots(*)$$

$$P(|Z| < k) = P(-k < Z < k)$$
$$= P(Z < k) - P(Z < -k)$$
$$= \Phi(k) - (1 - \Phi(k)) \text{ by } (*)$$
$$= 2\Phi(k) - 1$$

Long questions

1 a $1 = \int_{-\infty}^{\infty} f(x) \, dx$

$$= \int_0^5 5ax^2 - ax^3 \, dx$$

$$= \left[\frac{5ax^3}{3} - \frac{ax^4}{4} \right]_0^5$$

$$= a \left(\frac{5^4}{3} - \frac{5^4}{4} \right)$$

$$= \frac{5^4}{12} a$$

$$\Rightarrow a = \frac{12}{5^4} = \frac{12}{625} = 0.0192$$

b $E(X) = \int_{-\infty}^{\infty} x f(x) \, dx$

$$= \int_0^5 5ax^3 - ax^4 \, dx$$

$$= \left[\frac{5ax^4}{4} - \frac{ax^5}{5} \right]_0^5$$

$$= \left(\frac{5^5}{4} - \frac{5^5}{5} \right) a$$

$$= \left(\frac{5^5}{20} \right) a$$

$$= \frac{5^5}{20} \times \frac{12}{5^4}$$

$$= 3$$

$\text{Var}(X) = \int_{-\infty}^{\infty} x^2 f(x) \, dx - (E(X))^2$

$$= \int_0^5 5ax^4 - ax^5 \, dx - 9$$

$$= \left[\frac{5ax^5}{5} - \frac{ax^6}{6} \right]_0^5 - 9$$

$$= \left(\frac{5^6}{5} - \frac{5^6}{6} \right) a - 9$$

$$= \left(\frac{5^6}{30} \right) a - 9$$

$$= \frac{5^6}{30} \times \frac{12}{5^4} - 9$$

$$= 1$$

$\therefore$ standard deviation of X is 1.

c $P(X > 4) = \int_4^{\infty} f(x) \, dx$

$$= \int_4^5 5ax^2 - ax^3 \, dx$$

$$= \left[\frac{5ax^3}{3} - \frac{ax^4}{4} \right]_4^5$$

$$= a \left(\frac{5^4}{3} - \frac{5^4}{4} \right) - a \left(\frac{5}{3} 4^3 - 4^3 \right)$$

$$= \frac{5^4}{12} a - \frac{2}{3} \times 4^3 a$$

$$= \frac{5^4}{12} \times \frac{12}{5^4} - \frac{2}{3} \times 4^3 \times \frac{12}{5^4}$$

$$= 1 - \frac{2^9}{5^4}$$

$$= \frac{113}{625} = 0.1808$$

d Let $Y \sim N(\mu, \sigma^2)$

$E(Y) = E(X) \Rightarrow \mu = 3$

$P(Y > 4) = P(X > 4) = 0.1808$

$\therefore P(Y < 4) = 0.8192$

$$\frac{4 - \mu}{\sigma} = \Phi^{-1}(0.8192)$$

$$\frac{1}{\sigma} = 0.912$$

$$\Rightarrow \sigma = 1.10$$

24 Continuous distributions 337

2 a $1 = \int_{-\infty}^{\infty} f(x) \, dx$

$= \int_0^1 e - ke^{kx} \, dx$

$= \left[ex - e^{kx} \right]_0^1$

$= (e - e^k) - (0 - 1)$

$= e - e^k + 1$

$\Rightarrow e^k = e$

$\therefore k = 1$

b $P\left(\frac{1}{4} < X < \frac{1}{2}\right) = \int_{1/4}^{1/2} f(x) \, dx$

$= \int_{1/4}^{1/2} e - e^x \, dx$

$= \left[ex - e^x \right]_{1/4}^{1/2}$

$= \left(\frac{1}{2} e - e^{\frac{1}{2}} \right) - \left(\frac{1}{4} e - e^{\frac{1}{4}} \right)$

$= \frac{1}{4} e - \sqrt{e} + \sqrt[4]{e}$

c $E(X) = \int_{-\infty}^{\infty} x f(x) \, dx$

$= \int_0^1 ex - xe^x \, dx$

$= \left[\frac{ex^2}{2} \right]_0^1 - \int_0^1 xe^x \, dx$

$= \frac{e}{2} - \int_0^1 xe^x \, dx$

Integrating by parts, set:

$u = x \Rightarrow \frac{du}{dx} = 1$

$\frac{dv}{dx} = e^x \Rightarrow v = e^x$

$\int u \frac{dv}{dx} dx = uv - \int v \frac{du}{dx} dx$

$\therefore \int_0^1 xe^x \, dx = \left[xe^x \right]_0^1 - \int_0^1 e^x \, dx$

$= e - \left[e^x \right]_0^1$

$= e - (e - 1)$

$= 1$

So $E(X) = \frac{e}{2} - 1$

$\text{Var}(X) = \int_{-\infty}^{\infty} x^2 f(x) \, dx - (E(X))^2$

$= \int_0^1 ex^2 - x^2 e^x \, dx - \left(\frac{e}{2} - 1 \right)^2$

$= \left[\frac{ex^3}{3} \right]_0^1 - \int_0^1 x^2 e^x \, dx - \left(\frac{e^2}{4} - e + 1 \right)$

$= \frac{4e}{3} - \frac{e^2}{4} - 1 - \int_0^1 x^2 e^x \, dx$

Integrating by parts, set:

$w = x^2 \Rightarrow \frac{dw}{dx} = 2x$

$\frac{dv}{dx} = e^x \Rightarrow v = e^x$

$\int w \frac{dv}{dx} dx = wv - \int v \frac{dw}{dx} dx$

$\therefore \int_0^1 x^2 e^x \, dx = \left[x^2 e^x \right]_0^1 - 2 \int_0^1 xe^x \, dx$

$= (e - 0) - 2 \times 1$

$= e - 2$

(using the result $\int_0^1 xe^x \, dx = 1$ obtained above)

So $\text{Var}(X) = \frac{4e}{3} - \frac{e^2}{4} - 1 - (e - 2)$

$= 1 + \frac{e}{3} - \frac{e^2}{4}$

d 6 months $= \frac{1}{2}$ year

$P\left(X > \frac{1}{2}\right) = \int_{1/2}^{\infty} f(x) \, dx$

$= \int_{1/2}^{1} e - e^x \, dx$

$= \left[ex - e^x \right]_{1/2}^{1}$

$= (e - e) - \left(\frac{1}{2} e - e^{\frac{1}{2}} \right)$

$= \sqrt{e} - \frac{e}{2}$

$= 0.290 \, (3\text{SF})$

e Let Y be the number of batteries out of three which have failed at the end of six months.

$Y \sim B(3, 1-0.290) = B(3, 0.710)$

$P(Y=0) = 0.0243$

f $P(Y=1) = 0.179$

3 a i $E(X) = \int_{-\infty}^{\infty} xf(x)dx$

$= \int_0^2 \frac{2}{3}x^2 - \frac{1}{12}x^4 \, dx$

ii $E(X) = \int_0^2 \frac{2}{3}x^2 - \frac{1}{12}x^4 \, dx$

$= \left[\frac{2}{9}x^3 - \frac{1}{60}x^5\right]_0^2$

$= \frac{16}{9} - \frac{32}{60}$

$= \frac{56}{45}$

b i The median m is defined by $P(X < m) = 0.5$:

$\frac{1}{2} = \int_{-\infty}^{m} f(x)dx$

$= \int_0^m \frac{2}{3}x - \frac{1}{12}x^3 \, dx$

$= \left[\frac{1}{3}x^2 - \frac{1}{48}x^4\right]_0^m$

$= \frac{1}{3}m^2 - \frac{1}{48}m^4$

$\therefore 24 = 16m^2 - m^4$

$m^4 - 16m^2 + 24 = 0$

ii The above equation is a quadratic in m^2, with solutions

$m^2 = \frac{16 \pm \sqrt{16^2 - 4 \times 24}}{2} = 8 \pm 2\sqrt{10}$

Require $m \in [0, 2]$, so $m^2 = 8 - 2\sqrt{10}$

and hence $m = \sqrt{8 - 2\sqrt{10}}$

c The mode q is such that $f(x) \le f(q)$ for all x.

For stationary points of $f(x)$:

$f'(x) = \frac{2}{3} - \frac{1}{4}x^2 = 0$

$x^2 = \frac{8}{3}$

$x = \sqrt{\frac{8}{3}}$ (for $x \in [0, 2]$)

Compare value of f at stationary point and end points:

$f\left(\sqrt{\frac{8}{3}}\right) = 0.726$

$f(0) = 0$

$f(2) = \frac{8}{12} = 0.667$

$f\left(\sqrt{\frac{8}{3}}\right)$ is the greatest, so mode is $\sqrt{\frac{8}{3}}$

4 Let X (Pesos) be the monthly salary in Argentina.

$X \sim N(1500, \sigma^2)$

a $P(X > 2000) = 0.3$

$\Rightarrow P(X < 2000) = 0.7$

$\Rightarrow \frac{2000 - 1500}{\sigma} = \Phi^{-1}(0.7) = 0.524$

$\therefore \sigma = \frac{500}{0.524} = 953$ Pesos

b $P(X > 3000) = 1 - P(X < 3000)$

$= 1 - \Phi\left(\frac{3000 - 1500}{\sigma}\right)$

$= 0.0578$

24 Continuous distributions

c $P(X > 3000 | X > 2000) = \dfrac{P(X > 3000 \cap X > 2000)}{P(X > 2000)}$

$= \dfrac{P(X > 3000)}{P(X > 2000)}$

$= \dfrac{1 - P(X < 3000)}{0.3}$

$= 0.193$

d Let Y be the number of people in a random sample of three who earn less than 2000 Pesos a month.

$Y \sim B(3, 0.7)$

$P(Y \geq 2) = 1 - P(Y \leq 1)$

$= 1 - 0.216$

$= 0.784$

e The distribution of salaries is likely to be skewed rather than symmetrical, because there will be a small proportion of workers with very high salaries, while the majority have low salaries.

Moreover, if the normal model is used, the data here would suggest that $P(X < 0) = \Phi\left(-\dfrac{1500}{\sigma}\right) = 5.78\%$, i.e. around 6% of the population would have negative salaries!

25 Mathematical induction

Exercise 25B

1 $u_n = 2 \times 3^{n-1}$

Proposition: $S_n = \sum_{r=1}^{n} u_r = 3^n - 1$

Base case

For $n = 1$: $S_1 = u_1 = 2 = 3^1 - 1$

∴ the proposition is true for $n = 1$.

Inductive step

Assume the statement is true for $n = k$;
that is, $S_k = 3^k - 1$

Working towards: $S_{k+1} = 3^{k+1} - 1$

$S_{k+1} = S_k + u_{k+1}$
$= 3^k - 1 + 2 \times 3^k$
(using the formulae for S_k and u_{k+1})
$= 3 \times 3^k - 1$
$= 3^{k+1} - 1$

So if the statement is true for $n = k$ then it is also true for $n = k+1$.

The proposition is true for $n = 1$, and if true for $n = k$ it is also true for $n = k+1$. Therefore the proposition is true for all $n \in \mathbb{Z}^+$ by the principle of mathematical induction.

2 $u_n = n^2$

Proposition: $S_n = \sum_{r=1}^{n} u_r = \frac{n(n+1)(2n+1)}{6}$

Base case

For $n = 1$: $S_1 = u_1 = 1 = \frac{1(2)(3)}{6}$

∴ the proposition is true for $n = 1$.

Inductive step

Assume the statement is true for $n = k$;
that is,

$S_k = \frac{k(k+1)(2k+1)}{6}$

Working towards: $S_{k+1} = \frac{(k+1)(k+2)(2k+3)}{6}$

$S_{k+1} = S_k + u_{k+1}$
$= \frac{k(k+1)(2k+1)}{6} + (k+1)^2$
(using the formulae for S_k and u_{k+1})
$= \frac{(2k^3 + 3k^2 + k) + 6(k^2 + 2k + 1)}{6}$
$= \frac{2k^3 + 9k^2 + 13k + 6}{6}$
$= \frac{(k+1)(k+2)(2k+3)}{6}$

So if the statement is true for $n = k$ then it is also true for $n = k+1$.

The proposition is true for $n = 1$, and if true for $n = k$ it is also true for $n = k+1$. Therefore the proposition is true for all $n \in \mathbb{Z}^+$ by the principle of mathematical induction.

3 $u_n = n^3$

Proposition: $S_n = \sum_{r=1}^{n} u_r = \frac{n^2(n+1)^2}{4}$

Base case

For $n = 1$: $S_1 = u_1 = 1 = \dfrac{(1)^2(2)^2}{4}$

∴ the proposition is true for $n = 1$.

Inductive step

Assume the statement is true for $n = k$; that is,

$$S_k = \dfrac{k^2(k+1)^2}{4}$$

Working towards: $S_{k+1} = \dfrac{(k+1)^2(k+2)^2}{4}$

$S_{k+1} = S_k + u_{k+1}$

$= \dfrac{k^2(k+1)^2}{4} + (k+1)^3$

$= \dfrac{(k^4 + 2k^3 + k^2) + 4(k^3 + 3k^2 + 3k + 1)}{4}$

$= \dfrac{k^4 + 6k^3 + 13k^2 + 12k + 4}{4}$

$= \dfrac{(k^2 + 2k + 1)(k^2 + 4k + 4)}{4}$

$= \dfrac{(k+1)^2(k+2)^2}{4}$

So if the statement is true for $n = k$ then it is also true for $n = k+1$.

The proposition is true for $n = 1$, and if true for $n = k$ it is also true for $n = k+1$. Therefore the proposition is true for all $n \in \mathbb{Z}^+$ by the principle of mathematical induction.

4 $u_n = \dfrac{1}{n(n+1)}$

Proposition: $S_n = \sum_{r=1}^{n} u_r = \dfrac{n}{n+1}$

Base case

For $n = 1$: $S_1 = u_1 = \dfrac{1}{1(2)} = \dfrac{1}{2} = \dfrac{1}{1+1}$

∴ the proposition is true for $n = 1$.

Inductive step

Assume the statement is true for $n = k$; that is, $S_k = \dfrac{k}{k+1}$

Working towards: $S_{k+1} = \dfrac{k+1}{k+2}$

$S_{k+1} = S_k + u_{k+1}$

$= \dfrac{k}{k+1} + \dfrac{1}{(k+1)(k+2)}$

$= \dfrac{k(k+2)+1}{(k+1)(k+2)}$

$= \dfrac{k^2+2k+1}{(k+1)(k+2)}$

$= \dfrac{(k+1)^2}{(k+1)(k+2)}$

$= \dfrac{k+1}{k+2}$

So if the statement is true for $n = k$ then it is also true for $n = k+1$.

The proposition is true for $n = 1$, and if true for $n = k$ it is also true for $n = k+1$. Therefore the proposition is true for all $n \in \mathbb{Z}^+$ by the principle of mathematical induction.

5 $u_n = \dfrac{1}{(2n-1)(2n+1)}$

Proposition: $S_n = \sum_{r=1}^{n} u_r = \dfrac{n}{2n+1}$

Base case

For $n = 1$: $S_1 = u_1 = \dfrac{1}{1 \times 3} = \dfrac{1}{3} = \dfrac{1}{2 \times 1 + 1}$

∴ the proposition is true for $n = 1$.

Inductive step

Assume the statement is true for $n = k$; that is, $S_k = \dfrac{k}{2k+1}$

Working towards: $S_{k+1} = \dfrac{k+1}{2k+3}$

$S_{k+1} = S_k + u_{k+1}$

$= \dfrac{k}{2k+1} + \dfrac{1}{(2k+1)(2k+3)}$

$= \dfrac{k(2k+3)+1}{(2k+1)(2k+3)}$

$= \dfrac{2k^2+3k+1}{(2k+1)(2k+3)}$

$= \dfrac{(2k+1)(k+1)}{(2k+1)(2k+3)}$

$= \dfrac{k+1}{2k+3}$

So if the statement is true for $n = k$ then it is also true for $n = k+1$.

The proposition is true for $n = 1$, and if true for $n = k$ it is also true for $n = k+1$. Therefore the proposition is true for all $n \in \mathbb{Z}^+$ by the principle of mathematical induction.

6 $u_n = n \times n!$

Proposition: $S_n = \sum_{r=1}^{n} u_r = (n+1)! - 1$

<u>Base case</u>

For $n = 1$: $S_1 = u_1 = 1 \times 1! = 1 = 2! - 1$

$\therefore$ the proposition is true for $n = 1$.

<u>Inductive step</u>

Assume the statement is true for $n = k$;

that is, $S_k = (k+1)! - 1$

Working towards: $S_{k+1} = (k+2)! - 1$

$S_{k+1} = S_k + u_{k+1}$

$= (k+1)! - 1 + (k+1) \times (k+1)!$

$= (k+1)!(1+(k+1)) - 1$

$= (k+2)(k+1)! - 1$

$= (k+2)! - 1$

So if the statement is true for $n = k$ then it is also true for $n = k+1$.

The proposition is true for $n = 1$, and if true for $n = k$ it is also true for $n = k+1$. Therefore the proposition is true for all $n \in \mathbb{Z}^+$ by the principle of mathematical induction.

7 $u_n = (-1)^{n-1} n^2$

Proposition: $S_n = \sum_{r=1}^{n} u_r = (-1)^{n-1} \dfrac{n(n+1)}{2}$

<u>Base case</u>

For $n = 1$: $S_1 = u_1 = 1 = (-1)^0 \times \dfrac{1(2)}{2}$

$\therefore$ the proposition is true for $n = 1$.

<u>Inductive step</u>

Assume the statement is true for $n = k$;

that is, $S_k = (-1)^{k-1} \dfrac{k(k+1)}{2}$

Working towards: $S_{k+1} = (-1)^k \dfrac{(k+1)(k+2)}{2}$

$S_{k+1} = S_k + u_{k+1}$

$= (-1)^{k-1} \dfrac{k(k+1)}{2} + (-1)^k (k+1)^2$

$= (-1)^k (k+1) \left(-\dfrac{k}{2} + k + 1 \right)$

$= (-1)^k (k+1) \left(\dfrac{k}{2} + 1 \right)$

$= (-1)^k \dfrac{(k+1)(k+2)}{2}$

So if the statement is true for $n = k$ then it is also true for $n = k+1$.

The proposition is true for $n = 1$, and if true for $n = k$ it is also true for $n = k+1$. Therefore the proposition is true for all $n \in \mathbb{Z}^+$ by the principle of mathematical induction.

25 Mathematical induction

8 $u_n = n$

Proposition: $S_{2n} - S_n = \sum_{r=n+1}^{2n} u_r = \frac{1}{2}n(3n+1)$

> **COMMENT**
> Be alert for the opportunity to use a difference of two series to keep a formula simple.

Base case

For $n = 1$: $S_2 - S_1 = u_2 = 2 = \frac{1}{2} \times 1 \times (3+1)$

∴ the proposition is true for $n = 1$.

Inductive step

Assume the statement is true for $n = k$;

that is, $S_{2k} - S_k = \frac{1}{2}k(3k+1)$

Working towards:

$S_{2(k+1)} - S_{k+1} = \frac{1}{2}(k+1)(3k+4)$

$S_{2k+2} - S_{k+1} = S_{2k} - S_k + u_{2k+1} + u_{2k+2} - u_{k+1}$

$= \frac{1}{2}k(3k+1) + (2k+1) + (2k+2) - (k+1)$

(using the formulae for $S_{2k} - S_k$ and $u_{k+1}, u_{2k+1}, u_{2k+2}$)

$= \frac{3}{2}k^2 + \frac{7}{2}k + 2$

$= \frac{1}{2}(3k^2 + 7k + 4)$

$= \frac{1}{2}(k+1)(3k+4)$

So if the statement is true for $n = k$ then it is also true for $n = k+1$.

The proposition is true for $n = 1$, and if true for $n = k$ it is also true for $n = k+1$. Therefore the proposition is true for all $n \in \mathbb{Z}^+$ by the principle of mathematical induction.

9 $u_n = n \times 2^n$

Proposition: $S_n = \sum_{r=1}^{n} u_r = (n-1)2^{n+1} + 2$

Base case

For $n = 1$: $S_1 = u_1 = 1 \times 2^1 = 2 = (1-1)2^2 + 2$

∴ the proposition is true for $n = 1$.

Inductive step

Assume the statement is true for $n = k$; that is, $S_k = (k-1)2^{k+1} + 2$

Working towards: $S_{k+1} = k2^{k+2} + 2$

$S_{k+1} = S_k + u_{k+1}$

$= (k-1)2^{k+1} + 2 + (k+1) \times 2^{k+1}$

$= 2^{k+1}(k-1+k+1) + 2$

$= 2k2^{k+1} + 2$

$= k2^{k+2} + 2$

So if the statement is true for $n = k$ then it is also true for $n = k+1$.

The proposition is true for $n = 1$, and if true for $n = k$ it is also true for $n = k+1$. Therefore the proposition is true for all $n \in \mathbb{Z}^+$ by the principle of mathematical induction.

Exercise 25C

1 $u_{n+1} = 3u_n + 2$, $u_1 = 2$

Proposition: $u_n = 3^n - 1$

Base case

For $n = 1$: $u_1 = 2 = 3^1 - 1$

∴ the proposition is true for $n = 1$.

Inductive step

Assume the statement is true for $n = k$; that is, $u_k = 3^k - 1$

Working towards: $u_{k+1} = 3^{k+1} - 1$

$u_{k+1} = 3u_k + 2$

$= 3 \times (3^k - 1) + 2$

(using the recurrence relation and formula for u_k)

$= 3^{k+1} - 3 + 2$

$= 3^{k+1} - 1$

So if the statement is true for $n = k$ then it is also true for $n = k+1$.

The proposition is true for $n = 1$, and if true for $n = k$ it is also true for $n = k+1$. Therefore the proposition is true for all $n \in \mathbb{Z}^+$ by the principle of mathematical induction.

2 $U_{n+1} = 5U_n + 4$, $U_1 = 4$

Base case

For $n = 1$: $U_1 = 4 = 5^1 - 1$

$\therefore$ the proposition is true for $n = 1$.

Inductive step

Assume the statement is true for $n = k$; that is, $U_k = 5^k - 1$

Working towards: $U_{k+1} = 5^{k+1} - 1$

$U_{k+1} = 5U_k + 4$

$= 5(5^k - 1) + 4$

(using the recurrence relation and formula for U_k)

$= 5^{k+1} - 5 + 4$

$= 5^{k+1} - 1$

So if the statement is true for $n = k$ then it is also true for $n = k+1$.

The proposition is true for $n = 1$, and if true for $n = k$ it is also true for $n = k+1$. Therefore the proposition is true for all $n \in \mathbb{Z}^+$ by the principle of mathematical induction.

3 $U_{n+1} = 5U_n - 8$, $U_1 = 3$

Proposition: $U_n = 5^{n-1} + 2$

Base case

For $n = 1$: $U_1 = 3 = 5^0 + 2$

$\therefore$ the proposition is true for $n = 1$.

Inductive step

Assume the statement is true for $n = k$; that is, $U_k = 5^{k-1} + 2$

Working towards: $U_{k+1} = 5^k + 2$

$U_{k+1} = 5U_k - 8$

$= 5(5^{k-1} + 2) - 8$

(using the recurrence relation and formula for U_k)

$= 5^k + 10 - 8$

$= 5^k + 2$

So if the statement is true for $n = k$ then it is also true for $n = k+1$.

The proposition is true for $n = 1$, and if true for $n = k$ it is also true for $n = k+1$. Therefore the proposition is true for all $n \in \mathbb{Z}^+$ by the principle of mathematical induction.

4 $u_{n+1} = 3u_n + 1$, $u_1 = 1$

Proposition: $u_n = \dfrac{3^n - 1}{2}$

Base case

For $n = 1$: $u_1 = 1 = \dfrac{3^1 - 1}{2}$

$\therefore$ the proposition is true for $n = 1$.

Inductive step

Assume the statement is true for $n = k$; that is,

$u_k = \dfrac{3^k - 1}{2}$

25 Mathematical induction 345

Working towards: $u_{k+1} = \dfrac{3^{k+1}-1}{2}$

$u_{k+1} = 3u_k + 1$

$= 3 \times \dfrac{(3^k - 1)}{2} + 1$

(using the recurrence relation and formula for u_k)

$= \dfrac{3^{k+1} - 3 + 2}{2}$

$= \dfrac{3^{k+1} - 1}{2}$

So if the statement is true for $n = k$ then it is also true for $n = k+1$.

The proposition is true for $n = 1$, and if true for $n = k$ it is also true for $n = k+1$. Therefore the proposition is true for all $n \in \mathbb{Z}^+$ by the principle of mathematical induction.

5 $u_{n+2} = 5u_{n+1} - 6u_n$, $u_1 = 1$, $u_2 = 5$

Proposition: $u_n = 3^n - 2^n$

Base cases

For $n = 1$: $u_1 = 1 = 3^1 - 2^1$

For $n = 2$: $u_2 = 5 = 3^2 - 2^2$

∴ the proposition is true for $n = 1$ and for $n = 2$.

Inductive step

Assume the statement is true for $n = k$ and for $n = k+1$; that is,

$u_k = 3^k - 2^k$, $u_{k+1} = 3^{k+1} - 2^{k+1}$

Working towards: $u_{k+2} = 3^{k+2} - 2^{k+2}$

$u_{k+2} = 5u_{k+1} - 6u_k$

$= 5 \times (3^{k+1} - 2^{k+1}) - 6 \times (3^k - 2^k)$

(using the recurrence relation and formulae for u_{k+1} and u_k)

$5 \times (3^{k+1} - 2^{k+1}) - 2 \times 3^{k+1} + 3 \times 2^{k+1}$

$= (5-2) \times 3^{k+1} - (5-3) \times 2^{k+1}$

$= 3^{k+2} - 2^{k+2}$

So if the statement is true for $n = k$ and $n = k+1$, then it is also true for $n = k+2$.

The proposition is true for $n = 1$ and $n = 2$, and if true for $n = k$ and $n = k+1$ it is also true for $n = k+2$.
Therefore the proposition is true for all $n \in \mathbb{Z}^+$ by the principle of mathematical induction.

6 $u_{n+2} = 6u_{n+1} - 9u_n$, $u_1 = 3$, $u_2 = 36$

Proposition: $u_n = (3n-2)3^n$

Base cases

For $n = 1$: $u_1 = 3 = (3-2)3^1$

For $n = 2$: $u_2 = 36 = (6-2)3^2$

∴ the proposition is true for $n = 1$ and for $n = 2$.

Inductive step

Assume the statement is true for $n = k$ and for $n = k+1$; that is,

$u_k = (3k-2)3^k$, $u_{k+1} = (3k+1)3^{k+1}$

Working towards: $u_{k+2} = (3k+4)3^{k+2}$

$u_{k+1} = 6u_{k+1} - 9u_k$

$= 6(3k+1)3^{k+1} - 9(3k-2)3^k$

(using the recurrence relation and formulae for u_{k+1} and u_k)

$= 3^{k+2}(2(3k+1) - (3k-2))$

$= 3^{k+2}(3k+4)$

So if the statement is true for $n = k$ and $n = k+1$, then it is also true for $n = k+2$.

The proposition is true for $n=1$ and $n=2$, and if true for $n=k$ and $n=k+1$ it is also true for $n=k+2$.
Therefore the proposition is true for all $n \in \mathbb{Z}^+$ by the principle of mathematical induction.

7 $u_{n+2} = 5u_{n+1} - 6u_n$, $u_0 = -1$, $u_1 = -1$

Proposition: $u_n = 3^n - 2^{n+1}$

Base cases

For $n = 0$: $u_0 = -1 = 3^0 - 2^1$

For $n = 1$: $u_1 = -1 = 3^1 - 2^2$

∴ the proposition is true for $n = 0$ and for $n = 1$.

Inductive step

Assume the statement is true for $n = k$ and for $n = k+1$; that is,

$u_k = 3^k - 2^{k+1}$, $u_{k+1} = 3^{k+1} - 2^{k+2}$

Working towards: $u_{k+2} = 3^{k+2} - 2^{k+3}$

$u_{k+2} = 5u_{k+1} - 6u_k$
$= 5(3^{k+1} - 2^{k+2}) - 6(3^k - 2^{k+1})$

(using the recurrence relation and formulae for u_{k+1} and u_k)

$= 3^{k+1}(5-2) - 2^{k+2}(5-3)$
$= 3^{k+2} - 2^{k+3}$

So if the statement is true for $n = k$ and $n = k+1$, then it is also true for $n = k+2$.

The proposition is true for $n = 0$ and $n = 1$, and if true for $n = k$ and $n = k+1$ it is also true for $n = k+2$.
Therefore the proposition is true for all $n \in \mathbb{N}$ by the principle of mathematical induction.

8 $u_{n+1} = \dfrac{u_n}{u_n + 1}$, $u_1 = 1$

Proposition: $u_n = \dfrac{1}{n}$

Base case

For $n = 1$: $u_1 = 1 = \dfrac{1}{1}$

∴ the proposition is true for $n = 1$.

Inductive step

Assume the statement is true for $n = k$;

that is, $u_k = \dfrac{1}{k}$

Working towards: $u_{k+1} = \dfrac{1}{k+1}$

$u_{k+1} = \dfrac{u_k}{u_k + 1}$

$= \dfrac{\frac{1}{k}}{\frac{1}{k} + 1}$ (using the recurrence relation and formula for u_k)

$= \dfrac{\frac{1}{k}}{\frac{1}{k} + 1} \times \dfrac{k}{k}$

$= \dfrac{1}{1+k}$

So if the statement is true for $n = k$ then it is also true for $n = k+1$.

The proposition is true for $n = 1$, and if true for $n = k$ it is also true for $n = k+1$.
Therefore the proposition is true for all $n \in \mathbb{Z}^+$ by the principle of mathematical induction.

9 $u_{n+2} = u_{n+1} + u_n$, $u_1 = 1$, $u_2 = 1$

Proposition:

$u_n = \dfrac{1}{\sqrt{5}}\left(\left(\dfrac{1+\sqrt{5}}{2}\right)^n - \left(\dfrac{1-\sqrt{5}}{2}\right)^n\right)$

Let a and b be the roots of the equation $x^2 - x - 1 = 0$:

$a = \dfrac{1+\sqrt{5}}{2}$, $b = \dfrac{1-\sqrt{5}}{2}$

25 Mathematical induction 347

So the proposition becomes $u_n = \dfrac{1}{\sqrt{5}}(a^n - b^n)$

Note that $a+b=1$, $a-b=\sqrt{5}$ and hence $a^2 - b^2 = (a+b)(a-b) = \sqrt{5}$

> **COMMENT**
>
> If you are facing a question where a complicated value is likely to occur repeatedly, consider assigning the value as a constant for speed and clarity of working. If you can find a relationship which may be useful later, note it.

Base cases

For $n=1$: $u_1 = 1 = \dfrac{1}{\sqrt{5}}(a-b)$

For $n=2$: $u_2 = 1 = \dfrac{1}{\sqrt{5}}(a^2-b^2)$

∴ the proposition is true for $n=1$ and for $n=2$.

Inductive step

Assume the statement is true for $n=k$ and for $n=k+1$; that is,

$u_k = \dfrac{1}{\sqrt{5}}(a^k - b^k)$, $\quad u_{k+1} = \dfrac{1}{\sqrt{5}}(a^{k+1} - b^{k+1})$

Working towards: $u_{k+2} = \dfrac{1}{\sqrt{5}}(a^{k+2} - b^{k+2})$

$u_{k+2} = u_{k+1} + u_k$

$= \dfrac{1}{\sqrt{5}}(a^{k+1} - b^{k+1}) + \dfrac{1}{\sqrt{5}}(a^k - b^k)$

(using the recurrence relation and formulae for u_{k+1} and u_k)

$= \dfrac{1}{\sqrt{5}}(a^k(a+1) - b^k(b+1))$

But a and b are solutions of $x^2 - x - 1 = 0$, so $a+1 = a^2$ and $b+1 = b^2$. Hence

$u_{k+2} = \dfrac{1}{\sqrt{5}}(a^k(a^2) - b^k(b^2))$

$= \dfrac{1}{\sqrt{5}}(a^{k+2} - b^{k+2})$

So if the statement is true for $n=k$ and $n=k+1$, then it is also true for $n=k+2$.

The proposition is true for $n=1$ and $n=2$, and if true for $n=k$ and $n=k+1$ it is also true for $n=k+2$.
Therefore the proposition is true for all $n \in \mathbb{Z}^+$ by the principle of mathematical induction.

Exercise 25D

1. $y = \dfrac{1}{1-x}$

Proposition: $\dfrac{d^n y}{dx^n} = \dfrac{n!}{(1-x)^{n+1}}$

Base case

For $n=0$: $y = \dfrac{1}{1-x} = \dfrac{0!}{(1-x)^1}$

∴ the proposition is true for $n=0$.

Inductive step

Assume the statement is true for $n=k$; that is,

$\dfrac{d^k y}{dx^k} = \dfrac{k!}{(1-x)^{k+1}}$

Working towards: $\dfrac{d^{k+1} y}{dx^{k+1}} = \dfrac{(k+1)!}{(1-x)^{k+2}}$

$$\frac{d^{k+1}y}{dx^{k+1}} = \frac{d}{dx}\left(\frac{d^k y}{dx^k}\right)$$

$$= \frac{d}{dx}\left(\frac{k!}{(1-x)^{k+1}}\right)$$

(using the formula for $\frac{d^k y}{dx^k}$)

$$= \frac{d}{dx}\left(k!(1-x)^{-k-1}\right)$$

$$= -k!(-k-1)(1-x)^{-k-2}$$

$$= \frac{k!(k+1)}{(1-x)^{k+2}}$$

$$= \frac{(k+1)!}{(1-x)^{k+2}}$$

So if the statement is true for $n = k$ then it is also true for $n = k+1$.

The proposition is true for $n = 0$, and if true for $n = k$ it is also true for $n = k+1$. Therefore the proposition is true for all $n \in \mathbb{N}$ by the principle of mathematical induction.

2 $y = \dfrac{1}{1-3x}$

Proposition: $\dfrac{d^n y}{dx^n} = \dfrac{3^n n!}{(1-3x)^{n+1}}$

Base case

For $n = 0$: $y = \dfrac{1}{1-3x} = \dfrac{3^0 0!}{(1-3x)^1}$

∴ the proposition is true for $n = 0$.

Inductive step

Assume the statement is true for $n = k$; that is,

$$\frac{d^k y}{dx^k} = \frac{3^k k!}{(1-3x)^{k+1}}$$

Working towards: $\dfrac{d^{k+1} y}{dx^{k+1}} = \dfrac{3^{k+1}(k+1)!}{(1-3x)^{k+2}}$

$$\frac{d^{k+1}y}{dx^{k+1}} = \frac{d}{dx}\left(\frac{d^k y}{dx^k}\right)$$

$$= \frac{d}{dx}\left(\frac{3^k k!}{(1-3x)^{k+1}}\right)$$

(using the formula for $\frac{d^k y}{dx^k}$)

$$= \frac{d}{dx}\left(3^k k!(1-3x)^{-k-1}\right)$$

$$= 3^k k!(-k-1)(1-3x)^{-k-2}(-3)$$

$$= \frac{3 \times 3^k k!(k+1)}{(1-3x)^{k+2}}$$

$$= \frac{3^{k+1}(k+1)!}{(1-3x)^{k+2}}$$

So if the statement is true for $n = k$ then it is also true for $n = k+1$.

The proposition is true for $n = 0$, and if true for $n = k$ it is also true for $n = k+1$. Therefore the proposition is true for all $n \in \mathbb{N}$ by the principle of mathematical induction.

3 $y = xe^{2x}$

Proposition: $\dfrac{d^n y}{dx^n} = \left(2^n x + n2^{n-1}\right)e^{2x}$

Base case

For $n = 0$: $y = xe^{2x} = \left(2^0 x + 0\right)e^{2x}$

∴ the proposition is true for $n = 0$.

Inductive step

Assume the statement is true for $n = k$; that is,

$$\frac{d^k y}{dx^k} = \left(2^k x + k2^{k-1}\right)e^{2x}$$

Working towards:

$$\frac{d^{k+1}y}{dx^{k+1}} = \left(2^{k+1} x + (k+1)2^k\right)e^{2x}$$

$$\frac{d^{k+1}y}{dx^{k+1}} = \frac{d}{dx}\left(\frac{d^k y}{dx^k}\right)$$

$$= \frac{d}{dx}\left(\left(2^k x + k2^{k-1}\right)e^{2x}\right)$$

(using the formula for $\frac{d^k y}{dx^k}$)

$$= \frac{d}{dx}\left(2^k xe^{2x} + k2^{k-1}e^{2x}\right)$$

$$= 2^k e^{2x} + 2^k x \times 2e^{2x} + k2^{k-1} \times 2e^{2x}$$

$$= \left(2^k + 2^{k+1}x + k2^k\right)e^{2x}$$

$$= \left(2^{k+1}x + (k+1)2^k\right)e^{2x}$$

So if the statement is true for $n = k$ then it is also true for $n = k+1$.

The proposition is true for $n = 0$, and if true for $n = k$ it is also true for $n = k+1$. Therefore the proposition is true for all $n \in \mathbb{N}$ by the principle of mathematical induction.

4 $y = x\sin x$

Proposition: $\dfrac{d^{2n}y}{dx^{2n}} = (-1)^n (x\sin x - 2n\cos x)$

Base case

For $n = 0$: $y = x\sin x = (-1)^0(x\sin x - 0)$

∴ the proposition is true for $n = 0$.

Inductive step

Assume the statement is true for $n = k$; that is,

$$\frac{d^{2k}y}{dx^{2k}} = (-1)^k(x\sin x - 2k\cos x)$$

Working towards:

$$\frac{d^{2(k+1)}y}{dx^{2(k+1)}} = (-1)^{k+1}(x\sin x - 2(k+1)\cos x)$$

$$\frac{d^{2k+2}y}{dx^{2k+2}} = \frac{d^2}{dx^2}\left(\frac{d^k y}{dx^k}\right)$$

$$= \frac{d^2}{dx^2}\left((-1)^k(x\sin x - 2k\cos x)\right)$$

(using the formula for $\frac{d^k y}{dx^k}$)

$$= \frac{d}{dx}\left((-1)^k(\sin x + x\cos x + 2k\sin x)\right)$$

$$= \frac{d}{dx}\left((-1)^k(x\cos x + (2k+1)\sin x)\right)$$

$$= (-1)^k(\cos x - x\sin x + (2k+1)\cos x)$$

$$= (-1)^k(-x\sin x + (2k+2)\cos x)$$

$$= (-1)^{k+1}(x\sin x - 2(k+1)\cos x)$$

So if the statement is true for $n = k$ then it is also true for $n = k+1$.

The proposition is true for $n = 0$, and if true for $n = k$ it is also true for $n = k+1$. Therefore the proposition is true for all $n \in \mathbb{N}$ by the principle of mathematical induction.

5 $y = x^2 e^x$

Proposition: $\dfrac{d^n y}{dx^n} = \left(x^2 + 2nx + n(n-1)\right)e^x$ for $n \geq 2$

Base case

For $n = 2$:

$$\frac{d^2 y}{dx^2} = \frac{d^2}{dx^2}\left(x^2 e^x\right)$$

$$= \frac{d}{dx}\left(2xe^x + x^2 e^x\right)$$

$$= 2e^x + 2xe^x + 2xe^x + x^2 e^x$$

$$= \left(x^2 + 4x + 2\right)e^x$$

$$= \left(x^2 + 2(2)x + 2(1)\right)e^x$$

∴ the proposition is true for $n = 2$.

Inductive step

Assume the statement is true for $n = k$; that is,

$$\frac{d^k y}{dx^k} = \left(x^2 + 2kx + k(k-1)\right)e^x$$

Working towards:

$$\frac{d^{k+1} y}{dx^{k+1}} = \left(x^2 + 2(k+1)x + k(k+1)\right)e^x$$

$$\frac{d^{k+1} y}{dx^{k+1}} = \frac{d}{dx}\left(\frac{d^k y}{dx^k}\right)$$

$$= \frac{d}{dx}\left(\left(x^2 + 2kx + k(k-1)\right)e^x\right)$$

(using the formula for $\frac{d^k y}{dx^k}$)

$$= \frac{d}{dx}\left(x^2 e^x + 2kxe^x + k(k-1)e^x\right)$$

$$= 2xe^x + x^2 e^x + 2ke^x + 2kxe^x + k(k-1)e^x$$

$$= \left(x^2 + (2+2k)x + k(k-1+2)\right)e^x$$

$$= \left(x^2 + 2(k+1)x + k(k+1)\right)e^x$$

So if the statement is true for $n = k$ then it is also true for $n = k+1$.

The proposition is true for $n = 2$, and if true for $n = k$ it is also true for $n = k+1$. Therefore the proposition is true for all integers $n \geq 2$ by the principle of mathematical induction.

6 $y = uv$

Proposition: $\dfrac{d^n y}{dx^n} = \sum_{k=0}^{n} \binom{n}{k} \dfrac{d^k u}{dx^k} \dfrac{d^{n-k} v}{dx^{n-k}}$

Base case

For $n = 1$: by the product rule,

$$\frac{dy}{dx} = u\frac{dv}{dx} + v\frac{du}{dx} = \sum_{k=0}^{1} \binom{1}{k} \frac{d^k u}{dx^k} \frac{d^{n-k} v}{dx^{n-k}}$$

∴ the proposition is true for $n = 1$.

Inductive step

Assume the statement is true for $n = r$; that is,

$$\frac{d^r y}{dx^r} = \sum_{k=0}^{r} \binom{r}{k} \frac{d^k u}{dx^k} \frac{d^{r-k} v}{dx^{r-k}}$$

Working towards:

$$\frac{d^{r+1}y}{dx^{r+1}} = \sum_{k=0}^{r+1} \binom{r+1}{k} \frac{d^k u}{dx^k} \frac{d^{r+1-k}v}{dx^{r+1-k}}$$

$$\frac{d^{r+1}y}{dx^{r+1}} = \frac{d}{dx}\left(\frac{d^r y}{dx^r}\right)$$

$$= \frac{d}{dx}\left(\sum_{k=0}^{r} \binom{r}{k} \frac{d^k u}{dx^k} \frac{d^{r-k}v}{dx^{r-k}}\right)$$

(using the formula for $\frac{d^k y}{dx^k}$)

$$= \sum_{k=0}^{r} \binom{r}{k} \frac{d}{dx}\left(\frac{d^k u}{dx^k} \frac{d^{r-k}v}{dx^{r-k}}\right)$$

(derivative of a sum is the sum of the derivatives, so $\frac{d}{dx}$ can be taken inside the summation)

$$= \sum_{k=0}^{r} \binom{r}{k} \left(\frac{d^{k+1}u}{dx^{k+1}} \frac{d^{r-k}v}{dx^{r-k}} + \frac{d^k u}{dx^k} \frac{d^{r-k+1}v}{dx^{r-k+1}}\right)$$

(by the product rule)

$$= \sum_{k=0}^{r} \binom{r}{k} \left(\frac{d^{k+1}u}{dx^{k+1}} \frac{d^{r-k}v}{dx^{r-k}}\right) + \sum_{k=0}^{r} \binom{r}{k} \left(\frac{d^k u}{dx^k} \frac{d^{r+1-k}v}{dx^{r+1-k}}\right)$$

$$= \sum_{k=1}^{r+1} \binom{r}{k-1} \left(\frac{d^k u}{dx^k} \frac{d^{r-k+1}v}{dx^{r-k+1}}\right) + \sum_{k=0}^{r} \binom{r}{k} \left(\frac{d^k u}{dx^k} \frac{d^{r+1-k}v}{dx^{r+1-k}}\right)$$

(replacing the dummy variable k with $k-1$ in the first sum and adjusting the substituted values up by 1 to compensate)

$$= \sum_{k=0}^{r+1} \binom{r}{k-1} \left(\frac{d^k u}{dx^k} \frac{d^{r+1-k}v}{dx^{r+1-k}}\right) + \sum_{k=0}^{r+1} \binom{r}{k} \left(\frac{d^k u}{dx^k} \frac{d^{r+1-k}v}{dx^{r+1-k}}\right)$$

(changing the limits of the summations, allowing that $\binom{r}{-1} = 0 = \binom{r}{r+1}$)

$$= \sum_{k=0}^{r+1} \left(\binom{r}{k} + \binom{r}{k-1}\right) \frac{d^k u}{dx^k} \frac{d^{r+1-k}v}{dx^{r+1-k}}$$

(merging the two summations over the same substituted values)

$$= \sum_{k=0}^{r+1} \binom{r+1}{k} \frac{d^k u}{dx^k} \frac{d^{r+1-k}v}{dx^{r+1-k}}$$

(by the property that $\binom{n}{k} + \binom{n}{k-1} = \binom{n+1}{k}$)

So if the statement is true for $n = k$ then it is also true for $n = k+1$.

The proposition is true for $n = 1$, and if true for $n = k$ it is also true for $n = k+1$. Therefore the proposition is true for all $n \in \mathbb{Z}^+$ by the principle of mathematical induction.

Exercise 25E

1 Proposition: $5^n - 1$ is divisible by 4 for all $n \in \mathbb{N}$

Base case

For $n = 0$: $5^0 - 1 = 0 = 4 \times 0$

∴ the proposition is true for $n = 0$.

Inductive step

Assume the statement is true for $n = k$; that is, $5^k - 1 = 4A$ for some $A \in \mathbb{Z}$

Working towards: $5^{k+1} - 1 = 4B$ for some $B \in \mathbb{Z}$

$5^{k+1} - 1 = 5 \times (5^k - 1) + 4$

$= 5 \times 4A + 4$

(using the formula for $5^k - 1$)

$= 4 \times (5A + 1)$

$= 4B$ where $B = 5A + 1 \in \mathbb{Z}$

So if the statement is true for $n = k$ then it is also true for $n = k+1$.

The proposition is true for $n = 0$, and if true for $n = k$ it is also true for $n = k+1$. Therefore the proposition is true for all $n \in \mathbb{N}$ by the principle of mathematical induction.

2 Proposition: $4^n - 1$ is divisible by 3 for all $n \geq 1$

Base case

For $n = 1$: $4^1 - 1 = 3 = 3 \times 1$

∴ the proposition is true for $n = 1$.

Inductive step

Assume the statement is true for $n = k$; that is,

$4^k - 1 = 3A$ for some $A \in \mathbb{Z}$

Working towards: $4^{k+1} - 1 = 3B$ for some $B \in \mathbb{Z}$

$4^{k+1} - 1 = 4 \times (4^k - 1) + 3$

$= 4 \times 3A + 3$

(using the formula for $4^k - 1$)

$= 3 \times (4A + 1)$

$= 3B$ where $B = 4A + 1 \in \mathbb{Z}$

So if the statement is true for $n = k$ then it is also true for $n = k+1$.

The proposition is true for $n = 1$, and if true for $n = k$ it is also true for $n = k+1$. Therefore the proposition is true for all integers $n \geq 1$ by the principle of mathematical induction.

3 Proposition: $7^n - 3^n$ is divisible by 4 for all $n \in \mathbb{N}$

Base case

For $n = 0$: $7^0 - 3^0 = 0 = 4 \times 0$

∴ the proposition is true for $n = 0$.

Inductive step

Assume the statement is true for $n = k$; that is, $7^k - 3^k = 4A$ for some $A \in \mathbb{Z}$

Working towards: $7^{k+1} - 3^{k+1} = 4B$ for some $B \in \mathbb{Z}$

$7^{k+1} - 3^{k+1} = 7 \times (7^k - 3^k) + 4 \times 3^k$

$= 7 \times 4A + 4 \times 3^k$

(using the formula for $7^k - 3^k$)

$= 4 \times (7A + 3^k)$

$= 4B$ where $B = 7A + 3^k \in \mathbb{Z}$

So if the statement is true for $n = k$ then it is also true for $n = k+1$.

The proposition is true for $n = 0$, and if true for $n = k$ it is also true for $n = k+1$. Therefore the proposition is true for all $n \in \mathbb{N}$ by the principle of mathematical induction.

4 Proposition: $30^n - 6^n$ is divisible by 12 for all integers $n \geq 0$

Base case

For $n = 0$: $30^0 - 6^0 = 0 = 12 \times 0$

∴ the proposition is true for $n = 0$.

Inductive step

Assume the statement is true for $n = k$; that is,

$$30^k - 6^k = 12A \text{ for some } A \in \mathbb{Z}$$

Working towards: $30^{k+1} - 6^{k+1} = 12B$ for some $B \in \mathbb{Z}$

$30^{k+1} - 6^{k+1} = 30 \times (30^k - 6^k) + 24 \times 6^k$

$= 30 \times 12A + 24 \times 6^k$

(using the formula for $30^k - 6^k$)

$= 12(30A + 2 \times 6^k)$

$= 12B$ where $B = 30A + 2 \times 6^k \in \mathbb{Z}$

So if the statement is true for $n = k$ then it is also true for $n = k+1$.

The proposition is true for $n = 0$, and if true for $n = k$ it is also true for $n = k+1$. Therefore the proposition is true for all integers $n \geq 0$ by the principle of mathematical induction.

5 Proposition: $n^3 - n$ is divisible by 6 for all $n \geq 1$

Base case

For $n = 1$: $1^3 - 1 = 0 = 6 \times 0$

∴ the proposition is true for $n = 1$.

COMMENT

Although you could start at $n = 0$ as in previous examples, the question specifies that the proof is needed only for $n \geq 1$, so it is best to take this as the base case.

Inductive step

Assume the statement is true for $n = k$; that is, $k^3 - k = 6A$ for some $A \in \mathbb{Z}$

Working towards: $(k+1)^3 - (k+1) = 6B$ for some $B \in \mathbb{Z}$

$(k+1)^3 - (k+1) = k^3 + 3k^2 + 3k + 1 - k - 1$

$= k^3 - k + (3k^2 + 3k)$

$= 6A + 3k(k+1)$

(using the formula for $k^3 - k$)

Since one of k or $k+1$ must be even, their product is even, so $k(k+1) = 2C$ for some $C \in \mathbb{Z}$

∴ $(k+1)^3 - (k+1) = 6A + 6C$

$= 6B$ where $B = A + C \in \mathbb{Z}$

So if the statement is true for $n = k$ then it is also true for $n = k+1$.

The proposition is true for $n = 1$, and if true for $n = k$ it is also true for $n = k+1$. Therefore the proposition is true for all integers $n \geq 1$ by the principle of mathematical induction.

6 Proposition: $n^3 + 5n$ is divisible by 6 for all $n \geq 1$

Base case

For $n = 1$: $1^3 + 5 \times 1 = 6 = 6 \times 1$

∴ the proposition is true for $n = 1$.

Inductive step

Assume the statement is true for $n = k$; that is, $k^3 + 5k = 6A$ for some $A \in \mathbb{Z}$

Working towards: $(k+1)^3 + 5(k+1) = 6B$ for some $B \in \mathbb{Z}$

$(k+1)^3 + 5(k+1) = k^3 + 3k^2 + 3k + 1 + 5k + 5$
$= k^3 + 5k + 3k^2 + 3k + 6$
$= 6A + 3k(k+1) + 6$

(using the formula for $k^3 + 5k$)

Since one of k or $k+1$ must be even, their product is even, so $k(k+1) = 2C$ for some $C \in \mathbb{Z}$.

$\therefore (k+1)^3 + 5(k+1) = 6A + 6C + 6$
$= 6B$

where $B = A + C + 1 \in \mathbb{Z}$

So if the statement is true for $n = k$ then it is also true for $n = k+1$.

The proposition is true for $n = 1$, and if true for $n = k$ it is also true for $n = k+1$. Therefore the proposition is true for all integers $n \geq 1$ by the principle of mathematical induction.

7 Proposition: $7^n - 4^n - 3^n$ is divisible by 12 for all $n \in \mathbb{Z}^+$.

Base case

For $n = 1$: $7^1 - 4^1 - 3^1 = 0 = 12 \times 0$

$\therefore$ the proposition is true for $n = 1$.

Inductive step

Assume the statement is true for $n = k$; that is,

$7^k - 4^k - 3^k = 12A$ for some $A \in \mathbb{Z}$

Working towards: $7^{k+1} - 4^{k+1} - 3^{k+1} = 12B$ for some $B \in \mathbb{Z}$

$7^{k+1} - 4^{k+1} - 3^{k+1} = 7(7^k - 4^k - 3^k) + 3 \times 4^k + 4 \times 3^k$
$= 12A + 12 \times 4^{k-1} + 12 \times 3^{k-1}$

(using the formula $7^k - 4^k - 3^k = 12A$)

$= 12(A + 4^{k-1} + 3^{k-1})$
$= 12B$ where $B = 4^{k-1} + 3^{k-1} \in \mathbb{Z}$

So if the statement is true for $n = k$ then it is also true for $n = k+1$.

The proposition is true for $n = 1$, and if true for $n = k$ it is also true for $n = k+1$. Therefore the proposition is true for all $n \in \mathbb{Z}^+$ by the principle of mathematical induction.

8 Proposition: $3^{2n+2} - 8n - 9$ is divisible by 64 for all $n \in \mathbb{Z}^+$.

Since $3^{2n+2} = 3^{2(n+1)} = 9^{n+1}$, this is equivalent to the statement that $9^{n+1} - 8n - 9$ is divisible by 64 for all $n \in \mathbb{Z}^+$.

Base case

For $n = 1$: $9^2 - 8 \times 1 - 9 = 64 = 64 \times 1$

$\therefore$ the proposition is true for $n = 1$.

Inductive step

Assume the statement is true for $n = k$; that is,

$9^{k+1} - 8k - 9 = 64A$ for some $A \in \mathbb{Z}$

Working towards: $9^{k+2} - 8(k+1) - 9 = 64B$ for some $B \in \mathbb{Z}$

$9^{k+2} - 8(k+1) - 9 = 9 \times (9^{k+1} - 8k - 9) + 64k + 64$
$= 9 \times 64A + 64k + 64$

(using the formula $9^{k+1} - 8k - 9 = 64A$)

$= 64(9A + k + 1)$
$= 64B$ where $B = 9A + k + 1 \in \mathbb{Z}$

So if the statement is true for $n = k$ then it is also true for $n = k+1$.

25 Mathematical induction   355

The proposition is true for $n = 1$, and if true for $n = k$ it is also true for $n = k+1$. Therefore the proposition is true for all $n \in \mathbb{Z}^+$ by the principle of mathematical induction.

9 Proposition: $(n-1)^3 + n^3 + (n+1)^3$ is divisible by 9 for all $n \in \mathbb{Z}$

> **COMMENT**
> Note that the statement in the question is for all integers, but a proof for non-negative n can easily be converted into a proof for negative n using symmetry.

Start by proving the proposition for $n \in \mathbb{N}$.

Base case

For $n = 0$: $(-1)^3 + 0^3 + 1^3 = 0 = 9 \times 0$

∴ the proposition is true for $n = 0$.

Inductive step

Assume the statement is true for $n = k$; that is, $(k-1)^3 + k^3 + (k+1)^3 = 9A$ for some $A \in \mathbb{Z}$

Working towards:
$k^3 + (k+1)^3 + (k+2)^3 = 9B$ for some $B \in \mathbb{Z}$

$k^3 + (k+1)^3 + (k+2)^3$
$= k^3 + (k+1)^3 + k^3 + 6k^2 + 12k + 8$
$= k^3 + (k+1)^3 + k^3 - 3k^2 + 3k - 1 + 9k^2 + 9k + 9$
$= k^3 + (k+1)^3 + (k-1)^3 + 9k^2 + 9k + 9$
$= 9A + 9k^2 + 9k + 9$

(using the formula $(k-1)^3 + k^3 + (k+1)^3 = 9A$)

$= 9(A + k^2 + k + 1)$
$= 9B$ where $B = A + k^2 + k + 1 \in \mathbb{Z}$

So if the statement is true for $n = k$ then it is also true for $n = k+1$.

The proposition is true for $n = 0$, and if true for $n = k$ it is also true for $n = k+1$. Therefore the proposition is true for all $n \in \mathbb{N}$ by the principle of mathematical induction.

To extend the proof to all $n \in \mathbb{Z}$:

for $n < 0$ let $n = -m$ for $m \in \mathbb{Z}^+$; then

$(n-1)^3 + n^3 + (n+1)^3 = -(1-n)^3 - (-n)^3 - (-n-1)^3$
$= -(1+m)^3 - m^3 - (m-1)^3$
$= -\left((m-1)^3 + m^3 + (m+1)^3\right)$

Since $m > 0$, it has already been proved that $(m-1)^3 + m^3 + (m+1)^3$ is a multiple of 9.

∴ $(n-1)^3 + n^3 + (n+1)^3$ is also a multiple of 9.

Therefore the proposition is true for all $n \in \mathbb{Z}$.

Exercise 25F

1 Clearly for $n = 3$, $3^n = n^3$

Proposition: $3^n > n^3$ for all $n \geq 4$

Base case

For $n = 4$: $3^4 = 81 > 4^3 = 64$

∴ the proposition is true for $n = 4$.

Inductive step

Assume the statement is true for $n = k$ where $k \geq 4$; that is, $3^k > k^3$

Working towards: $3^{k+1} > (k+1)^3$

$3^{k+1} = 3 \times 3^k > 3k^3$ (using $3^k > k^3$)

$3k^3 = k^3 + k^3 + k^3$

For $k > 3$, $k^3 > 3k^2 > 3k + 1$

∴ $3k^3 > k^3 + 3k^2 + 3k + 1 = (k+1)^3$ for $k > 3$

Hence $3^{k+1} > (k+1)^3$ for $k > 3$

So if the statement is true for $n = k$ then it is also true for $n = k+1$.

The proposition is true for $n = 4$, and if true for $n = k \geq 4$ it is also true for $n = k+1$. Therefore the proposition is true for all integers $n \geq 4$ by the principle of mathematical induction.

2 Proposition: $2^n > 1 + n$ for all $n > 1$

Base case

For $n = 2$: $2^2 = 4 > 1 + 2 = 3$

$\therefore$ the proposition is true for $n = 2$.

Inductive step

Assume the statement is true for $n = k$ where $k \geq 2$; that is, $2^k > k+1$

Working towards: $2^{k+1} > k+2$

$2^{k+1} = 2 \times 2^k > 2(k+1)$ (using $2^k > k+1$)

$\therefore 2^{k+1} > 2k+2 > k+2$ for $k \geq 2$

So if the statement is true for $n = k$ then it is also true for $n = k+1$.

The proposition is true for $n = 2$, and if true for $n = k \geq 2$ it is also true for $n = k+1$. Therefore the proposition is true for all integers $n > 1$ by the principle of mathematical induction.

3 Proposition: $2^n > n^2$ for all $n > 4$

Base case

For $n = 5$: $2^5 = 32 > 5^2 = 25$

$\therefore$ the proposition is true for $n = 5$.

Inductive step

Assume the statement is true for $n = k$ where $k \geq 5$; that is, $2^k > k^2$

Working towards: $2^{k+1} > (k+1)^2$

$2^{k+1} = 2 \times 2^k > 2k^2$ (using $2^k > k^2$)

$k^2 - 2k - 1 = 0$ has roots $1 \pm \sqrt{2}$, so $k^2 - 2k - 1 > 0$ for $k \geq 5$
i.e. $k^2 > 2k+1$ for $k \geq 5$

Hence $2k^2 = k^2 + k^2$
$> k^2 + 2k + 1 = (k+1)^2$ for $k \geq 5$

$\therefore 2^{k+1} > (k+1)^2$

So if the statement is true for $n = k$ then it is also true for $n = k+1$.

The proposition is true for $n = 5$, and if true for $n = k \geq 5$ it is also true for $n = k+1$. Therefore the proposition is true for all integers $n > 4$ by the principle of mathematical induction.

4 Proposition: $n! > 2^n$ for all $n \geq 4$

Base case

For $n = 4$: $4! = 24 > 2^4 = 16$

$\therefore$ the proposition is true for $n = 4$.

Inductive step

Assume the statement is true for $n = k$ where $k \geq 4$; that is, $k! > 2^k$

Working towards: $(k+1)! > 2^{k+1}$

$(k+1)! = (k+1) \times k! > (k+1) \times 2^k$ (using $k! > 2^k$)

$k+1 > 2$ for $k \geq 4$

$\Rightarrow (k+1) \times 2^k > 2^{k+1}$ for $k \geq 4$

$\therefore (k+1)! > 2^{k+1}$ for $k \geq 4$

So if the statement is true for $n = k$ then it is also true for $n = k+1$.

The proposition is true for $n = 4$, and if true for $n = k \geq 4$ it is also true for $n = k+1$. Therefore the proposition is true for all integers $n \geq 4$ by the principle of mathematical induction.

5 Proposition: $\sum_{i=1}^{n}\frac{1}{\sqrt{i}} > \sqrt{n}$ for all $n > 1$

Base case

For $n = 2$: $\frac{1}{\sqrt{1}} + \frac{1}{\sqrt{2}} \approx 1.707 > 1.414 \approx \sqrt{2}$

$\therefore$ the proposition is true for $n = 2$.

Inductive step

Assume the statement is true for $n = k$

where $k \geq 2$; that is, $\sum_{i=1}^{k}\frac{1}{\sqrt{i}} > \sqrt{k}$

Working towards: $\sum_{i=1}^{k+1}\frac{1}{\sqrt{i}} > \sqrt{k+1}$

$\sum_{i=1}^{k+1}\frac{1}{\sqrt{i}} = \sum_{i=1}^{k}\frac{1}{\sqrt{i}} + \frac{1}{\sqrt{k+1}} > \sqrt{k} + \frac{1}{\sqrt{k+1}}$

(using $\sum_{i=1}^{k}\frac{1}{\sqrt{i}} > \sqrt{k}$)

$\sqrt{k} + \frac{1}{\sqrt{k+1}} = \sqrt{k+1}\left(\frac{\sqrt{k}}{\sqrt{k+1}} + \frac{1}{k+1}\right)$

$0 < \frac{k}{k+1} < 1$

$\therefore \sqrt{\frac{k}{k+1}} > \frac{k}{k+1}$

Hence

$\sqrt{k+1}\left(\sqrt{\frac{k}{k+1}} + \frac{1}{k+1}\right) > \sqrt{k+1}\left(\frac{k}{k+1} + \frac{1}{k+1}\right)$

$= \sqrt{k+1}$

$\therefore \sum_{i=1}^{k+1}\frac{1}{\sqrt{i}} > \sqrt{k+1}$

So if the statement is true for $n = k$ then it is also true for $n = k+1$.

The proposition is true for $n = 2$, and if true for $n = k$ it is also true for $n = k+1$. Therefore the proposition is true for all integers $n > 1$ by the principle of mathematical induction.

6 Proposition: $\sum_{i=1}^{n}\frac{1}{\sqrt{i}} > 2(\sqrt{n+1} - 1)$ for all $n \geq 1$

Base case

For $n = 1$: $\frac{1}{\sqrt{1}} = 1 > 2(\sqrt{2} - 1) \approx 0.828$

$\therefore$ the proposition is true for $n = 1$.

Inductive step

Assume the statement is true for $n = k$;

that is, $\sum_{i=1}^{k}\frac{1}{\sqrt{i}} > 2(\sqrt{k+1} - 1)$

Working towards: $\sum_{i=1}^{k+1}\frac{1}{\sqrt{i}} > 2(\sqrt{k+2} - 1)$

$\sum_{i=1}^{k+1}\frac{1}{\sqrt{i}} = \sum_{i=1}^{k}\frac{1}{\sqrt{i}} + \frac{1}{\sqrt{k+1}} > 2(\sqrt{k+1} - 1) + \frac{1}{\sqrt{k+1}}$

(using $\sum_{i=1}^{k}\frac{1}{\sqrt{i}} > 2(\sqrt{k+1} - 1)$)

$2(\sqrt{k+1} - 1) + \frac{1}{\sqrt{k+1}} = \frac{2(k+1) + 1}{\sqrt{k+1}} - 2$

$= \frac{2k+3}{\sqrt{k+1}} - 2$

$= \sqrt{\frac{4k^2 + 12k + 9}{k+1}} - 2$

$= \sqrt{\frac{4k^2 + 12k + 8 + 1}{k+1}} - 2$

$= \sqrt{\frac{4(k^2 + 3k + 2) + 1}{k+1}} - 2$

$= \sqrt{\frac{4(k+1)(k+2)}{k+1} + \frac{1}{k+1}} - 2$

$= \sqrt{4(k+2) + \frac{1}{k+1}} - 2$

$> \sqrt{4(k+2)} - 2$

$= 2\sqrt{k+2} - 2$

$= 2(\sqrt{k+2} - 1)$

$\therefore \sum_{i=1}^{k+1}\frac{1}{\sqrt{i}} > 2(\sqrt{k+2} - 1)$

So if the statement is true for $n = k$ then it is also true for $n = k+1$.

The proposition is true for $n = 1$, and if true for $n = k$ it is also true for $n = k+1$. Therefore the proposition is true for all integers $n \geq 1$ by the principle of mathematical induction.

7 $3^0 = 1 = 0! = 0!$

$3^1 = 3 > 1! = 1$

$3^2 = 9 > 2! = 2$

$3^3 = 27 > 3! = 6$

$3^4 = 81 > 4! = 24$

$3^5 = 243 > 5! = 120$

$3^6 = 729 > 6! = 720$

$3^7 = 2187 < 7! = 5040$

$\therefore N = 7$

Proposition: $3^n < n!$ for all $n \geq 7$

Base case

For $n = 7$: $3^7 = 2187 < 7! = 5040$

$\therefore$ the proposition is true for $n = 7$.

Inductive step

Assume the statement is true for $n = k$ where $k \geq 7$; that is, $3^k < k!$

Working towards: $3^{k+1} < (k+1)!$

$3^{k+1} = 3 \times 3^k < 3k!$ (using $3^k < k!$)

$k \geq 7 \Rightarrow k+1 > 3$

$\Rightarrow 3k! < (k+1)k! = (k+1)!$

$\therefore 3^{k+1} < (k+1)!$

So if the statement is true for $n = k$ then it is also true for $n = k+1$.

The proposition is true for $n = 7$, and if true for $n = k \geq 7$ it is also true for $n = k+1$. Therefore the proposition is true for all integers $n \geq 7$ by the principle of mathematical induction.

8 Proposition: $(1+x)^n \geq 1+nx$ for all $n \in \mathbb{N}$ and $x \in \mathbb{R}$

Base case

For $n = 0$: $(1+x)^0 = 1 = 1+0x$

$\therefore$ the proposition is true for $n = 0$, irrespective of the value of x.

Inductive step

Assume the statement is true for $n = k$; that is, $(1+x)^k \geq 1+kx$ for all $x \in \mathbb{R}$

Working towards: $(1+x)^{k+1} \geq 1+(k+1)x$ for all $x \in \mathbb{R}$

$(1+x)^{k+1} = (1+x)(1+x)^k \geq (1+x)(1+kx)$

(using $(1+x)^k \geq 1+kx$)

$(1+x)(1+kx) = 1+(k+1)x+kx^2$

$\geq 1+(k+1)x$

(since $x^2 \geq 0$ and $k \geq 0$)

$\therefore (1+x)^{k+1} \geq 1+(k+1)x$ for all $x \in \mathbb{R}$

So if the statement is true for $n = k$ then it is also true for $n = k+1$.

The proposition is true for all values of x when $n = 0$, and if true for $n = k$ it is also true for $n = k+1$, for all values of x. Therefore the proposition is true for all $n \in \mathbb{N}$ and $x \in \mathbb{R}$ by the principle of mathematical induction.

Mixed examination practice 25

Short questions

1 $u_n = n(n+1)$

Proposition: $S_n = \sum_{r=1}^{n} u_r = \frac{1}{3}n(n+1)(n+2)$

Base case

For $n = 1$: $S_1 = u_1 = 1 \times 2 = \frac{1}{3}(1)(2)(3)$

$\therefore$ the proposition is true for $n = 1$.

Inductive step

Assume the statement is true for $n = k$;

that is, $S_k = \frac{1}{3}k(k+1)(k+2)$

Working towards:

$S_{k+1} = \frac{1}{3}(k+1)(k+2)(k+3)$

$S_{k+1} = S_k + u_{k+1}$

$= \frac{1}{3}k(k+1)(k+2) + (k+1)(k+2)$

(using the formulae for S_k and u_{k+1})

$= (k+1)(k+2)\left(\frac{k}{3}+1\right)$

$= \frac{1}{3}(k+1)(k+2)(k+3)$

So if the statement is true for $n = k$ then it is also true for $n = k+1$.

The proposition is true for $n = 1$, and if true for $n = k$ it is also true for $n = k+1$. Therefore the proposition is true for all $n \in \mathbb{Z}^+$ by the principle of mathematical induction.

2 Proposition: $3^{2n} + 7$ is divisible by 8 for all $n \in \mathbb{N}$

Base case

For $n = 0$: $3^0 + 7 = 8 = 8 \times 1$

∴ the proposition is true for $n = 0$.

Inductive step

Assume the statement is true for $n = k$; that is, $3^{2k} + 7 = 8A$ for some $A \in \mathbb{Z}$

Working towards: $3^{2k+2} + 7 = 8B$ for some $B \in \mathbb{Z}$

$3^{2k+2} + 7 = 9 \times (3^{2k} + 7) - 56$

$= 9 \times 8A - 56$

(using the formula $3^{2k} + 7 = 8A$)

$= 8 \times (9A - 7)$

$= 8B$ where $B = 9A - 7 \in \mathbb{Z}$

So if the statement is true for $n = k$ then it is also true for $n = k+1$.

The proposition is true for $n = 0$, and if true for $n = k$ it is also true for $n = k+1$. Therefore the proposition is true for all $n \in \mathbb{N}$ by the principle of mathematical induction.

3 $u_n = \dfrac{n}{(n+1)!}$

Proposition: $S_n = \sum_{r=1}^{n} u_r = \dfrac{(n+1)! - 1}{(n+1)!}$

Base case

For $n = 1$: $S_1 = u_1 = \dfrac{1}{2!} = \dfrac{1}{2} = \dfrac{2! - 1}{2!}$

∴ the proposition is true for $n = 1$.

Inductive step

Assume the statement is true for $n = k$;

that is, $S_k = \dfrac{(k+1)! - 1}{(k+1)!}$

Working towards: $S_{k+1} = \dfrac{(k+2)! - 1}{(k+2)!}$

$S_{k+1} = S_k + u_{k+1}$

$= \dfrac{(k+1)! - 1}{(k+1)!} + \dfrac{k+1}{(k+2)!}$

(using the formulae for S_k and u_{k+1})

$= \dfrac{(k+2)(k+1)! - (k+2) + (k+1)}{(k+2)!}$

$= \dfrac{(k+2)! - 1}{(k+2)!}$

So if the statement is true for $n = k$ then it is also true for $n = k+1$.

The proposition is true for $n = 1$, and if true for $n = k$ it is also true for $n = k+1$. Therefore the proposition is true for all $n \in \mathbb{Z}^+$ by the principle of mathematical induction.

4 Proposition: $11^{n+2} + 12^{2n+1}$ is divisible by 133 for all $n \in \mathbb{N}$

Base case

For $n = 0$: $11^2 + 12^1 = 133 = 133 \times 1$

$\therefore$ the proposition is true for $n = 0$.

Inductive step

Assume the statement is true for $n = k$; that is, $11^{k+2} + 12^{2k+1} = 133A$ for some $A \in \mathbb{Z}$

Working towards: $11^{k+3} + 12^{2k+3} = 133B$ for some $B \in \mathbb{Z}$

$11^{k+3} + 12^{2k+3} = 11(11^{k+2} + 12^{2k+1}) + (12^2 - 11) \times 12^{2k+1}$

$= 11 \times 133A + 133 \times 12^{2k+1}$

(using the formula $11^{k+2} + 12^{2k+1} = 133A$)

$= 133(11A + 12^{2k+1})$

$= 133B$ where $B = 11A + 12^{2k+1} \in \mathbb{Z}$

So if the statement is true for $n = k$ then it is also true for $n = k+1$.

The proposition is true for $n = 0$, and if true for $n = k$ it is also true for $n = k+1$. Therefore the proposition is true for all $n \in \mathbb{N}$ by the principle of mathematical induction.

5 $u_n = \dfrac{n}{2^n}$

Proposition:

$S_n = \sum_{r=0}^{n} u_r = 2 - \left(\dfrac{1}{2}\right)^n (n+2)$ for $n \in \mathbb{N}$

Base case

For $n = 0$: $S_0 = u_0 = 0 = 2 - \left(\dfrac{1}{2}\right)^0 \times 2$

$\therefore$ the proposition is true for $n = 0$.

Inductive step

Assume the statement is true for $n = k$; that is, $S_k = 2 - \left(\dfrac{1}{2}\right)^k (k+2)$

Working towards: $S_{k+1} = 2 - \left(\dfrac{1}{2}\right)^{k+1} (k+3)$

$S_{k+1} = S_k + u_{k+1}$

$= 2 - \left(\dfrac{1}{2}\right)^k (k+2) + \dfrac{k+1}{2^{k+1}}$

(using the formulae for S_k and u_{k+1})

$= 2 + \left(\dfrac{1}{2}\right)^{k+1} (-2(k+2) + (k+1))$

$= 2 + \left(\dfrac{1}{2}\right)^{k+1} (-k-3)$

$= 2 - \left(\dfrac{1}{2}\right)^{k+1} (k+3)$

So if the statement is true for $n = k$ then it is also true for $n = k+1$.

The proposition is true for $n = 0$, and if true for $n = k$ it is also true for $n = k+1$. Therefore the proposition is true for all $n \in \mathbb{N}$ by the principle of mathematical induction.

6 $u_n = \sin((2n-1)\theta)$

Proposition:

$S_n = \sum_{r=1}^{n} u_r = \dfrac{\sin^2(n\theta)}{\sin \theta}$ for $n \in \mathbb{Z}^+$

Base case

For $n = 1$: $S_1 = u_1 = \sin \theta = \dfrac{\sin^2 \theta}{\sin \theta}$

$\therefore$ the proposition is true for $n = 1$.

Inductive step

Assume the statement is true for $n = k$;

that is, $S_k = \dfrac{\sin^2(k\theta)}{\sin\theta}$

Working towards: $S_{k+1} = \dfrac{\sin^2((k+1)\theta)}{\sin\theta}$

$S_{k+1} = S_k + u_{k+1}$

$= \dfrac{\sin^2(k\theta)}{\sin\theta} + \sin((2k+1)\theta)$

(using the formulae for S_k and u_{k+1})

$= \dfrac{\sin^2(k\theta) + \sin\theta\sin((2k+1)\theta)}{\sin\theta}$

To proceed further, we will use the formula
$\sin(A+B)\sin(A-B) = \sin^2 A - \sin^2 B$

Proof of this result:

$\sin(A+B)\sin(A-B)$
$= (\sin A\cos B + \sin B\cos A)$
$\quad(\sin A\cos B - \sin B\cos A)$
$= \sin^2 A\cos^2 B - \sin^2 B\cos^2 A$
$= \sin^2 A(1 - \sin^2 B) - \sin^2 B(1 - \sin^2 A)$
$= \sin^2 A - \sin^2 B$

Taking $A = (k+1)\theta$ and $B = k\theta$, so that $A+B = (2k+1)\theta$ and $A-B = \theta$, the formula gives

$S_{k+1} = \dfrac{\sin^2(k\theta) + \left(\sin^2(k+1)\theta - \sin^2(k\theta)\right)}{\sin\theta}$

$= \dfrac{\sin^2((k+1)\theta)}{\sin\theta}$

So if the statement is true for $n = k$ then it is also true for $n = k+1$.

The proposition is true for $n = 1$, and if true for $n = k$ it is also true for $n = k+1$. Therefore the proposition is true for all $n \in \mathbb{Z}^+$ by the principle of mathematical induction.

---

7 $u_n = 4n - 2$

Proposition: $P_n = \displaystyle\prod_{r=1}^{n} u_r = \dfrac{(2n)!}{n!}$ for $n \in \mathbb{Z}^+$

> **COMMENT**
>
> The large Pi notation represents a product of terms, in the same way that the large Sigma notation indicates the sum of terms. This notation is used here to keep the working easy to read, but you would not be required to use it in an examination; any clear notation, including use of ... to indicate continuation of a pattern, is acceptable.

Base case

For $n = 1$: $P_1 = u_1 = 2 = \dfrac{(2\times 1)!}{1!}$

$\therefore$ the proposition is true for $n = 1$.

Inductive step

Assume the statement is true for $n = k$;

that is, $P_k = \dfrac{(2k)!}{k!}$

Working towards: $P_{k+1} = \dfrac{(2k+2)!}{(k+1)!}$

$P_{k+1} = P_k \times u_{k+1}$

$= \dfrac{(2k)!}{k!} \times (4k+2)$

(using the formulae for P_k and u_{k+1})

$= \dfrac{(2k)! \times 2 \times (2k+1)}{k!}$

$= \dfrac{(2k+1)! \times 2}{k!}$

$= \dfrac{(2k+1)! \times 2 \times (k+1)}{(k+1)!}$

$= \dfrac{(2k+1)! \times (2k+2)}{(k+1)!}$

$= \dfrac{(2k+2)!}{(k+1)!}$

So if the statement is true for $n = k$ then it is also true for $n = k+1$.

The proposition is true for $n = 1$, and if true for $n = k$ it is also true for $n = k+1$.
Therefore the proposition is true for all $n \in \mathbb{Z}^+$ by the principle of mathematical induction.

8 $u_n = \cos(2^{n-1} x)$

$$\cos x \times \cos 2x \times \cos 4x \times \cos 8x \times \ldots \times \cos(2^{n-1} x) = \prod_{r=1}^{n} u_r$$

Proposition:

$$P_n = \prod_{r=1}^{n} u_r = \frac{\sin(2^n x)}{2^n \sin x} \text{ for } n \in \mathbb{Z}^+$$

Base case

For $n = 1$:

$$P_1 = u_1 = \cos(2^0 x) = \cos x$$

$$= \frac{2 \sin x \cos x}{2 \sin x} = \frac{\sin(2x)}{2 \sin x} = \frac{\sin(2^1 x)}{2^1 \sin x}$$

∴ the proposition is true for $n = 1$.

Inductive step

Assume the statement is true for $n = k$;

that is, $P_k = \dfrac{\sin(2^k x)}{2^k \sin x}$

Working towards: $P_{k+1} = \dfrac{\sin(2^{k+1} x)}{2^{k+1} \sin x}$

$P_{k+1} = P_k \times u_{k+1}$

$$= \frac{\sin(2^k x)}{2^k \sin x} \times \cos(2^k x) \text{ (using the formulae for } P_k \text{ and } u_{k+1})$$

$$= \frac{2 \sin(2^k x) \cos(2^k x)}{2^{k+1} \sin x}$$

$$= \frac{\sin(2 \times 2^k x)}{2^{k+1} \sin x}$$

$$= \frac{\sin(2^{k+1} x)}{2^{k+1} \sin x}$$

So if the statement is true for $n = k$ then it is also true for $n = k+1$.

The proposition is true for $n = 1$, and if true for $n = k$ it is also true for $n = k+1$.
Therefore the proposition is true for all $n \in \mathbb{Z}^+$ by the principle of mathematical induction.

Long questions

1 a Proposition:
$(\cos\theta + i\sin\theta)^n = \cos(n\theta) + i\sin(n\theta)$ for $n \in \mathbb{Z}^+$

<u>Base case</u>

For $n = 1$:

$(\cos\theta + i\sin\theta)^1 = \cos\theta + i\sin\theta = \cos(1\theta) + i\sin(1\theta)$

∴ the proposition is true for $n = 1$.

<u>Inductive step</u>

Assume the statement is true for $n = k$;

that is, $(\cos\theta + i\sin\theta)^k = \cos(k\theta) + i\sin(k\theta)$

Working towards:

$(\cos\theta + i\sin\theta)^{k+1} = \cos((k+1)\theta) + i\sin((k+1)\theta)$

$$\begin{aligned}(\cos\theta + i\sin\theta)^{k+1} &= (\cos\theta + i\sin\theta)^k (\cos\theta + i\sin\theta) \\ &= (\cos(k\theta) + i\sin(k\theta))(\cos\theta + i\sin\theta) \\ &\quad \text{(using the inductive assumption)} \\ &= \cos(k\theta)\cos\theta - \sin(k\theta)\sin\theta \\ &\quad + i(\sin\theta\cos(k\theta) + \cos\theta\sin(k\theta)) \\ &= \cos(k\theta + \theta) + i\sin(k\theta + \theta) \\ &= \cos((k+1)\theta) + i\sin((k+1)\theta)\end{aligned}$$

So if the statement is true for $n = k$ then it is also true for $n = k+1$.

The proposition is true for $n = 1$, and if true for $n = k$ it is also true for $n = k+1$. Therefore the proposition is true for all $n \in \mathbb{Z}^+$ by the principle of mathematical induction.

b Proposition: $(\cos\theta + i\sin\theta)^{-n} = \cos(n\theta) - i\sin(n\theta)$ for $n \in \mathbb{Z}^+$

<u>Base case</u>

For $n = 1$:

$$(\cos\theta + i\sin\theta)^{-1} = \frac{(\cos\theta + i\sin\theta)^*}{|(\cos\theta + i\sin\theta)|^2}$$

$$= \frac{\cos\theta - i\sin\theta}{1}$$

$$= \cos(1\theta) - i\sin(1\theta)$$

∴ the proposition is true for $n = 1$.

Inductive step

Assume the statement is true for $n = k$; that is,

$$(\cos\theta + i\sin\theta)^{-k} = \cos(k\theta) - i\sin(k\theta)$$

Working towards:

$$(\cos\theta + i\sin\theta)^{-k-1} = \cos((k+1)\theta) - i\sin((k+1)\theta)$$

$$\begin{aligned}(\cos\theta + i\sin\theta)^{-k-1} &= (\cos\theta + i\sin\theta)^{-k}(\cos\theta + i\sin\theta)^{-1} \\ &= (\cos(k\theta) - i\sin(k\theta)) \times (\cos\theta - i\sin\theta)\end{aligned}$$

(using the inductive assumption and the base case)

$$\begin{aligned}&= \cos(k\theta)\cos\theta - \sin(k\theta)\sin\theta \\ &\quad - i(\sin\theta\cos(k\theta) + \cos\theta\sin(k\theta)) \\ &= \cos(k\theta + \theta) - i\sin(k\theta + \theta) \\ &= \cos((k+1)\theta) - i\sin((k+1)\theta)\end{aligned}$$

So if the statement is true for $n = k$ then it is also true for $n = k+1$.

The proposition is true for $n = 1$, and if true for $n = k$ it is also true for $n = k+1$.
Therefore the proposition is true for all $n \in \mathbb{Z}^+$ by the principle of mathematical induction.

c $|2i - 2| = \sqrt{2^2 + 2^2} = 2\sqrt{2}$

$$\arg(2i - 2) = \arctan\left(\frac{2}{-2}\right) = \frac{3\pi}{4}$$

(argument in second quadrant as $\text{Re}(z) < 0, \text{Im}(z) > 0$)

d $z^3 = 2i - 2$

Let $z = r\text{cis}\theta$; then the equation becomes

$$r^3 \text{cis}(3\theta) = 2\sqrt{2}\,\text{cis}\left(\frac{3\pi}{4}\right)$$

$$\therefore r^3 = 2\sqrt{2}, \quad 3\theta = \frac{3\pi}{4}, \frac{11\pi}{4}, \frac{19\pi}{4}$$

$$\Rightarrow r = \sqrt{2}, \quad \theta = \frac{\pi}{4}, \frac{11\pi}{12}, \frac{19\pi}{12}$$

$$\cos\left(\frac{\pi}{4}\right) = \sin\left(\frac{\pi}{4}\right) = \frac{1}{\sqrt{2}}$$

$$\cos\left(\frac{11\pi}{12}\right) = -\frac{\sqrt{2}+\sqrt{6}}{4}, \quad \sin\left(\frac{11\pi}{12}\right) = \frac{\sqrt{6}-\sqrt{2}}{4}$$

$$\cos\left(\frac{19\pi}{12}\right) = \frac{\sqrt{6}-\sqrt{2}}{4}, \quad \sin\left(\frac{19\pi}{12}\right) = \frac{-\sqrt{2}-\sqrt{6}}{4}$$

So the solutions are

$$z_1 = \sqrt{2}\operatorname{cis}\left(\frac{\pi}{4}\right) = 1+i$$

$$z_2 = \sqrt{2}\operatorname{cis}\left(\frac{11\pi}{4}\right) = -\frac{2+2\sqrt{3}}{4} + i\frac{2\sqrt{3}-2}{4} = -\frac{1+\sqrt{3}}{2} + i\frac{\sqrt{3}-1}{2}$$

$$z_3 = \sqrt{2}\operatorname{cis}\left(\frac{19\pi}{4}\right) = \frac{2\sqrt{3}-2}{4} + i\frac{-2-2\sqrt{3}}{4} = \frac{\sqrt{3}-1}{2} - i\frac{1+\sqrt{3}}{2}$$

> **COMMENT**
>
> There is no requirement that you know the sine and cosine of all multiples of $\frac{\pi}{12}$, though it can be useful. A calculated answer to 3SF would be acceptable for z_2 and z_3. If you were asked for the exact form, always remember that you can use the double angle formula to split the cosine and sine of $\frac{\pi}{6}$, which you should know. This gives exact values, albeit in a slightly different form.

2 a $\binom{n}{r} = \frac{n!}{r!(n-r)!}$

$$= \frac{n(n-1)!}{r!(n-r)!}$$

$$= \frac{r(n-1)! + (n-r)(n-1)!}{r!(n-r)!}$$

$$= \frac{r(n-1)!}{r!(n-r)!} + \frac{(n-r)(n-1)!}{r!(n-r)!}$$

$$= \frac{r(n-1)!}{r(r-1)!(n-r)!} + \frac{(n-r)(n-1)!}{r!(n-r)(n-1-r)!}$$

$$= \frac{(n-1)!}{(r-1)!(n-r)!} + \frac{(n-1)!}{r!(n-1-r)!}$$

$$= \binom{n-1}{r-1} + \binom{n-1}{r}$$

366 Mixed examination practice 25

b Proposition:

$$\sum_{r=1}^{n-1} \binom{n}{r} = 2^n - 2 \text{ for } n \geq 2$$

Base case

For $n = 2$: $\binom{2}{1} = 2 = 2^2 - 2$

∴ the proposition is true for $n = 2$.

Inductive step

Assume the statement is true for $n = k$;

that is, $\sum_{r=1}^{k-1} \binom{k}{r} = 2^k - 2$

Working towards: $\sum_{r=1}^{k} \binom{k+1}{r} = 2^{k+1} - 2$

$$\sum_{r=1}^{k} \binom{k+1}{r} = \sum_{r=1}^{k} \left(\binom{k}{r-1} + \binom{k}{r} \right) \text{ using (a)}$$

$$= \sum_{r=1}^{k} \binom{k}{r-1} + \sum_{r=1}^{k} \binom{k}{r}$$

$$= \sum_{r=0}^{k-1} \binom{k}{r} + \binom{k}{k} + \sum_{r=1}^{k-1} \binom{k}{r}$$

$$= \binom{k}{0} + \sum_{r=1}^{k-1} \binom{k}{r} + \binom{k}{k} + \sum_{r=1}^{k-1} \binom{k}{r}$$

$$= \binom{k}{0} + \binom{k}{k} + 2\sum_{r=1}^{k-1} \binom{k}{r}$$

$$= 1 + 1 + 2(2^k - 2)$$

(using the inductive assumption)

$$= 2^{k+1} - 2$$

So if the statement is true for $n = k$ then it is also true for $n = k+1$.

The proposition is true for $n = 2$, and if true for $n = k$ it is also true for $n = k+1$. Therefore the proposition is true for all integers $n \geq 2$ by the principle of mathematical induction.

3 a $u_n = ar^{n-1}$

Proposition: $S_n = \sum_{i=1}^{n} u_i = \dfrac{a(r^n - 1)}{r - 1}$

Base case

For $n = 1$: $S_1 = u_1 = a = \dfrac{a(r^1 - 1)}{(r-1)}$

∴ the proposition is true for $n = 1$.

Inductive step

Assume the statement is true for $n = k$;

that is, $S_k = \dfrac{a(r^k - 1)}{r - 1}$

Working towards: $S_{k+1} = \dfrac{a(r^{k+1} - 1)}{r - 1}$

$S_{k+1} = S_k + u_{k+1}$

$$= \dfrac{a(r^k - 1)}{r - 1} + ar^k$$

(using the formulae for S_k and u_{k+1})

$$= \dfrac{a}{r-1}(r^k - 1 + r^k(r - 1))$$

$$= \dfrac{a(r^{k+1} - 1)}{r - 1}$$

So if the statement is true for $n = k$ then it is also true for $n = k+1$.

The proposition is true for $n = 1$, and if true for $n = k$ it is also true for $n = k+1$. Therefore the proposition is true for all $n \in \mathbb{Z}^+$ by the principle of mathematical induction.

b $a = 1.2$, $r = 0.5$

Require $S_n > 2.399$

$$\frac{1.2(0.5^n - 1)}{0.5 - 1} > 2.399$$

$$2.4 \times (1 - 0.5^n) > 2.399$$

$$1 - 0.5^n > \frac{2.399}{2.4}$$

$$0.5^n < \frac{0.001}{2.4}$$

$$n \ln(0.5) < \ln\left(\frac{0.001}{2.4}\right)$$

$$n > \ln\left(\frac{0.001}{2.4}\right) \div \ln(0.5)$$

$$n > 11.2$$

∴ the least such n is 12

4 a Proposition: $n^5 - n$ is divisible by 5 for all $n \in \mathbb{Z}^+$

Base case

For $n = 1$: $1^5 - 1 = 0 = 5 \times 0$

∴ the proposition is true for $n = 1$.

Inductive step

Assume the statement is true for $n = k$; that is, $k^5 - k = 5A$ for some $A \in \mathbb{Z}$

Working towards: $(k+1)^5 - (k+1) = 5B$ for some $B \in \mathbb{Z}$

$(k+1)^5 - (k+1) = k^5 + 5k^4 + 10k^3 + 10k^2 + 5k$
$\qquad + 1 - k - 1$
$= k^5 - k + 5(k^4 + 2k^3 + 2k^2 + k)$
$= 5A + 5(k^4 + 2k^3 + 2k^2 + k)$
(using the formula for $k^5 - k$)
$= 5(A + k^4 + 2k^3 + 2k^2 + k)$
$= 5B$ where $B = A + k^4 + 2k^3$
$\qquad + 2k^2 + k \in \mathbb{Z}$

So if the statement is true for $n = k$ then it is also true for $n = k + 1$.

The proposition is true for $n = 1$, and if true for $n = k$ it is also true for $n = k + 1$. Therefore the proposition is true for all $n \in \mathbb{Z}^+$ by the principle of mathematical induction.

b $n^5 - n = n(n^4 - 1)$
$= n(n-1)(n^3 + n^2 + n + 1)$
$= n(n-1)(n+1)(n^2 + 1)$

One of n, $n-1$ and $n+1$ must be a multiple of 3, since these are three consecutive numbers.

One of n and $n+1$ must be even, since these are two consecutive numbers.

Hence the product $(n-1)n(n+1)$ must be a multiple of 6, and so $n^5 - n$ is divisible by 6.

c From (a) and (b), since $n^5 - n = n(n-1)(n+1)(n^2+1)$ is divisible by 5 and 6, it is a multiple of 30.

If n is even, then $(n-1)$, $(n+1)$ and (n^2+1) are all odd and their product will be odd.

Therefore, unless n is a multiple of 4 or odd, $n(n-1)(n+1)(n^2+1)$ will not be a multiple of 4 and therefore not a multiple of 60.

Using this reasoning, we expect that $n^5 - n$ will not be a multiple of 60 for $n = 6$.

As expected, $6^5 - 6 = 7770$ is not a multiple of 60.

Therefore the proposition is not true for all $n \geq 3$.

> **COMMENT**
> Proving that something is not true requires only a counter-example. If you have concrete reasoning for why a proposition is false, you should also be able to devise a counter-example. An alternative approach here would be to try proving the proposition by induction. If you succeed, then the proposition is proved and you are finished. If you find you cannot make a logical deduction to complete the proof, then the point at which you fail may give you insight into what value of n could provide a counter-example.

5 a $\cos\left(x + \dfrac{\pi}{2}\right) = \cos x \cos\left(\dfrac{\pi}{2}\right) - \sin x \sin\left(\dfrac{\pi}{2}\right)$

$\qquad\qquad\qquad = \cos x \times 0 - \sin x \times 1$

$\qquad\qquad\qquad = -\sin x$

b Proposition:

$\dfrac{d^n}{dx^n}(\cos x) = \cos\left(x + \dfrac{n\pi}{2}\right)$ for $n \in \mathbb{Z}^+$

<u>Base case</u>

For $n = 1$:

$\dfrac{d}{dx}(\cos x) = -\sin x = \cos\left(x + \dfrac{\pi}{2}\right)$ by the result from (a)

$\therefore$ the proposition is true for $n = 1$.

<u>Inductive step</u>

Assume the statement is true for $n = k$;

that is, $\dfrac{d^k}{dx^k}(\cos x) = \cos\left(x + \dfrac{k\pi}{2}\right)$

Working towards:

$\dfrac{d^{k+1}}{dx^{k+1}}(\cos x) = \cos\left(x + \dfrac{(k+1)\pi}{2}\right)$

$\dfrac{d^{k+1}}{dx^{k+1}}(\cos x) = \dfrac{d}{dx}\left(\dfrac{d^k}{dx^k}(\cos x)\right)$

$\qquad\qquad\quad = \dfrac{d}{dx}\left(\cos\left(x + \dfrac{k\pi}{2}\right)\right)$ (using the inductive assumption)

$\qquad\qquad\quad = -\sin\left(x + \dfrac{k\pi}{2}\right)$

$\qquad\qquad\quad = \cos\left(x + \dfrac{k\pi}{2} + \dfrac{\pi}{2}\right)$ (using the result from (a) again)

$\qquad\qquad\quad = \cos\left(x + \dfrac{(k+1)\pi}{2}\right)$

25 Mathematical induction 369

So if the statement is true for $n = k$ then it is also true for $n = k+1$.

The proposition is true for $n = 1$, and if true for $n = k$ it is also true for $n = k+1$.
Therefore the proposition is true for all $n \in \mathbb{Z}^+$ by the principle of mathematical induction.

6 For n lines in the plane subject to the conditions given, let I_n be the number of intersection points and R_n the number of regions formed.

Proposition 1: $I_n = \dfrac{n(n-1)}{2}$

Base case

For $n = 1$: a single line produces no intersections, so

$$I_1 = 0 = \dfrac{1(1-1)}{2}$$

∴ the proposition is true for $n = 1$.

Inductive step

Assume the statement is true for $n = k$;

that is, $I_k = \dfrac{k(k-1)}{2}$

Working towards: $I_{k+1} = \dfrac{(k+1)k}{2}$

The $(k+1)$th line is not parallel to any of the preceding k lines, so it must have a single intersection with each of them. Since no three lines pass through any single point, each of these k intersections occurs at a different point.

∴ $I_{k+1} = I_k + k$

$= \dfrac{k(k-1)}{2} + k$ (using the inductive assumption)

$= \dfrac{k}{2}(k-1+2) = \dfrac{k(k+1)}{2}$

So if the statement is true for $n = k$ then it is also true for $n = k+1$.

The proposition is true for $n = 1$, and if true for $n = k$ it is also true for $n = k+1$.
Therefore the proposition is true for all $n \in \mathbb{Z}^+$ by the principle of mathematical induction.

Proposition 2: $R_n = \dfrac{n(n+1)}{2} + 1$

Base case

For $n = 1$: a single line divides the plane into two regions,

so $R_1 = 2 = \dfrac{1(1+1)}{2} + 1$

∴ the proposition is true for $n = 1$.

<u>Inductive step</u>

Assume the statement is true for $n = k$;

that is, $R_k = \dfrac{k(k+1)}{2} + 1$

Working towards: $R_{k+1} = \dfrac{(k+1)(k+2)}{2} + 1$

The $(k+1)$th line is not parallel to any of the preceding k lines, so it must have a single intersection with each of them.

The k intersection points divide the $(k+1)$th line into $k+1$ line segments.

Each line segment divides a region which was previously undivided into two parts.

The $(k+1)$th line must therefore increase the number of regions by $k+1$.

$\therefore R_{k+1} = R_k + k + 1$

$= \dfrac{k(k+1)}{2} + 1 + (k+1)$ (using the inductive assumption)

$= \dfrac{(k+1)}{2}(k+2) + 1$

$= \dfrac{(k+1)(k+2)}{2} + 1$

So if the statement is true for $n = k$ then it is also true for $n = k+1$.

The proposition is true for $n = 1$, and if true for $n = k$ it is also true for $n = k+1$.
Therefore the proposition is true for all $n \in \mathbb{Z}^+$ by the principle of mathematical induction.

26 Questions crossing chapters

Short questions

1 The number of possible outcomes is 36.
An average of 3 is equivalent to a sum of 6, and the outcomes that give this result are
1+5
2+4
3+3
4+2
5+1
Each has probability $\frac{1}{36}$, so the total probability is $\frac{5}{36}$

2 The first two terms are:
$u_1 = S_1$
$= 3(1) + 2(1)^2$
$= 5$
$u_2 = S_2 - S_1$
$= 3(2) + 2(2)^2 - 5$
$= 14 - 5$
$= 9$
$\therefore d = u_2 - u_1 = 4$

> **COMMENT**
> We can also find d by comparing the given expression for S_n to the general formula
> $S_n = \frac{n}{2}(2u_1 + (n-1)d)$
> $= \left(u_1 - \frac{1}{2}d\right)n + \frac{1}{2}dn^2$
> Comparing the coefficient of n^2 in the general formula with that in $S_n = 3n + 2n^2$, we get $\frac{1}{2}d = 2 \Rightarrow d = 4$.

3 $|x| = \begin{cases} x & \text{for } x \geq 0 \\ -x & \text{for } x < 0 \end{cases}$

The graph is composed of two straight lines, one for positive x with gradient 1 and one for negative x with gradient -1, which meet at the origin.

Therefore the graph of $f'(x)$ is as follows:

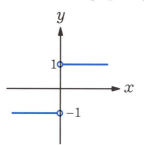

Figure 26S.3

4 Mode at $x = 4 \Rightarrow f'(4) = 0$
$f(x) = abx - ax^2 \Rightarrow f'(x) = ab - 2ax$
$f'(4) = 0$
$\Rightarrow ab - 8a = 0$
$\therefore b = 8$
($a \neq 0$, otherwise $f(x)$ would be zero.)
$f(x)$ is a pdf, so $\int_0^b f(x)\,dx = 1$
$\int_0^8 8ax - ax^2\,dx = 1$
$\left[4ax^2 - \frac{1}{3}ax^3\right]_0^8 = 1$
$\frac{256}{3}a = 1$
$\therefore a = \frac{3}{256}$

5 $u \cdot v = x^2 + x + x$
$= x^2 + 2x$
$\therefore \int u \cdot v \, dx = \int x^2 + 2x \, dx$
$= \dfrac{1}{3}x^3 + x^2 + c$

6 Average $= \dfrac{S_n}{n}$
$= \dfrac{1}{n}\left(\dfrac{a(1-r^n)}{1-r}\right)$
$= \dfrac{a(1-r^n)}{n(1-r)}$

7 $f(g(x)) = ((ax+b)-1)^2 + 3$
$\therefore (ax+b-1)^2 + 3 = 16x^2 - 16x + 7$
$a^2x^2 + 2a(b-1)x + (b-1)^2 + 3 = 16x^2 - 16x + 7$

Equating coefficients:
$x^2 : a^2 = 16$ ...(1)
$x^1 : 2a(b-1) = -16$ ...(2)
$x^0 : (b-1)^2 + 3 = 7$ ...(3)

From (1), $a = 4$ or -4

From (2), if $a = 4$ then
$8(b-1) = -16$
$b = -1$

If $a = -4$ then
$-8(b-1) = -16$
$b = 3$

Check these values in (3):
$(-1-1)^2 + 3 = 7$ and $(3-1)^2 + 3 = 7$

Both are valid,
$\therefore a = 4, b = -1$ or $a = -4, b = 3$

8 From GDC:

Figure 26S.8 Graph of $y = |1 + \tan 2x|$

$y < 1$ when
$1.02 < x < 1.57$ or $2.59 < x < 3.14$

9 If $X \sim B(10, p)$ then $\mathrm{Var}(X) = 10p(1-p)$

Graph of $\mathrm{Var}(X)$ versus p is a negative quadratic with roots at $p = 0$ and 1:

Figure 26S.9 Graph of $y = 10p(1-p)$

From the symmetry of the graph, the maximum is at $p = 0.5$.

10 Variance $s^2 = E(X^2) - E(X)^2$

$$= \frac{3^2 + 7^2 + x^2}{3} - \left(\frac{3+7+x}{3}\right)^2$$

$$= \frac{58 + x^2}{3} - \frac{100 + 20x + x^2}{9}$$

$$= \frac{2x^2 - 20x + 74}{9}$$

$$= \frac{2(x-5)^2 + 24}{9}$$

$$= \frac{2}{9}(x-5)^2 + \frac{8}{3}$$

∴ minimum value of s^2 is $\frac{8}{3} = 2.67$ (3SF)

11 $\sum_x P(X = x) = 1$

$\ln k + \ln 2k + \ln 3k + \ln 4k = 1$

$\ln(24k^4) = 1$

$24k^4 = e$

$\Rightarrow k = \sqrt[4]{\frac{e}{24}}$

12 $\binom{n}{2} = k$

$\frac{n(n-1)}{2} = k$

$n^2 - n - 2k = 0$

$n = \frac{1 + \sqrt{1 + 8k}}{2}$

(Take the positive sign because $n > 0$.)

13 The graph of $y = 1 - \cos x$ is obtained from the graph of $y = \cos x$ by reflection in the x-axis followed by translation one unit up:

Compared with $y = \sin x$, this graph is translated $\frac{\pi}{2}$ units to the right and 1 unit up, so the single transformation is the translation with vector $\begin{pmatrix} \frac{\pi}{2} \\ 1 \end{pmatrix}$.

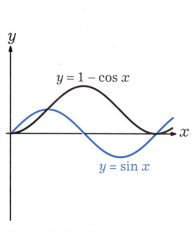

Figure 26S.13

374 Questions crossing chapters

14 a Using the binomial expansion:
$$(\cos\theta + i\sin\theta)^5 = \cos^5\theta + 5\cos^4\theta(i\sin\theta) + 10\cos^3\theta(i\sin\theta)^2$$
$$+ 10\cos^2\theta(i\sin\theta)^3 + 5\cos\theta(i\sin\theta)^4 + (i\sin\theta)^5$$
$$= \cos^5\theta + 5i\cos^4\theta\sin\theta - 10\cos^3\theta\sin^2\theta$$
$$- 10i\cos^2\theta\sin^3\theta + 5\cos\theta\sin^4\theta + i\sin^5\theta$$

b By De Moivre: $(\cos\theta + i\sin\theta)^5 = \cos 5\theta + i\sin 5\theta$
Comparing imaginary parts:
$$\sin 5\theta = 5\cos^4\theta\sin\theta - 10\cos^2\theta\sin^3\theta + \sin^5\theta$$
$$= 5(1-\sin^2\theta)^2\sin\theta - 10(1-\sin^2\theta)\sin^3\theta + \sin^5\theta$$
$$= 5\sin\theta - 10\sin^3\theta + 5\sin^5\theta - 10\sin^3\theta + 10\sin^5\theta + \sin^5\theta$$
$$= 5\sin\theta - 20\sin^3\theta + 16\sin^5\theta$$

15 a The horizontal stretch scale factor is $\dfrac{1}{k}$.

b $y = \ln kx$
$= \ln k + \ln x$
So the vertical translation is by $\ln k$ units up.

16 a This is a geometric series with first term $0.5^0 = 1$ and common ratio 0.5.
There are 11 terms in the sum. So
$$S_{11} = \frac{1(1 - 0.5^{11})}{1 - 0.5}$$
$$= \frac{1 - \frac{1}{2^{11}}}{\frac{1}{2}} = \frac{2^{11} - 1}{2^{10}}$$
$$= \frac{2047}{1024}$$

b $\ln u_r = \ln 0.5^r$
$= r\ln 0.5$
$$\sum_{r=0}^{10} \ln(u_r) = \ln 0.5 \sum_{r=0}^{10} r$$
$$= \ln 0.5(0 + 1 + \ldots + 10)$$
$$= 55\ln 0.5$$
$$= 55\ln 2^{-1}$$
$$= -55\ln 2$$

26 Questions crossing chapters

17 **a** $\arg(i) = \dfrac{\pi}{2}$

b $|i| = 1$, so $i = e^{i\frac{\pi}{2}}$

c $i^i = \left(e^{i\frac{\pi}{2}}\right)^i = e^{i\frac{\pi}{2} \times i} = e^{-\frac{\pi}{2}}$

18 $y = x^{\sin x}$

$\ln y = \ln\left(x^{\sin x}\right)$

$\ln y = \sin x \ln x$

Using implicit differentiation with respect to x:

$\dfrac{1}{y}\dfrac{dy}{dx} = \cos x \ln x + \dfrac{1}{x}\sin x$

$\dfrac{dy}{dx} = y\left(\cos x \ln x + \dfrac{1}{x}\sin x\right)$

$= x^{\sin x}\left(\cos x \ln x + \dfrac{1}{x}\sin x\right)$

19 When $n = 1$: $\left(a^x\right)^1 = a^x = a^{1x}$

∴ the statement is true for $n = 1$.

Assume that the statement is true for $n = k$: $\left(a^x\right)^k = a^{kx}$

Then for $n = k+1$:

$\left(a^x\right)^{k+1} = \left(a^x\right)^k\left(a^x\right)$

$= a^{kx}a^x$

$= a^{kx+x}$

$= a^{(k+1)x}$

Thus the statement is true for $n = k+1$.

The statement is true for $n = 1$, and if true for $n = k$ then it is also true for $n = k+1$. Therefore the statement is true for all $n \in \mathbb{Z}^+$ by the principle of mathematical induction.

20 **a** $\dfrac{dy}{dx} = \lambda e^{\lambda x}$

$\dfrac{d^2 y}{dx^2} = \lambda^2 e^{\lambda x}$

b Substituting $y = e^{\lambda x}$ into

$\dfrac{d^2 y}{dx^2} + 5\dfrac{dy}{dx} - 6y = 0$:

$\lambda^2 e^{\lambda x} + 5\lambda e^{\lambda x} - 6e^{\lambda x} = 0$

$e^{\lambda x}\left(\lambda^2 + 5\lambda - 6\right) = 0$

$\lambda^2 + 5\lambda - 6 = 0$ (as $e^{\lambda x} \neq 0$)

$(\lambda - 1)(\lambda + 6) = 0$

∴ $\lambda = 1$ or -6

21 $P(X = k) = \dfrac{e^{-m}m^k}{k!}$

$P(X = 7) = P(X = 8) + P(X = 9)$

∴ $\dfrac{e^{-m}m^7}{7!} = \dfrac{e^{-m}m^8}{8!} + \dfrac{e^{-m}m^9}{9!}$

$\dfrac{m^7}{7!} = \dfrac{m^8}{8!} + \dfrac{m^9}{9!}$

$1 = \dfrac{m}{8} + \dfrac{m^2}{9 \times 8}$ (multiplying by $\dfrac{7!}{m^7}$)

$m^2 + 9m - 72 = 0$

$m = \dfrac{-9 \pm \sqrt{81 + 4 \times 72}}{2}$

$m = 5.10$ (reject negative value)

22 $f(g(x)) = 3(ax^2 - x + 5) + 1$

$= 3ax^2 - 3x + 16$

$f(g(x)) = 0$ is a quadratic equation and has equal roots when $\Delta = 0$:

$(-3)^2 - 4 \times 3a \times 16 = 0$

$9 - 192a = 0$

$a = \dfrac{3}{64}$

23 **a**

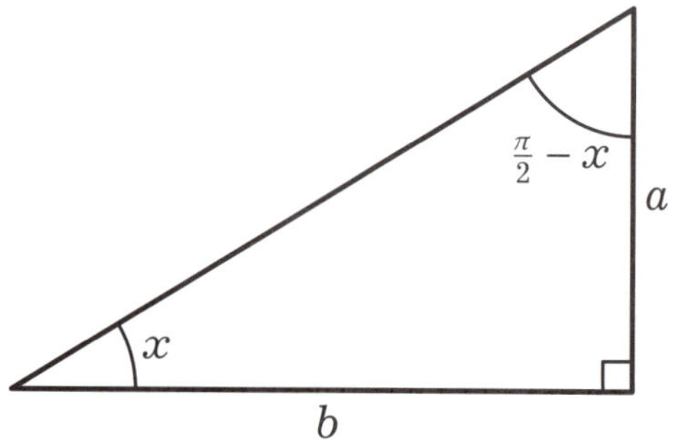

Figure 26S.23

376 Questions crossing chapters

From the diagram:

$\tan x = \dfrac{a}{b}$, $\tan\left(\dfrac{\pi}{2} - x\right) = \dfrac{b}{a}$

$\therefore \tan\left(\dfrac{\pi}{2} - x\right) = \dfrac{1}{\tan x}$

b From (a),

$\tan\left(\dfrac{\pi}{2} - \alpha\right) = \dfrac{1}{\tan\alpha}$, $\tan\left(\dfrac{\pi}{2} - \beta\right) = \dfrac{1}{\tan\beta}$

The roots of $ax^2 + bx + c = 0$ satisfy

$$\begin{cases} \tan\alpha + \tan\beta = -\dfrac{b}{a} \\ \tan\alpha \tan\beta = \dfrac{c}{a} \end{cases}$$

Then

$\dfrac{1}{\tan\alpha} + \dfrac{1}{\tan\beta} = \dfrac{\tan\alpha + \tan\beta}{\tan\alpha \tan\beta}$

$= \dfrac{\left(-\dfrac{b}{a}\right)}{\dfrac{c}{a}} = -\dfrac{b}{c}$

$\dfrac{1}{\tan\alpha} \dfrac{1}{\tan\beta} = \dfrac{1}{\tan\alpha \tan\beta}$

$= \dfrac{1}{c/a} = \dfrac{a}{c}$

Therefore an equation with roots $\dfrac{1}{\tan\alpha}$ and $\dfrac{1}{\tan\beta}$ is

$x^2 + \dfrac{b}{c}x + \dfrac{a}{c} = 0$

or $cx^2 + bx + a = 0$

24 a Substituting $y = ax^2 + bx$ into $y^2 - 2y - 24 = 0$:

$(ax^2 + bx)^2 - 2(ax^2 + bx) - 24 = 0$

$a^2x^4 + 2abx^3 + b^2x^2 - 2ax^2 - 2bx - 24 = 0$

$a^2x^4 + 2abx^3 + (b^2 - 2a)x^2 - 2bx - 24 = 0$

Comparing coefficients with
$x^4 + 10x^3 + 23x^2 - 10x - 24 = 0$:

$x^4 : a^2 = 1$ ...(1)

$x^3 : 2ab = 10$ ...(2)

$x^2 : b^2 - 2a = 23$ ...(3)

$x^1 : -2b = -10$ ...(4)

From (1), $a = 1$ or -1

From (2), when $a = 1$:

$2b = 10$

$\Rightarrow b = 5$

When $a = -1$

$-2b = 10$

$\Rightarrow b = -5$

But from (4), $b = 5$

$\therefore b = -5, a = -1$ is not a solution.

Check $a = 1, b = 5$ in (3):

$5^2 - 2 = 23$

$\therefore a = 1$ and $b = 5$

b $x^4 + 10x^3 + 23x^2 - 10x - 24 = 0$ is equivalent to

$y^2 - 2y - 24 = 0$

$(y-6)(y+4) = 0$

$y = 6$ or -4

$\therefore x^2 + 5x = 6$

$x^2 + 5x - 6 = 0$

$(x-1)(x+6) = 0$

$x = 1$ or -6

or $x^2 + 5x = -4$

$x^2 + 5x + 4 = 0$

$(x+1)(x+4) = 0$

$x = -1$ or -4

Therefore the solutions are $x = -6, -4, -1, 1$

25 $\int_0^y x^2 + 1 \, dx = 4$

$\left[\dfrac{1}{3}x^3 + x\right]_0^y = 4$

$\dfrac{1}{3}y^3 + y - 4 = 0$

From GDC: $y = 1.86$

26 This is a geometric series with common ratio $r = x^2 - x$. It converges when $|x^2 - x| < 1$.

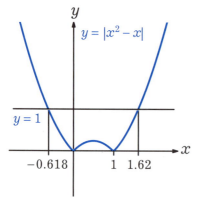

Figure 26S.26 Graphs of $y = |x^2 - x|$ and $y = 1$

From the figure, $|x^2 - x| < 1$ for $-0.618 < x < 1.62$.

27 $\mathbf{u} \cdot \mathbf{v} = 0 \Rightarrow \mathbf{u}$ and $\mathbf{v}$ are perpendicular.

The vector $\mathbf{u} - \mathbf{v}$ is the diagonal of the rectangle, as shown in Figure 26S.27.

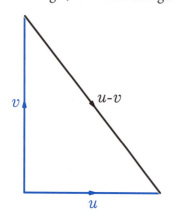

Figure 26S.27

$|\mathbf{u} - \mathbf{v}|^2 = \sqrt{|\mathbf{u}|^2 + |\mathbf{v}|^2} = \sqrt{2^2 + 3^2} = \sqrt{13}$

378 Questions crossing chapters

COMMENT

We can also solve the problem algebraically, using $|\mathbf{v}|^2 = \mathbf{v} \cdot \mathbf{v}$:

$|\mathbf{u} - \mathbf{v}|^2 = (\mathbf{u} - \mathbf{v}) \cdot (\mathbf{u} - \mathbf{v})$

$= |\mathbf{u}|^2 + |\mathbf{v}|^2 - 2\mathbf{u} \cdot \mathbf{v}$

$= 2^2 + 3^2 - 2 \times 0$

$= 13$

$\therefore |\mathbf{u} - \mathbf{v}| = \sqrt{13}$

---

28 Let $S_n = u_1 + \ldots + u_n$ where $u_k = 2r^{k-1}$.

When $n = 1$:

$S_1 = u_1 = 2$

and $S_1 = \dfrac{2(1 - r^1)}{1 - r} = 2$

So the statement is true for $n = 1$.

Assume that the statement is true for

$n = k$: $S_k = \dfrac{2(1 - r^k)}{1 - r}$.

Then for $n = k + 1$:

$S_{k+1} = S_k + u_{k+1}$

$= \dfrac{2(1 - r^k)}{1 - r} + 2r^k$

$= \dfrac{2 - 2r^k + 2r^k - 2r^{k+1}}{1 - r}$

$= \dfrac{2 - 2r^{k+1}}{1 - r}$

$= \dfrac{2(1 - r^{k+1})}{1 - r}$

Hence the statement is true for $n = k + 1$.

The statement is true for $n = 1$, and if true for $n = k$ then it is also true for $n = k + 1$. Therefore it is true for all $n \in \mathbb{Z}^+$ by the principle of mathematical induction.

29 a $P(\text{first 6 on third roll}) = P(6') \times P(6') \times P(6)$

$$= \frac{5}{6} \times \frac{5}{6} \times \frac{1}{6} = \frac{25}{216}$$

b To get the first six on the rth roll, he rolls $r-1$ non-sixes followed by a six:

$$p_r = \left(\frac{5}{6}\right)^{r-1}\left(\frac{1}{6}\right) = \frac{5^{r-1}}{6^r}$$

c $\sum_{r=1}^{\infty} p_r = \sum_{r=1}^{\infty} \left(\frac{1}{6}\right)\left(\frac{5}{6}\right)^{r-1}$

This is a geometric series with first term $\frac{1}{6}$ and common ratio $\frac{5}{6}$, so

$$S_\infty = \frac{\frac{1}{6}}{1-\frac{5}{6}}$$

$$= \frac{1/6}{1/6}$$

$$= 1$$

30 The sum is a geometric series with first term 1 and common ratio x (which converges because $0 < x < 1$), so $\sum_{r=1}^{\infty} x^r = \frac{1}{1-x}$

Hence $\int_0^{1/2} \sum_{r=1}^{\infty} x^r \, dx = \int_0^{1/2} \frac{1}{1-x} \, dx$

$$= \left[-\ln(1-x)\right]_0^{1/2}$$

$$= -\ln\left(\frac{1}{2}\right) + \ln(1)$$

$$= \ln 2$$

31 a $\sin[(A+B)x] - \sin[(A-B)x]$

$= \sin(Ax + Bx) - \sin(Ax - Bx)$

$= (\sin Ax \cos Bx + \sin Bx \cos Ax) - (\sin Ax \cos Bx - \sin Bx \cos Ax)$

$= 2 \sin Bx \cos Ax$

b Let $A = 5$ and $B = 3$ in part (a); then

$$\sin 3x \cos 5x = \frac{1}{2}(\sin 8x - \sin 2x)$$

$$\therefore \int \sin 3x \cos 5x \, dx = \int \frac{1}{2}\sin 8x - \frac{1}{2}\sin 2x \, dx$$

$$= -\frac{1}{16}\cos 8x + \frac{1}{4}\cos 2x + c$$

26 Questions crossing chapters 379

32 Need to express the volume V of the cone in terms of θ.

From the diagram, the height is $h = l\cos\theta$ and the radius of the base is $l\sin\theta$.

$$\therefore V = \frac{1}{3}\pi(l\sin\theta)^2(l\cos\theta)$$

$$= \frac{1}{3}\pi l^3 \sin^2\theta \cos\theta$$

From GDC, the maximum value of $y = \sin^2\theta\cos\theta$ for $\theta \in \left(0, \dfrac{\pi}{2}\right)$ is 0.385, so the maximum possible volume is

$$V_{max} = \frac{1}{3} \times 0.385\pi l^3 = 0.403 l^3$$

> **COMMENT**
>
> Although the maximum value can be found directly using a GDC, differentiation can also be used: local maximum of V occurs where $\dfrac{dV}{d\theta} = 0$:
>
> $$\frac{1}{3}\pi l^3 \left(2\sin\theta\cos^2\theta - \sin^3\theta\right) = 0$$
>
> $$\sin\theta\left(2\cos^2\theta - \sin^2\theta\right) = 0$$
>
> $$\sin\theta = 0 \text{ or } 2\cos^2\theta - \sin^2\theta = 0$$
>
> $$\therefore 2\cos^2\theta - \sin^2\theta = 0 \quad \left(\text{as } \sin\theta \neq 0 \text{ for } 0 < \theta < \frac{\pi}{2}\right)$$
>
> $$\frac{\sin^2\theta}{\cos^2\theta} = 2$$
>
> $$\tan^2\theta = 2$$
>
> $$\therefore \tan\theta = \sqrt{2} \quad \left(\tan\theta > 0 \text{ for } 0 < \theta < \frac{\pi}{2}\right)$$
>
> Hence, using a $1, \sqrt{2}, \sqrt{3}$ right-angled triangle, $\cos\theta = \dfrac{1}{\sqrt{3}}$.
> At this value of θ,
>
> $$V = \frac{1}{3}\pi l^3 \tan^2\theta \cos^3\theta$$
>
> $$= \frac{1}{3}\pi l^3 \times 2 \times \frac{1}{3\sqrt{3}}$$
>
> $$= \frac{2\pi l^3}{9\sqrt{3}} = 0.403 l^3$$

33 a $y = \arcsin x \Rightarrow x = \sin y$

b $\dfrac{dx}{dy} = \cos y$

c $\dfrac{dy}{dx} = \dfrac{1}{\cos y}$

$= \dfrac{1}{\sqrt{1-\sin^2 y}}$

$= \dfrac{1}{\sqrt{1-x^2}}$

34 a $\csc x = \dfrac{1}{\sin x}$

$\dfrac{d}{dx}(\csc x) = \dfrac{d}{dx}\left[(\sin x)^{-1}\right]$

$= -(\sin x)^{-2} \cos x$

$= -\dfrac{\cos x}{\sin^2 x}$

$= -\dfrac{1}{\sin x} \dfrac{\cos x}{\sin x}$

$= -\csc x \cot x$

b $\dfrac{dy}{dx} = 2\sqrt{3}$

$\therefore -\csc x \cot x = 2\sqrt{3}$

$-\dfrac{\cos x}{\sin^2 x} = 2\sqrt{3}$

$-\cos x = 2\sqrt{3}(1-\cos^2 x)$

$2\sqrt{3}\cos^2 x - \cos x - 2\sqrt{3} = 0$

$\cos x = \dfrac{1 \pm \sqrt{1+48}}{4\sqrt{3}}$

$= \dfrac{1 \pm 7}{4\sqrt{3}}$

$= \dfrac{2\sqrt{3}}{3}$ or $-\dfrac{\sqrt{3}}{2}$

As $|\cos x| \leq 1$, only one solution is valid:

$\cos x = -\dfrac{\sqrt{3}}{2}$

$\therefore x = \dfrac{5\pi}{6}$

$y = \dfrac{1}{\sin\left(\dfrac{5\pi}{6}\right)} = 2$

$\therefore$ the coordinates are $\left(\dfrac{5\pi}{6}, 2\right)$

35 a ϕ is an interior angle of a parallelogram, so $\phi = \pi - 2\theta$

b $\mathbf{a}+\mathbf{b}$ is the longer diagonal of the parallelogram.

Using the cosine rule in the triangle with sides $|\mathbf{a}|$ and $|\mathbf{b}|$ and angle ϕ between them,

$|\mathbf{a}+\mathbf{b}|^2 = |\mathbf{a}|^2 + |\mathbf{b}|^2 - 2|\mathbf{a}||\mathbf{b}|\cos\phi$

$= 1 + 1 - 2\cos(\pi - 2\theta)$

$= 2 - 2(-\cos 2\theta)$

$= 2 + 2(2\cos^2\theta - 1)$

$= 4\cos^2\theta$

36 a $f(x) = \ln(x^2 - 9) - \ln(x+3) - \ln x$

$= \ln\left(\dfrac{x^2-9}{x(x+3)}\right)$

$= \ln\left(\dfrac{(x-3)(x+3)}{x(x+3)}\right)$

$= \ln\left(\dfrac{x-3}{x}\right)$

> **COMMENT**
> It is possible to cancel the fraction since $x \neq -3$ for the function to be defined.

b $y = \ln\left(\dfrac{x-3}{x}\right)$

$\dfrac{x-3}{x} = e^y$

$xe^y - x = -3$

$x(e^y - 1) = -3$

$\Rightarrow x = \dfrac{-3}{e^y - 1}$

$\therefore f^{-1}(x) = \dfrac{3}{1-e^x}$

37 **a** Let $z = x+iy$ and $w = a+ib$; then

$$(zw)^* = ((x+iy)(a+ib))^*$$
$$= (xa+ixb+iya-yb)^*$$
$$= xa-yb-ixb-iya$$
$$= (x-iy)(a-ib)$$
$$= z^* w^*$$

b When $n = 1$: $(z^1)^* = z^* = (z^*)^1$

So the statement is true for $n = 1$.

Assume that the statement is true for $n = k$: $(z^k)^* = (z^*)^k$.

Then for $n = k+1$:

$$(z^{k+1})^* = (z^k z)^*$$
$$= (z^k)^* z^*$$
$$= (z^*)^k z^*$$
$$= (z^*)^{k+1}$$

So the statement is true for $n = k+1$.

The statement is true when $n = 1$, and if true for $n = k$ then it is also true for $n = k+1$. Therefore it is true for all integers $n \geq 1$ by the principle of mathematical induction.

38 **a** $p = P(X=1) + P(X=2)$

$$= \frac{e^{-\mu}\mu^1}{1} + \frac{e^{-\mu}\mu^2}{2}$$
$$= \left(\mu + \frac{1}{2}\mu^2\right)e^{-\mu}$$

b Maximum of p occurs when $\frac{dp}{d\mu} = 0$:

$$(1+\mu)e^{-\mu} - \left(\mu + \frac{1}{2}\mu^2\right)e^{-\mu} = 0$$

$$\left(1 - \frac{1}{2}\mu^2\right)e^{-\mu} = 0$$

$$1 - \frac{1}{2}\mu^2 = 0 \quad (\text{as } e^{-\mu} \neq 0)$$

$$\therefore \mu = \sqrt{2} \quad (\text{choose positive root})$$

39 **a** $\int e^{kx} \, dx = \dfrac{1}{k} e^{kx} + c$

b $e^{(1+3i)x} = e^x e^{i(3x)}$

$= e^x (\cos 3x + i \sin 3x)$

$\therefore \operatorname{Im}\left(e^{(1+3i)x}\right) = e^x \sin 3x$

c $\int e^x \sin 3x \, dx = \int \operatorname{Im}\left(e^{(1+3i)x}\right) dx$

$= \operatorname{Im}\left(\int e^{(1+3i)x} \, dx\right)$

$= \operatorname{Im}\left(\dfrac{1}{1+3i} e^{(1+3i)x}\right) + c$

$= \operatorname{Im}\left(\dfrac{e^x (\cos 3x + i \sin 3x)}{1+3i}\right) + c$

$= \operatorname{Im}\left(\dfrac{e^x (\cos 3x + i \sin 3x)(1-3i)}{10}\right) + c$

$= \dfrac{e^x (\sin 3x - 3\cos 3x)}{10} + c$

40 **a** $\ln(x^2) = 2\ln x$, so the transformation is a vertical stretch with scale factor 2.

b $\log_{10} x = \dfrac{\ln x}{\ln 10}$, so the transformation is a vertical stretch with scale factor $\dfrac{1}{\ln 10}$.

41 **a** $\dfrac{d}{dx}(\ln y) = \dfrac{1}{y} \dfrac{dy}{dx}$

b $y = \dfrac{x^4}{(2+5x)\sqrt{x^2+1}}$

$\ln y = \ln(x^4) - \ln(2+5x) - \ln\left(\sqrt{x^2+1}\right)$

$= 4\ln x - \ln(2+5x) - \dfrac{1}{2}\ln(x^2+1)$

c $\dfrac{d}{dx}(\ln y) = \dfrac{4}{x} - \dfrac{5}{2+5x} - \dfrac{x}{x^2+1}$

i.e. $\dfrac{1}{y}\dfrac{dy}{dx} = \dfrac{4}{x} - \dfrac{5}{2+5x} - \dfrac{x}{x^2+1}$

$\therefore \dfrac{dy}{dx} = y\left(\dfrac{4}{x} - \dfrac{5}{2+5x} - \dfrac{x}{x^2+1}\right)$

$= \dfrac{x^4}{(2+5x)\sqrt{x^2+1}}\left(\dfrac{4}{x} - \dfrac{5}{2+5x} - \dfrac{x}{x^2+1}\right)$

42 a $e^{\frac{i\pi}{4}} = \cos\left(\frac{\pi}{4}\right) + i\sin\left(\frac{\pi}{4}\right)$

$= \frac{\sqrt{2}}{2} + \frac{i\sqrt{2}}{2}$

b $e^{\frac{i\pi}{4}} = \frac{\sqrt{2}}{2}(1+i)$

$\Rightarrow 1+i = \sqrt{2}\,e^{\frac{i\pi}{4}}$

$\therefore \ln(1+i) = \ln\left(\sqrt{2}\,e^{\frac{i\pi}{4}}\right)$

$= \ln\sqrt{2} + \ln e^{\frac{i\pi}{4}}$

$= \ln\sqrt{2} + i\frac{\pi}{4}$

43 $\sum_{x=0}^{\infty} P(X=x) = 1$

$\therefore \frac{4}{5} + \frac{4}{5}p + \frac{4}{5}p^2 + \ldots = 1$

This is a geometric series with $u_1 = \frac{4}{5}$ and $r = p$, so $S_\infty = \frac{\frac{4}{5}}{1-p}$

$S_\infty = 1$

$\frac{\frac{4}{5}}{1-p} = 1$

$1 - p = \frac{4}{5}$

$\therefore p = \frac{1}{5}$

44 $P(-1 < x < 1) = \int_{-1}^{1} \frac{1}{\pi(x^2+3)}\,dx$

$= \frac{1}{\pi}\left[\frac{1}{\sqrt{3}}\arctan\left(\frac{x}{\sqrt{3}}\right)\right]_{-1}^{1}$

$= \frac{1}{\pi\sqrt{3}}\left[\arctan\left(\frac{1}{\sqrt{3}}\right) - \arctan\left(-\frac{1}{\sqrt{3}}\right)\right]$

$= \frac{1}{\pi\sqrt{3}}\left[\frac{\pi}{6} - \left(-\frac{\pi}{6}\right)\right]$

$= \frac{1}{\pi\sqrt{3}}\cdot\frac{\pi}{3}$

$= \frac{1}{3\sqrt{3}}$

45 Using the binomial expansion:

$$\sum_{r=0}^{n}\binom{n}{r}p^r(1-p)^{n-r} = (p+(1-p))^n = 1^n = 1$$

46 a Setting $x = 1$ in $(1+x)^n$ gives

$$(1+1)^n = \sum_{r=0}^{n}\binom{n}{r}1^{n-r}1^r$$

$$\sum_{r=0}^{n}\binom{n}{r} = 2^n$$

b Setting $x = -1$ in $(1+x)^n$:

$$(1-1)^n = \sum_{r=0}^{n}\binom{n}{r}1^{n-r}(-1)^r$$

$$\Rightarrow \sum_{r=0}^{n}(-1)^r\binom{n}{r} = 0$$

47 The length of an arc is $l = r\theta$, so need to find the radii OB, OB$_1$, OB$_2$ etc.

OA = OB = 1

OA$_1$ = OB$_1$
 = OA cosθ
 = 1 × cosθ
 = cosθ

OA$_2$ = OB$_2$
 = OA$_1$ cosθ
 = cosθ × cosθ
 = cos$^2\theta$

Hence the radii form a geometric sequence with first term 1 and common ratio cosθ.

$AB + A_1B_1 + A_2B_2 + \ldots$
$= 1 \times \theta + (\cos\theta)\theta + (\cos^2\theta)\theta + \ldots$
$= (1 + \cos\theta + \cos^2\theta + \ldots)\theta$
$= \left(\dfrac{1}{1-\cos\theta}\right)\theta$ (using formula for S_∞)
$= \dfrac{\theta}{1-\cos\theta}$

48 $\tan\left(\dfrac{\pi}{3}\right) = \sqrt{3}$

$\therefore \tan^{2r}\left(\dfrac{\pi}{3}\right) = \left(\sqrt{3}^2\right)^r = 3^r$

Using the binomial expansion:

$$\sum_{r=0}^{n}\binom{n}{r}\tan^{2r}\left(\dfrac{\pi}{3}\right) = \sum_{r=0}^{n}\binom{n}{r}3^r$$

$= (1+3)^n$
$= 4^n$

49 a $f(x) = d$
$\Rightarrow |f(x+1)| - 1 = d$
$\therefore |d| - 1 = d$
(since $f(x) = d \Rightarrow f(x+1) = d$)
$|d| = d + 1$

d cannot be positive, as this would mean that $d = d + 1$,
$\therefore d < 0$
and so $|d| = -d$
Hence $-d = d + 1$
$2d = -1$
$d = -\dfrac{1}{2}$

COMMENT

Note that this says that reflecting the line $y = -\dfrac{1}{2}$ in the x-axis is the same as translating it up by one unit.

b The equation $f(x) = |f(x+1)| - 1$ says that translation by one unit to the left, followed by reflecting the negative part in the x-axis, followed by translation of one unit down, returns the graph to the original function.

By the same argument as in (a), the centre line must be at $y = -\frac{1}{2}$, so $c = -\frac{1}{2}$. The translation to the left must be by half a period, so $b = \pi$. Finally, to retain the same shape after reflection in the x-axis, the original graph must not cross the x-axis, hence $|a| \leq \frac{1}{2}$.

50 **a** $y = e^x + \frac{1}{e^x}$

$\Rightarrow e^{2x} - ye^x + 1 = 0$

This is a quadratic equation in e^x; using the quadratic formula,

$e^x = \frac{y \pm \sqrt{y^2 - 4}}{2}$

$\Rightarrow x = \ln\left(\frac{y \pm \sqrt{y^2 - 4}}{2}\right)$

b Summing these two roots,

$x_1 + x_2 = \ln\left(\frac{y + \sqrt{y^2 - 4}}{2}\right) + \ln\left(\frac{y - \sqrt{y^2 - 4}}{2}\right)$

$= \ln\left(\frac{(y + \sqrt{y^2 - 4})(y - \sqrt{y^2 - 4})}{4}\right)$

$= \ln\left(\frac{y^2 - (y^2 - 4)}{4}\right)$

$= \ln\left(\frac{4}{4}\right)$

$= 0$

51 **a** Let $z = a + ib$; then

$z^2 = (a + ib)^2 = a^2 + 2iab - b^2$

$\therefore a^2 + 2iab - b^2 = i - 1$

Comparing real and imaginary parts:

Re: $a^2 - b^2 = -1$ ...(1)

Im: $2ab = 1$ ...(2)

From (2), $b = \frac{1}{2a}$

Substituting into (1):

$a^2 - \frac{1}{4a^2} = -1$

$4a^4 + 4a^2 - 1 = 0$

$a^2 = \frac{-4 \pm \sqrt{32}}{8}$

$= \frac{-1 + \sqrt{2}}{2}$ (as $a^2 \geq 0$)

$a = \pm\sqrt{\frac{\sqrt{2} - 1}{2}}$

Then

$b = \frac{1}{2a}$

$= \pm\frac{1}{\sqrt{2}\sqrt{\sqrt{2} - 1}}$

$= \pm\sqrt{\frac{\sqrt{2} + 1}{2}}$

$\therefore z = \pm\left(\sqrt{\frac{\sqrt{2} - 1}{2}} + i\sqrt{\frac{\sqrt{2} + 1}{2}}\right)$

b $w^2 + 2iw - i = 0$

$w = \frac{-2i \pm \sqrt{-4 + 4i}}{2}$

$= -i \pm \sqrt{i - 1}$

$= -i \pm \left(\sqrt{\frac{\sqrt{2} - 1}{2}} + i\sqrt{\frac{\sqrt{2} + 1}{2}}\right)$

52 **a** Reflection in the line $y = x$ exchanges x and y, so the coordinates become (y, x). The reflection in the y-axis makes the x-coordinate negative, so the new coordinates are $(-y, x)$.

b Reflection in the line $y = x$ followed by reflection in the y-axis results in 90° anticlockwise rotation about the origin. This maps (x, y) to $(-y, x)$, so the new equation is

$x = f(-y)$

$f^{-1}(x) = -y$

i.e. $y = -f^{-1}(x)$

> **COMMENT**
>
> The fact that $f(x)$ is one-to-one is needed in order for the inverse function to exist.

53 a To obtain the graph of $y=||x|-1|$, the graph of $y=|x|$ is translated down by one unit and then the negative part is reflected in the x-axis:

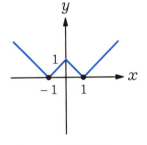

Figure 26S.53.1

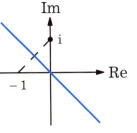

Figure 26S.54 Points in the Argand plane that satisfy $|z-i|=|z+1|$, i.e. the line $y=-x$

> **COMMENT**
>
> We can also think about this problem geometrically by noting that $|z-i|$ is the distance between the points representing complex numbers z and i in the Argand diagram, and $|z+1|$ is the distance between z and -1. So the points satisfying the equation are those which are at the same distance from points i and -1.

b

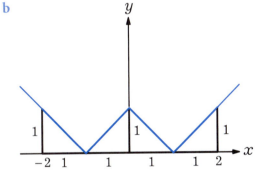

Figure 26S.53.2

From the graph:

$$\int_{-2}^{2}||x|-1|\,dx = 4\left(\frac{1\times 1}{2}\right) = 2$$

54 a $|z-i| = |x+iy-i|$

$\qquad = |x+i(y-1)|$

$\qquad = \sqrt{x^2+(y-1)^2}$

b $|z-i|=|z+1|$

$\sqrt{x^2+(y-1)^2} = \sqrt{(x+1)^2+y^2}$

$x^2+y^2-2y+1 = x^2+2x+1+y^2$

$-2y = 2x$

$y = -x$

55

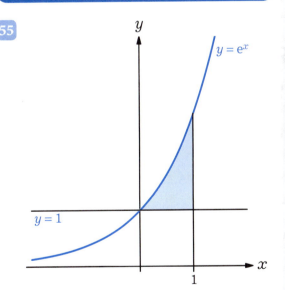

Figure 26S.55

This is the same volume as when the region between the graph of $y=e^x-1$, the x-axis and the line $x=1$ is rotated around the x-axis. (The whole picture is just translated vertically by one unit.)

$$\therefore V = \pi \int_0^1 (e^x - 1)^2 \, dx$$

$$= \pi \int_0^1 e^{2x} - 2e^x + 1 \, dx$$

$$= \pi \left[\frac{1}{2} e^{2x} - 2e^x + x \right]_0^1$$

$$= \pi \left[\left(\frac{1}{2} e^2 - 2e + 1 \right) - \left(\frac{1}{2} e^0 - 2e^0 + 0 \right) \right]$$

$$= \pi \left(\frac{1}{2} e^2 - 2e + \frac{5}{2} \right)$$

56 Probability has to be between 0 and 1, so need all integers x for which

$$0 \le \frac{1}{7}(x^2 - 14x + 38) \le 1$$

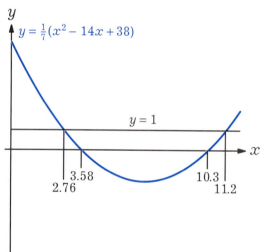

Figure 26S.56

From the graph on the GDC, the inequality is satisfied for
$2.76 \le x \le 3.68$ and $10.3 \le x \le 11.2$.

Hence the possible integer values of x are 3 and 11.

57 a This is a geometric series with first term $x^0 = 1$ and common ratio x.
There are $n+1$ terms in the sum.

$$\therefore \sum_{k=0}^{n} x^k = S_{n+1} = \frac{1 - x^{n+1}}{1 - x}$$

b Notice that the required expression is the derivative of the sum from (a). So

$$1 + 2x + 3x^2 + \ldots + nx^{n-1} = \frac{d}{dx}\left(1 + x + x^2 + \ldots + x^n\right)$$

$$= \frac{d}{dx}\left(\frac{1 - x^{n+1}}{1 - x}\right)$$

$$= \frac{-(n+1)x^n(1-x) - (-1)(1 - x^{n+1})}{(1-x)^2}$$

$$= \frac{-(n+1)x^n + (n+1)x^{n+1} + 1 - x^{n+1}}{(1-x)^2}$$

$$= \frac{1 + nx^{n+1} - (n+1)x^n}{(1-x)^2}$$

$$= \frac{1 - x^n}{(1-x)^2} - \frac{nx^n(1-x)}{(1-x)^2}$$

$$= \frac{1 - x^n}{(1-x)^2} - \frac{nx^n}{1 - x}$$

58 This is a geometric series with first term 1 and common ratio ω. The sum of the first n terms is

$$S_n = \frac{\omega^n - 1}{\omega - 1}, \omega \neq 1$$

Since the ω is the solution of $z^n = 1$, we know that $\omega^n - 1 = 0$ and hence $S_n = 0$.

But $\omega = 1$ is also a solution of $z^n = 1$, and in that case $S_n = n$.

So the possible values of the sum are 0 and n.

59 a $(2+i)(3+i) = 6 + 2i + 3i - 1 = 5 + 5i$

b By considering the position of $5 + 5i$ in the Argand diagram,

$$\tan \theta = \frac{5}{5} = 1$$

$$\therefore \arg(5 + 5i) = \arctan(1) = \frac{\pi}{4}$$

c $\arg(2+i) = \arctan\frac{1}{2}$, $\arg(3+i) = \arctan\frac{1}{3}$

$\arg z_1 + \arg z_2 = \arg(z_1 z_2)$

$$\therefore \arctan\left(\frac{1}{2}\right) + \arctan\left(\frac{1}{3}\right) = \arg(2+i)(3+i) = \frac{\pi}{4}$$

> **COMMENT**
>
> As all the complex numbers in this question are in the first quadrant of the Argand plane, drawing a diagram isn't really necessary. However, do beware! When numbers are not in the first quadrant, it is always worth drawing a diagram as the argument won't be as obvious.

60 a $\dfrac{1+e^{2i\theta}}{1-e^{2i\theta}} = \dfrac{1+\cos 2\theta + i\sin 2\theta}{1-\cos 2\theta - i\sin 2\theta}$

$= \dfrac{(1+\cos 2\theta + i\sin 2\theta)(1-\cos 2\theta + i\sin 2\theta)}{(1-\cos 2\theta - i\sin 2\theta)(1-\cos 2\theta + i\sin 2\theta)}$

$= \dfrac{(1+i\sin 2\theta)^2 - (\cos 2\theta)^2}{(1-\cos 2\theta)^2 + \sin^2 2\theta}$

$= \dfrac{(1-\cos^2 2\theta - \sin^2 2\theta) + i(2\sin 2\theta)}{1 - 2\cos 2\theta + \cos^2 2\theta + \sin^2 2\theta}$

$= \dfrac{(1-1) + i(2\sin 2\theta)}{1 - 2\cos 2\theta + 1}$

$= 0 + i\dfrac{\sin 2\theta}{1-\cos 2\theta}$

$\therefore \text{Re}\left(\dfrac{1+e^{2i\theta}}{1-e^{2i\theta}}\right) = 0, \quad \text{Im}\left(\dfrac{1+e^{2i\theta}}{1-e^{2i\theta}}\right) = \dfrac{\sin 2\theta}{1-\cos 2\theta}$

b Using double angle formulae:

$\dfrac{1+e^{2i\theta}}{1-e^{2i\theta}} = \dfrac{i(2\sin\theta\cos\theta)}{1-(1-2\sin^2\theta)} = \dfrac{i\cos\theta}{\sin\theta} = i\cot\theta$

61 When $n = 1$:

$\dfrac{\sin 2\theta}{2\sin\theta} = \dfrac{2\sin\theta\cos\theta}{2\sin\theta} = \cos\theta$

So the statement is true for $n = 1$.

Suppose that the statement is true for some $n = k$: $\cos\theta + \cos 3\theta + \ldots + \cos(2k-1)\theta = \dfrac{\sin(2k\theta)}{2\sin\theta}$

Then for $n = k+1$:

$\cos\theta + \cos 3\theta + \ldots + \cos(2k-1)\theta + \cos(2(k+1)-1)\theta$

$= \dfrac{\sin(2k\theta)}{2\sin\theta} + \cos(2k+1)\theta$

$= \dfrac{\sin(2k\theta) + 2\sin\theta\cos(2k+1)\theta}{2\sin\theta}$

$= \dfrac{\sin(2k\theta) + 2\sin\theta(\cos 2k\theta\cos\theta - \sin 2k\theta\sin\theta)}{2\sin\theta}$

$= \dfrac{\sin(2k\theta)(1-2\sin^2\theta) + \cos 2k\theta(2\sin\theta\cos\theta)}{2\sin\theta}$

$= \dfrac{\sin 2k\theta\cos 2\theta + \cos 2k\theta\sin 2\theta}{2\sin\theta}$

$= \dfrac{\sin(2k\theta + 2\theta)}{2\sin\theta}$

$= \dfrac{\sin(2(k+1)\theta)}{2\sin\theta}$

So the statement is true for $n = k+1$.

The statement is true for $n = 1$, and if true for $n = k$ then it is also true for $n = k+1$. Therefore it is true for all $n \in \mathbb{Z}^+$ by the principle of mathematical induction.

> **COMMENT**
>
> It may not have been clear at first how to proceed in the proof, so it is worth thinking about what you want to get to: in this case $\sin(2k+2)\theta$ in the numerator. Expanding this and working backwards should then show you the way ahead!

62 a Convert to modulus–argument form in order to raise to a power:

$$|2+i| = \sqrt{5}, \ \arg(2+i) = \arctan\left(\frac{1}{2}\right)$$

So, by De Moivre,

$$(2+i)^n = \left(\sqrt{5}\right)^n \operatorname{cis}\left(n \arctan\left(\frac{1}{2}\right)\right)$$

$$\therefore \operatorname{Re}\left((2+i)^n\right) = \left(\sqrt{5}\right)^n \cos\left(n \arctan\left(\frac{1}{2}\right)\right)$$

b $2+i = \sqrt{5} \operatorname{cis} \theta$ and $2-i = \sqrt{5} \operatorname{cis}(-\theta)$, where $\theta = \arctan\left(\frac{1}{2}\right)$

By De Moivre,

$$(2-i)^n = \left(\sqrt{5}\right)^n \operatorname{cis}(-n\theta)$$

So $(2+i)^n + (2-i)^n = \left(\sqrt{5}\right)^n (\cos n\theta + i \sin n\theta) + \left(\sqrt{5}\right)^n (\cos n\theta - i \sin n\theta)$

$$= 2\left(\sqrt{5}\right)^n \cos n\theta$$

which is always real.

> **COMMENT**
>
> In general, $(z*)^n = (z^n)^*$, so $z^n + (z*)^n = 2\operatorname{Re}(z^n)$.

63 a $\binom{n}{1} = \dfrac{n!}{(n-1)!\,1!}$

$$= \frac{n(n-1)!}{(n-1)!}$$

$$= n$$

b $(x+h)^n = x^n + \binom{n}{1}x^{n-1}h + \binom{n}{2}x^{n-2}h^2 + \ldots + h^n$

c Differentiating from first principles:

$$\frac{d}{dx}(x^n) = \lim_{h \to 0} \frac{(x+h)^n - x^n}{h}$$

$$= \lim_{h \to 0} \frac{x^n + \binom{n}{1}x^{n-1}h + \binom{n}{2}x^{n-2}h^2 + \ldots + h^n - x^n}{h}$$

$$= \lim_{h \to 0} \frac{nx^{n-1}h + \binom{n}{2}x^{n-2}h^2 + \ldots + h^n}{h}$$

$$= \lim_{h \to 0} \left(nx^{n-1} + \binom{n}{2}x^{n-2}h + \ldots + h^{n-1} \right)$$

$$= nx^{n-1}$$

64 Using $a \cdot a = |a|^2$, $b \cdot b = |b|^2$ and $|a| = |b| = x$:

$(a+b) \cdot (a+b) = 6x$

$a \cdot a + b \cdot b + 2a \cdot b = 6x$

$2x^2 + 2a \cdot b = 6x$

$a \cdot b = 3x - x^2$

Then, using $a \cdot b = |a||b|\cos\theta$,

$3x - x^2 = |a||b|\cos\theta$

$= x \times x \cos\theta$

$= x^2 \cos\theta$

Since $-1 \leq \cos\theta \leq 1$, $-x^2 \leq x^2 \cos\theta \leq x^2$

and so $-x^2 \leq 3x - x^2 \leq x^2$

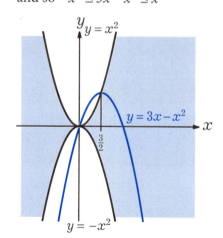

Figure 26S.64.1

From the graph on the GDC, $y = 3x - x^2$ is between $y = -x^2$ and $y = x^2$ for $x \geq \frac{3}{2} = 1.5$, so the smallest possible value of x is 1.5.

> **COMMENT**
>
> It is also possible to solve these inequalities without a calculator:
>
> $-x^2 \leq 3x - x^2 \leq x^2$
>
> $\Leftrightarrow 3x \geq 0$ and $2x^2 - 3x \geq 0$
>
> $\Leftrightarrow 3x \geq 0$ and $x(2x-3) \geq 0$
>
> $\Leftrightarrow x \geq 0$ and $x \leq 0$ or $x \geq \frac{3}{2}$
>
> $\therefore x = 0$ or $x \geq \frac{3}{2}$
>
> so $x \geq \frac{3}{2}$ (as $x \neq 0$ is given)
>
> Nonetheless, it is still advisable to consult a graph when solving a quadratic inequality (such as $2x^2 - 3x \geq 0$); without a calculator, a sketch of the quadratic is needed.
>
> If it is unclear where the solutions to two inequalities both hold (such as $x \geq 0$ and $x \leq 0$ or $x \geq \frac{3}{2}$), highlight them on a number line and look for the region where they overlap:
>
>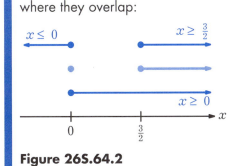
>
> Figure 26S.64.2

Long questions

1 a i By the product rule,
$f'(x) = pe^{px}(x+1) + e^{px}(1)$
$= e^{px}(p(x+1)+1)$

ii Part (i) shows that the statement is true for $n = 1$.
Assume that it is true for some $n = k$:
$f^{(k)}(x) = p^{k-1}e^{px}(p(x+1)+k)$

Then for $n = k+1$:
$f^{(k+1)}(x) = \dfrac{d}{dx}f^{(k)}(x)$
$= p^{k-1}pe^{px}(p(x+1)+k) + p^{k-1}e^{px}(p)$
$= p^k e^{px}(p(x+1)+k+1)$

So the statement is true for $n = k+1$.

The statement is true for $n = 1$, and if true for $n = k$ then it is also true for $n = k+1$.

Therefore it is true for all $n \in \mathbb{Z}^+$ by the principle of mathematical induction.

b i Minimum point when $f'(x) = 0$:
$e^{px}(p(x+1)+1) = 0$
$px + p + 1 = 0$
$x = -\dfrac{p+1}{p}$
$= -\dfrac{\sqrt{3}+1}{\sqrt{3}}$

ii Point of inflexion when $f''(x) = 0$:
$pe^{px}(p(x+1)+2) = 0$
$px + p + 2 = 0$
$x = -\dfrac{p+2}{p}$
$= -\dfrac{\sqrt{3}+2}{\sqrt{3}}$

c The graph of $y = e^{\frac{x}{2}}(x+1)$ crosses the x-axis at $x = -1$. It is negative for $-2 \leq x \leq -1$.

$\therefore$ Area $= -\displaystyle\int_{-2}^{-1} e^{\frac{x}{2}}(x+1)\,dx + \int_{-1}^{2} e^{\frac{x}{2}}(x+1)\,dx$

Using integration by parts:

$u = x+1 \Rightarrow \dfrac{du}{dx} = 1$

$\dfrac{dv}{dx} = e^{\frac{x}{2}} \Rightarrow v = 2e^{\frac{x}{2}}$

$$\text{Area} = \left[-2e^{\frac{x}{2}}(x+1)\right]_{-2}^{-1} + \int_{-2}^{-1} 2e^{\frac{x}{2}}\,dx + \left[2e^{\frac{x}{2}}(x+1)\right]_{-1}^{2} - \int_{-1}^{2} 2e^{\frac{x}{2}}\,dx$$

$$= \left[0 + 2e^{-1}(-1)\right] + \left[4e^{\frac{x}{2}}\right]_{-2}^{-1} + \left[2e(3) - 0\right] - \left[4e^{\frac{x}{2}}\right]_{-1}^{2}$$

$$= -2e^{-1} + \left(4e^{-\frac{1}{2}} - 4e^{-1}\right) + 6e - \left(4e - 4e^{-\frac{1}{2}}\right)$$

$$= 2e + 8e^{-\frac{1}{2}} - 6e^{-1}$$

$$= 8.08 \ (3\text{SF})$$

> **COMMENT**
>
> It is always a good idea to sketch the function to be integrated on the GDC to check whether it goes below the x-axis, rather than just integrating between the limits blindly. If the function is negative for part of the interval, then the integration needs to be split up as shown here.

2 a $P(\text{Daniel gets heads}) = \dfrac{1}{5}$

  b $P(\text{Daniel gets tails, then Theo gets heads})$

   $= \dfrac{4}{5} \times \dfrac{1}{5} = \dfrac{4}{25}$

  c $P(\text{D gets tails, then T gets tails, then D gets heads})$

   $= \dfrac{4}{5} \times \dfrac{4}{5} \times \dfrac{1}{5} = \dfrac{16}{125}$

  d The probabilities of Daniel winning on a particular throw are:

   $P(\text{first throw}) = \dfrac{1}{5}$

   $P(\text{second throw}) = \left(\dfrac{4}{5}\right)^2 \left(\dfrac{1}{5}\right)$

   $P(\text{third throw}) = \left(\dfrac{4}{5}\right)^4 \left(\dfrac{1}{5}\right)$

   $\vdots$

   This is a geometric sequence with first term $\dfrac{1}{5}$ and common ratio $\left(\dfrac{4}{5}\right)^2$.

   The probability of Daniel winning is S_∞, which is

   $$S_\infty = \dfrac{\dfrac{1}{5}}{1 - \left(\dfrac{4}{5}\right)^2} = \dfrac{5}{25 - 16} = \dfrac{5}{9}$$

e   P(Theo wins) = 1 − P(Daniel wins)

$$= 1 - \frac{5}{9}$$

$$= \frac{4}{9}$$

> **COMMENT**
>
> Here we are assuming that the game eventually ends, so that there is no possibility of a draw. This is the case because the probability of no one winning after n throws is $P(n \text{ tails}) = \left(\frac{4}{5}\right)^n$, which tends to zero as $n \to \infty$.

f   Let $P(\text{head}) = p$

Using the same argument as above, $P(\text{Daniel wins}) = \dfrac{p}{1-(1-p)^2}$.

If $P(\text{D wins}) = 2\,P(\text{T wins})$, then since the two probabilities must add up to 1, $P(\text{D wins}) = \dfrac{2}{3}$

$$\therefore \frac{p}{1-(1-p)^2} = \frac{2}{3}$$

$$3p = 2 - 2(1 - 2p + p^2)$$

$$2p^2 - p = 0$$

$$p(2p - 1) = 0$$

So $p = \dfrac{1}{2}$ (as $p \neq 0$)

3 a If $f(x)$ is a continuous function on a single domain interval, then $f(x)$ is not one-to-one because the gradient changes from positive to negative and back. A one-to-one function must be either increasing or decreasing throughout its domain.

b $f(3) = 4$

$$\therefore f(f(3)) = f(4) = 6$$

c Translation by $\begin{pmatrix} 2 \\ 3 \end{pmatrix}$ results in the function $f(x-2) + 3$, and reflection in the x-axis gives

$$g(x) = -f(x-2) - 3$$

$$\therefore g'(x) = -f'(x-2)$$

and so $g'(2) = -f'(0) = -7$

4 a De Moivre's theorem: $(\cos\theta + i\sin\theta)^n = \cos(n\theta) + i\sin(n\theta)$

Proof by induction:

When $n = 1$:
$(\cos\theta + i\sin\theta)^1 = \cos(\theta) + i\sin(\theta) = \cos(1\theta) + i\sin(1\theta)$

So the statement is true for $n = 1$.

26 Questions crossing chapters

Assume that the statement is true for some $n = k$:
$$(\cos\theta + i\sin\theta)^k = \cos(k\theta) + i\sin(k\theta)$$

Then for $n = k+1$:
$$(\cos\theta + i\sin\theta)^{k+1} = (\cos(k\theta) + i\sin(k\theta))(\cos\theta + i\sin\theta)$$
$$= \cos(k\theta)\cos\theta - \sin(k\theta)\sin\theta + i\sin(k\theta)\cos\theta + i\cos(k\theta)\sin\theta$$
$$= \cos(k\theta + \theta) + i\sin(k\theta + \theta)$$
$$= \cos(k+1)\theta + i\sin(k+1)\theta$$

Hence the statement is true for $n = k+1$.

The statement is true for $n = 1$, and if true for $n = k$ then it is also true for $n = k+1$. Therefore it is true for all $n \in \mathbb{Z}^+$ by the principle of mathematical induction.

b $r = -\dfrac{1}{2}e^{i\theta}$

$|r| = \dfrac{1}{2}|e^{i\theta}| = \dfrac{1}{2} < 1$

c $u_1 = 1$, $r = -\dfrac{1}{2}e^{i\theta}$, and the sum to infinity exists since $|r| < 1$.

$$S_\infty = \dfrac{1}{1 - \left(-\dfrac{1}{2}e^{i\theta}\right)} = \dfrac{2}{2 + e^{i\theta}}$$

d The required expression is the real part of the series from (b):

$$1 - \dfrac{1}{2}\cos\theta + \dfrac{1}{4}\cos 2\theta - \dfrac{1}{8}\cos 3\theta + \ldots$$

$$= \text{Re}\left(1 - \dfrac{1}{2}e^{i\theta} + \dfrac{1}{4}e^{2i\theta} - \dfrac{1}{8}e^{3i\theta} + \ldots\right)$$

$$= \text{Re}\left(\dfrac{2}{2 + e^{i\theta}}\right)$$

$$= \text{Re}\left(\dfrac{2}{2 + \cos\theta + i\sin\theta}\right)$$

$$= \text{Re}\left(\dfrac{2(2 + \cos\theta - i\sin\theta)}{(2 + \cos\theta)^2 + \sin^2\theta}\right)$$

$$= \text{Re}\left(\dfrac{4 + 2\cos\theta - 2i\sin\theta}{4 + 4\cos\theta + \cos^2\theta + \sin^2\theta}\right)$$

$$= \text{Re}\left(\dfrac{4 + 2\cos\theta - 2i\sin\theta}{5 + 4\cos\theta}\right)$$

$$= \dfrac{4 + 2\cos\theta}{5 + 4\cos\theta}$$

5 a $f'(x) = 2 + \cos 2x - \sec^2 x$

b Stationary points occur where $f'(x) = 0$:

$$2 + \cos 2x - \sec^2 x = 0$$

$$2 + (2\cos^2 x - 1) - \left(\frac{1}{\cos x}\right)^2 = 0$$

$$2\cos^2 x + 2\cos^4 x - \cos^2 x - 1 = 0$$

$$2\cos^4 x + \cos^2 x - 1 = 0$$

c Solving the above equation for $\cos x$:

$$(2\cos^2 x - 1)(\cos^2 x + 1) = 0$$

$$\therefore 2\cos^2 x - 1 = 0 \quad (\text{as } \cos^2 x + 1 \neq 0)$$

$$\Rightarrow \cos x = \pm \frac{1}{\sqrt{2}}$$

$$\therefore x = \pm \frac{\pi}{4} \quad \text{as } x \in \left[-\frac{\pi}{2}, \frac{\pi}{2}\right]$$

i.e. there are two stationary points.

6 a The series converges because the common ratio is $r = \cos x$, so $|r| < 1$. (Note that $\cos x \neq \pm 1$ because $x \neq 0, \pi$.)

b With $u_1 = 1$, $r = \cos x$:

$$S_\infty = \frac{1}{1 - \cos x} = \frac{1}{1 - \left(1 - 2\sin^2 \frac{x}{2}\right)} = \frac{1}{2\sin^2 \frac{x}{2}} = \frac{1}{2}\csc^2 \frac{x}{2}$$

c Using the answer from (b):

$$\int_{\pi/3}^{\pi/2} \left(1 + \cos x + \cos^2 x + \cos^3 x + \ldots\right) dx = \int_{\pi/3}^{\pi/2} \frac{1}{2}\csc^2\left(\frac{x}{2}\right) dx$$

$$= \left[-\cot\left(\frac{x}{2}\right)\right]_{\pi/3}^{\pi/2}$$

$$= -\cot\left(\frac{\pi}{4}\right) + \cot\left(\frac{\pi}{6}\right)$$

$$= \sqrt{3} - 1$$

7 a $\left(z - \dfrac{1}{z}\right)^5 = z^5 - \dfrac{5z^4}{z} + \dfrac{10z^3}{z^2} - \dfrac{10z^2}{z^3} + \dfrac{5z}{z^4} - \dfrac{1}{z^5}$

$$= z^5 - 5z^3 + 10z - \frac{10}{z} + \frac{5}{z^3} - \frac{1}{z^5}$$

b Let $z = \cos\theta + i\sin\theta$

Then $\dfrac{1}{z} = z^{-1} = \cos\theta - i\sin\theta$

$\therefore z - \dfrac{1}{z} = \cos\theta + i\sin\theta - (\cos\theta - i\sin\theta) = 2i\sin\theta$

Similarly, $z^k = \cos k\theta + i\sin k\theta$

and $\dfrac{1}{z^k} = \cos k\theta - i\sin k\theta$

$\therefore z^k - \dfrac{1}{z^k} = 2i\sin k\theta$

Regrouping the terms from (a):

$$\left(z - \dfrac{1}{z}\right)^5 = z^5 - \dfrac{1}{z^5} - 5z^3 + \dfrac{5}{z^3} + 10z - \dfrac{10}{z}$$

$$= \left(z^5 - \dfrac{1}{z^5}\right) - 5\left(z^3 - \dfrac{1}{z^3}\right) + 10\left(z - \dfrac{1}{z}\right)$$

Then, substituting in the above expressions for $z^k - \dfrac{1}{z^k}$:

$(2i\sin\theta)^5 = 2i\sin 5\theta - 5(2i\sin 3\theta) + 10(2i\sin\theta)$

$32i\sin^5\theta = 2i\sin 5\theta - 10i\sin 3\theta + 20i\sin\theta$

$32\sin^5\theta = 2\sin 5\theta - 10\sin 3\theta + 20\sin\theta$

c $\displaystyle\int_0^{\pi/2} \sin^5\theta\, d\theta = \int_0^{\pi/2} \dfrac{1}{16}(\sin 5\theta - 5\sin 3\theta + 10\sin\theta)\, d\theta$

$= \dfrac{1}{16}\left[-\dfrac{\cos 5\theta}{5} + \dfrac{5\cos 3\theta}{3} - 10\cos\theta\right]_0^{\pi/2}$

$= \dfrac{1}{16}\left[-\dfrac{1}{5}\cos\dfrac{5\pi}{2} + \dfrac{5}{3}\cos\dfrac{3\pi}{2} - 10\cos\dfrac{\pi}{2}\right] - \dfrac{1}{16}\left[-\dfrac{1}{5} + \dfrac{5}{3} - 10\right]$

$= \dfrac{1}{16}(0) - \dfrac{1}{16}\left(\dfrac{-3 + 25 - 150}{15}\right)$

$= \dfrac{128}{240}$

$= \dfrac{8}{15}$

8 a $\sum_{r=0}^{\infty} a^r = 1 + a + a^2 + \ldots$

This is a geometric series with first term 1 and common ratio a.

$S_{\infty} = 1.5$

$\dfrac{1}{1-a} = 1.5$

$1.5 - 1.5a = 1$

$1.5a = 0.5$

$\therefore a = \dfrac{1}{3}$

b $1 - x + x^2 - x^3 + \ldots$ is a geometric series with first term 1 and common ratio $-x$.

Using the formula for S_{∞},

$1 - x + x^2 - x^3 + \ldots = \dfrac{1}{1-(-x)} = \dfrac{1}{1+x}$

The series converges for $|-x| < 1$, i.e. $|x| < 1$

$\therefore k = 1$

c $\dfrac{d}{dx} \ln(1+x) = \dfrac{1}{1+x}$

$\therefore \ln(1+x) = \int \dfrac{1}{1+x} dx$

$= \int (1 - x + x^2 - x^3 + \ldots) dx$ from (b)

$= x - \dfrac{1}{2}x^2 + \dfrac{1}{3}x^3 - \dfrac{1}{4}x^4 + \ldots + c$

When $x = 0$,

$\ln(1) = c$

$\therefore c = 0$

Hence $\ln(1+x) = x - \dfrac{1}{2}x^2 + \dfrac{1}{3}x^3 - \dfrac{1}{4}x^4 + \ldots$

> **COMMENT**
>
> This assumes that an infinite series can be integrated term by term, which is true in this case, although the formal proof requires techniques of mathematical analysis that are usually introduced at undergraduate level.

d Set $x = 0.1$ in (c):

$\ln 1.1 \approx 0.1 - \dfrac{0.01}{2} + \dfrac{0.001}{3} - \dfrac{0.0001}{4}$

$= (0.1 + 0.00033\ldots) - (0.005 + 0.000025)$

$= 0.10033\ldots - 0.005025$

$= 0.0953\ldots$

$= 0.095$ (3DP)

9 a

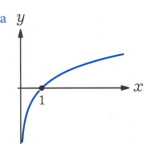

Figure 26L.9.1 Graph of $y = \ln x$

b $\dfrac{dy}{dx} = \dfrac{1}{x}$

Equation of the tangent at $x = p$ is

$y - \ln p = \dfrac{1}{p}(x - p)$

Given that this tangent passes through $(0, 0)$:

$-\ln p = \dfrac{1}{p}(0 - p)$

$\ln p = 1$

$\therefore p = e$

c

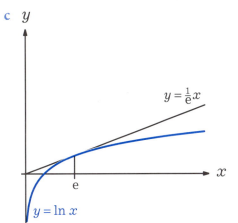

Figure 26L.9.2 Graph of $y = \ln x$ and the tangent line $y = \dfrac{1}{e}x$

The line $y = kx$ intersects the graph of $y = \ln x$ twice when its gradient k is smaller than the gradient $\dfrac{1}{e}$ of the tangent from (b); but the gradient needs to be positive, otherwise there will be only one intersection.

$$\therefore 0 < k < \dfrac{1}{e}$$

10 a Dividing $f(x)$ by $(x-a)^2$ gives a polynomial $g(x)$ and the remainder which cannot be divided. If the remainder is linear, $R = mx + c$, then

$$\dfrac{f(x)}{(x-a)^2} = g(x) + \dfrac{mx+c}{(x-a)^2}$$

$$\therefore f(x) = g(x)(x-a)^2 + mx + c$$

b Using the product rule,
$$f'(x) = g'(x)(x-a)^2 + 2(x-a)g(x) + m$$

c Substituting $x = a$ into the equations from (a) and (b):

$$f(a) = ma + c \quad \ldots(1)$$
$$f'(a) = m \quad \ldots(2)$$

Substituting (2) into (1):
$$f(a) = af'(a) + c$$
$$c = f(a) - af'(a)$$

So the remainder is
$$mx + c = f'(a)x + f(a) - af'(a)$$
$$= f'(a)(x-a) + f(a)$$

d If $(x-a)^2$ is a factor of $f(x)$, then the remainder $mx + c$ equals zero for all x.

This means that $m = c = 0$

$$\therefore f'(a) = 0 \text{ and } f(a) = af'(a) = 0$$

11 a For independent events, $P(A \cap B) = P(A)P(B)$. But

$$P(A) \times P(B) = 0.85 \times 0.60 = 0.51$$

$$P(A \cap B) = 0.55$$

so A and B are not independent events.

b Require the probability that the building will not be completed on time (B') given that the materials arrive on time (A):

$$P(B' | A) = 1 - P(B | A)$$

$$= 1 - \dfrac{P(A \cap B)}{P(A)}$$

$$= 1 - \dfrac{0.55}{0.85}$$

$$= 0.353 \text{ (3SF)}$$

c The total number of possible selections is $\binom{10}{5} = 252$.

The number of selections with two electricians, one plumber and two others is $\binom{3}{2}\binom{2}{1}\binom{5}{2} = 60$.

The required probability is
$$\dfrac{60}{252} = \dfrac{5}{21} = 0.238.$$

d Let X be the number of hours worked by a random team member; then

$$X \sim N(42, \sigma^2)$$

First we need to find σ.

Let $Z = \dfrac{X - 42}{\sigma} \sim N(0,1)$

$P(X > 48) = 10\%$

$$\Rightarrow P\left(Z > \dfrac{48 - 42}{\sigma}\right) = 0.1$$

$$\dfrac{6}{\sigma} = 1.2816 \text{ (from GDC)}$$

$$\therefore \sigma = 4.682$$

Then

$P(\text{both plumbers work more than 40 hours})$
$= P(X > 40) \times P(X > 40)$
$= 0.665^2$ (from GDC)
$= 0.443$

> **COMMENT**
>
> We assume that time is a continuous random variable, so 'more than 40 hours' means $X > 40$ rather than, say, $X \geq 41$ or $X > 40.5$.

12 a This is a geometric series with $u_1 = e^{i\theta}$ and $r = e^{2i\theta}$.

The sum of the first n terms is

$$S_n = \frac{e^{i\theta}\left(1 - e^{2ni\theta}\right)}{1 - e^{2i\theta}}$$

b This is the real part of the series from (a).

$\cos\theta + \cos 3\theta + \cos 5\theta + \ldots + \cos(2n-1)\theta$

$= \text{Re}\left(\dfrac{e^{i\theta}\left(1 - e^{2ni\theta}\right)}{1 - e^{2i\theta}}\right)$

$= \text{Re}\left(\dfrac{1 - e^{2ni\theta}}{e^{-i\theta} - e^{i\theta}}\right)$

$= \text{Re}\left(\dfrac{1 - \cos 2n\theta - i\sin 2n\theta}{(\cos\theta - i\sin\theta) - (\cos\theta + i\sin\theta)}\right)$

$= \text{Re}\left(\dfrac{1 - \cos 2n\theta - i\sin 2n\theta}{-2i\sin\theta}\right)$

$= \text{Re}\left(\dfrac{i(1 - \cos 2n\theta) + \sin 2n\theta}{2\sin\theta}\right)$

$= \dfrac{\sin 2n\theta}{2\sin\theta}$

> **COMMENT**
>
> Notice that dividing by $e^{i\theta}$ at the beginning results in a nicer expression in the denominator to work with.

c Setting $n = 3$ in the result of (b):

$\cos\theta + \cos 3\theta + \cos 5\theta = \dfrac{\sin 6\theta}{2\sin\theta}$

$\therefore \dfrac{\sin 6\theta}{2\sin\theta} = 0$

$\sin 6\theta = 0$ for $0 < 6\theta < 6\pi$

$6\theta = \pi, 2\pi, 3\pi, 4\pi, 5\pi$

$\theta = \dfrac{\pi}{6}, \dfrac{\pi}{3}, \dfrac{\pi}{2}, \dfrac{2\pi}{3}, \dfrac{5\pi}{6}$

> **COMMENT**
>
> Note that the result from (b) applies only when $\sin\theta \neq 0$, which is the case for all five solutions above, so they are all valid.

13 a Using several times the fact that $k!(k+1) = (k+1)!$,

$\binom{n}{r} + \binom{n}{r+1} = \dfrac{n!}{r!(n-r)!} + \dfrac{n!}{(r+1)!(n-r-1)!}$

$= \dfrac{n!(r+1) + n!(n-r)}{(r+1)!(n-r)!}$

$= \dfrac{n!(r+1+n-r)}{(r+1)!(n-r)!}$

$= \dfrac{n!(n+1)}{(r+1)!(n-r)!}$

$= \dfrac{(n+1)!}{(r+1)!(n-r)!}$

$= \binom{n+1}{r+1}$

b When $n = 1$:

$\dfrac{d}{dx}(uv) = u\dfrac{dv}{dx} + \dfrac{du}{dx}v$ by the product rule

$= \binom{1}{0}u\dfrac{dv}{dx} + \binom{1}{1}\dfrac{du}{dx}v$

so the statement is true for $n = 1$.

Assume that the statement is true for some $n = k$:

$$\frac{d^k}{dx^k}(uv) = \sum_{i=0}^{k} \binom{k}{i} \frac{d^i}{dx^i}(u) \frac{d^{k-i}}{dx^{k-i}}(v)$$

Then, for $n = k+1$, use the product rule for each term in the sum to differentiate this expression with respect to x:

$$\frac{d^{k+1}}{dx^{k+1}}(uv) = \sum_{i=0}^{k} \binom{k}{i} \left(\frac{d^i}{dx^i}(u) \frac{d^{k-i+1}}{dx^{k-i+1}}(v) + \frac{d^{i+1}}{dx^{i+1}}(u) \frac{d^{k-i}}{dx^{k-i}}(v) \right)$$

$$= \binom{k}{0}\left(u \frac{d^{k+1}v}{dx^{k+1}} + \frac{du}{dx} \frac{d^k v}{dx^k} \right) + \binom{k}{1}\left(\frac{du}{dx} \frac{d^k v}{dx^k} + \frac{d^2 u}{dx^2} \frac{d^{k-1}v}{dx^{k-1}} \right)$$

$$+ \binom{k}{2}\left(\frac{d^2 u}{dx^2} \frac{d^{k-1}v}{dx^{k-1}} + \frac{d^3 u}{dx^3} \frac{d^{k-2}v}{dx^{k-2}} \right) + \ldots$$

$$= \binom{k}{0} u \frac{d^{k+1}v}{dx^{k+1}} + \left(\binom{k}{0} + \binom{k}{1} \right) \frac{du}{dx} \frac{d^k v}{dx^k} + \left(\binom{k}{1} + \binom{k}{2} \right) \frac{d^2 u}{dx^2} \frac{d^{k-1}v}{dx^{k-1}} + \ldots$$

Now, using the result from (a) and noticing that $\binom{k}{0} = \binom{k+1}{0} = 1$:

$$\frac{d^{k+1}}{dx^{k+1}}(uv) = \binom{k+1}{0} u \frac{d^{k+1}v}{dx^{k+1}} + \binom{k+1}{1} \frac{du}{dx} \frac{d^k v}{dx^k} + \binom{k+1}{2} \frac{d^2 u}{dx^2} \frac{d^{k-1}v}{dx^{k-1}} + \ldots$$

which is of the required form with $n = k+1$.

Hence, if the statement is true for $n = k$ then it is also true for $n = k+1$.
As it is true for $n = 1$, it follows that it is true for all integers $n \geq 1$ by the principle of mathematical induction.

14 If $t = \tan\frac{x}{2}$, then

$$\frac{1-t^2}{1+t^2} = \frac{1 - \frac{\sin^2\frac{x}{2}}{\cos^2\frac{x}{2}}}{1 + \frac{\sin^2\frac{x}{2}}{\cos^2\frac{x}{2}}} = \frac{\cos^2\frac{x}{2} - \sin^2\frac{x}{2}}{\cos^2\frac{x}{2} + \sin^2\frac{x}{2}} = \frac{\cos x}{1} = \cos x$$

and

$$\frac{2t}{1+t^2} = \frac{2\frac{\sin\frac{x}{2}}{\cos\frac{x}{2}}}{1 + \frac{\sin^2\frac{x}{2}}{\cos^2\frac{x}{2}}} = \frac{2\sin\frac{x}{2}\cos\frac{x}{2}}{\cos^2\frac{x}{2} + \sin^2\frac{x}{2}} = \frac{\sin x}{1} = \sin x$$

where the double angle formulae have been used to simplify the numerators.

b $\dfrac{dt}{dx} = \dfrac{1}{2}\sec^2 \dfrac{x}{2} = \dfrac{1}{2}\left(\tan^2 \dfrac{x}{2} + 1\right) = \dfrac{t^2+1}{2}$

$\Rightarrow dx = \dfrac{2\,dt}{t^2+1}$

$\therefore \displaystyle\int \dfrac{\sin x}{1+\cos x}\,dx = \int \dfrac{\dfrac{2t}{1+t^2}}{1+\dfrac{1-t^2}{1+t^2}} \cdot \dfrac{2\,dt}{1+t^2} = \int \dfrac{2t}{1+t^2+1-t^2} \cdot \dfrac{2\,dt}{1+t^2} = \int \dfrac{2t}{1+t^2}\,dt$

c Change the limits:

when $x = 0,\ t = \tan 0 = 0$

when $x = \dfrac{\pi}{2},\ t = \tan \dfrac{\pi}{4} = 1$

$\displaystyle\int_0^{\frac{\pi}{2}} \dfrac{\sin x}{1+\cos x}\,dx = \int_0^1 \dfrac{2t}{1+t^2}\,dt = \left[\ln(1+t^2)\right]_0^1 = \ln 2 - \ln 1 = \ln 2$

15 a Given that $X = 4$, $Z = 5$ is equivalent to $Y = 1$

$\therefore P(Z = 5 \mid X = 4) = P(Y = 1 \mid X = 4)$

But X and Y are independent, so

$P(Y = 1 \mid X = 4) = P(Y = 1) = \dfrac{e^{-n} n^1}{1!} = n e^{-n}$

b If $Z = k$, there are the following possibilities:

$X = 0,\ Y = k$

or $X = 1,\ Y = k-1$

or $X = 2,\ Y = k-2$

$\vdots$

or $X = k,\ Y = 0$

These are mutually exclusive outcomes, so their probabilities can be added; and since X and Y are independent, the probabilities in each term can be multiplied.

$\therefore P(Z = k) = P(X = 0)P(Y = k) + P(X = 1)P(Y = k-1) + \ldots$

$= \displaystyle\sum_{r=0}^{k} P(X = r)P(Y = k-r)$

26 Questions crossing chapters

c $P(X=r)P(y=k-r) = \dfrac{e^{-m}m^r}{r!} \times \dfrac{e^{-n}n^{k-r}}{(k-r)!}$

$\qquad\qquad\qquad = \dfrac{e^{-(m+n)}m^r n^{k-r}}{r!(k-r)!}$

The factor $e^{-(m+n)}$ is common to all terms in the sum from (b), so

$P(Z=k) = e^{-(m+n)} \displaystyle\sum_{r=0}^{k} \dfrac{m^r n^{k-r}}{r!(k-r)!}$

$\qquad\quad = \dfrac{e^{-(m+n)}}{k!} \displaystyle\sum_{r=0}^{k} \dfrac{k!}{r!(k-r)!} m^r n^{k-r}$

$\qquad\quad = \dfrac{e^{-(m+n)}}{k!} \displaystyle\sum_{r=0}^{k} \binom{k}{r} m^r n^{k-r}$

$\qquad\quad = \dfrac{e^{-(m+n)}}{k!} (m+n)^k$

This is the correct expression for $P(Z=k)$ if $Z \sim \text{Po}(m+n)$.

COMMENT

In this kind of proof, it is worth keeping in mind what we are working towards. We need to get an expression involving $(m+n)^k = \displaystyle\sum_{r=0}^{k} \binom{k}{r} m^r n^{k-r}$, and since $\binom{k}{r} = \dfrac{k!}{r!(k-r)!}$, we were just missing a *k!*, which is why we multiplied numerator and denominator by *k!*.

---

404 Questions crossing chapters